ENVIRONMENTAL ASPECTS OF NUCLEAR POWER

Geoffrey G. Eichholz
Professor of Nuclear Engineering
Georgia Institute of Technology

ANN ARBOR SCIENCE
PUBLISHERS INC
P.O. BOX 1425 • ANN ARBOR, MICH. 48106

Second Printing, 1977

Copyright © 1976 by Ann Arbor Science Publishers, Inc.
230 Collingwood, P.O. Box 1425, Ann Arbor, Michigan 48106

Library of Congress Catalog Card No. 76-22243
ISBN   0-250-40138-X

PREFACE

Seldom has any current issue attracted as much public notice and emotion as the question of the environmental effects of the growing nuclear industry. Unfortunately many of the participants in this struggle make little attempt at placing all the technical aspects in any reasonable perspective and few members of the public feel qualified to do so. This book is an attempt to collect many of the essential technical facets of the subject, in order to help train and educate scientists and engineers in this field. The viewpoint taken is that of a responsible nuclear engineer; that is, the conservation and protection of the environment is an important responsibility of any citizen who at the same time accepts the need for development of additional power sources to meet present and future energy needs in an increasingly industrial and urbanized society.

To a large extent this book is an outgrowth of lecture courses and discussion groups conducted at Georgia Tech over the past few years. The selection of material and emphasis employed are necessarily determined by the author's preferences. However, it is assumed and hoped that the presentation adopted is sound and logical and will continue to be useful, even in the face of rapidly changing technological developments inherent in nuclear engineering at this time and continuing changes in the regulatory and licensing procedures in all countries.

While the book leans heavily on current usage and experience in the United States, it is reasonable to expect that similar problems and engineering solutions can be anticipated in other industrial countries. Most of the material presented here will be of general validity and it is hoped that it will form a firm base for the education and training of nuclear engineers and a convenient reference source for others interested in this important aspect of nuclear technology.

Since the material has been drawn from many sources covering a wide range of scientific fields and engineering practice, it has not been found practical to convert all data into a common system of units. Both metric and English units have been retained where this seemed expedient without loss of meaning or general intent; conversion tables have been provided in the Appendix.

I wish to acknowledge with gratitude the advice and comments received from my friends and colleagues, Wayne L. Britz, Fisher Craft, Thomas W. Jackson, Donald G. Jacobs, Bernd Kahn, Paul G. Mayer, Karl Z. Morgan, John W. Poston, C. J. Roberts, A. Schneider and C. L. Wakamo who reviewed various portions of the manuscript.

Geoffrey G. Eichholz
Atlanta, Georgia

CONTENTS

LIST OF FIGURES

Figure

Figure

LIST OF TABLES

Table

Table

Table

CHAPTER 1

INTRODUCTION

One of the basic tenets of the proponents of the Industrial Revolution was that "Progress is Inevitable" and that the fruits of technical progress are invariably beneficial. The smoking chimneys of industry have been symbols of prosperity to most industrial societies and it is only in the past two decades that some of the underlying technical and sociological assumptions associated with the development of an increasingly industrialized and urbanized society have been seriously questioned. Nowhere has this questioning become more evident and insistent than in considerations of the effect of the growth of industry and of metropolitan cities on the "quality of life," a term frequently equated with the access to and enjoyment of unspoiled countryside by the average citizen. For many, this desire to preserve a natural environment that seems to slip rapidly beyond our grasp has led to attempts to recreate the more pastoral life and bucolic charms of a bygone age, without many of the evident drawbacks, or to an emotional, unreasoned reaction against all symbols and monuments of our technical society. At heart, many of these latter-day Luddites know that at the present rate of population growth and with the need to feed the world's population, which can only be met by improved technical methods of agriculture and distribution, the steady urbanization of society is probably inevitable and irreversible.

Yet we have a responsibility to future generations to husband the world's resources and preserve, as much as possible, all that is beautiful and desirable in our world. To do this and to reconcile many conflicting demands on our inherently limited resources, it is essential to have enough people available who can competently assess all of the social, technical and ecological factors involved in developing these resources. Most of the apparent problems have technically feasible solutions; however, in most cases, a balanced assessment of the costs and benefits is needed to determine the optimum solution (1).

1

The nuclear industry is caught in the cross fire of environmental questions at this time both because it is an obvious target for the antitechnologists on emotional grounds and also because it happens to be at a stage of growth where the size of economic investment at stake as well as the degree of extrapolation from a limited base of engineering experience are unprecedented (2,3).

In the following chapters the various factors involved in the environmental aspects of the nuclear power industry will be discussed. In view of the growing emphasis on reviewing the impact of nuclear power as a whole, all stages of the nuclear fuel cycle will be considered. For the sake of conciseness, little attention will be paid to the conventional aspects of electrical power generation and distribution, except to the extent that they interact with nuclear technology.

For this purpose, this chapter will briefly set the stage by considering the reasons for the growing adoption of nuclear energy as a primary energy source, some of the social implications of steadily increasing energy consumption, those aspects of our environment likely to be affected by nuclear technology, and the nature of the effects being considered. Regulatory and legal matters, which are in a considerable state of flux at this time, will not be discussed in this book in any detail in order to avoid undue obsolescence of the material presented.

THE NEED FOR POWER

The modern technological society in its vital aspects is based on readily available energy sources for the production of goods, their distribution, all forms of transportation and communication, the heating of homes, the preparation of food, entertainment, and waste disposal. The amount of energy consumed has grown significantly with the rising population in the world, but more importantly the amount consumed per capita has increased appreciably in the past decades and is often accepted as one appropriate indicator of the standard of living in a country. Based on projected world population growth and extrapolations of the rate of increase in per capita consumption of energy, one obtains predictions for an almost exponential increase in energy demand for the next few decades (4-6). There are several points on which one might argue with such projections, such as the assumption that a rising standard of living or a high rate of personal energy consumption can be maintained in the face of rapidly growing population densities, increasing fuel costs, and greater competition for existing energy resources, particularly in developing nations. Nevertheless, over the short term, i.e., the next 20-30 years, such projections may well be justified and they immediately pose a number of important questions.   One relates to the type of fuel to be used, and the existence and reasonably assured availability of that type of fuel.   Another is the effectiveness of present uses of energy resources;   it is unlikely, for instance, that the present pattern of individual automobile travel, with its inefficient use of raw materials and highway capacity can be extrapolated indefinitely to larger populations or should really be emulated in developing countries with high population densities.   Then there is the question of the efficiency of conversion of the raw fuel material into usable form, a subject that attracts and deserves much additional research and development.   Improvements in access to and distribution of power generating facilities also will have a major effect on energy utilization.   Some of these factors will be discussed briefly in the following sections;   additional discussion may be found in publications listed in the bibliography at the end of this chapter.

THE NATURE OF ENERGY DEMAND

The increase in the living standard of mankind has been accompanied by a steady growth of the utilization of energy in all forms.   After early use of raw manpower and animal draught strength, other energy sources, such as coal for fuel, wind and water power, steam and finally electricity were pressed into service as societies attained a more complex organization and as the needs of the individual and the group society became more refined.   Of late, it has become customary to measure the standard of living, a rather elusive quality, in terms of the per capita consumption of energy.   Figure 1, based on Hubbert (7,8), illustrates the kind of correlation obtained.   It is evident that many of the developing countries are characterized by low energy consumption and excessive dependence on animal and manpower for food production, transportation and manufacture.

A similar correlation is obtained if following Salvetti one uses an "ecological index" based on energy consumption per unit area.   In the United States and Western Europe, total energy consumption has increased almost exponentially (10) over the past hundred years.   This increase has been dependent on the availability of new and plentiful energy sources, resulting in a shift from traditional, local fuels to oil and gas, based on a world-wide fuel economy (7).   Since fossil fuels are not in unlimited supply and the politics and economics of importation of petroleum from less-developed countries appear to be uncertain over the long haul, the search for alternative energy sources is gathering momentum.   These considerations result in an increasing proportion of new generating capacity being based on nuclear power.   For the United States this development is shown in Figure 2, as predicted by the U.S. Atomic Energy Commission (11).

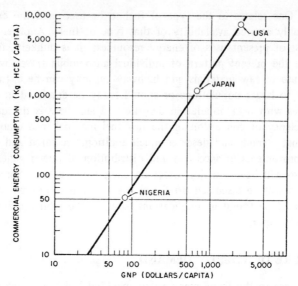

Figure 1. Per capita energy use and gross national product in selected countries (7,9).

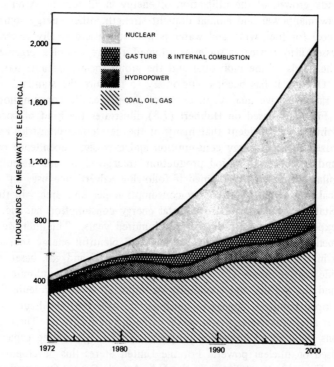

Figure 2. Projected installed generating capacity in the United States [1974 estimate] (11).
Source: 1972 Actual Data, FPC, Forecast Data, AEC

Clearly nuclear power will play an increasingly important part as the availability of other conventional power sources diminishes.

It is generally assumed that the growth of nuclear power in the United States will be closely paralleled by similar growth in Western Europe and the U.S.S.R., both in the rate of development and the installed capacity (Figure 3) (12). Most predictions have assumed that in the developed

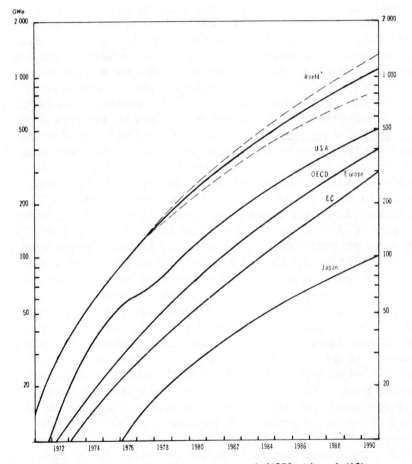

Figure 3. Net world nuclear power growth [1973 estimate] (12).

countries demand will follow population growth at the present per capita consumption, or even at an increased demand rate, reaching at least the U.S. consumption level for all Western countries. In the developing countries a slower trend, but one directed towards the same goal, is similarly assumed, leading to a world-wide exponential growth in power demand, much of which will be nuclear (5). One may question whether such a trend is desirable or realistic, leading as it does to an exhaustion of the readily accessible fossil and nuclear fuels within the next two or three generations, a situation often termed irresponsible but for which no ready remedy has been suggested short of a complete reordering of Western society. Some of the more persistent critics of nuclear power have pressed this point, but their many valid arguments have often tended to be overshadowed by their shrill vilification of the Establishment and by overbiased presentation (13-16). Only very rarely has there been a realistic assessment of the economic and social cost of not having an adequate availability of power. However, many sober voices are reviewing the prospects for effective energy conservation and improved utilization, without necessarily stopping the clock (17-21). Impending exhaustion of readily accessible fossil fuel resources over the next hundred years at the current rate of production and the necessarily long time scale of development for new approaches, based on nuclear fusion or solar power, make it evident that over the next few decades there are no foreseeable alternatives to power production based on nuclear fission (5). The supply cycles for coal and oil production have been discussed by Hubbert (8) who showed that they may peak-out around 2100 and 2000, respectively (Figure 4). Even if one did not accept all the details of these projections, it would not greatly affect the overriding need for nuclear power in the immediate future; however, recent economic developments have shown that rising fuel costs and a shortage of investment capital may rather effectively slow down the expansion of the power economy.

In terms of the overall impact on the world we live in, the effects of the exploration, production and utilization of energy sources are just one of the many facets of the growth of the world's population, its increasing urbanization and industrialization, and of the ever-expanding occupation by man and his works of much of the Earth's surface. To single out nuclear power as a major factor in the changing environment is seeing this development out of focus; in fact, most of the changes we can observe do not depend greatly on the particular source of energy used, except on a purely local scale, and would not be much affected if we could continue using conventional fuels at the rate and on the scale demanded. The purely environmental factors will be discussed in the next sections; however, short of devastating wars or pestilence, even with "zero population growth" now, the present rise in the world's population would not be effectively stabilized for at least another century. In the meantime the relation between man and nature has changed to an unprecedented extent from man's total dependence on the forces of his

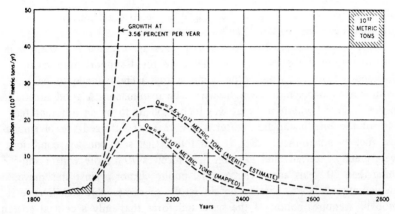

A

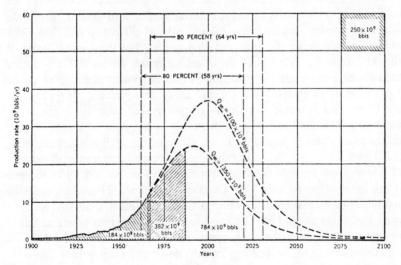

B

Figure 4. Complete cycle of world production for two estimates of total available resources (Q$_\infty$) A coal B crude oil (8,22).

natural environment to one where the existence of wildlife, plant growth and even the weather tend to be judged and controlled purely by their supposed beneficial role in man's economy.

Most of the changes that have taken place are irreversible; there is no way to restore the grassy prairies that once could support millions of buffaloes. However, a few other changes that were assumed to be natural phenomena have been shown to be man-made. A good illustration is the famous London fog that was long thought to be an inherent feature of the British climate. After a few years of enforced use of smokeless fuel by government edict, it was found that soot and fog conditions were greatly reduced, and London now receives 70% more winter sunshine than 20 years ago. The original public clamor against this move, incidentally, indicates that to obtain results of comparable significance it probably requires action of the scale and force that only a central government can command.

Many of the environmental problems associated with new power-producing facilities arise from the larger size of the current generation of plants. Where not long ago a plant generating 100-300 megawatts (MW) of electricity was considered "large," present plans for new plants, both fossil-fueled and nuclear, include plants with capacities of 1000-4000 MW at one site. This results in environmental problems of greater magnitude in several respects, such as heat release, exhaust plumes, power line right-of-way requirements, and transportation needs. For the United States, if the AEC forecasts listed in Figure 2 are accepted, the most likely nuclear capacity projection for the year 2000 of an equivalent capacity of about 1200 reactors of current design would imply that virtually everyone in the United States would live within several tens of miles from one or more power stations and be subject to exposure by the effluents (11). In the more densely populated industrial countries of Western Europe and Japan, there will be a correspondingly greater environmental impact of power plant construction and operation than in the U.S. For example, it has been estimated that West Germany must build 21 nuclear generating stations with a total capacity of 18.8 GWe by 1980 if Germany is to meet its energy needs (24). It is evident that any action taken at this time that will control the discharge of radionuclides, heat, and other contaminants to the surrounding countryside can greatly reduce future exposure of the population to such effects.

For these reasons a fair and forward-looking assessment must be made of present and future power plant locations, their design and requirements. Almost inevitably, plant locations are likely to be clustered on sites that must be carefully selected and identified well in advance of actual construction. This will help to lay the groundwork for adequate land use planning, baseline measurements needed for environmental assessments, and psychological preparation of any citizen groups likely to be affected by the proposed plant operation.

ALTERNATIVE ENERGY SOURCES

As Figure 2 indicated, nuclear power will supplement not replace alternative energy sources for a long time to come, although an increasing fraction of new installed capacity is likely to be nuclear-powered. In reviewing the choice among prime power sources at the time a new generating plant is planned, all potential alternatives must be considered. Considering the state of technology and the limited sites available for their exploitation, solar energy, geothermal energy, tidal, and even hydroelectric power are not serious contenders for the production of large blocks of power in most countries at this time, though, of course, they may be suitable in specific circumstances. Very few countries have sizable unexploited hydroelectric potential left and with few exceptions such sites are either so remote as to make transmission costs excessive or too limited in capacity to compete in the range of installed capacity over 300 MWe that is of principal interest now.

Tidal power has been discussed widely and offers possibilities in a few places around the world, though with a huge capital investment. There is an operating plant in the Rance estuary in France and there have been recurrent feasibility studies for the Passamaquoddy estuary in the Bay of Fundy and there may well be other possible locations. Table 1, after Hubbert, shows the distribution of possible tidal power sites that have been evaluated. However, for most utilities this does not constitute a practical alternative.

Table 1. Tidal-Power Sites and Maximum Potential Power (9)

Locality or Region	Average Potential Power (10^3 kW)	Potential Annual Energy Production (10^6 kWh)
North America, Bay of Fundy (Nine sites)	29,027	255,020
South America, Argentina, San Jose	5,870	51,500
Europe, England, Severn	1,680	14,700
France (Nine sites)	11,149	97,811
USSR (Four sites)	16,049	140,452
Totals	63,775	599,483

Solar power has shown its value for space power in satellites and has been utilized in small-scale demonstrations (25), but nobody appears to have accomplished the design of a sufficiently compact system capable of large-scale continuous energy production at competitive cost (26,27). The sunlight that falls on a small percentage of the land area of the U.S. would satisfy most of the energy needs of the country in the year 2000 if converted to electricity with an efficiency of 12% (8). Serious problems of economic energy storage and distribution still have to be solved.

Geothermal energy has been tapped on a small scale in some places, notably Iceland, Italy and New Zealand, but is clearly not an alternative that is available just anywhere. Table 2 shows the world potential for geothermal power (8). Proposals for a hydrogen fuel system (28) must be considered rather speculative at this time; such systems await considerable development work before they can be considered technically and economically attractive.

In general, the real choice will lie between fossil fuels (oil, gas and coal) and nuclear power. Interestingly enough, to a large extent the factors that determine such a decision have shifted in relative importance in recent years. They include:

1. plant size contemplated
2. fuel type
3. fuel availability
4. capital investment required
5. operating cost
6. effluent treatment needed
7. environmental factors (fuel transport, waste disposal)
8. cooling requirements.

Table 3, adapted from Starr (5), shows the distribution of economically recoverable fuel supplies for the world and the U.S. It is evident that the supply situation will become exceedingly critical unless nuclear breeder reactors can be brought into use at an early date or consumption is cut back drastically. Of all the energy needs projected to the end of the century, nonelectric uses such as transportation, space heating and industrial processes account for about two-thirds of the total. Short of a wholesale abandonment of the internal combustion engine such uses presumably will continue to preempt much of the available fossil fuel, requiring the power industry to rely increasingly on nuclear fuels.

With regard to hydroelectric generating facilities, their environmental impact is, of course, not negligible. Substantial land areas are inundated by the impounded lake behind the dam. The dams themselves are large structures requiring substantial quantities of concrete and steel, and these structures may impede the migration of certain fish species. The impoundment of water itself will usually increase the surface area exposed to

Table 2. Developed and Planned Geothermal Electric Power Installations (5)

Country and Locality	Installed Capacity 1969 (MW)	Planned Additional Capacity (MW)	Total Capacity by Early 1970s (MW)	Date of Earliest Installation
Italy				
Larderello	370		370	1904
Monte Amiata	19		19	ca 1962
Total	389	-	389	
U.S.A.				
The Geysers, California	82	318	400	1960
New Zealand				
Wairakei	290		290	Nov.1958
Mexico				
Pathé	3.5		3.5	ca 1958
Cerro Prieto (Mexicali)		75	75	ca 1971
Total	3.5	75	78.5	
Japan				
Matsukawa	20	40	60	Oct. 1966
Otake	13	47	60	Aug.1967
Goshogate		10	10	
Total	33	97	130	
Iceland	(Geothermal			
Hveragerdi	energy for house and greenhouse heating)	17	17	1960
USSR				
Kamchatka				
Pauzhetsk	5	7.5	12.5	1966
Paratunka	0.75		0.75	1968
Bolshiye Bannyye	25		25	1968
Total	30.75	7.5	38.25	
Grand Total	828.25	514.5	1342.75	

sunlight and it is not uncommon to find waters in reservoirs warmed several degrees above those temperatures found naturally in the unconfined stream. Also, flow of water through the turbine and its discharge to the stream below the dam may create changes in the dissolved gas content, which in turn lead to specific diseases in fish (29).

Fossil-fuel steam electric plants, which burn carbonaceous materials to release heat energy, provide the bulk of electric generation capacity in

Table 3. U.S. and World Energy Sources (5)
A. Economically Recoverable Fuels (1971 estimates)
B. Continuous or Renewable Energy Supply

A	Depletable Supply (10^{12} Watt-Years)	World	U.S.
	Coal	670-1000	160-230
	Petroleum	100-200	20-35
	Gas	70-170	20-35
	Subtotal	840-1370	200-300
	Nuclear (ordinary reactor)	$\sim$ 3,000	$\sim$ 300
	Nuclear (breeder reactor)	$\sim$300,000	$\sim$30,000
	Cumulative Demand 1960 to Year 2000 (10^{12} Watt-Years)	350-700	100-140

B	Continuous Supply (10^{12} Watts)	World Maximum	World Possible by 2000	U.S. Maximum	U.S. Possible by 2000
	Solar Radiation	28,000		1,600	
	Fuel wood	3	1.3	0.1	0.05
	Farm waste	2	0.6	0.2	0.00
	Photosynthesis fuel	8	0.01	0.5	0.001
	Hydropower	3	1	0.3	0.1
	Wind power	0.1	0.01	0.01	0.001
	Direct conversion	?	0.01	?	0.001
	Space heating	0.6	0.006	0.01	0.001
	Nonsolar				
	Tidal	1	0.06	0.1	0.06
	Geothermal	0.06	0.006	0.01	0.006
	Total	18+	3	1.2	0.2
	Annual demand Year 2000 (10^{12} Watts)	$\sim$15		$\sim$5 - 6	

the world at the present time. Approximately 11 tons of coal are required to produce 1 megawatt-day (MWd) of electric energy; about 44 barrels of oil and 250,000 cubic feet of gas are burned per megawatt-day of electric energy. The conversion of each of these fuels to electric energy results in some environmental pollution, which varies with the different fuels used (30).

Looking first at coal, many kinds of environmental effects can be identified as results from coal use. The first is the effect on land use, including the ecological effects of mining operations, the commitment of substantial areas of land for power plant sites, rail delivery facilities, coal storage areas, and the requirements for disposal of solid wastes. Second, there are effects on air quality stemming from the discharge into the atmosphere of combustion products including nitrogen oxides, sulfur oxides, and particulates. Effects on water quality from the discharges of waste heat and some chemicals, and effects on land use associated with transmission rights-of-way are essentially common to all thermal generating stations and are not necessarily specific to the use of coal as a fuel.

The world's resources of coal are extremely large. However, coal is not uniform in quality, varying quite widely in both physical and chemical properties and in its cost to mine and transport. For example, the bulk of the low-sulfur and inexpensive coals found in the western part of the United States are located quite far from power demand centers, which are predominantly in the East. Coals that are nearer to load centers are generally rather high in sulfur or are expensive to mine.

The sulfur content of U.S. coals ranges from 0.2% to 7% by weight. The ash content normally ranges from 5% to 20% by weight. The bulk of the coal burned by U.S. utilities had a sulfur content in the range of 2% to 4%. Increasingly, electric utilities are required by air quality constraints to burn fuel having a sulfur content of 1% or less and, in some instances, the allowable limit has been set below 0.5% sulfur. The supplies of coal available with sulfur contents as low as 1% are very limited in relation to the enormous needs of the power industry. Substantial reserves of this coal are either dedicated to the metallurgical and export markets or are directly owned by steel companies.

The burning of coal results in the emission of particulates and of oxides of sulfur and nitrogen; the disposal of the ash resulting from the burning of coal constitutes a major environmental problem. For example, in 1970 the 320 million tons of coal burned by U.S. electric utilities resulted in the generation of 30-35 million tons of ash requiring disposal as solid waste, and approximately 24 million tons of sulfur dioxide assuming an average of 3.5% sulfur in the coal burned.

From a technological point of view, the control of particulate emission from coal-burning plants is well established. Techniques include the installation of mechanical collectors and electrostatic precipitators with nominal removal efficiencies in excess of 99% from the flue gas streams at these plants. The control of sulfur emissions from coal-fired stations is not nearly as well developed as the control technology for particulate emissions. The present status of these processes can at best be described as developmental; that is, there are a number of processes that have demonstrated a reasonable capability in small-scale experiments or pilot plant operations and are now in the testing phase at a number of different utility plants around the country to determine their effectiveness when installed at generating plants under realistic operating conditions. The U.S. Environmental Protection Agency has generally stated that technology has been demonstrated for the sulfur removal processes; most utilities and the Federal Power Commission disagree with this position.

A final consideration that relates to this problem and has an environmental impact of its own is the fact that almost all of the low-sulfur coal available in the western United States would be extracted using strip mining techniques. Although reclamation practices are available for dealing with strip mining problems, there has been appreciable resistance by some mining companies to employ them in a meaningful fashion.

Since fuel oil typically contains less sulfur than does coal, it has been utilized in many instances as a replacement for coal where air quality regulations no longer permitted coal to be burned. Assuming an average content of 1.5% sulfur for the 335 million barrels of oil burned in 1970 by U.S. electric stations, the resulting emissions of sulfur oxides in that year totalled approximately 200,000 tons, or less than 1% of the total contributed by utility coal combustion in that same period. Although processes are available for desulfurizing oil, the capacity of existing refineries does not now (nor will it for several years) match the demand for fuel oil with a low sulfur content.

There are other environmental problems associated with the use of oil as a fuel. These include the potential contamination of ocean waters during the exploration, recovery, and transportation stages associated with the development of oil resources. Certainly the controversy about the Alaskan pipeline or the periodic episodes of oil spills from seabottom wells or from tanker mishaps in coastal waters have not escaped public notice. Additionally, construction of the marine terminal facilities required for docking and unloading the so-called "supertankers" has encountered substantial opposition in many locations due to concern about adverse environmental effects.

As the experience of recent years has shown, the availability of natural gas for power plants is severely restricted by geographical factors and pipeline capacity. The supply of fuel oil is increasingly dependent on political factors and a rapidly rising cost pattern. This leaves coal as the principal fossil fuel available for future development for large power plants. Coal-fired plants require a lower capital investment than nuclear plants, but pose operational problems related to fuel transportation and labor costs. However, a bigger problem has arisen due to the pollution potential of large fossil-fuel plants from their emission of fly ash, nitrous oxides and sulfur oxides.

Table 4 shows 1968 emission figures for American fossil-fueled power generating stations (31) compared with other effluent sources. Table 5 shows comparable, more detailed data for Canada (32,33). With the current emphasis on the reduction of sources of pollution, advocates of nuclear power have found themselves suddenly on the side of the angels by promoting a "clean" fuel. In addition, most coal contains small amounts of uranium and radium, so that it has been claimed that in some cases coal- and oil-burning plants discharged to the atmosphere a greater fraction of the maximum permissible concentration of radioactive dust than certain types of nuclear reactors (20). This point is illustrated in Figure 5, which compares the radioactivity in fly ash and soil at three plant locations. We will return to this matter in Chapter 13.

Table 4. Estimated U.S. Discharges of Airborne Pollutants, 1968 (31)
(million tons per year)

	Carbon Monoxide	Particulate Matter	Sulfur Oxides	Hydro-Carbons	Nitrogen Oxides	Total
Power plants	0.1	5.6	16.8	Neg.	4.0	26.5
Other fuel combustion in stationary sources	1.8	3.3	7.6	0.7	6.0	19.4
Transportation	63.8	1.2	0.8	16.6	8.1	90.5
Industrial processes	9.7	7.5	7.3	4.6	0.2	29.3
Solid waste disposal	7.8	1.1	0.1	1.6	0.6	11.2
Miscellaneous	16.9	9.6	0.6	8.5	1.7	37.3
Total	100.0	28.3	33.2	32.0	20.6	214.2

Note: Sulfur oxides expressed as tons of sulfur dioxide and nitrogen oxides as tons of nitrogen dioxide.

While scrubbing of stack gases is technically feasible, it is expensive and controversial in many respects. Low-sulfur coal is in relatively short supply and often mined in places far removed from the consumer. High-sulfur coal and oil are more plentiful, but it is claimed that most commercial systems for the removal of sulfur from stack gases are ineffective and unreliable. This means that enforcement of strict environmental emission standards severely reduces the coal reserves that are usable and available to the power industry. Consequently, tables of recoverable fossil fuel reserves, of the type shown in Table 6, after Hubbert (8,9), tend to

Table 5. Canadian Inventory of Air Pollution from Fuel Combustion - 1971 (32)

		Weight of Pollutant Emitted, lb x 10^6				
	Total	NO	CO	Particulates	SO_2	SO_3
Domestic Fuels						
LPG	7541	9.426	3.91	–	–	–
Natural gas	25670	8.588	15.41	–	–	–
Light furnace oil	86110	35.110	43.68	68.26	374.6	9.558
Stove oil	8565	3.500	4.36	6.80	16.0	0.41
Subtotals	127886	56.624	67.36	75.06	390.6	9.968

Total   128485.612x10^6 lb = 64.24x10^6 short tons = 58.28x10^6
metric tons

Commercial Fuels						
Natural gas	17230	24.930	10.34	–	–	–
Diesel fuel	40260	192.800	119.90	16.56	153.20	–
Motor gasoline	91430	846.800	19470.00	16440.	23.55	–
Aircraft turbine	8975	14.450	21.87	29.30	16.73	0.43
Can. bit. coal	1339	2.716	0.564	3.972	8.208	0.201
Can. sub-bit. coal	684	1.320	0.274	2.235	2.869	0.072
Can. lignite	207	0.395	0.082	0.951	1.101	0.028
Imported anthracite	607	1.128	0.234	1.948	1.364	0.034
Imported bit. coal	2855	5.772	1.197	5.955	11.120	0.278
Subtotals	163587	1090.311	19624.461	16500.921	218.142	1.043

Total   201021.878x10^6 lb = 100.51x10^6 short tons = 91.18x10^6
metric tons

Industrial Fuels						
Natural gas	48710	70.490	29.240	–	–	–
Coke oven gas	3085	5.200	2.160	–	–	–
Blast furnace gas	831	0.230	0.215	–	–	–
Refinery fuel	5259	7.260	1.613	4.170	52.476	2.316
Heavy fuel oil	106400	249.800	51.810	84.610	1692.000	42.300
Can. bit. coal	22930	46.520	9.648	34.010	140.500	3.512
Can. sub-bit. coal	9477	18.270	3.790	15.470	39.730	0.993
Can. lignite	6044	11.540	2.393	13.890	32.160	0.804
Imported anthracite	1784	3.313	0.687	2.861	4.006	0.100
Imported bit. coal	92920	187.900	38.960	96.920	361.900	9.046
Subtotals	297440	600.523	140.516	251.931	2322.772	59.071

Total   300814.813x10^6 lb = 150.41x10^6 short tons = 136.45x10^6
metric tons

Additional Major Pollution						
Smelting	–	–	–	440.0	9600.0	–
Cement & lime	20000	–	–	–	–	–
Subtotals	20000	–	–	440.0	9600.0	–

Total   30040x10^6 lb = 15.02x10^6 short tons = 13.62x10^6 metric tons

Grand Total   660362.935x10^6 lb = 330.182x10^6 short tons = 299.536x10^6
metric tons

% from motor gasoline: 19.4   % from commercial fuels less gasoline: 11.0
% from domestic fuels: 19.5   % from industrial fuels: 45.6
% from other industrial sources: 4.5%

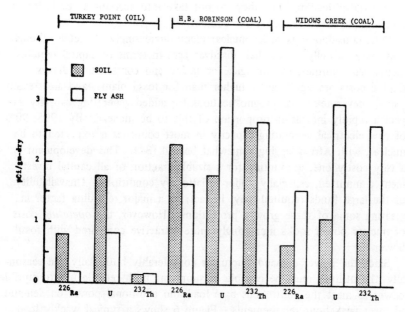

Figure 5. Natural radioactivity in soil and fly ash near three fossil-fueled power stations (20).

Table 6. Approximate Magnitudes and Energy Contents of Recoverable World Fossil Fuels (8)

Fuel	Quantity	Energy Content		Percent
		10^{21} thermal joules	10^{15} thermal kWh	
Coal and lignite	2.35×10^{12} metric tons	53.2	14.80	63.78
Petroleum liquids	2400×10^{9} bbls	14.2	3.95	17.03
Natural gas	$12,000 \times 10^{12}$ ft^3	13.1	3.64	15.71
Tar-sand oil	300×10^{9} bbls	1.8	0.50	2.16
Shale oil	190×10^{9} bbls	1.1	0.31	1.32
Total		83.4	23.20	100.00

be a little misleading since they do not take into account accessibility of fuel to the consumer or sulfur content.

Such considerations make nuclear plants increasingly attractive, though they only partially offset their disadvantages in terms of capital requirements, legal hurdles, psychological obstacles, and construction delays. Capital costs are significantly higher than for fossil plants, and are steadily rising; nevertheless, all prognostications for added generating capacity expect a rapidly increasing proportion of this to be nuclear. By 1995, 90% of new electrical generating capacity in most countries is expected to be nuclear, with Africa lagging somewhat behind (34). This development is a very costly one, accounting for a sizable fraction of all capital investment committed, especially under inflationary conditions. Unavailability of the large funds required may, in fact, be a major retarding factor in meeting some of these growth projections. However, the *operating* costs of nuclear power looks increasingly more attractive compared with fossil power.

Since the power demand fluctuates considerably both daily and seasonally, the utilities must be capable of supplying both reliable uninterruptible power to meet the continuous base load and additional power on demand to meet peak hour requirements. Figure 6 shows a typical weekly load curve. Evidently it is most economical to meet the base load by

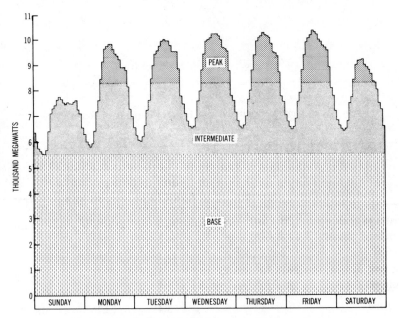

Figure 6. Typical weekly load curve and source of power.

minimum-cost generating capacity such as nuclear power plants in which the heavy capital investment puts a premium on continuous operation. In contrast, peak loads are best met by smaller generating plants, typically oil- or gas-fired, with a low standby cost and somewhat higher operating costs, that can be turned on or off at short notice. Hence, for a long time to come, most larger power companies will be operating a mixed system of generating plants. Furthermore, some utility networks are proposing pumped storage systems that pump water up to a high lake at night, when spare power is available, and allow it to run back through the generators during periods of high demand in order to smooth out the variation between peak load and base load.

In addition to daily fluctuations, there are also seasonal fluctuations. Until the advent of air conditioning, in North America the seasonal peak occurred generally in the Christmas season in the late afternoons, when a combination of industry-commercial use, holiday lighting and domestic heating coincided to produce the maximum demand on every utility. The increasing acceptance of air conditioning in the U.S. has shifted the maximum demand to the summer season in a rather unpredictable fashion depending on the existence of hot, humid weather over large sections of the country. To meet such peak loads, most utilities try to maintain a reserve capacity of 15-20%. This capacity also may depend on the size of the largest unit that may suffer emergency shutdown and on the availability of electricity from adjoining grid networks.

THE NUCLEAR FUEL CYCLE

The flowsheet of operations of a fossil-fueled power plant is fairly straightforward: fuel is transported from mine or well to the plant, conveyed to the burners, and the ashes or residues are disposed of or dispersed, with or without prior treatment. In a nuclear power plant this flowsheet is much more complex, as shown in Figure 7. The basic steps in this operation are essentially the same whether the fuel is uranium, thorium, plutonium, or a mixture of them. The details of each step depend on the particular type of reactor employed and are characterized by the fuel used, the type of coolant (light water, heavy water, gas, organic liquid, or liquid metal), the neutron energy, and, where appropriate, any breeder characteristics.

The energy-producing stage is the fission of uranium (or plutonium) nuclei under bombardment by neutrons. This releases about 200 MeV $(= 3.2 \times 10^{-11}$ joule) per fission, in the form of two heavy nuclei, called "fission products," and several energetic neutrons, at least one of which has to interact with another fissile uranium nucleus to continue the process.

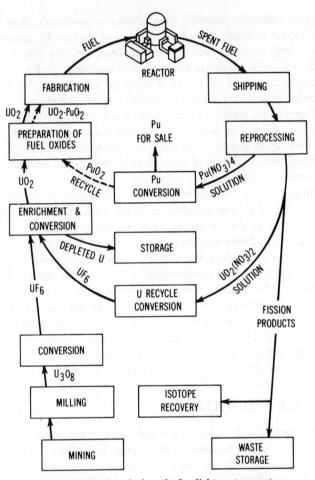

Figure 7. Nuclear fuel cycle for light water reactors.

The unique feature of a fuel "cycle" arises from the fact that complete combustion of all available fissile fuel in the reactor core is not practical because the fission products, as they build up, absorb neutrons parasitically and make it progressively more difficult and inefficient to maintain a critical mass for continued fission to proceed. For this and related reasons, it is customary to withdraw the fuel after about 25-30% of the fissile fuel has been consumed. After some months of storage to allow decay of most of the induced radioactivity, the used fuel is then "reprocessed" to remove the unwanted fission products and to recover and purify the remaining fissile fuel material for reuse. This recycle of fuel gives rise to the characteristic nuclear fuel cycle shown in Figure 7.

To assess the economic and environmental contributions of the various stages of the fuel cycle, it is expedient to break it down into its components:

1. mining and milling of uranium ore
2. refining
3. conversion
4. enrichment of ^{235}U content
5. fuel preparation
6. fuel fabrication
7. power generation
8. spent fuel transportation
9. reprocessing and waste recovery
10. waste storage and disposal.

The distribution of costs between the major components of the fuel cycle, other than actual power generation, is shown in Table 7. To provide some idea of the quantities required for a "model" 1000 MWe light-water reactor in terms of annual fuel needs, typical enrichment ratios, separative (enrichment) work required, and thermal efficiencies, Table 8 is provided for reference.

Table 7. Estimated Fuel Generation Cost of LWR Nuclear Fuel and Comparative Costs for 1000 MWe Power Plants (1981 costs in 1973 dollars) (11)

Cost Component	Cost mills/kWh
Mining & milling ($10/lb U_3O_8)	0.54
Conversion to UF_6 ($1.35/lb U)	0.07
Enrichment ($42/kg SWU)	0.76
Reconversion & fabrication ($70/kg U)	0.33
Spent fuel shipping ($5/kg U)	0.02
Reprocessing ($35/kg U)	0.14
Waste management	0.04
Plutonium credit ($8/g)	(0.22)
Subtotal	1.68
Fuel inventory carrying charge (at 12%)	0.82
Total	2.50

Cost Component	LWR	Coal	Oil
Capital	11.70	10.90	8.00
Fuel	2.50	5.50	24.60
O&M	1.00	1.60	0.80
Total	15.20	18.00	33.40

Assumptions: (1) Annual generation of 7.0 x 10^9 kWh (80% utilization)
(2) Annual fixed charge of 15% used to translate unit capital costs to annual capital costs.

Table 8. Characteristics of "Model" Nuclear Plants, 1000 MWe Capacity (11)[a]

Plant Starting Commercial Operation in Period:	Boiling Water		Pressurized Water	
	1976-80	1981-85	1976-80	1981-85
Thermal efficiency (%)	34	34	33	33
Specific power (MWt/tonne U)	22	26	37	41
Initial core (average)				
Irradiation level (MWDt/tonne U)	21,000	24,000	26,000	26,000
Fresh fuel assay (wt % U-235)	2.2	2.4	2.8	2.7
Spent fuel assay (wt % U-235)	0.8	0.7	0.9	0.8
Fissile Pu discharged (kg/tonne U)	5.1	5.4	6.0	6.0
Feed required (ton U_3O_8/MWe)[b]	0.680	0.635	0.545	0.480
Separative work required (kg/MWe)[b]	345	340	320	275
Replacement loadings (steady state)				
Irradiation level (MWDt/tonne U)	27,000	33,000	33,000	33,000
Fresh fuel assay (wt % U-235)	2.6	2.8	3.3	3.2
Spent fuel assay (wt % U-235)	0.8	0.7	0.9	0.8
Fissile Pu discharged (kg/tonne U)	5.6	5.9	6.7	6.7
Feed required (ton U_3O_8/MWe/yr)[b]	0.145	0.140	0.165	0.165
Separative work required (kg/MWe/yr)[b]	105	100	130	125

[a]See Appendix for symbols used.

[b]Based on operation of enriching facilities at a tails assay of 0.2% U-235 and on no recycle of plutonium. For replacement loadings, the required feed and separative work are net in that they allow for the use of uranium recovered from spent fuel.

Those factors having particular bearing on the environmental impact of the various operations will be discussed in later chapters. However, since there is a growing tendency to review the effects of the whole fuel cycle whenever a nuclear power plant is planned, it may be appropriate to discuss the main features of the fuel cycle here in general terms.

Mining and Milling of Uranium

Uranium is a fairly common constituent of the earth's crust, with an average abundance in rocks of 1-2 ppm. To be commercially significant, uranium ores must contain at least 0.05% U_3O_8 by weight; such material which once had a market value of $8-12/lb U_3O_8 in refined ("yellowcake") form had risen to a value of $40/lb in 1975. Lower-grade ores are exploited under special conditions, such as in South African gold-bearing ores where the cost of production is largely allocated to the gold production, making it possible to extract a low-grade uranium at highly competitive prices. In general, the extraction of uranium from low-grade sources such as phosphates and shales will inevitably require much higher prices and is not seriously contemplated at present. At the present time the main producers of uranium concentrate in commercial quantities are the United States, the Soviet Union, Canada, South Africa, and Australia. France, Spain, and the Niger Republic are among other smaller though significant producers.

Apart from the environmental impact normally associated with any hard-rock mining operation and the dumping of mill tailings, the principal, peculiar health hazard associated with uranium mining is connected with the emanation of radon gas from uranium-bearing rocks. This aspect will be discussed in detail in Chapter 11.

Purification and Conversion

The uranium concentrate is next refined to reactor-grade purity to eliminate all gross impurities (Si, Fe, S, Th, Co, V and others) and to reduce all neutron-absorbing elements, such as cadmium, boron, hafnium and various rare earths, to acceptable trace concentrations. The refined product is then converted to uranium tetrafluoride (UF_4, green salt), which is the starting material for subsequent processes. The green salt can either be reduced to uranium metal, converted to uranium oxide, or to uranium hexafluoride (UF_6), which is the operating form for all enrichment processes.

All of these processes do not differ substantially from comparable processes in the chemical industry, and the associated environmental impact is essentially related to the plant effluents that have to be treated to remove fluorides and acids, as well as some trace metallic constituents.

Enrichment

Although there are a number of natural-uranium, heavy-water moderated reactors, by far the greatest number of power reactors use enriched uranium. This is obtained by artificially increasing the proportion of the uranium isotope ^{235}U from its natural abundance of 0.07% within the major constituent ^{238}U to a much higher concentration that may range as high as 95% by weight. Most commercial reactors use enriched fuel containing 2-3% ^{235}U. To obtain such an enrichment level the natural uranium must be processed, as the UF_6 gas, in an enrichment plant. So far most uranium fuel is enriched by means of the gaseous diffusion process (24), though other processes such as the gas centrifuge (35) are under active development.

The environmental impact of gaseous diffusion plants is due mainly to the gigantic size of the plants and their enormous energy consumption.

This is shown in Table 9 for existing U.S. operations (36), which lists the power requirements and acreage of the existing plants.

Table 9. Size and Cost Data for U.S. Enrichment Plants (36)

Plant	Process Building	Separative Stages	Ground Coverage (Acres)	Cost (Million $)	Comp. Date	Full Power[b] (megawatts)
Oak Ridge	K-25	2,304[a]	40.0	347	Aug. 1945	
	K-27	540	8.6	61	Jan. 1946	
	K-29	300	6.7	43	Jan. 1951	
	K-31	600	17.1	97	Dec. 1951	
	K-33	640	32.3	267	Jun. 1954	
		4,384[b]	104.7	815		1700
Paducah	C-310	60[a]	1.3	15	Jan. 1953	
	C-331	400	11.8	99	Feb. 1953	
	C-335	400	11.8	101	Apr. 1954	
	C-333	480	24.4	314	Nov. 1953	
	C-337	472	24.4	226	Dec. 1954	
		1,812	73.7	755		2550
Portsmouth	X-326	2,340[a]	28.3	248	Feb. 1956	
	X-330	1,100	32.0	237	Jul. 1955	
	X-333	640	32.4	271	Nov. 1955	
		4,080	92.7	756		1750

[a]Includes purge stages.
[b]Before upgrading improvement program.

Fabrication

The enriched fuel material is next converted to the desired form and fabricated into fuel elements. Metallic fuels consist of uranium alloys, such as U-Mo and U-Zr, in which the minor alloying component primarily serves to promote dimensional stability. Most nuclear reactors now utilize uranium oxide fuel, which is isotropic in its properties and can operate safely at a much higher temperature. The oxide fuel is sintered into rods or may be in the form of small pellets or small carbon-coated spheres. The fuel is clad with a thin metallic coating most commonly made of Zircaloy-4 or stainless steel; a few older reactors still use aluminum or magnesium alloy. The fuel elements are assembled into fuel assemblies, bundles of fuel elements in a carefully adjusted configuration.

None of these operations differ in their environmental impact from any other specialized metallurgical plant of comparable complexity except that quality control and dust collection procedures are superior. Special precautions apply whenever plutonium is incorporated into any fuel, since the high toxicity of plutonium imposes special requirements for remote handling and effluent purification (37).

Power Generation

The use of the fuel in the nuclear reactor forms the central and only productive phase of the nuclear fuel cycle. It represents the justification for the whole fuel cycle operation and, therefore, is subject to the most detailed technological assessment. Furthermore, whereas mines, fabrication and enrichment plants are relatively limited in number and often located in remote places, the number of nuclear power plants is steadily growing; hence their environmental effects constitute the central feature of this book and will be analyzed in detail in subsequent chapters.

Reprocessing and Waste Disposal

After partial burnup of the contained fissile material in the reactor, the fuel assemblies are withdrawn, stored to allow decay of most of the radioactivity, and shipped to a reprocessing plant. The high level of radioactivity of the "spent" fuel being shipped makes the transportation step one of major significance in terms of total environmental impact. This is reviewed in Chapter 10.

At the reprocessing plant the fuel is dissolved, the remaining fissile materials, both uranium and plutonium, are recovered and concentrated, and the waste fission products are separated, stored, and converted to a form most suitable for disposal or burial. This stage in the fuel cycle involves the largest quantities of unconfined radioactive material and, although the number of reprocessing plants is small, particular attention must be paid to their location and the impact of any effluents. These factors are discussed in Chapter 11.

Finally, the long-lived fission products must be stored or buried in a place where they can decay safely over many centuries without any risk of their reemergence into the biosphere. The selection of a procedure that will guarantee such long-range disposal is clearly difficult and receives considerable public scrutiny. These problems are considered in Chapter 12.

CONCLUSION

Overall consideration of the nuclear fuel cycle is clearly important, but it may impose an undue burden on the utility industry to account for the environmental impact of all the subsidiary portions of the fuel cycle, which in any case are under detailed government supervision.

In practice the environmental impact of nuclear power resolves itself into three major components: the effect of a large amount of heat released at a given location on the surrounding ecosystem; the hazards associated with the contained radioactive materials in reactor fuel and the population exposure to low-levels of radioactive contaminants from plant effluents; and the overall impact on neighboring areas of selecting a specific power plant site in terms of land use, powerline locations, transportation routes, esthetic and recreational factors, and economic benefits.

The balancing and assessment of these various factors is still difficult and subject to many attempts to improve procedures and to develop acceptable criteria. An effort will be made to explain the problems involved and some of the current procedures in use.

REFERENCES

1. Jarrett, H., Ed. *Environmental Quality in a Growing Economy* (Baltimore, Maryland: Resources for the Future, Inc., Johns Hopkins Press, 1966).
2. Katz, M. "Decision Making in the Production of Power," *Scient. Amer.* **224** (3), 191 (1971).
3. Maddox, J. *The Doomsday Syndrome* (New York: McGraw-Hill, 1972).
4. Frejka, T. "Prospects for a Stationary World Population," *Scient. Amer.* **228** (3), 15 (1973).
5. Starr, C. "Energy and Power," *Scient. Amer.* **224**(3), 37 (1971).
6. U.S. Congress. "Environmental Effects of Producing Electric Power," Hearings, Joint Committee on Atomic Energy, Congress of the United States (Washington, D.C.: U.S. Government Printing Office, 1969).
7. Cook, E. "The Flow of Energy in an Industrial Society," *Scient. Amer.* **224**(3), 135 (1971).
8. Hubbert, M. K. "Energy Resources," in *Resources and Man* (San Francisco: Freeman, 1969); also *Bull. Canad. Inst. Mining Met.* **66**(735), 37 (1973).
9. "Environmental Aspects of Nuclear Power Stations," *Proceedings of the New York Symposium 1970.* (Vienna: International Atomic Energy Agency, 1971).
10. Singer, S. F. "Human Energy Production as a Process in the Biosphere," *Scient. Amer.* **223**(3), 174 (1970).

11. "The Nuclear Industry - 1974," AEC Report WASH 1174-74 (Washington, D.C.: U.S. Atomic Energy Commission, 1974).
12. "Uranium-Production and Short-Term Demand," (Paris: Organization for Economic Cooperation and Development, 1969).
13. Curtis, R. and E. Hogan. *Perils of the Peaceful Atom*. (New York: Doubleday, 1969).
14. Gofman, J. W. and A. R. Tamplin. *Poisoned Power* (Emmaus, Pennsylvania: Rodale Press, 1971).
15. Novick, S. *The Careless Atom* (Boston: Houghton, Mifflin , 1969).
16. Shepherd, J. "The Nuclear Threat Inside America," *Look* 34(25), 21 (1970).
17. Cambel, A. B. "Impact of Energy Demands," *Physics Today* 23, 38 (1970).
18. Commoner, B. *Science and Survival* (New York: Viking Press, 1966).
19. Freeman, D. "Toward a Policy of Energy Conservation," *Bull. Atomic Scientists* 27(8), 8 (1971).
20. Martin. J. E., E. D. Harward and D. T. Oakley. "Comparison of Radioactivity from Fossil Fuel and Nuclear Power Plants," HEW Report, U.S. Public Health Service (1969).
21. Netschert, B. C. "How Much Power Do We Really Need?" AIF Atomic Conference, Bal Harbor, Florida (October 1971).
22. Hubbert, M. K. "The Energy Resources of the Earth," *Scient. Amer.* 224 (3), 61 (1971).
23. "Proposed Rule Making Action: Numerical Guides for Design Objectives and Limiting Conditions for Operation to Meet the Criterion 'As Low as Practicable' for Radioactive Material in Light-Water-Cooled Nuclear Power Reactor Effluents," Draft Environmental Statement (Washington, D.C.: U.S. Atomic Energy Commission, 1973).
24. *Nuclear News* 17(4), 52 (1974).
25. Meinel, A. B. and M. P. Meinel. "Physics Looks at Solar Energy," *Physics Today* 25(2), 44; 25(5), 15 (1972).
26. Clark, W. "How to Harness Sunpower and Avoid Pollution," *Smithsonian* 2(8), 14 (1971).
27. Gast, P. L. "Solar Economics," *Physics Today* 25(5), 15 (1972).
28. Gregory, D. P. "The Hydrogen Economy," *Scient. Amer.* 228(1), 13 (1973).
29. Berkowitz, D. A. and A. M. Squires, Eds. *Power Generation and Environmental Change* (Cambridge, Mass.: MIT Press, 1971).
30. Energy Policy Staff. *Electric Power and the Environment*. (Washington, D.C.: Office of Science and Technology, 1970).
31. Federal Power Commission. *The 1970 National Power Survey*. (Washington, D.C.: Federal Power Commission, 1971).
32. Mitchell, E. R. "Only People Pollute," Information Circular IC 268 (Ottawa, Canada: Mines Branch, Dept. of Energy, Mines and Resources, 1971).
33. Mitchell, E. R. "Impact on the Environment of Fuel Combustion and Pollution Abatement Measures," Mines Branch Report FRC

72/58-CCRL. (Ottawa, Canada: Department of Energy, Mines and Resources, 1972).

34. Spinrad, B. I. "The Role of Nuclear Power in Meeting World Energy Needs," *Environmental Aspects of Nuclear Power Stations* (Vienna: International Atomic Energy Agency, 1971).

35. Olander, D. R. "Technical Basis of the Gas Centrifuge," *Adv. Nucl. Sci. Technol.* **6**, 106 (1972).

36. "U.S. AEC Gaseous Diffusion Plant Operations," U.S. Atomic Energy Commission Report ORO-658 (1968).

37. Holden, A. N., Ed. *High-Temperature Nuclear Fuels* (New York: Gordon and Breach, 1967).

CHAPTER 2

THE PASSIVE ENVIRONMENT

THE PASSIVE ENVIRONMENT AND
THE GENERATION OF ELECTRICITY

Man lives in an environment, confined to a narrow layer at the Earth's surface, that has been shaped by geological, climatic and biological factors through the eons to achieve a delicate balance that is increasingly disturbed by man's presence and activities. This balanced environment is in a process of continuous if slow dynamic changes, and any consideration of "preserving" it must allow for these natural changes. Such changes are evident if one regards such phenomena as the glaciation of the polar caps, the cyclic predator-prey relationships and the variations in wild animal population following forest fires, natural droughts, or overgrazing. More subtle changes may be associated with sunspot cycles, shifts in the jet stream or changes in atmospheric carbon dioxide levels. Such changes cannot be considered inherently "good" or "bad" and it seems frivolous to consider environmental conditions purely in relation to momentary human benefits or primarily as resources for the maintenance and comfort of mankind.

In the present context "the environment" enters into the picture in two respects: (1) as a passive framework or stage within which large-scale industrial processes, including electric power generation, occur and whose characteristics are modified in some discernible fashion, and (2) as a dynamic system whose movements and interconnections diffuse and extend the effects of an otherwise highly localized occurrence. It is convenient to discuss the effects of power generation and distribution in terms of the localized "passive" environment, and the movement of any contaminants and their consequent long-range impact in terms of "dynamic" environmental systems, even though in some respects such a distinction may seem artificial.

29

One way of assessing power plant impact on a passive environment is by considering first those effects that are essentially short-range in character, or confined to primary interaction events. Such factors are usually pin-pointed more easily than are long-range effects or possible accumulative effects, although that does not imply that their importance or societal costs are necessarily more readily determined. Among factors in this immediate category the following deserve particular mention:

1. land use and land requirements
2. industrial build-up
3. local pollution effects
4. transportation effects
5. changes in ecology
6. distribution line effects
7. cooling water needs and heat release
8. recreational factors and general amenities
9. aesthetic factors
10. employment opportunities
11. raw materials production
12. noise

Obviously, some of these factors are more than purely "environmental" in character, but they are advanced here because of their significance in cost-benefit assessments, which usually cannot be confined solely to the traditional environment; in other words, we must recognize that our largely man-made environment is already severely dependent on socio-economic factors. Transportation and heat release effects will be discussed in detail in later chapters and are mentioned here solely for completeness. Mention must also be made of transient environmental effects, such as vibration, dust and noise, associated with the construction phase of any large power plant. Although the importance of this to neighboring residents should not be underestimated, it is an inevitable evil that can be minimized by care and good planning but should not be classed with the more permanent effects associated with the other factors listed.

Land Use and Land Requirements

A typical large power plant in the 500-2000 MWe* power range will probably be located in a rural location not too far from a major industrial or population center, on a stream or lake or occasionally near an ocean beach. In general, to keep land costs reasonable previous land use may include marginal farming, wood lots or open pasture. As experience is gained in the reliable operation of large nuclear plants, new power plants may be located closer to cities, and this prospective land use may

---

*MWe = megawatts (electric); unit of power generated for distribution.

compete with commercial surburban land development. On the average, coal-fired plants require a minimum of about 60 acres (24 hectares) for each 1000 MW of capacity. The use of cooling towers and ponds would add to this requirement. A 3000 MW plant would require a minimum of 180 acres (72.85 hectares). These acreages do not include ash disposal areas, which must be found near the plant and, in general, require about 250,000 cubic feet per MW, or about 180 acres of land utilized to a depth of 100 feet (30 m) for the lifetime of operation of a 3000 MW plant (1). Severe environmental problems are associated with stability and seepage problems in such disposal tips and underscore the value of developing industrial uses for fly ash and bottom ash. Table 10 indicates the quantities of ash collected and utilized in the U.S. in 1968 (1); of the total 29 million tons of ash collected only 17.5% found by-product uses at that time.

For nuclear-fueled plants, land requirements are determined almost entirely by exclusion area criteria and may vary from plant to plant depending on the situation of the plant with regard to open water, population areas, the type of reactor, and meteorological conditions. Typical values for site requirements have been 300-600 acres (120-240 ha) for 2200 MW plants comprising two units each; smaller plants do not necessarily require proportionally less land (Chapter 9). In all cases, additional land is needed for the switchyard and transmission lines; the latter will be discussed below.

Not all of the land tied up in the exclusion areas is necessarily unproductive. While access limitations at some plants have turned unneeded land into wildlife refuges, in other cases limited farming and cattle grazing has been permitted, provided adequate radiation monitoring was maintained. Such farming uses are likely to expand as the density of plant location increases and land values go up. Similarly at shore locations the utilization of heated water will encourage use of some of the sequestered shore for fish farming and related activities.

One of the long-range land-use problems relates to the need to stabilize population patterns near large nuclear power plants. Environmental impact assessments attempt to allow for long-term population trends; however, nuclear site selection usually assumes that low population areas near a plant will retain that character even though the availability of large amounts of power and the need for waste heat utilization may tend to attract additional industries and their employees to the neighborhood. This trend requires long-range land-use planning, a notion that is resisted fiercely by American municipalities and land developers alike. Failing this, the effect of large power plants in accelerating urbanization of adjoining land tracts may be the most significant impact factor of them all.

Table 10.   Major Methods of Utilizing Ash in the U.S. in 1968 (1)

	Fly Ash (tons)	Bottom Ash (tons)	Boiler Slag (if Separated from Bottom Ash) (tons)
1. Total ash collected	19,813,747	7,259,212	2,554,560
2. Ash utilized (ash utilized includes ash sold as well as used by company for its own use in concrete, road base stabilization, etc. Does not include ash hauled or pumped by company to a disposal or fill area):			
A. Mixed with cement clinker or mixed with cement (pozzolan cement)	23,458	—	—
B. Mixed with raw material before forming cement clinker	107,666	17,510	23,025
C. Stabilizer for road bases, parking areas, etc.	141,142	17,009	69,046
D. Partial replacement of cement in:			
1. Concrete products (blocks, bricks, pipe, etc.)	113,790	–	17
2. Structural concrete	106,244	–	–
3. Dams and other mass concrete	92,411	–	–
E. Lightweight aggregate	190,192	32,387	–
F. Fill material for roads, construction sites, etc.	205,757	583,094	964,220
G. Filler in asphalt mix	103,827	13,820	63,604
H. Miscellaneous	132,258	448,445	256,136
Total item No. 2	1,216,745	1,112,265	1,376,048
3. Ash removed from plant sites at no cost to utility but not covered in categories listed under "Ash Utilization"	686,738	705,788	96,432
Total utilized items No. 2 and No. 3	1,903,483	1,818,053	1,472,480

Some experiences regarding the impact of large installations on nearby areas have been described in Reference 2.

As operating experience with large nuclear reactors is accumulated and progressively better data are available on the reliability of nuclear power systems, one may confidently expect the land requirements of such plants to shrink. However, such a development lies obviously in the future, at least 8-10 years ahead. The same applies to plans to locate nuclear power plants in densely populated areas close to the load centers in order to

reduce distribution costs. Before the general population will feel sufficiently reassured to accept urban locations for nuclear installations, far
more tangible experience to establish safety and reliability of nuclear
plants will be needed than is available at this time.

Because of the long-range impact of power generating plants and the
relative scarcity of sites that are acceptable to all concerned parties, most
countries are beginning to set aside certain areas for such developments
or are obligating utilities to do so to ensure the availability of suitable
sites in the foreseeable future. In the United States, a start has been
made in this direction through the Power Plant Siting Act of 1973 (3).
Failing this or similar legislation, one can only expect an increasing clus-
tering of power plants at existing sites, ultimately leading to a cumulation
of impact effects under most categories (4).

Air and Water Pollution

Although air and water pollution from industrial sources affect wide
areas through their dispersion in the hydrologic and atmospheric systems,
there is often an immediate highly localized impact zone, as anybody
who has passed through the almost sterilized regions near such places as
large copper smelters or paper mills can attest. Both air and water pollution are foremost in everyone's mind when the question of environmental
damage from power plants is discussed, whether it refers to the now greatly
cleaned-up stacks of the Battersea Power Station in London, the factories
of the Ruhr, or the Four Corners coal-fired power plant at Farmington,
New Mexico. In fossil power plants, sulfur dioxide, nitrogen oxides and
carbon monoxides account for the bulk of the gaseous emissions, and
strenuous efforts are being made in most industrialized countries to reduce
their quantities. Tables 4 and 5 showed the distribution of airborne pollutants by sources in the U.S. and Canada. It is evident that the power
generating industry was a major source of SO_2 emissions; since that time,
a move has been made to restrict the use of high-sulfur fuels by utilities,
resulting in serious fuel shortages and a significant rise in power costs.
Other forms of air pollution, principally over larger cities, can cause
severe enough changes in the atmosphere to result in observable climatic
changes (Table 11).

Attempts to remove sulfur from stack gases have not been notably
successful, and commercial stack-scrubber installations have been notori-
ously unreliable. Better results have been obtained in controlling fly ash
emissions, and Figure 8 indicates the trend towards improved performance
(1) with efficiencies in the 97-99% range for all but the finest ($<0.05\ \mu$)
particles. The substitution of taller stacks to ensure better dispersion of

Table 11. Climatic Changes Produced by Large Cities

Element	Comparison with Rural Environs
Contaminants	
Dust particles	10 times more
Sulfur dioxide	5 times more
Carbon dioxide	10 times more
Carbon monoxide	25 times more
Radiation	
Total on horizontal surface	15 to 20% less
Ultraviolet, winter	30% less
Ultraviolet, summer	5% less
Cloudiness	
Clouds	5 to 10% more
Fog, winter	100% more
Fog, summer	30% more
Precipitation	
Amounts	5 to 10% more
Days with 0.2 in.	10% more
Temperature	
Annual mean	1 to 1.5°F more
Winter minima	2 to 3°F more
Relative humidity	
Annual mean	6% less
Winter	2% less
Summer	8% less
Wind speed	
Annual mean	20 to 30% less
Extreme gusts	10 to 20% less
Calms	5 to 20% less

NO_2 and other pollutants can only be considered an interim solution pending the development of satisfactory removal techniques. Figure 9 shows the trend towards taller stacks; nobody can seriously argue that the environmental impact of a 400-1000-foot tall stack by itself is negligible in any respect! In contrast, the effluents from the stacks of nuclear power plants are relatively clean; the significance of residual discharges of low concentrations of radioactive effluents will be discussed in Chapter 8.

Water pollution is an ever-present by-product of modern civilization, and particularly so in highly industrialized areas. However, neither fossil

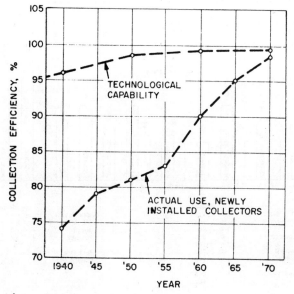

A.

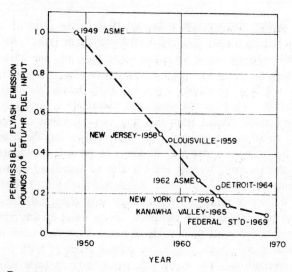

B.

Figure 8. Trends in control of fly ash emission from coal-fired power plants (1).
   A. Growth in efficiency level of fly ash collectors capability versus
   actual use
   B. Trend of emission limits for large units

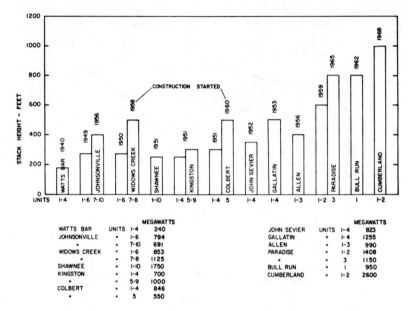

		MEGAWATTS				MEGAWATTS
WATTS BAR	UNITS 1-4	240	JOHN SEVIER	UNITS 1-4	823
JOHNSONVILLE	" 1-6	794	GALLATIN	" 1-4	1255
"	" 7-10	691	ALLEN	" 1-3	990
WIDOWS CREEK	" 1-6	853	PARADISE	" 1-2	1408
"	" 7-8	1125	"	" 3	1150
SHAWNEE	" 1-10	1750	BULL RUN	" 1	950
KINGSTON	" 1-4	700	CUMBERLAND	" 1-2	2600
"	" 5-9	1000			
COLBERT	" 1-4	846			
"	" 5	550			

Figure 9. Historical trend in TVA power plant stack heights (5).

nor nuclear power plants create water pollution in the form of untreated
sewage or industrial wastes discharged into streams or lakes. All steam-
electric plants, however, must discharge waste heat into the environment;
this waste heat, given up by steam in the power plant condensers, is first
absorbed by the flow of cooling water and then usually discharged into
a lake, stream or river near the plant, and eventually is largely dissipated
into the atmosphere. Apart from this, low concentrations of some chemicals such as chlorine and various wood preservatives may be discharged
(6). Since the effluent water may be lower in bacteria and biota than
the bulk of the receiving waters, there is also an associated environmental
effect in terms of reduced nutrients. Similarly, diversion of substantial
portions of the water through a power plant may change the presence or
concentration of essential trace elements. Since water quality is to some
extent defined in terms of permissible bacterial levels and trace element
concentrations, it is important to know normal levels of both under all
regular flow conditions. For many U.S. rivers, data on trace element
content have been reported in the National Water Quality Network (7)
and by Kopp and Kroner (8). Similar information has been compiled in
Canada by the Department of Energy, Mines and Resources. Table 11
summarizes the recommended water quality criteria for public water

Table 12. Surface Water Criteria for Public Water Supplies (9)

Constituent or Characteristic	Permissible Criteria	Desirable Criteria
Physical		
Color (color units)	75	$<$10
Odor	Narrative	Virtually absent
Temperature[a]	Narrative	Narrative
Turbidity	Narrative	Virtually absent
Microbiological		
Coliform organisms	10,000/100 ml[b]	$<$100/100 ml[b]
Fecal coliforms	2,000/100 ml[b]	$<$20/100 ml[b]
Inorganic chemicals	(mg/l)	(mg/l)
Alkalinity	Narrative	Narrative
Ammonia	0.5 (as N)	$<$0.01
Arsenic[a]	0.05	Absent
Barium[a]	1.0	Absent
Boron[a]	1.0	Absent
Cadmium[a]	0.01	Absent
Chloride[a]	250	$<$25
Chromium,[a] hexavalent	0.05	Absent
Copper[a]	1.0	Virtually absent
Dissolved oxygen	$\geq$4 (monthly mean) $\geq$3 (individual sample)	Near saturation
Fluoride[a]	Narrative	Narrative
Hardness[a]	Narrative	Narrative
Iron (filterable)	0.3	Virtually absent
Lead[a]	0.05	Absent
Manganese[a] (filterable)	0.05	Absent
Nitrates plus nitrites[a]	10 (as N)	Virtually absent
pH (range)	6.0-8.5	Narrative
Phosphorus[a]	Narrative	Narrative
Selenium[a]	0.01	Absent
Silver[a]	0.05	Absent
Sulfate[a]	250	$<$50
Total dissolved solids[a] (filterable residue)	500	$<$200
Uranyl ion[a]	5	Absent
Zinc[a]	5	Virtually absent
Organic chemicals		
Carbon chloroform extract[a] (CCE)	0.15	$<$0.04
Cyanide[a]	0.20	Absent
Methylene blue active substances[a]	0.5	Virtually absent
Oil and grease[a]	Virtually absent	Absent
Pesticides		
Aldrin[a]	0.017	Absent
Chlordane[a]	0.003	Absent
DDT[a]	0.042	Absent
Dieldrin[a]	0.017	Absent
Endrin[a]	0.001	Absent
Heptachlor[a]	0.018	Absent
Heptachlor epoxide[a]	0.018	Absent
Lindane[a]	0.056	Absent
Methoxychlor[a]	0.035	Absent
Organic phosphates plus carbamates[a]	0.1[c]	Absent
Toxaphene[a]	0.005	Absent
Herbicides		
2,4-D plus 2,4,5-T, plus 2,4,5-TP[a]	0.1	Absent
Phenols[a]	0.001	Absent

Table 12, continued

Constituent or Characteristic	Permissible Criteria	Desirable Criteria
Radioactivity	(pCi/l)	(pCi/l)
Gross beta[a]	1,000	$\leq$100
Radium-226[a]	3	$<$ 1
Strontium-90[a]	10	$<$ 2

[a]The defined treatment process has little effect on this constituent.

[b]Microbiological limits are monthly arithmetic averages based upon an adequate number of samples. Total coliform limit may be relaxed if fecal coliform concentration does not exceed the specified limit.

[c]As parathion in cholinesterase inhibition. It may be necessary to resort to even lower concentrations for some compounds or mixtures.

supplies (9). As in similar connection elsewhere in this book, in using such data it is important to appreciate the difference between complying with a stated criterion, such as maximum desirable concentrations, and achieving minimum impact by altering preexisting conditions, whatever they were, to the least extent. Many of the U.S. states have incorporated similar water quality criteria into their regulatory guidelines.

NATURAL ECOSYSTEMS

In considering the environmental impact of industrial activities, one immediately thinks of the animal and plant life in the area; there is a natural concern to preserve their status quo. In actual fact, most ecosystems are themselves in a continuous state of flux and subject to even minor changes in the many ecological cycles required to maintain them. Since life is supported by a finite amount of solar energy that is fixed by green plants, any factors changing the amount of solar energy available or the type and character of plants in a given region will change the life cycles of organisms and indirectly of man there. Forests, which cover about a tenth of the earth's surface, fix almost half of the biosphere's total energy. Massive destruction of woods and forests changes the type of vegetation possible and by facilitating land erosion often reduces the amount of fertile soil available to other plants. It is possible to describe the natural and agricultural ecosystems in terms of energy cycles in which any diversion of energy available to any one species has its ineluctable consequences (10). Any variation in the quantity of nutrients along the cycle sets off a chain reaction, such as the cycle of arctic foxes and lemmings, or the northerly shift of herrings and sharks following a fractional rise in ocean temperatures.

Compared with such large-scale ecological changes the impact of any single industrial plant is obviously small and only of purely local significance. There can be no doubt that placing a power plant on a previously deserted ocean beach may affect the egg laying of turtles or grunnions, just as the location of such a plant in the desert or on a quiet lake will affect the survival of existing animals there. Some animals readily adapt to new conditions such as urbanization, and some animals multiply in the less competitive "no-hunting" woods of many national research laboratories to the extent that overpopulation becomes a problem.

Given the decision that additional power plants are needed, and not everybody is willing to concede this at this moment, the problem then becomes how to select a site having the least deleterious impact on the natural environment. Clearly, this becomes a matter of judgment; is the "value" of a given species related to its rarity? Are trout automatically more desirable than catfish? It has been suggested that diversity of species is a more useful index of ecological impact than the fate of any selected "valuable" species. Such value questions highlight the enormous task of analyzing the ecosystem of any locality and the food chain of each species to determine how such an impact can be minimized. This is an almost impossible undertaking, yet the environmental laws in various countries impose a requirement to determine the suitability of alternative sites in terms of their relative impact on flora and fauna. Common sense demands avoiding areas of obvious scenic beauty, historic sites, wildlife refuges and recognized nesting grounds of rare birds, unique geographic features and important spawning grounds, but beyond such considerations it is often barely possible to stay ahead of the free-enterprise real estate developer who is untrammeled by law or environmental considerations. Though aiming to preserve existing environmental conditions, it must be realized that the continued presence of man itself in most places irrevocably changes conditions for plant and animal life.

In justifying the selection of a specific site, U.S. Nuclear Regulatory Commission (NRC) Guidelines require an applicant "to identify the important local flora and fauna, their habitats and distribution as well as the relationship between species and their environments. A species, whether animal or plant, is important if it is commercially or recreationally valuable, if it is rare or endangered, if it is of specific scientific interest, or if it is necessary to the well-being of some significant species (*e.g.*, a food chain component) or to the balance of the ecological system" (11).

Such a catalog of local organisms should include a discussion of terrestrial vertebrate species that migrate through the area or use it for breeding grounds, the distribution of principal plant communities, and of plant and animal populations within the aquatic environments. The discussion must include descriptions of area usage (*e.g.*, habitat and breeding), life histories of important regional animals, their normal population fluctuations and habitat requirements such as thermal tolerance ranges, and identification of food chains and interspecies relationship. Correlating these relationships with predictions or evaluations of the environmental impact of a nuclear power plant, or any other industrial complex, on the regional biota poses the greatest difficulty to a conscientious applicant because only rarely does our current understanding extend fully to the role of the lower-order biota in many of the more complex food chains.

In view of the potential complexity of supplying a complete ecological inventory, a plant operator can only identify the most prominent species and subsequently provide adequate environmental surveillance programs to verify predictions of low impact. As an example of a very detailed ecological analysis, the Environmental Statement for the Watts Bar Nuclear Plant (12) includes listings of seven terrestrial game species, 32 zooplankton and 43 phytoplankton species, and 37 fishes observed in the vicinity of the site. Five endangered bird species are mentioned and there are ten pages listing mammals, birds, amphibians and reptiles whose ranges include the plant site. The Environmental Statement for the Zion Nuclear Power Station (13) includes 11 pages listing such vertebrates under the various headings and 19 pages listing vascular plants observed in the vicinity. Though required by the NRC, it is hard to see that such lists by themselves aid greatly in the evaluation of true environmental impact.

Site assessment should include a vegetation survey, supplemented by a review of shrub cover, major plant communities and a description of a field test program to observe any changes in the plant cover of the site area. Ideally, a survey should be made at intervals over a five-mile radius from the plant; in practice the work involved forces concentration on a relatively small number of indicator plants and animals.

Aesthetic and Recreational Factors

Power plants vary widely in the visual impression they make on the viewer. Depending on one's point of view their appearance may be judged as blending harmoniously with their surroundings or as an insult to an otherwise attractive countryside. Objectively, suitable architectural treatment utilizing colors and textures, reflections in cooling ponds and a distribution of masses can do much to enhance the plants' appearance. In the writer's experience, the French have been most concerned and successful in providing appropriate settings for nuclear facilities.

There has been some objection to the appearance of tall hyperbolic cooling towers. To the observer who does not cherish their stark graceful lines they may appear objectionably obtrusive. Attempts to camouflage them by painting them or surrounding them with a shroud rarely have been successful; where their appearance is considered unacceptable, alternative means of providing comparable cooling capacity may have to be explored.

The aesthetic impact of high voltage transmission lines is constantly under attack. At the voltages in common use, 230 kV and 500 kV, underground lines are not feasible now. The prospect of superconducting transmission is still a long way from practical realization and is not a feasible alternative at present (14). On the other hand, as Figure 10 shows, the cost of long distance transmission is encouraging a move to higher transmission voltages. At the moment, the only thing a utility can do is to run high tension lines along the least objectionable route, and such routes are getting harder to find.

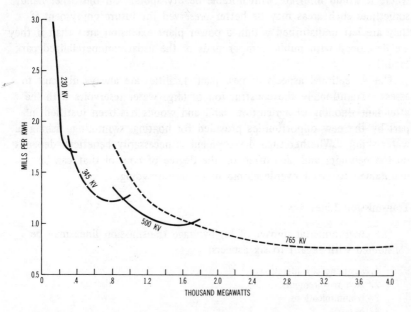

Figure 10. Cost of electric energy transmission for 200 miles in 1970 (1).

In addition to the reactor building the general appearance of a power plant encompasses the turbine building, cooling towers, water treatment and waste storage tanks, and the switch yard. While the switch yard can rarely be aesthetically pleasing, there is no reason why the grounds cannot be landscaped and planted in a way to improve their appearance.

Suitable placing of plants and shrubbery may also help to reduce noise effects.

The plant location and power line routing should not impinge on areas valued for their historic or natural significance. Only an industry very insensitive to the historical past will permit the immediate juxtaposition of a power plant and an historic monument or the mining of coal up to the walls of historic buildings, as happened at Temple Newsham in England. On the other hand, an historic site, shrine, battlefield or monument does not rule out locating a nuclear power plant in a well-screened location in the vicinity, as long as the visitors to such a site are taken into account in assessing and minimizing population radiation exposure.

Siting near areas of natural significance, wildlife refuges, beauty spots, and the like clearly requires a rational approach and a careful screening of alternative sites. It is obviously inappropriate to locate a power plant where it would interfere with notable beauty spots. On the other hand, sometimes such areas may be better preserved for future enjoyment if they are left undisturbed within a power plant exclusion area than if they are developed with public campgrounds or the usual commercialized park facilities.

The recreational aspects of new plant facilities are always difficult to assess. Traditionally the construction of large water reservoirs with the attendant flooding of agricultural land and woods has been justified in part by the new opportunities provided for boating, swimming, fishing or water skiing. Whether such development is necessarily beneficial depends on the beholder and also often on the degree of control that can be maintained to avoid overdevelopment or overcrowding.

Transmission Lines

The environmental impact of high-voltage transmission lines may be considered with regard to six aspects:

1. aesthetic considerations
2. land requirements
3. communications
4. hazards
5. ozone
6. habitat effects

There has been some reference already to the aesthetic objections to high-voltage transmission lines. Since underground lines at this time are limited to 250 kV and to short-range distribution systems, high-voltage transmission lines will remain part of our environmental scene for a considerable time. Efforts have been made to design more attractive towers and to select cables permitting longer free distances between towers. Routing can take advantage of topological features to reduce visibility, and use of higher transmission voltages will reduce the number of lines radiating from major power centers. General guidelines for good practice in routing, designing and clearing for transmission lines were published recently (15,16).

Land requirements will vary with the number of lines and the height of towers, but the right-of-way for major transmission lines may be 100-400 feet (30-120 m) wide. In general, use of the land below the lines is restricted to pasturing or low-intensity farming, though in most cases the utility may be sole owner of the right-of-way, may keep it fenced and sheared, and restrict access. Such restrictions will be governed by state law and federal criteria (15). In any case, it is evident that the land demand for easements associated with any new power station that may be 40 miles or more from major load centers will greatly exceed the land area of the plant itself and of the exclusion area, which would be 600-1000 acres (240-400 hectares) for a 1200 MWe plant. For the Watts Bar plant, for example, the plant site is 967 acres, and transmission lines require easements on 3165 acres; for the Maine Yankee plant of 855 MW, the plant site is 125 acres plus a 130-acre pond, and the two extra transmission rights-of-way are 3.5 miles long, 350 feet and 170 feet wide, respectively, or 441 acres in area.

Regarding communication problems, high tension lines may cause some interference with nearby radio and television reception and may introduce fluctuation on signal strengths on windy days or under icy conditions. As a rule they are less subject to icing than low power distribution lines and are designed to withstand strong winds and earthquakes. Tall towers and multiple lines may pose a hazard to air traffic, particularly under conditions of poor visibility, and major air lanes need to be considered when selecting a route. Land erosion can be minimized by rapid replacement of ground cover and a policy of minimum soil disturbance. Near beach sites, salt encrustation on power lines and insulators may cause arcing problems.

Ozone may be generated by any corona or electrical discharge in air or other oxygen medium. The quantities produced depend on the severity of the discharge and the quantity of oxygen in the affected volume. Corona discharges can increase as a result of abrasions, corrosion effects, foreign particles or sharp points on electrical conductors, or incorrect design that produces excessively high potential gradients.

Ozone is not considered injurious to vegetables, animals or humans, unless concentrations exceed about 0.05 ppm over prolonged periods (17). Extremely sensitive varieties of tobacco can be injured after about eight hours of exposure of 0.05 ppm ozone or one hour exposure of 0.07 ppm. Most people generally experience discomfort from ozone's odor by the time concentrations approach 0.05 ppm, and visual acuity may be impaired by prolonged exposures of 0.20 to 0.50 ppm. The Community Air Quality Guide (17) recommends an upper concentration limit of 0.05 ppm for not more than 1-2 hours per day to protect very sensitive plants and an exposure limit of 0.1 ppm/hr/day on the average during any year if human health is not to be significantly impaired during a lifetime of exposure.

Experimental work has been conducted to determine the added ozone production around a 1000 kV high-voltage test line. The results showed an insignificant increase in ozone concentration over that produced by sunlight. Consequently, current types of high tension lines are not expected to produce harmful ozone concentrations (12).

Increased public pressure to minimize obtrusion of transmission lines and the economic incentives to make maximum use of rights-of-way and to transmit large blocks of power reliably over long distances with the minimum number of lines have accelerated the trend toward higher transmission voltages. Many transmission lines now operate at 525 and 765 kV, with consideration being given to 1000 kV and above. Adoption of voltage levels above 765 kV is complicated and requires detailed economic and technical studies before decisions can be reached. For example, the limits of practical air insulation may not be too far above the 1300 kV level and could be the determining factor in establishing a transmission voltage level.

The capability of a transmission circuit increases approximately as the square of its rated voltage. This higher voltage transmission becomes intrinsically more economical if the transmission circuit can be built for higher voltages at a cost that increases less than its transmission capability. Concentrating generating capacity in single units and plants has created a similar requirement for concentration of transmission capability on a single transmission corridor.

Actual costs of several 500-kV overhead transmission lines prior to 1969 averaged $120,000 per mile including right-of-way, clearance, and line construction. In urban areas, not only are right-of-way costs higher, but line costs are also higher because of the large number of angle towers, increased clearance requirements, foundation problems, and aesthetic considerations.

Neglecting the problems associated with system stability and thermal capacity of conductors, one 765-kV line is equivalent in load-carrying capability to five 345-kV lines or thirty 138-kV lines. Spacewise, a 765-kV line requires only about 200 feet (66 m) of right-of-way, as compared with about 750 feet (240 m) needed for five 345-kV lines and about 3,000 feet (1 km) needed for thirty 138-kV lines. Due to the scarcity of available land for rights-of-way, EHV transmission is essential except in metropolitan areas where overhead transmission lines are prohibited. In this latter case, underground cable must be used.

The interface or "edge" between two diverse plant communities will often produce or attract more kinds and numbers of animals than would occur in either habitat type alone. Power line right-of-way has great potential as a wildlife habitat because of the "edge effect" that it creates and the high food and cover productivity of emerging vegetation. Many power lines are good producers of wildlife without special management because of these two factors.

Shear clearing through heavily forested areas is not inconsistent with good forestry and wildlife management practices. A common management practice in large sections of unbroken forest land is to open the tract by means of small, evenly spaced clearings. The rationale for this practice is to provide diversity and food in the forest environment and to create edge. Wildfires originally provided this type of habitat. Power line right-of-way creates long linear forest openings that are indefinitely maintained to prevent power outages. The sunlight penetrating the forest via the right-of-way stimulates understory growth adjacent to the power line. Periodic line maintenance may perpetuate these beneficial wildlife habitat conditions.

Where transmission line right-of-way crosses wooded areas, the utility may perform the necessary clearing as its part of a cooperative arrangement with county agents, state and federal park commissions, soil conservation agencies, sportsmen groups, and other interested agencies that propose compatible uses for wooded land within easement areas that will meet the goals of the interested parties. Under such an arrangement, forest development that allows growing of small trees such as Christmas trees and nursery stock can be implemented. Also, buckwheat, Korean and Kobe Lespedeza, and other low-growing seed crops and grasses beneficial to a small game habitat can be planted for the establishment of shooting preserves. Right-of-way not totally cleared can be utilized for production of many low-growing forest products.

Many additional multiple right-of-way uses can be identified. If the landowner desires to use the right-of-way for the establishment of playgrounds, athletic fields, golf courses, parks, picnic areas or trails for hiking and horseback riding, such use would usually be permissible.

Raw Materials Production

The impact of fuel material extraction from the ground is significant and very evident. Of the various fuels, coal mining is notorious for its hazardous working conditions and its attendant damage to the countryside. Strip mining is practiced widely and often leaves the country scarred and devastated; only recently have mining companies made any attempt at partial restoration. Underground mining frequently has led to serious subsidences and to build-up of large coal tips and waste pits that are unsightly and occasionally unstable. Acid mine waters can cause serious water pollution.

Uranium is usually mined underground in hard rock and considerably less gangue material is produced. The total quantities of ore handled are small compared with coal mines, ranging from 500 to 5000 tons per day. Some open pit mining is done for uranium, including the famous Shinkolobwe mine in Katanga and some of the Australian mines, with the usual surface effects. The special problems associated with radon in mines and with uranium mill tailings will be discussed in Chapter 11.

Oil and gas production has comparatively little environmental impact. What there is is primarily associated with the big pipelines that cross the country or with off-shore drilling. Recent debates on the effects of the Trans-Alaska pipeline are an indication of the type of problem encountered. There are some aesthetic effects associated with derricks and well pumps, but few people quarrel with the associated benefits. Pollution of ocean waters and beaches by tankers and from wrecks has been the most serious environmental effect. The special problems connected with gas stimulation by underground nuclear explosions will not be discussed here.

Other areas of environmental concern and their economic cost will be discussed in their appropriate context elsewhere.

REFERENCES

1. *The 1970 National Power Survey* (Washington, D.C.: Federal Power Commission, 1971).
2. Breese, G., R. J. Klingenmeier, H. P. Cahill, J. E. Whelan, A. E. Church and D. E. Whiteman. *The Impact of Large Installations on Nearby Areas* (Beverly Hills, California: Sage Publications, Inc., 1965).
3. United States Congress. "Power Plant Siting Act of 1973," HR 180, 93rd U.S. Congress, Washington, D.C. (1973).
4. U.S. Senate. "Land Use Policy and Planning Assistance Act," (S.268), Hearings, Committee of Interior and Insular Affairs, U.S. Senate (1973).

5. Gartrell, F. E., G. F. Stone and T. A. Wojtalik. "Environmental Quality Protection: Large Steam-Electric Power Stations," In *Environmental Aspects of Nuclear Power Stations* (Vienna: International Atomic Energy Agency, 1971).

6. Becker, C. B. and T. O. Thatcher. "Toxicity of Power Plant Chemicals to Aquatic Life," U.S. AEC Report WASH-1249 (Washington, D.C.: U.S. Atomic Energy Commission, 1973).

7. National Water Quality Network, 1962. Public Health Service, Publ. No. 663 (Washington, D.C.: U.S. Dept. of Health, Education and Welfare, 1963).

8. Kopp, J. F. and R. C. Kroner. "Trace Metals in Waters of the United States," (Cincinnati, Ohio: Federal Water Pollution Control Administration, U.S. Dept. of Interior, 1967).

9. "Water Quality Criteria," (Washington, D.C.: Federal Water Pollution Control Administration, 1968).

10. Woodwell, G. M. "The Energy Cycle of the Biosphere," *Scient. Amer.* **223** (3), 64 (1970).

11. "Guide to the Preparation of Environmental Reports for Nuclear Power Plants," (Washington, D.C.: U.S. Atomic Energy Commission, 1973).

12. "Environmental Statement, Watts Bar Nuclear Plant, Units 1 and 2," TVA Report OHES-EIS-72-9. (Chattanooga, Tennessee: Tennessee Valley Authority, 1972).

13. "Environmental Statement. Zion Nuclear Power Station, Units 1 and 2," Commonwealth Edison Company, U.S. Atomic Energy Commission (1972).

14. Snowden, D. P. "Superconductors for Power Transmission," *Scient. Amer.* **224** (10), 84 (1972).

15. "Environmental Criteria for Electric Transmission Systems." (Washington, D.C.: U.S. Dept. of the Interior, 1970).

16. Electric Power Research Institute. "Transmission Line Reference Book: 345 kV and Above," Palo Alto, California (1975).

17. "Community Air Quality Guides, Ozone," *J. Amer. Hyg. Assoc.* **29**, 299 (1968).

CHAPTER 3

THE DYNAMIC ENVIRONMENT

Apart from immediate, purely local effects, widespread environmental changes may result through action of transport mechanisms in the biosphere. Such changes may be long-range or even global in character, resulting primarily from atmospheric circulation of air and airborne matter, surface and subsurface circulation of water, and the interrelation of plant and animal matter through the various food chains. With regard to nuclear power plants, these mechanisms may be responsible for the dispersion, intentional or accidental, of radioactive contaminants from their point of release and in some cases for any subsequent reconcentration. Thanks to the very extensive studies on fallout distribution from nuclear explosions that have been carried out over the past three decades and to the use of radioactive tracer methods to detect the mobility and uptake of the more important fission products, we now have a qualitative understanding of the mechanisms involved, even though the complexity of the phenomena, the scale of effects and the large number of variables at this time still preclude accurate prediction, especially of atmospheric effects.

In addition, the various dynamic chains interact with each other, and the physical and chemical forms of the contaminants may change in different surroundings. An attempt will be made to present here those facets of the major chains having the greatest influence on the prospective movement of radioactive matter in the environment; details on specific materials will be covered in Chapter 8.

THE AQUATIC ENVIRONMENT

Viewed as a cyclic system, water circulates from the vast reservoir of the oceans by solar evaporation into the atmosphere where it may condense into clouds. Under wind action some of the clouds are moved to land areas where the water precipitates as rain or snow. Subsequent run-off may pass through subsurface aquifers, as well as by surface runoff, returning the water ultimately to the oceans (1). Since precipitation may occur close to the source or thousands of miles away, the cycle residence time may vary from a few hours to several weeks; a general average is nine or ten days. The cycle is also strongly affected by seasonal variations in solar radiation, wind conditions, air temperature and the presence of condensation nuclei and dust particles in the air. Water quality and quantity in underground water storage and water table conditions are especially susceptible to interference from human activities, particularly by the pumping out of well water, by changes in surface drainage patterns and by impoundment of streams. The movement of pollutants in the water cycle in general will be at low concentrations and may be influenced by the presence of other materials, both organic or inorganic, in the water. This includes dissolved ions, variations in pH, suspended insoluble matter or aggregates, and silt and clay particles, which may attract soluble ions by ion exchange. Table 13 lists the various processes affecting the movement of material in surface waters.

Table 13. Transport Vector Categories

   I. Physical processes
      A. Transport by water current
      B. Dispersion and dilution by turbulent and convective mixing
      C. Deposition of solids on bottom
      D. Resuspension of deposited material
  II. Chemical processes
      A. Precipitation of dissolved materials
      B. Dissolution of solid materials
      C. Simple combination and complexing
      D. Oxidation or reduction
      E. Sorption of solutes by solids due to ion exchange, fixation, attraction, etc.
 III. Biological processes
      A. Plants
         1. Absorption, adsorption
         2. Metabolism, respiration, photosynthesis
         3. Release of materials upon death
         4. Decomposition and dissolution
      B. Animals
         1. Absorption
         2. Metabolism, respiration
         3. Excretion
         4. Transport by motile species

The main aquatic transport phenomena relevant to nuclear power plants are entrainment and diffusion processes in lakes and streams. They determine the dissipation of heat in water or the dispersion of radioactive contaminants, with adsorption and precipitation processes aiding the removal or reconcentration of effluent materials and the movement of contaminated fluids through permeable media. These processes differ fundamentally from each other and will be reviewed in turn. Special conditions are superimposed on them in tidal estuaries and near shallow beaches, and these, too, will be discussed briefly.

Dispersion Processes in Lakes and Streams

When pollutants are discharged into a stream, the parameter of interest is the concentration variation downstream attending a well-localized discharge of an identifiable effluent into a moving stream. The pollutant may consist of soluble contaminants, fine particulates, or heated water, and it is often desirable to predict and measure rates of dilution or entrainment, the affected water volumes, recirculation patterns and any possible reconcentration by sediment transport or organic vectors.

In the absence of currents the elementary diffusion law provides an estimate of the concentration distribution:

$$C_{(r,t)} = QC_0 \left(\frac{1}{2\sqrt{\pi D t}} \right)^3 e^{-\frac{r^2}{4Dt}} \tag{1}$$

where r is the distance between the observation point and the pollutant outlet point, t is the time interval the pollutant takes to travel the distance r, and D is the diffusion coefficient (assumed constant irrespective of the diffusion direction).

If the effluent material is released into a stationary or a slowly moving body of water, it approaches equilibrium as described by an exponential relation of the form

$$C_{(r,t)} = \frac{Q}{R} \left(1 - e^{-kt} \right) \tag{2}$$

where $C_{(r,t)}$ is the total concentration of effluent material (or total heat in appropriate units) present in the lake, over and above the normal background concentration, at time t, Q is the rate of plant effluent addition to the lake, and k is the rate constant determined by the rate of decay or removal from the lake, whether it is by outflow from the lake or by dissipation or deposition at the surface or the bottom of the lake.

$$k = a + \frac{f}{V} \tag{3}$$

where a is the coefficient of excess heat loss, f is the flow rate through the lake, and V is the effective lake volume.

The flow rate can be calculated for a given lake by measuring the outflow across a well-defined bed cross section or a weir. In the more general case of a stream with irregular bed contours, measurement or prediction of flow velocities and mixing rates is much more complicated. For open channels with approximately trapezoidal cross section, any of the well-known hydraulic equations may be applied. For uniform flow and a moderate slope, the most frequently used formula is the Chézy-Manning formula:

$$V = \frac{1.49}{n} R^{2/3} S^{1/2} \tag{4}$$

where R is the hydraulic radius (cross-sectional area divided by wetted perimeter), S is the slope of the energy grade line (equal to the water surface slope for uniform flow), and n is the Manning roughness coefficient, a property of the channel bed surface. Table 14 tabulates values of the roughness coefficient n for different channel linings. If the roughness is not uniform across the channel width, an average value of n must be selected, or the channel may be treated as two or more contiguous channels. This formula serves mainly as a first approximation under idealized conditions.

Actually, the value of n is highly variable, and when selecting a proper value of n it is useful to be aware of the factors determining it. The primary characteristic in bed surface roughness is size and shape of the grains of the material forming the wetted perimeter, with coarse grains resulting in a relatively high value of n and fine grains in a low value. In addition, erosion and flow obstructions may impede flow significantly. Stage height, *i.e.*, depth of water exposing any nonuniform characteristics of the stream bottom, may significantly affect the effective value of n (2). Figure 11 shows a nomograph to solve Equation 4 for simple cases. For more detailed calculations the reader is referred to standard texts on hydraulics.

It should be noted that use of the Manning formula is questionable for very wide streams where effects of surface roughness become less significant. It also does not consider the effect of suspended matter and bed loads. In the absence of any natural or artificial source of turbulence, very little lateral mixing may occur even at moderate stream velocities, and dilution of effluents may proceed relatively slowly. Figure 12 shows

Table 14. Hydraulic Roughness Coefficients for Different Channel Linings in
Clean, Straight Channels (2)

Type of Lining	Condition	n (Kutter and Manning)	γ (Bazin)
Glazed coating or enamel	In perfect order	0.010	
Timber	Planed boards, carefully laid	0.010	0.06
	Planed boards, inferior workmanship or aged	0.012	
	Unplaned boards, carefully laid	0.012	0.16
	Unplaned boards, inferior workmanship or aged	0.014	
Metal	Smooth	0.010	0.06
	Riveted	0.015	0.30
	Slightly tuberculated	0.020	
Masonry	Neat cement plaster	0.010	0.06
	Sand and cement plaster	0.012	
	Concrete, steel troweled	0.012	
	Concrete, wood troweled	0.013	
	Brick, in good condition	0.013	0.16
	Brick, rough	0.015	0.30
	Masonry in bad condition	0.020	
Stonework	Smooth, dressed ashlar	0.013	0.16
	Rubble set in cement	0.017	0.46
	Fine, well-packed gravel	0.020	
Earth	Regular surface in good condition	0.020	0.85
	In ordinary condition	0.0225	1.30
	With stones and weeds	0.025	1.75
	In poor condition	0.035	
	Partially obstructed with debris or weeds	0.050	

an example of the horizontal mixing pattern in the Columbia River; even there, conditions will vary from point to point and with momentary flow conditions. For precise predictions of mixing patterns, detailed horizontal and vertical measurements must be made under all representative flow rate and volume conditions.

Mixing processes near shore and in deep waters in the marine environment show significant divergences from the simple Fick diffusion law. This has been discussed in detail in a 1971 report of the National Academy of Sciences (4).

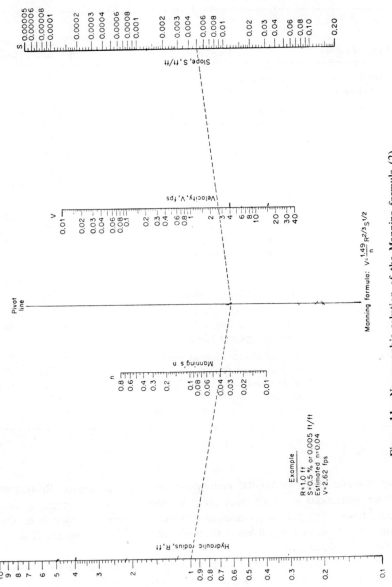

Figure 11. Nomographic solution of the Manning formula (2).

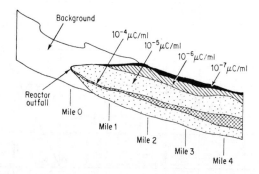

Figure 12. Horizontal mixing of radioactive reactor effluents in the Columbia River, 1958 (3).

Effect of Suspended Silt and Solids

Every stream carries some suspended sediment and moves larger solids along the stream bed as "bed load." Since the specific gravity of soil materials is about 2.65, the particles of suspended sediment tend to settle to the channel bottom, but upward currents in turbulent flow tend to counteract the gravitational settling. The smallest particles may remain in suspension for a long time and some may pass dams and weirs with water discharged through sluiceways, turbines, or over the spillway.

The relation between total sediment transport Q and stream flow q is often expressed by the equation

$$Q = kq^{n'} \tag{5}$$

where n' commonly varies between 2 and 3, and k is usually small. A sediment rating curve such as Figure 13 may be used to estimate total sediment transport from the continuous record of streamflow (5). The settling rate and sediment volume vary significantly with particle size distribution; appreciable fluctuations in particle size distribution may be encountered in a single stream over a range of flow conditions (6).

The sediments play an important role in aquatic ecology by serving as a repository for radioactive substances and soluble chemical pollutants that they may transport over considerable distances and may pass to the higher trophic by way of bottom-feeding biota. The main mechanism of removal of dissolved matter is ion exchange on sediment surfaces; particulates with good ion exchange properties, such as most clay minerals, act

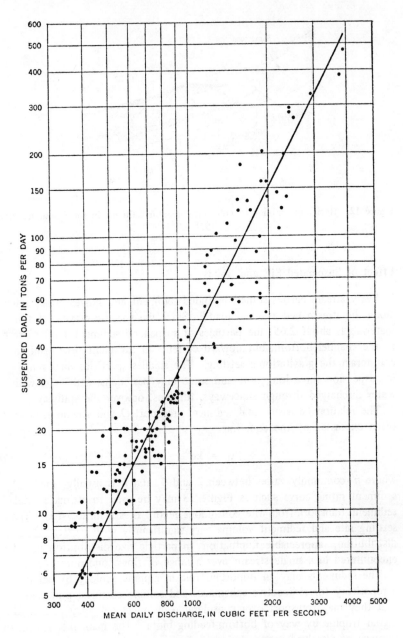

Figure 13. Suspended-sediment transport curve for Ichawaynochaway Creek at Milford, Georgia (6).

as efficient scavengers and may serve to purify the water from the more readily adsorbed ions. Kennedy (7) has reported the ion exchange properties of many fluvial sediments. Since the exchange properties of different elements in solution vary markedly, this process results in significant modification of the proportion and concentrations of radionuclides released in effluents to streams carrying an appreciable sediment load (8). This fractionation is shown in Figure 14 which shows the effect of clay minerals and suspended organic materials in changing the proportion of trace elements remaining dissolved in water. More recently Duursma and Gross (4) have studied the sorption and desorption of various radionuclides on marine sediments in the Dutch Wadden Zee, the Irish Sea and the Mediterranean. As might be suspected, the smallest size fractions (below 16μ) account for the main adsorption removal. Sorption rates will also depend on oxygen levels; anaerobic conditions are rare in open waters but may be found in interstitial waters of sediments, especially those deposited in highly productive nearshore ocean areas. The scavenging process through sediment adsorption for radionuclides from fallout and reactor effluents is illustrated in Figure 15, for the cesium-137 levels measured in the upper reaches of the Hudson River (9).

Water Quality

The impact of any industrial operation on local water supplies and water quality can be determined in two ways: by measuring the change in specific characteristics in existing streams and reservoirs resulting from plant operations, or by establishing set water quality standards that must be met by effluents or receiving streams regardless of prior conditions. The first approach leads to difficulties in enforcement since baseline data are often unavailable over a long enough period to establish such factors as diurnal or seasonal variations in water levels, temperature, concentrations of trace elements or pesticides. In the U.S., analyses of trace elements for many major rivers have been published for the National Water Quality Network (10); however, many of the more exotic elements of interest in nuclear technology, such as zirconium, ruthenium or cesium, are rarely reported. Kopp and Kroner have published analyses for trace elements in selected rivers and lakes (11) indicating concentrations that may be expected in various mineral provinces, but again many elements of potential importance are not referred to.

In many cases it will be found that the effluent released by a plant has been demineralized or sterilized and thus is cleaner than the intake water. That type of environmental impact may be just as objectionable as the release of potentially toxic trace elements even at low concentrations.

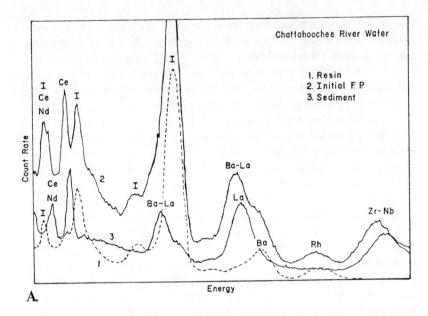

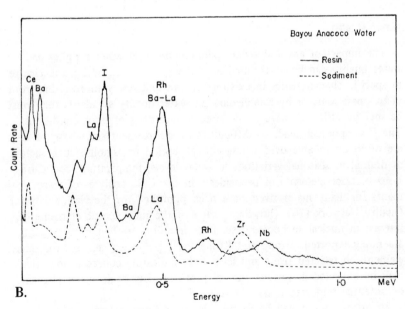

Figure 14. Fractionation of fission products: Gamma-ray spectra of fission products in solution and adsorbed on sediment (8).

A: river water with clay sediments;   B: swamp water with high organic content.

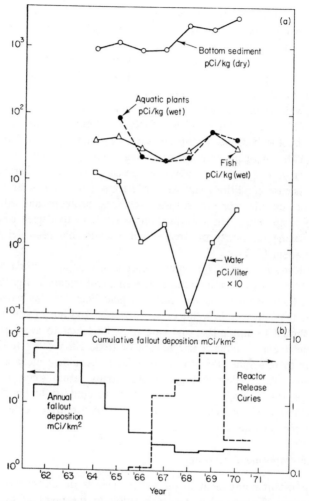

Figure 15. Cesium-137 activities in the upper reaches of the Hudson River, 1962-1971 (9).

Typically, biocides may be added to cooling water to inhibit algal growth, wood preservatives and trace minerals may appear in blowdown water from cooling towers, and metallic ions may be added from pipe corrosion. The effect of these additions is hard to assess and the cost of removing them may be out of proportion to the potential damage they may cause. On the other hand, fish kills due to excessive flow rates at intake trash racks can be readily minimized by proper screen design.

In view of these many variables it is often preferable to set certain water quality standards that must be met regardless of baseline conditions. Typically such criteria would establish acceptable ranges of hardness, turbidity, pH, odor, color, temperature, sediment concentration, concentration of selected inorganic and organic tracer materials, COD or biological oxygen demand (BOD), and radioactivity. In many cases, state regulations insist on water purity in plant effluents equivalent to or better than that required for public water supplies. Table 12 presented a list of criteria proposed by the National Technical Advisory Committee on Water Quality Criteria (12). In many practical cases, condition of the intake water needs to be taken into account. Criteria on radioactivity are further qualified by plant license conditions and will be discussed in detail later. Similarly, temperature criteria represent a severe technical problem and will be discussed in Chapter 6. The classification of rivers as to degree of cleanness and the testing for purity of streams and estuaries has been described in detail by Klein (13).

When seawater is used for cooling purposes, separate criteria have to be established to cause minimum interference with marine organisms. At this time, there seem to be few accepted guidelines for this purpose. It is evident that changes in salinity, pH, dissolved oxygen or temperature changes should be minimized and care must be taken to avoid release of oil and petroleum products, raw sewage or toxic chemicals. It is also important to remember that the marine food chain may include many detrital and filter feeders that may preferentially return some of the more insoluble elements, such as zinc or mercury, in contrast to the terrestrial food chain through herbivores, which tend to concentrate some of the more soluble pollutants.

Subsurface Movement of Water

An important component in the water cycle is the subsurface movement of water from the area of surface deposition to the point of release as well water or spring water. The basic mechanism of percolation through a porous medium is well understood and is thoroughly discussed in texts on hydrology (1,14). "Pure" water will flow under diffusion-controlled conditions through any permeable medium under the influence of gravity or pressure gradients, given continuity of geological layers that can serve as aquifers. This type of dispersion has been discussed in detail by Baetslé (15,16) who provides dispersion coefficients for various flow velocities and particle sizes. The intersection of such aquifers with the soil-atmosphere interface and its physical condition will govern the release of free water to surface streams. For the simplest case of seepage of a soluble material through a permeable, nonadsorbing medium one can apply the Darcy equation:

$$Q' = \mu \, i \, A \qquad (5)$$

where Q' = seepage rate
μ = permeability
i = hydraulic gradient
A = cross-sectional area

For a typical case, a storage pond of 12.5 acres, A = 12.5 x 43,560 sq. ft = 544,500 sq. ft, i = 1, μ = 10^{-5} ft/min for silty clay soil; hence the infiltration rate, Q', was 40 gallons per minute. Any faulting of water-bearing strata may trap large bodies of underground water or conversely provide a more rapid pathway for water movement along cracks and fissures. It is this imperfect nature of subsurface pathways that introduces so much uncertainty into the postulated movement of liquid wastes and waterborne pollutants.

Most actual water movement involves water containing both dissolved material in trace concentrations and submicron-size suspended material. These are the substances of prime interest to the soil scientist and the geochemist because they provide nourishment to plants and are responsible for the distribution and formation of minerals in rocks and soils. Among the dissolved components are the essential trace elements required by plants and animals to sustain vital metabolic functions. They may be leached out of soil minerals or may have to be supplied artificially in fertilizers. Many pesticides and industrial waste materials may be transported through the subsoil in solution, as may any soluble radioactive effluents. The distance traveled by such materials depends on their concentration and chemical form.

In many cases bacterial action or the presence of organic complexing agents, such as humic acids, may modify the chemical form of the contaminant and thereby its solubility and ionic character. This is of importance since one of the main impediments to long-range movement of dissolved contaminants is their removal by adsorption or ion exchange on clay minerals and soil particles. Most clay minerals are strong cation exchangers and remove many simple dissolved cations such as those of cesium or barium; however, any complexing effects by organic matter may convert such cations to anions that are less strongly adsorbed (17-20). The behavior of ruthenium, which is particularly prone to the formation of various complexes, has been studied by a number of people (21-23).

This removal by ion exchange is not necessarily permanent; although fixation by soil particles is an effective means of reducing the spread of cationic contaminants, and therefore is important in assessing the hazard of leakage from tanks storing radioactive wastes, it was shown by Champlin (24) that such adsorbed cations in a sand bed could be re-mobilized by the injection of a strong NaCl solution or a phosphate detergent. The migration distance and exchange rate, therefore, depend on the accessible ion exchange sites, the total concentration of exchange-able ions in the water and the degree of saturation of the mineral surfaces in relation to the competitive adsorption characteristics of the various solutes.

The penetration of sorbable solutes through a permeable bed may be expressed as

$$D\left(\frac{\partial^2 C}{\partial y^2}\right) - u_y \frac{\partial C}{\partial y} = \frac{\partial C}{\partial t} + k(K_m C - m) \qquad (6)$$

where D = diffusivity of solute through interstices
$\quad\quad\quad C$ = liquid concentration at time t and distance beneath surface y
$\quad\quad\quad u_y$ = vertical convective velocity
$\quad\quad\quad m$ = sorbed concentration at t and y on a volumetric basis
$\quad\quad\quad k$ = mass-transfer coefficient
$\quad\quad\quad K_m$ = dimensionless distribution coefficient

As long as the surface is saturated and absorption has not occurred, the sorbed concentration becomes

$$C' = C'_0 \exp(-py) \qquad (7)$$

where C' = sorbed concentration on a mass basis (M/M) at a depth y below the surface
$\quad\quad\quad C'_0$ = concentration at surface
$\quad\quad\quad p$ = linear uptake coefficient

The effective depth is simply the reciprocal linear sorption coefficient. The mass of radionuclides beneath a unit surface area or the surface con-centration becomes

$$C_s = K_d C_e d_s W_s \qquad (8)$$

where C_s = surface concentration
$\quad\quad\quad K_d$ = slurry distribution coefficient
$\quad\quad\quad C_e$ = activity concentration in overlying water at equilibrium
$\quad\quad\quad d_s$ = effective sorbing depth of bed
$\quad\quad\quad W_s$ = *in situ* unit weight of bed

Rooted plants also provide surfaces and volumes for sorbing and storing solutes, thus increasing the detention capacity attributable to a unit area of the system.

Model river and aquaria experiments have shown that a large portion of plant uptake must be due to absorption, since uptake is very rapid. Also, it has been demonstrated that desorption is rapid and nearly complete, so that macroplants do not have a significant detention effect on discrete releases of radionuclides; however, under continuous release and considerable plant growth conditions, they may provide a significant storage reservoir for radionuclides.

On an overall basis, equilibrium sorption by plants may be quantitatively expressed as follows:

$$K_c = \frac{C'_e}{C_e} \tag{9}$$

where C'_e = equilibrium concentration in plant, mass basis
C_e = equilibrium concentration in water, volume basis

Sorption and retention by plants are functions of growth rate and stage, so K_c varies if ecological parameters change. For the most part K_c has been found to be between 10^3 and 10^4 ml/g on a dry-weight basis for several macroplants and several nuclides, including zinc, cobalt, ruthenium, strontium and cesium.

A second variable in migration characteristics is introduced by the presence of submicron particulates in the subsurface water. If these particles are too small to be filtered out by the porous medium serving as the aquifer, they will be carried along by the water flow and may serve in turn as carriers of contaminants. Since many of the suspended particles may be clay minerals they may absorb cations in solution and prevent their deposition on the soil matrix. Such an effect may account for the unexpectedly long migration distances observed for some pollutants (22,24). Another type of suspended material capable of serving as a carrier of contaminants, possibly as organic complexes, has been identified by Champlin as bacteria (25), either live or dead. More recently such bacterial movement through fractured bedrock has been studied by Morrison and Allen (26), who showed that coliform bacteria can travel rapidly through shallow ground water.

To some extent the presence of such particulate carriers negates the surface adsorption effects of the aquifer matrix and results in a further fractionation effect on the trace element composition of the ground water: some components are rapidly removed and fixed in the soil, others are removed but carried over long ranges by minute particulates. Both processes are of concern in evaluating radioactive effluent migration and the safe containment of long-lived waste materials.

ATMOSPHERIC TRANSPORT

The atmosphere surrounding the earth's surface is the most pervading dynamic medium available for the transport of matter. Its average composition is given in Table 15; its density varies with height and temperature, being about 0.0013 g/cm^3 at the surface of the earth where the pressure is 1 atmosphere (= 0.1 MPa). At 50 km above sea level the atmospheric pressure drops to 10^{-3} atm, and at 100 km to 10^{-6} atm. The total mass of the dry atmosphere is about 5 x 10^{18} kg, to which are added about 1.5 x 10^{17} kg of water vapor, the most variable constituent and the one governing many of its thermodynamic characteristics.

Table 15. The Average Composition of the Atmosphere (3)

Gas	Composition by Volume (ppm)	Composition by Weight (ppm)	Total Mass (x 10^{22} g)
N_2	780,900	755,100	38.648
O_2	209,500	231,500	11.841
A	9,300	12,800	0.655
CO_2	300	460	0.0233
Ne	18	12.5	0.000636
He	5.2	0.72	0.000037
CH_4	1.5	0.9	0.000043
Kr	1	2.9	0.000146
N_2O	0.5	0.8	0.000040
H_2	0.5	0.03	0.000002
O_3	0.4	0.6	0.000031
Xe	0.08	0.36	0.000018

The mixing characteristics of the atmosphere are determined largely by its vertical temperature gradient, which changes abruptly with increasing altitude. The lowest region, extending to a height of 11 km in the temperature zone, is known as the troposphere. In this region the temperature normally decreases at a rate of about -6.5°C/km (-3.5°F per 1000 ft), called the "lapse rate of temperature." The troposphere contains most of the atmospheric water vapor content in the form of clouds and moisture. Above the troposphere, extending from about 11 to 32 km in altitude, lies an isothermal region called the stratosphere, with a mean temperature of -55°C. It is essentially dry and cloudless and is separated from the troposphere by an imaginary boundary layer known as the tropopause. The height of the tropopause varies with latitude and the season of the year. Above the stratosphere lies the mesosphere, which consists of increasingly thinner air whose temperature rises with height. This extends upward to the ionosphere, a strongly ionized region of low molecular density that begins at altitudes greater than 60 km.

The troposphere also contains natural and man-made aerosols that originate from many sources. Many other suspended solids are introduced by volcanic activities, dust storms, forest fires and ocean spray as well as air pollutants produced by human activities. Any aerosol or gas vapor introduced into the atmosphere is diluted by molecular or turbulent diffusion. Since the troposphere is the scene of most meteorological phenomena, such as rain, cloud movement, storms and other wind effects, most transport effects are due to turbulence processes with diffusivity coefficients ranging from 0.2 cm^2/sec to 10^{11} cm^2/sec. Thus atmospheric motions contributing to the mixing processes vary in scale from almost microscopic eddies to large cyclonic storms extending over hundreds of kilometers. The transport processes in the atmosphere form a central subject in the study of meteorology and are described in most texts on that subject. In the present context the reader is referred to the books by Stern (27), Pasquill (28), Slade (29) and Reiter (30) for more detailed discussions. Here, we will present only the more basic concepts that are needed to facilitate the discussion in later chapters.

Turbulent Diffusion

The principal interest in atmospheric effects as related to nuclear power is less concerned with global phenomena and major weather systems than with the more local diffusion processes governing the dispersion and diffusion of airborne pollutants, be they gaseous or particulate, under all conceivable meteorological conditions around a plant site for distances up to 100 miles (160 km). This implies a detailed knowledge of local weather patterns, topography, wind speeds and directions, humidity and precipitation conditions, turbulence and temperature distributions, and diurnal and seasonal variations. In view of the imperfect status of general weather prediction methodology, in most cases a probabilistic approach has been taken, based on available meteorological data, to predict dispersion patterns and precipitation ("fallout") probabilities.

The movement of air in any given locality at lower levels of the troposphere depends on ground contours, wind speeds imposed by global weather systems, and local temperature gradients. Figure 16 illustrates the variations in airflow in the vicinity of topographic obstacles. These may include tall buildings; for instance, the presence of tall cooling towers may change the flow pattern from plant stacks in certain directions. Wind speed information and the relative frequency of given wind

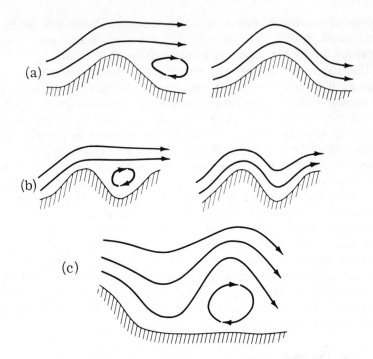

Figure 16. Flow patterns over topographic obstacles
A. over hills; **B.** over valleys; **C.** a rotor in the lee of a mountain (29).

directions is frequently displayed in the form of a wind rose, illustrated in Figure 17 (29). The usual significant difference between day and night conditions may be indicated in the wind rose, as was done in Figure 17b. Other diurnal variations of local significance can be displayed in a similar fashion. Since low-level wind conditions are heavily affected by local topology, it is often found that wind data, usually obtained at the nearest airport, may not apply several miles away in hilly terrain and varies from hilltop to valley. In some cases, long-term observations can lead to a relationship for wind data at moderately close-spaced locations.

The mixing characteristics of the atmosphere depend on its turbulence, which itself is governed in large measure by the vertical temperature gradient. Under normal daytime conditions, the earth's surface absorbs more solar radiation than it reflects or is removed by evaporation or near-surface convection. Hence it becomes warmer than the air above. The lapse rate then becomes superadiabatic, *i.e.*, the temperature decreases at a rate exceeding that which would occur if a volume of air

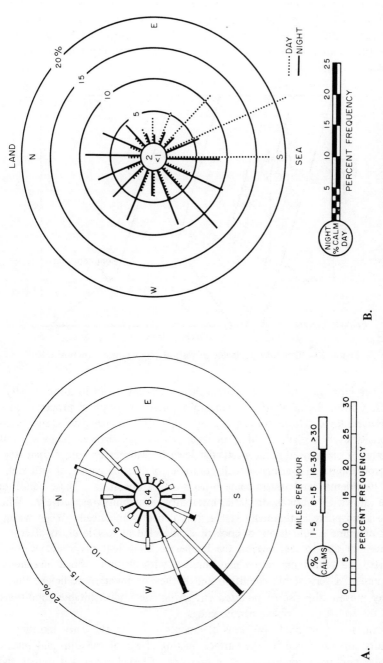

Figure 17. Wind rose presentations. A. wind speed information; B. day-night wind rose showing the diurnal effect of a sea breeze (29).

was raised and allowed to expand adiabatically as the pressure decreased (Figure 18). Under these conditions all vertical motions are accelerated because a mass of air that is forced to rise is warmer and less dense than the surrounding air and thus continues to rise. The dry adiabatic or "neutral" lapse rate is normally about -1°C/100 m or -5.4°F/1000 ft.

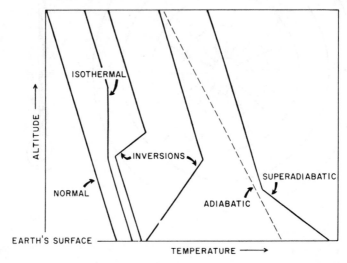

Figure 18. Examples of low-level vertical temperature structure (29).

The lapse rate in the real atmosphere over any given location usually differs from the adiabatic. Owing to a variety of physical processes the observed lapse rate in the atmosphere will differ in magnitude, and occasionally even in sign, in different layers above the surface. Thus a lifted parcel of air may at some particular level be warmer or cooler than its surroundings. If warmer, it will continue to rise; if cooler, it will sink to its original level unless some other force is operating. These conditions are known as thermal, or static, instability or stability respectively. When the atmosphere is thermally stable, turbulence is enhanced. When warm air overruns a cold layer or cool air is advected at low level, as after sunset or due to sea breezes, the temperature gradient may become inverted, *i.e.*, the temperature may increase with altitude. Such inversions represent a very stable condition and if they occur at some height they may prevent the rise of polluted air, giving rise under suitable conditions to the well-known phenomenon, smog.

Surface radiational inversions commonly occur in the early morning hours, are destroyed by the surface heating effect of the sun, and return under nocturnal conditions. The presence of cloud cover will reduce the intensity of the radiation inversion. Seasonal differences in solar radiation will favor longer inversion periods during cold months (31). Since it is easier to measure vertical temperature gradients than atmospheric turbulence, the former is often used as an index of expected turbulence conditions. Another visual indicator of turbulence conditions is the plumes of stack gases. The varying temperature gradients and accompanying turbulence conditions are exemplified by the five plume formations shown in Figure 19. It is evident that the concentration of a contaminant at ground level and the strike range of the plume depend drastically on turbulence conditions and, of course, the stack height.

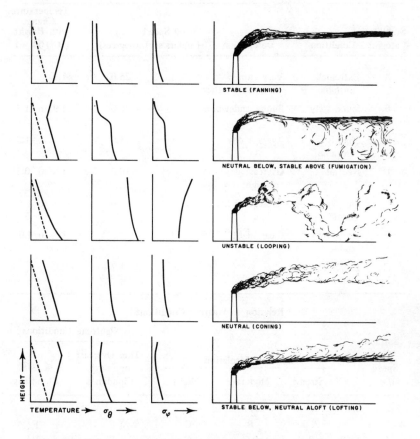

Figure 19. Types of smoke-plume patterns observed in the atmosphere (29). (The dashed curves in the left-hand column of diagrams show the adiabatic lapse rate, and the solid lines are the observed profiles.)

Turbulence is characterized by highly nonregular motions, which have thus far escaped exact mathematical definition, though some improved calculational models have been developed for ideal cases (30). In nature atmospheric mixing is the result of turbulent diffusion, and the usual theoretical approach is based on a development of Fick's diffusion law. In choosing appropriate numerical quantities, it is customary to classify turbulence conditions by means of Pasquill's classification (28), which is shown in Table 16 with the corresponding wind speeds and weather conditions (29).

Table 16. Atmospheric Stability Categories

Stability Category	Condition	Description	Wind Speed (meters/sec)	σ_θ (degrees)	Temperature Change with Height ($^\circ$C/100 m)
A	Extremely unstable	Very sunny summer weather	1	25.0	-1.9
B	Moderately unstable	Sunny and warm	2	20.0	-1.9 to -1.7
C	Slightly unstable	Average day	5	15.0	-1.7 to -1.5
D	Neutral	Overcast day or night	5	10.0	-1.5 to -0.5
E	Slightly stable	Average night	3	5.0	-0.5 to 1.5
F	Moderately stable	Clear night	2	2.5	1.5 to 4.0
G	Extremely stable		-	1.7	>4.0

Relation to Weather Conditions

Surface Wind Speed m/sec	Daytime Insolation			Nighttime Conditions	
				Thin Overcast or $\geqslant 4/8$	$\leqslant 3/8$
	Strong	Moderate	Slight	Cloudiness	Cloudiness
<2	A	A-B	B		
2	A-B	B	C	E	F
4	B	B-C	C	D	E
6	C	C	D	D	D
>6	C	C	D	D	D

To obtain concentration distributions around a point source of effluents, one must calculate the concentration at any given point (x,y) on the ground for various atmospheric diffusion conditions. The development of such diffusion theories has been presented in detail by F. A. Gifford (29) and only the major results will be quoted here. If one assumes a normal, Gaussian distribution for the atmospheric diffusion in the plume, then Fick's law for an instantaneous point-source of material under isotropic conditions takes the form

$$\chi(x,y,z,t) = \frac{Q}{(2\pi\sigma_y^2)^{3/2}} \ \exp\left(\frac{-r^2}{2\sigma_y^2}\right) \tag{10}$$

where Q = source strength, in grams or curies
   χ = concentration at point (x,y,z)   [note $\chi \neq x$]
   t = time of travel
   $\bar{u}$ = mean wind speed; hence $x = \bar{u}t$
   σ_y^2 = variance of the horizontal plume spread
   r^2 = $[(x - \bar{u}t)^2 + y^2 + z^2]$.

For isotropic distributions

$$\sigma_x^2 = \sigma_y^2 = \sigma_z^2$$

Under stable meteorological conditions or in the presence of boundary effects it is usually assumed that diffusion takes place independently in three coordinate directions. Then

$$\chi(x,y,z) = \frac{Q}{(2\pi)^{3/2}\sigma_x\sigma_y\sigma_z} \ \exp\left[-\left(\frac{(x - \bar{u}t)^2}{2\sigma_x^2} + \frac{y^2}{2\sigma_y^2} + \frac{z^2}{2\sigma_z^2}\right)\right] \tag{11}$$

This is the statistical form of the well-known Sutton law.

If the source is at a stack elevation h above ground and the receptor at the ground level, $z = 0$, then

$$\chi(x,y) = \frac{Q'}{\pi\sigma_y\sigma_z\bar{u}} \ \exp\left[-\left(\frac{y^2}{2\sigma_y^2} + \frac{h^2}{2\sigma_z^2}\right)\right] \tag{12}$$

where Q' is the continuous source strength, in g/sec, or Ci/sec. This is the form of the Gaussian plume model most frequently used, resulting in a concentration pattern like that shown in Figure 20.

In order to calculate distributions of χ, numerical values of the diffusion coefficients σ_y and σ_z must be found. These will vary according to

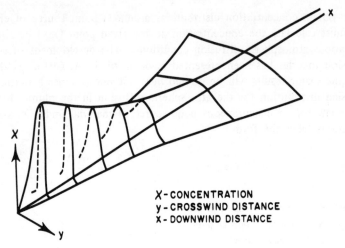

X-CONCENTRATION
y-CROSSWIND DISTANCE
X-DOWNWIND DISTANCE

Figure 20. Surface-concentration pattern downwind from an elevated source (29).

stability conditions, wind shear and roughness of terrain. However, for practical applications it is sufficient to use the curves derived from experimental tests (29), which are shown in Figure 21, for each of the six Pasquill turbulence types. Pasquill's stability categories can also be described in terms of measured values of σ_θ, which are defined as the standard deviation of horizontal wind direction fluctuation over a period of 15 minutes to 1 hour as shown in Table 16.

Pasquill's method of estimating diffusion is well suited to field use because a simple recording wind vane and anemometer erected at a proposed site can, when used with the wind-direction range theory, furnish climatologically useful estimates of σ_θ rapidly. In practice, dispersion patterns for airborne pollutants have to be computed for various turbulence conditions. An example of the type of computer program used is described in the paper by Plato, Menker and Dauer (32).

It has been found that the values predicted by the diffusion equations will be higher than observed values by a factor that will vary depending on the length of the sampling period and on topographical factors (33). During stable inversion conditions plumes have been observed to remain aloft indefinitely with no evidence of ground-level contamination for more than 25 miles downwind from a 400-ft stack. For the noble gases, however, Equation 12 seems to represent a reasonable approximation, at least over level ground.

To predict the maximum ground-level concentration that will occur downwind from the stack, Equation 12 for χ_{max}.

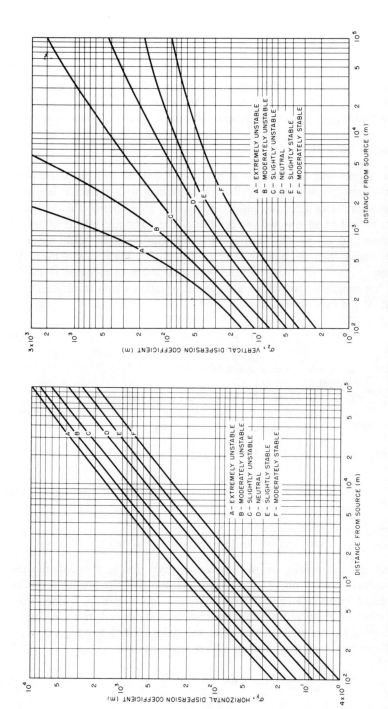

Figure 21. Graphs of diffusion coefficients σ_y and σ_z at downwind locations for Pasquill's turbulence types (29).

$$\chi_{max} = \frac{2Q'}{e\pi\bar{u}h^2} \frac{\sigma_z}{\sigma_y} \tag{13}$$

where e is the base of the natural logarithms. The downwind distance from the stack to χ_{max} is usually of the order of 15-30 stack heights. Note that χ_{max} is inversely proportional to h^2 and the wind speed.

Effective Stack Height

In formulating Sutton's law it is assumed that the effluent is released to the atmosphere at a well-defined h coinciding with the stack height. In practice, stack gases are often discharged with a considerable vertical velocity and at an elevated temperature. The combined effects of temperature and gas velocity then result in an effective stack height somewhat higher than the nominal height. Briggs (34) has reviewed the various methods of calculating the buoyancy effect of a heated plume and has proposed the following empirical equation which seems to fit most of the experimental data:

$$\Delta h = 1.6 \; \frac{F^{1/3} \; x^{2/3}}{\bar{u}} \tag{14}$$

where Δh = effective stack height increment (m)
F = $3.7 \times 10^{-5} \; Q_H$
Q_H = heat emission due to stack gas efflux (cal/sec)
x = downwind distance to a maximum value of 10 stack heights (m)
$\bar{u}$ = mean horizontal wind speed (m/sec).

Another widely used formula is Holland's equation

$$\Delta h = \frac{1.5 \; w_s d + 4 \cdot 10^{-5} \; Q_H}{\bar{u}} \tag{15}$$

where d = stack diameter (m)
w_s = stack exit velocity (m/sec)

Equation 15 is usually regarded as conservative, that is, it underestimates the actual value of Δh. Holland suggested also that the predicted rise be increased by 10-20% in unstable conditions. More recently, Briggs (35) has reviewed these and other plume rise calculations and Yeh (36) has attempted to resolve discrepancies of the order of 10% in χ_{max}.

Briggs proposed the following plume rise equations:

A. Neutral

$$\Delta h = 400 \; \frac{F}{\bar{u}^3} + 3r \; \frac{w}{\bar{u}} \tag{16a}$$

B. Stable with wind

$$\Delta h = 2.6 \left(\frac{F}{\bar{u}s} \right)^{1/3}$$ (16b)

C. Stable and calm

$$\Delta h = 5.1 \; F^{1/4} \; s^{-3/8}$$ (16c)

where F = buoyancy coefficient = $3.8 \times 10^{-5} \; Q_H \; m^4/sec^3$

   s = stability parameter = $\dfrac{g}{T} \dfrac{\partial \theta}{\partial z}$

   $\dfrac{\partial \theta}{\partial z} = \dfrac{\partial T}{\partial z} + 9.8^{\circ}C/km$

   r = inside stack radius

Effect of Buildings

Most of the dispersion formulas assume unobstructed dispersion in air with only minimal turbulence, and that of a large-scale nature, due to topographic features. In practice, the source of effluent release, be it a stack, a roof vent or a cooling tower, tends to be in close proximity to other buildings that may modify the flow pattern close to the source by introducing a downwash or a wake on the lee side of any sizable building. This effect has been discussed by Halitsky (29) and is illustrated in Figure 22, which shows the flow pattern around a steep-sided building with a round dome. The initial streamline pattern is broken up into a displacement zone, a cavity region and the wake. Within the cavity zone there exists a reverse current eddy flow that may result in a condition of high concentration. The wake is a region of high turbulence with rapid mixing characteristics. From wind tunnel tests it has been found (3) that the area of the wake may be taken as twice the projected area of the building, A, and the average wind velocity is $\bar{u}/2$. Assuming that a contaminant is mixed uniformly across the area of the wake

$$\chi = \frac{Q'}{Au}$$ (17)

In nuclear plant assessment, one of the major concerns is with the consequences of any accidental releases from leaks in containment buildings. Halitsky (29) has described wind tunnel tests to predict the building configuration effect on any possible concentration of released effluents in the cavity region. The effects have been determined in terms of a non-dimensional concentration coefficient K_c, which is the ratio of the actual

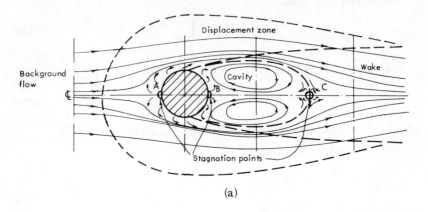

(a)

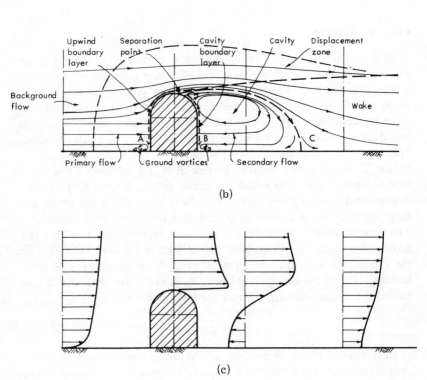

(b)

(c)

Figure 22. Flow patterns around a rounded building. **A.** Flow in a horizontal plane near the ground; **B.** flow in the longitudinal center plane; **C.** velocity profiles in the longitudinal center plane (29).

concentration X to the ideal isotropic concentration, or

$$K_c = \frac{\overline{X} \, L^2 u}{Q'} \tag{18}$$

where L is the length or distance parameter. The K_c isopleths obtained are very complex and depend on the release point assumed. It is evident that for any specific building comparable data must be developed to provide relevant information for accident analysis for different postulated release points, both upwind and downwind.

When the wind speed is high, effluent may be drawn into the low pressure area on the leeward side of the chimney, a condition known as downwash. In practice downwash can be minimized as long as stack velocities exceed the windspeed at which the worst pollution is expected. Another way to avoid downwash when the efflux velocity is low is to install a horizontal disk around the chimney top about one chimney diameter in width. Specific problems of pollution effects over cities have been reviewed by Gifford (37).

Cloud Rise

The rise of an instantaneous cloud, such as that originating from an explosion or a filter failure, is similar to the rise of a plume, except that the cloud diffuses in three dimensions. The manner in which dispersion occurs depends not only on meteorological parameters, but also on the temperature of the cloud and the total time during which the release occurs. Morton, Taylor and Turner (38) have proposed a relation for the rise rate of clouds:

$$\Delta h = 2.66 \left(\frac{Q}{C_p \rho \, \frac{\partial \theta}{\partial z}} \right)^{1/4} = 2.66 \, F_i^{1/4} \, s^{1/4} \tag{19}$$

where C_p = specific heat at constant pressure of ambient air
ρ = air density
$F_i = \dfrac{gQ}{C_p \rho T}$ = buoyant force imparted to entrained air

s = stability parameter, as defined in Equation 16

This equation has been confirmed on observations of clouds from nuclear detonations. The subsequent movement of a cloud is assumed to follow the dispersion law described previously and any ground dose calculations for radioactive clouds are based on computations of the airborne concentration integrated over any significant volume. It is important to distinguish deposition rates and patterns on the ground, the exposure to an individual immersed in the skirt of a passing cloud, and the ground effects due to a suspended cloud volume passing at some distance overhead. Some of these factors will be reviewed briefly in a later chapter.

Movement of Particulates

In many cases stack effluents may contain particulates and liquid droplets of varying composition and physical characteristics. Such material may be comprised of dust and grit, smokes or fumes, aerosols, and soluble molecules entrained in vapor droplets. Their ultimate fate in the air will depend on their vapor pressure, susceptibility to photochemical reactions, tendency to coalesce, density, and surface properties. In the end, most of the material that is not truly volatile at ambient temperature will be deposited on the ground and, if radioactive, is referred to as fallout. The rate of fallout will depend on the nature of the material being deposited, the turbulence of the air, the initial altitude reached in the atmosphere and the effective deposition mechanism. In a more specific sense, fallout is often taken to refer as the deposition of debris from nuclear explosions, where one distinguishes between local and global fallout (39). In the case of nuclear power plants neither the quantity of fallout materials nor the mechanism of atmospheric injection would be of comparable scale, and one is primarily concerned with small quantities of entrained material from the plant stack or the cooling tower drift even under design-base accident conditions.

Gravitational Settling

The rate of deposition is essentially a process of gravitational settling determined by the size of the particles and wind flow conditions. For streamline motion, Reynolds number Re between 10^{-4} and 10, the settling velocity is given by Stokes' law:

$$v_s = \frac{2gr^2(\rho_p - \rho_a)}{9\eta} \tag{20}$$

where r = particle radius
η = atmospheric dynamic viscosity
g = gravitational acceleration
ρ_p, ρ_a = density of particle and air, respectively.

The effect of actual particle shape upon fall velocity is, on the average, to reduce the velocity by about two-thirds from that of a smooth sphere. Equation 20 is applicable for a range of particle diameters from a few microns to about 400μ. At fall velocities less than about 1 cm/sec the effect of sedimentation is negligible and vertical movement of the particle is largely controlled by the larger turbulent and mean air movements. As an empirical rule of thumb, a 10 micron mineral particle of density 2.5 (quartz) falls at a rate of about 1 ft/min (30 cm/min) at sea level. The fall velocity of other particles can be approximated from the relationship $v = d^2/100$ for streamline motion. Figure 23 shows the relation of fall speed versus particle size.

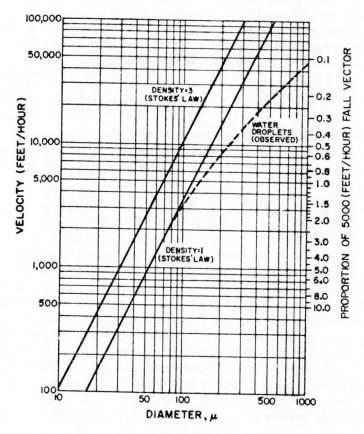

Figure 23. Fall speed *versus* size and density of spherical particles in air.

When a particle becomes so small that its diameter is less than the mean free path of gas molecules, λ, or at high altitudes, Cunningham's modification of Stokes' law may be applied:

$$v_{corr} = v_s \left(1 + \frac{0.43\lambda}{r} \right) \tag{21}$$

where $\lambda = 10^{-5}$ cm at sea level.

Dry Deposition

In addition to gravitational settling, small particles may be deposited on the ground by attachment on suspended dust particles by surface impaction, electrostatic attraction, adsorption, and chemical interaction. These mechanisms may remove volatile materials such as iodine and highly ionized molecules. For particle sizes below 20μ this effect can be described in terms of a deposition velocity v_g defined as

$$v_g = \frac{\text{total amount deposited per unit area}}{\text{time integral of volumetric concentration above surface (cloud dosage)}}$$

The net effect of this deposition is to reduce the source term Q_o' in Sutton's equation by an effective depleted-source term Q_x', the residual source at x meters downwind. The surface deposition at (x,y) is then

$$\omega_{(x,y)} = v_g \bar{\chi}(x,y,o) = \frac{v_g Q_x'}{\pi \sigma_y \sigma_2 \bar{u}} \exp \left[-\frac{1}{2} \left(\frac{y^2}{\partial \sigma_y^2} + \frac{h^2}{2\sigma_z^2} \right) \right] \tag{22}$$

Van der Hoven (29) has given examples of measurements of deposition rates and depletion ratios Q_x'/Q_o', and has shown their dependence on turbulence conditions. The effect of this source depletion effect on calculated values of χ_{max} has been discussed by Milman and Tadmor (40). Chemically active aerosols, such as iodine, deposit more readily than inactive materials, such as cesium or nonradioactive fluorescent particles. For iodine, deposition velocities are probably between 0.1 and 1.0 and Otway (41) suggests a value $v_g = 0.3$ cm/sec as a reasonable value for estimating the whole-body dose from surface-deposited material.

Precipitation Scavenging

A third process resulting in removal of particulates and chemically active materials from a cloud or plume is known as precipitation scavenging. Under the proper conditions precipitation can remove nearly all the radioactivity from the cloud, a fact long known to northern uranium prospectors, who found unusually large amounts of radon daughter activity on the ground after heavy snows, and to early fallout observers during periods of atmospheric nuclear testing. The scavenging process involves three separate events: (1) transport of the material to the scavenging site, (2) in-cloud scavenging by cloud elements, "rainout," or "snowout," and (3) below-cloud scavenging by precipitation, "washout." These effects have been discussed in detail by Junge (42) and Engelmann (29). Since the scavenging process consists of repeated exposures of particles and gases to cloud or precipitation elements with some chance of collection on the element on each exposure, it is an exponential process of the form

$$\chi = \chi_0 \exp(-\Lambda t) \tag{23}$$

where Λ may be the rainout or washout coefficient. The scavenging efficiency ϵ is a function of particle size and drop diameter D, or

$$\epsilon = f(r^2, \rho, D) \tag{24}$$

Figure 24 shows the effect of washout versus the rate of rainfall for various aerosol particle sizes. Figure 25 relates scavenging efficiencies and drop diameters for various particle sizes (29). Snow should be about as efficient as rain for removing radioactivity from a cloud.

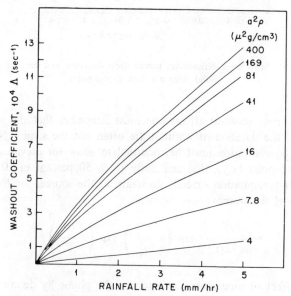

Figure 24. Washout coefficients for unit density particles as a function of rainfall and particle size (29).

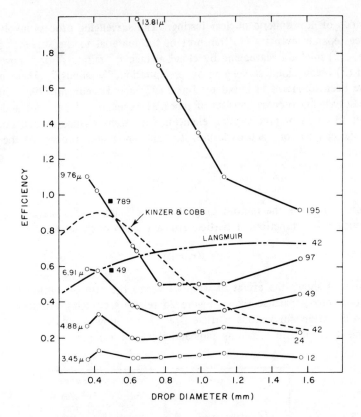

Figure 25. Scavenging efficiencies *versus* drop diameter and particle size (29). Solid lines are ZnS experiments.

In calculating washout effects one must remember that the length of release and the duration of washout are often not the same. Normally the cloud concentration must be modified to allow for previous scavenging upstream of point (x,y), that can easily reach 50 percent in 10 miles of rainfall. If precipitation extends upstream to the source, for the Gaussian plume model the washout rate becomes

$$\omega = \frac{\Lambda Q'}{\bar{u}\sigma_y(2\pi)^{1/2}} \exp\left(-\frac{y^2}{2\sigma_y^2}\right) \exp\left(-\frac{\Lambda x}{\bar{u}}\right) \tag{25}$$

The net effect of progressive depletion of the plume by deposition of particulates is a fractionation in the effective range of plume components and the consequent composition in the ground level concentration. This is illustrated in Figure 26.

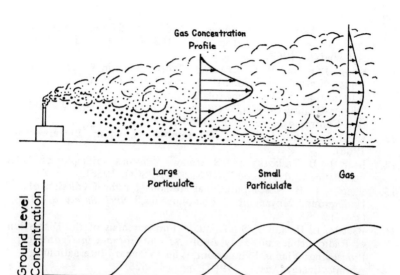

Figure 26. Differential composition and deposition profiles for gaseous and particulate plumes at a given wind speed.

REFERENCES

1. Penman, H. L. "The Water Cycle," *Scient. Amer.* **223** (3), 99 (1970).
2. Chow, V. T. *Open-Channel Hydraulics.* (New York: McGraw-Hill 1959).
3. Eisenbud, M. *Environmental Radioactivity*, 2nd ed. (New York: McGraw-Hill, 1973).
4. "Radioactivity in the Marine Environment," (Washington, D.C.: National Academy of Sciences, 1971).
5. Linsley, R. K. and J. B. Franzini. *Elements of Hydraulic Engineering* (New York: McGraw-Hill Book Co., 1955).
6. Kennedy, V. C. "Sediment Transported by Georgia Streams," Geological Survey Water Supply Paper 1668 (Washington, D.C.: U.S. Government Printing Office, 1964).
7. Kennedy, V. "Mineralogy and Exchange Capacity of Modern Fluvial Sediments," U.S. Atomic Energy Commission Report TID-7664, 71-81 (1963).
8. Eichholz, G. G., T. F. Craft and A. N. Galli. "Trace Element Fractionation by Suspended Matter in Water," *Geochim. Cosmochim. Acta* **31**, 737 (1967).
9. Wrenn, M. E., J. W. Lentsch, M. Eisenbud, J. Lauer and G. P. Howells. "Radiocesium Distribution in Water, Sediment and Biota in the Hudson River Estuary from 1964 through 1970," *Proceedings 3rd National Symposium of Radioecology,* Oak Ridge (1971).

10. National Water Quality Network, 1962; Public Health Service Publ. No. 663, (Washington, D.C.: U.S. Dept. of Health, Education and Welfare, 1963).

11. Kopp, J. F. and R. C. Kroner. "Trace Metals in Waters of the United States," (Cincinnati: Federal Water Pollution Control Administration, U.S. Dept. of Interior, 1967).

12. "Water Quality Criteria," (Washington, D.C.: Federal Water Pollution Control Administration, 1968).

13. Klein, L. *River Pollution, III. Control* (London: Butterworths, 1966).

14. Bear, J., D. Zaslavsky and S. Irmay. "Physical Principles of Water Percolation and Seepage," (Paris: UNESCO, 1968).

15. Baetslé, L. H. "Computational Methods for the Prediction of Underground Movement of Radionuclides," *Nucl. Safety* **8**, 576 (1967).

16. Baetslé, L. H. and J. Souffriau. "Fundamentals of the Dispersion of Radionuclides in Sandy Aquifers," in *Isotopes in Hydrology*, Proceedings Vienna Symposium 1966 (Vienna: International Atomic Energy Agency, 1967), pp. 617-628.

17. Borak, T. B., M. Awschalom, W. Fairman, F. Iwami and J. Sedlet. "The Underground Migration of Radionuclides Produced in Soil near High-Energy Proton Accelerators," *Health Phys.* **23**, 679 (1972).

18. Fowler, E. B., Ed. *Radioactive Fallout: Soils, Plants, Food, Man* (Amsterdam: Elsevier, 1965).

19. Garrett, A. R., S. C. Cummings and J. E. Regnier. "Accumulation of ^{137}Cs and ^{85}Sr by Florida Forages in a Uniform Environment," *Health Physics* **21**, 67 (1971).

20. Jordan, C. F., J. R. Kline and D. S. Sasscer. "A Simple Model of Strontium and Manganese Dynamics in a Tropical Rain Forest," *Health Phys.* **24**, 477 (1973).

21. Bolch, W. E. and E. F. Gloyna. "Radioactivity Transport in Water—Behavior of Ruthenium in Algal Environments," Env. Health Eng. Res. Lab. Techn. Rept. 4, University of Texas (1963).

22. Champlin, J. B. F. and G. G. Eichholz. "The Movement of Radioactive Sodium and Ruthenium through a Simulated Aquifer," *Water Resources* **4**, 147 (1968).

23. Lomenick, T. F. "Movement of Ruthenium in the Bed of White Oak Lake," *Health Phys.* **9**, 835 (1963).

24. Champlin, J. B. F. "The Transport of Radioisotopes by Fine Particulate Matter in Aquifers," Thesis, Georgia Institute of Technology, Atlanta, Georgia (1969).

25. Champlin, J. B. F. "The Movement of Micron-Size Particles through a Sand Bed," M.S. Thesis, Georgia Institute of Technology, Atlanta, Georgia (1967).

26. Morrison, S. M. and M. J. Allen. "Bacterial Movement through Fractured Bedrock," Completion Report Series No. 32, Environmental Resources Center, Colorado State University, Fort Collins, Colorado (1972).

27. Stern, A. C. *Air Pollution*, 2nd ed. (New York: Academic Press, 1968).

28. Pasquill, F. *Atmospheric Diffusion.* (New York: Van Nostrand, 1962).

29. Slade, D. H., Ed. *Meteorology and Atomic Energy–1968.* (Washington, D.C.: U.S. Atomic Energy Commission, 1968).

30. Reiter, E. R. "Atmospheric Transport Processes, Part I," Energy Transfer and Transformations, U.S. Atomic Energy Commission (1969).

31. Hosler, C. R. "Low-Level Inversion Frequency in the Contiguous United States," *Monthly Weather Rev.* **89**, 319 (1961).

32. Plato, P. A., D. F. Menker and M. Dauer. "Computer Model for the Prediction of the Dispersion of Airborne Radioactive Pollutants," *Health Phys.* **13**, 1105 (1967).

33. Whaley, H. "The Derivation of Plume Dispersion Parameters from Measured Three-Dimensional Data," Mines Branch Research Rept. R254, Canada. (Ottawa, Canada: Dept. of Energy, Mines and Resources, 1972).

34. Briggs, G. A. "Plume Rise," Critical Review Series, TID-25075 (Washington, D.C.: U.S. Atomic Energy Commission, 1969).

35. Briggs, G. A. "Plume Rise: A Recent Critical Review," *Nucl. Safety* **12**, 15 (1971).

36. Yeh, C. S. *The Effect of Plume Rise on the Parameters Determining the Maximum Ground Concentration of Effluents.* (Oradell, N.J.: Burns & Roe, Inc., 1973).

37. Gifford, F. A., Jr. "Atmospheric Transport and Dispersion over Cities," *Nucl. Safety* **13**, 391 (1972).

38. Morton, B. R., G. I. Taylor and J. S. Turner. "Turbulent Gravitational Convection from Maintained and Instantaneous Sources," *Proc. Roy. Soc.* (London) **A-234**, 1 (1956).

39. Brode, H. L. "Review of Nuclear Weapons Effects," *Ann. Rev. Nucl. Sci.* **18**, 153 (1968).

40. Milman, Y. and J. Tadmor. "The Maxima of Sutton's Equation Corrected for Deposition," *Health Phys.* **13**, 739 (1967).

41. Otway, H. J. "The Application of Risk Allocation to Reactor Siting and Design," U.S. Atomic Energy Commission Rept. LA-4316, Los Alamos Scientific Lab (1969).

42. Junge, C. E. *Air Chemistry and Radioactivity.* (New York: Academic Press, 1963).

CHAPTER 4

RADIATION EFFECTS

BASIC CONCEPTS

The one characteristic that distinguishes the nuclear industry and its impact on the surroundings from other industrial processes is the presence of sources of ionizing radiation. Such radiations, at high intensities, are known to cause severe damage to biological tissue; at low intensity levels any radiation effects may be less readily detected and may, in fact, be fairly innocuous. Since the levels of radiation associated with most stages of the nuclear fuel cycle are actually very low, it becomes important to determine how a low level can reasonably be maintained at all times and what level of radiation exposure can be considered "safe" or "innocuous" to the general population. This is a subject of considerable debate among experts and, owing to the intangible nature of the radiations involved, a subject of much psychological impact on the lay population. In general, any significant increase in radiation exposure may be harmful and should be avoided. However, such significance may have to be viewed in relation to other hazards and this discussion will attempt to focus on those issues that have relevance to any assessment of the environmental impact of nuclear plants as indirect sources of radiation.

The radiations involved here are primarily of nuclear origin, *i.e.*, they arise from the excitation of nuclear energy levels, usually caused by the capture of charged or uncharged particles by a nucleus or by the radioactive decay or fission of an unstable nuclide. The radiation types may be categorized as charged or uncharged particles, as particulates (electrons, alpha particles, protons, tritons, fission products) or as electromagnetic radiation (gamma rays, Bremsstrahlung, X-rays). Table 17 summarizes the basic characteristics of the more common radiation types encountered. Other, higher-energy radiations associated with the use of high-energy accelerators or found in cosmic radiation will not be considered here.

Table 17. Characteristics of Nuclear Radiations

Radiation	Rest Mass	Charge	Typical Energy Range	Path Length (Order of Magnitude)		General Comments
				Air	Solid	
α	4.00 amu	2+	4-10 MeV	5-10 cm	25-40 μm	Identical to ionized He nucleus
β (negatron)	5.48×10^{-4} amu 0.51 MeV	–	0-4 MeV	0-1 m	0-1 cm	Identical to electron
Positron (β positive)	5.48×10^{-4} amu 0.51 MeV	+	–	0-1 m	0-1 cm	Identical to electron except for charge
Proton	938.26 MeV 1.0073 amu	+	–	–	–	–
Neutron	1.0086 amu 939.55 MeV	0	0-15 MeV	0-100 m	0-100 cm	Free half life: 16 min
X (e.m. photon)	–	0	eV-100 keV	0.1-10 m[a]	0-1 m[a]	Photons from electron transitions
γ (e.m. photon)	–	0	10 keV-3 MeV	0.1-10 m[a]	1 mm-1 m	Photons from nuclear transitions

[a]Exponential attenuation in the case of electromagnetic radiation.

The radiations will interact with materials, that is, they will lose kinetic energy to any solid, liquid or gas through which they pass by a variety of interaction mechanisms that depend on the energy of the incident radiation, the density and atomic number of the absorbing medium, and the relative probability for one or another process to take place. These probabilities are called cross sections and are usually expressed in barns (1 barn = 10^{-24} cm^2 = 10^{-28} m^2). The result of these interaction processes is a gradual slowing down of any incident particle until it is brought to rest or "stopped" at the end of its range. Although the terminology varies among the nuclear physicist, the health physicist, the radiobiologist, and the radiation chemist, they all are interested in the rate of energy loss or transfer ("dE/dx, LET") as the incident particle is slowed down, the total energy lost or transferred by various secondary processes, and the consequent effect per unit mass or unit volume of the target material. The terminology is subject to periodic revisions by the International Commission on Radiation Units and Measurements (ICRU), which attempts to refine and redefine the terms and units in common use at frequent intervals (1). The concepts that follow are of general usefulness.

Radiation Field

The intensity of a radiation field can be described in several ways. The simplest is to describe it in terms of a particle flux or flux density, *i.e.*, the number of particles passing unit area per unit time, with all particle paths projected normally onto the reference plane. This is a convenient way of expressing field intensities for parallel beams of particles or for fields created by a point source. It is the customary way of describing neutron fields around a reactor where one also needs to distinguish between neutrons in different energy ranges. *Flux density* is usually expressed in number of particles/cm^2-sec.

For X-rays and gamma-rays, a more experimentally oriented approach is customary in which the field intensity is expressed in terms of the number of ions produced in a known mass or volume of detector. Such a measure of the ionization produced in air by X- or gamma radiation is called the *exposure*. The unit of exposure is the Roentgen (R). It has been defined as that quantity of X- or gamma radiation that would, through associated corpuscular radiation, produce 1 statcoulomb of charge of either sign in one cubic centimeter of dry air at STP.

$$1 \text{ R} = 2.58 \times 10^{-4} \text{ coulomb/kg of air}$$

The energy absorbed by that air volume represents an equivalent of 87.7 ergs per gram of air or 8.77×10^{-9} joule/kg; this quantity is useful in comparing exposure values and dose values, but it is important to maintain the distinction.

Source Strength

Radiation will originate from natural or artificial sources, which may be naturally decaying radioactive materials, nuclear reactors, astronomic sources or high voltage accelerators of various types. The source strength is a measure of the rate of emission of radiation. More than one kind of radiation and more than one energy range of a given type of radiation may be emitted, and this fact needs to be considered in describing the source characteristics. Consequently there are several ways in which a radiation source can be described.

Emergent Flux

For machine sources and reactor cores it may often be simplest to describe source strength in terms of total number of particles or photons emitted per unit time into the total solid angle (4π). This procedure is not appropriate if the emission is not isotropic or if there is significant conversion of the radiation within the source volume. In reactors or machines any measurement of this quantity must allow for reflection or backscattering effects and, in the case of neutrons, for spectral shift. The result is usually a description of the particle or photon flux at a given external point near the source, which is assumed to be concentrated at a small volume.

In reactors the equivalent source strength of the core region may involve both prompt fission effects and delayed fission product activity; hence it is rarely practical to describe ambient radiation fields in terms of an equivalent source strength.

Radioactive Sources

For radioactive materials it is customary to describe the source strength in terms of the source activity, which is defined as the number of disintegrations per unit time occurring in a given quantity of this material. For any pure radioactive substance, the rate of decay is usually described by its half life τ, i.e., the time it takes for a specified source material to decay to half its initial activity. The activity A can be written as

$$A = -\lambda N = \frac{\log_e 2}{\tau} N = \frac{0.693\, N}{\tau} \tag{26}$$

where λ = decay constant (sec^{-1}) and N = total number of atoms of the radioactive species.

The unit of activity is the *curie*, which is defined as:

$$1 \text{ curie (Ci)} = 3.70 \times 10^{10} \text{ disintegrations/second.}$$

The SI unit of activity is the *becquerel*: 1 becquerel = 1 disint./sec. Since activity is proportional to the number of atoms of the radioactive material, the quantity of any radioactive material is usually expressed in curies, regardless of its purity or concentration. The usual metric multiples apply to the curie. Since many nuclides have complex decay schemes, the activity does not in itself indicate the emergent flux from any source.

To calculate the radiation field for a given radioactive source it is necessary to know the decay scheme of the material, any branching ratios and the composition of the emitted radiation. Also note that the activity of a given source is a function of time since the number of radioactive atoms decreases exponentially:

$$N = N_o e^{-\lambda t} = N_o e^{\frac{-0.693t}{\tau}} \tag{27}$$

where N_o = number of atoms at time t = 0. Hence,

$$A = A_o e^{\frac{-0.693t}{\tau}} \tag{28}$$

In the case of radioactive materials contained in living tissue an additional allowance has to be made for the reduction in observed activity due to regular processes of elimination of the respective chemical or biochemical substance from the organism. This introduces a rate constant called the *biological half life*, which is approximately the same for both stable and radioactive isotopes of a given element.

Under such conditions the time required for a radioactive element in any living organism to be halved as a result of the combined action of radioactive decay and biological elimination is the *effective half life*:

$$\tau_{eff} = \frac{\tau_{biol} \times \tau_{RA}}{\tau_{biol} + \tau_{RA}} \tag{29}$$

Table 18 presents some effective half lives of particular interest.

Absorbed Dose

Since different types of radiation interact differently with any material through which they pass, any attempt to assess their effect on the human body or on plants and animals should take into account these differences.

Table 18. Half-Lives of Some Radionuclides in Body Organs

Radionuclide	Critical Organ	Half-Life[a]		
		Physical	Biological	Effective
Hydrogen-3[b] (Tritium)	Whole body	12.3 y	12 d	11.97 d
Iodine-131	Thyroid	8 d	138 d	7.6 d
Strontium-90	Bone	28 y	50 y	18 y
Plutonium-239	Bone	24,400 y	200 y	198 y
	Lung	24,400 y	500 d	500 d
Cobalt-60	Whole body	5.3 y	99.5 d	9.5 d
Iron-55	Spleen	2.7 y	600 d	388 d
Iron-59	Spleen	45.1 d	600 d	41.9 d
Manganese-54	Liver	303 d	25 d	23 d
Cesium-137	Whole body	30 y	70 d	70 d

[a]d = days, y = years

[b]Mixed in body water as tritiated water.

The quantity of interest is the *absorbed dose*, D, which is defined as the energy imparted by the incident radiation to unit mass of the target material, or

$$D = \frac{\Delta E_D}{\Delta m} \tag{30}$$

The unit of absorbed dose is the rad; 1 rad = 100 erg/gram = 0.01 J/kg. The rate of energy absorption is called the absorbed dose rate $\Delta D/\Delta t$ and is expressed in rad/hr, rad/min, rad/day, as appropriate.

The physical meaning of dose has concerned health physicists for many years and has led to many attempts at clarification. The problem arises from two sources: one, to measure internally the amount of energy actually transferred to an organic material and to correlate any observed effects with this energy deposition, and two, to account and predict secondary processes, such as collision effects or biologically triggered effects, that are an indirect consequence of the primary interaction event.

For this reason such interaction processes are often broken up into two parts: the rate of energy loss by the incident particle, which presumably is reflected by primary energy absorption effects, and, the rate of biological effect production, which is essentially a characteristic property of the irradiated tissue material and related to its sensitivity to radiation.

Energy Transfer

The mechanism by which an incident particle will lose energy in a medium depends on the type of interaction possible, the energy of the moving particle, and the density and nuclear charge of the medium. For charged particles ionization and excitation predominate, with little atomic displacement, especially for the lighter particles.

For Rutherford scattering the differential scattering cross section is

$$\frac{d\sigma}{d\theta} = \frac{0.8139 \ z^2}{E^2} \ \frac{\sin\theta}{\sin^4(\theta/2)} \ \text{x} \ 10^{-2} \ \text{barn per nucleus} \tag{31}$$

where: θ = scattering angle
E = energy of incident particle (MeV)
Z = atomic number of target material
ze = charge of incident particle

The energy loss due to ionization is given by Bethe's formula

$$-\frac{dE}{dx} = \frac{4\pi n_0 \ z^2 e^4}{mv^2} \left[\log_e \frac{2mv^2}{I(1-\beta^2)} - \beta^2 \right] \tag{32}$$

where n_0 = electron density per cm^3
β = v/c (relativistic velocity of incident particle)
m = mass of electron
I = average ionization potential of target material
I = 9.1 Z $(1 + 1.9Z^{-2/3})$ electron volts (eV).

The quantity dE/dx is variously referred to as the linear stopping power or the linear energy transfer (LET). The range, or maximum distance traveled, for 4-8 MeV alpha particles in tissue is of the order of 40-90 microns. Figure 27 shows the variation of stopping power with energy for charged particles. For nonrelativistic electrons or beta particles we have

$$-\frac{dE}{dx} = \frac{4\pi e^4 n_0}{m_0 v^2} (\log_e \frac{m_0 v^2}{2I} + 0.153) \tag{33}$$

The range of electrons can be calculated by means of empirical relations, such as

$$R_\beta = 412 \ E_\beta^{1.265 \ - \ 0.0954 \ \log_e E_\beta} \tag{34}$$

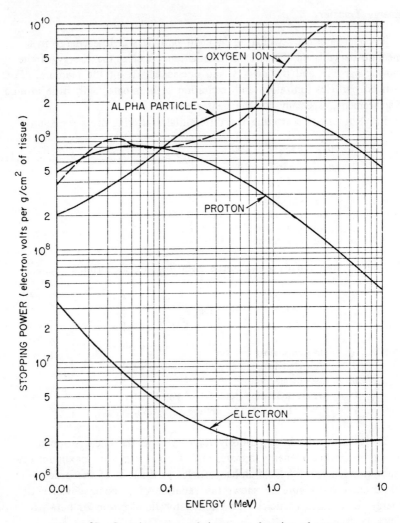

Figure 27. Stopping power of tissue as a function of energy.

or

$$\log_e R_\beta = 6.02 + (1.265 - 0.0954 \log_e E_\beta) \log_e E_\beta \qquad (35)$$

where R_β = range of beta particles (in mg/cm^2)
         E_β = nominal (maximum) energy of β particles (MeV)

For beta energies above 600 keV, Feather's rule may be used

$$R_\beta = 542E - 133 \qquad\qquad (36)$$

Note that these ranges are expressed in mg/cm^2, not in distance units directly; for a given material, equivalent thicknesses can be calculated by dividing the range by the density, in appropriate units.

For gamma rays, energy transfer occurs mainly by three processes: Compton effect, photoelectric effect and pair production. Since the cross section for each of the effects varies in a different way with photon energy and target composition, the LET for gamma rays and X-rays is a complicated function of these parameters. For photons of a given energy the rate of energy loss follows an exponential law

$$I'_x = I'_o \, e^{-\mu x} \qquad\qquad (37)$$

where I'_x and I'_o =   the beam intensities of the incident and emergent beam

               x =   thickness of absorber

               μ =   linear attenuation coefficient.

The attenuation coefficient μ is directly related to the linear stopping power and has been tabulated for many materials of interest for a wide range of photon energies (2,3). It is proportional to the sum of the cross sections for the three interaction processes at that energy; *i.e.,* $\mu = \sigma_{pe} + \sigma_{compton} + \sigma_{pp}$. Since for electron absorption, the *mass absorption coefficient* $\mu' = \mu/\rho$ is found to be relatively independent of the composition of the absorber, it is frequently convenient to express gamma-ray absorption also in terms of μ'. Figure 28 shows the dependence of μ' on photon energy in air and water (4).

Equation 37 can also be used to estimate the effective range of a photon of a given energy in a medium of attenuation coefficient μ. It can be shown that the mean range R_{av} is given by

$$R_{av} = \frac{1}{\mu} \qquad\qquad (38)$$

and we can define a half-value layer $x_{1/2}$ so that the incident intensity I'_o is reduced to half its initial value:

$$\text{half-value layer } x_{1/2} = \frac{\log_e 2}{\mu} = \frac{0.693}{\mu} \qquad\qquad (39)$$

This is useful in making rough estimates of shielding thicknesses required around a given radiation source.

Since the energy of an incident photon decreases progressively as it undergoes Compton scattering, the mean free path between interaction

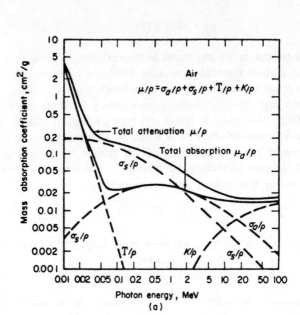

(a)

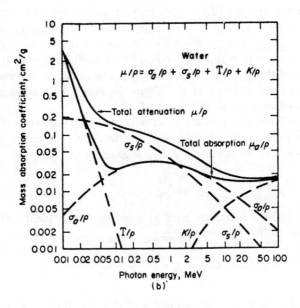

(b)

Figure 28. Mass attenuation coefficients for photons in air and water (4).

sites will decrease steadily and so does the amount of energy lost per collision (5). Each primary interaction can transfer enough energy to the Compton electrons to give rise to secondary events. It is also evident that the mean free path between collisions is enormously longer for photons than for charged particles, giving rise to a much lower LET. Consequently photons cause radiation damage over a longer range in tissue but with a lower energy transfer efficiency, whereas charged particles cause highly local damage, transferring most of their energy over a fairly restricted path length.

Summarizing then, for a given exposure, which determines the amount of radiation entering a given volume of material, we may get an amount of energy transferred to that volume that depends on the nature of the incident radiation, its linear energy transfer, and the density and composition of the target material. Table 19 lists some LET values for tissue.

Table 19. Representative LET and RBE Values

Radiation	Energy (MeV)	Av. LET (keV/μ)	RBE	Quality Factor
X-rays, 200 kVp	0.01-0.2	3.0	1.00	1
Gamma rays	1.25	0.3	0.7	1
	4	0.3	0.6	1
Electrons (β)	0.1	0.42	1.0	1
	0.6	0.3	1.3	1
	1.0	0.25	1.4	–
Protons	0.1	90	–	6
	2.0	16	2	10
	5.0	8	2	10
Alpha particle	0.1	260	–	–
	5.0	95	10-20	10
Heavy ions	10-30	~150	~25	20
Neutrons	thermal		4-5	3
	1.0	20	2-10	10

These values are general and approximate. RBE and QF values vary widely with different measures of biological injury.

We shall not discuss the direct radiation effects of neutrons here, since they will occur essentially only in close proximity to a nuclear reactor, well within the containment building, and their environmental effect in the present context is minimal.

Dose Equivalent

The interaction of the incident particle or photon with target material atoms produces secondary projectiles that may carry enough energy to cause further collisional interactions, resulting in a cascade of displaced or ionized atoms around the initial collision site. By means of photographic emulsions or electron microscopy one can show that the path of a heavily ionizing particle like a fission product atom is surrounded by a short tubular damage region with a high density of affected atoms. The weaker interacting charged particles have similar but less pronounced damage tracks. In contrast, a fast neutron with a long mean free path between interactions gives rise to widely spaced drop-like damage regions along its track. Since an incident nuclear particle may carry a kinetic energy in the multi-MeV range and it takes only a few electron volts to break a molecular bond, a higher density of displaced atoms implies a greater number of damaged molecules within an affected tissue volume. Since a certain fraction of broken bonds may recombine, the probability of lasting cell damage increases steeply as the number of damaged molecules in a given affected volume increases. As a consequence we find that some types of radiation are more damaging than others, even though both transfer the same amount of energy to the target tissue. This difference is described as biological effectiveness or *quality* of the incident radiation.

In practice it is difficult to measure accurately the energy actually deposited in a given small tissue volume; however, one can determine the relative number of cells seriously damaged by the radiation. We thus have two distinct concepts: the *relative biological effectiveness* (RBE), comparing the magnitude of observed effects produced in tissue by a comparable quantity of energy deposited (dose) by different types of radiation, and the radiation sensitivity of a given tissue, describing the degree of destruction of cells by a comparable received dose. Table 19 presents approximate RBE values, which by convention are expressed in comparison to effects of 200 kVp X-rays. It is seen that there is obviously some relationship between LET and RBE values, but it is not close. To assess the biological effect of various radiations more precisely and to predict likely radiation damage, it therefore is customary to determine the effective or tissue dose for different radiations. This is obtained by multiplying the absorbed dose, expressed in rads, by the appropriate RBE value. This tissue dose equivalent is expressed in terms of a unit called the rem, so that

$$1 \text{ rem} = 1 \text{ rad} \times \text{RBE}$$

More recently it has become customary to replace the term RBE in radiation protection applications by a related quantity called the Quality Factor (*QF* or *Q*), which expresses on a common scale for all ionizing radiations the effectiveness of the absorbed dose. Then

$$1 \text{ rem} = 1 \text{ rad} \times Q.$$

The dose equivalent, in rems, can express the total biologically effective dose regardless of the spectrum and composition of the incident radiation. It is used also to establish population doses where a large number of individuals is exposed to various sources of radiation. In that case the total dose is obtained by multiplying the average dose by the exposed population, yielding a population dose expressed in *man-rems*.

The man-rem notation is probably most useful to those attempting to relate doses received by large population groups to the possible effects of radiation, if any, on the group. It has the advantage of relating doses received to real people instead of hypothetical people; when the man-rem is used, it means that the actual number of people prospectively exposed to the radiation has been taken into consideration for specified areas.

The radiation sensitivity of various types of tissue is a difficult quantity to establish. The most sensitive function of a cell is cellular division; this fact is the basis of the Law of Bergonie and Tribondeau, which states that the greater their mitotic activity the more sensitive are cells to radiation. The mitotic activity depends on the number of cells entering mitosis and the length of time each cell spends in mitosis. The law is not absolute, but many cells showing much mitotic activity, including the cells of blood-forming tissues, germinal cells, and cells of the intestinal epithelium, are in fact radiosensitive, whereas other cells that show little mitotic activity, such as the cells of connective tissue, nerve tissue and fat, are relatively insensitive to radiation. Radiation sensitivity can also be affected by the presence or absence of oxygen, and this *oxygen effect* is of importance in radiation treatment of certain malignant tumors.

Although at high levels of irradiation the effects observed due to a certain tissue dose seem to be independent of dose rate, this may not necessarily be true at very low dose rate levels once the rate of destruction of cells by radiation becomes comparable or significantly smaller than the normal regeneration rate of those cells. This is sometimes referred to as *recovery*, but may be simply due to different rate-controlling processes in normal body metabolism.

For the same reason distinctions have to be made between acute and chronic exposure conditions, and whether exposure is to the whole body or only to specific parts of the body.

Exposure Conditions

In view of the greatly varying radiation sensitivity of different organs and tissues, the location and direction of a radiation field are significant. In particular one distinguishes between whole-body exposure and localized exposure. In whole-body exposure, all of the body is assumed to be bathed in a uniform radiation field and to receive a specified average dose to all tissues. Such exposure can be assumed to have more serious consequences for the survival chances of the body than a comparable dose confined to a single organ. In fact, in cancer therapy the local dose often given to the malignant tissue would be fatal to the patient if all of his body were exposed. Similarly, in view of the lesser sensitivity of the extremities, a high level of exposure confined to the hands or feet may be tolerated when it might cause serious damage to more sensitive areas. Hence, it is important in medical radiology and in the handling of radioactive materials to shield all portions of the body not directly involved in the radiation application. This applies particularly to such radiation-sensitive organs as the gonads, the lens of the eye and the bone marrow.

Radiation exposure may be *chronic* or *acute*. Chronic exposure implies continued exposure over longs periods of time leading to a given total dose value. Acute exposure refers to sudden, perhaps massive, radiation exposure such as might arise in the case of a radiation accident or a nuclear explosion. At high radiation levels the rate of exposure is usually assumed to be unimportant and only the total absorbed dose is considered. Nevertheless, it is important, particularly in accident situations, to reconstruct the dose received and the duration of exposure.

Another important distinction arises from the limited range of some of the radiations, especially the heavier charged particles. If the source of radiation is external to the body, the dose to internal organs depends on the attenuation of the radiation by intervening tissue and the dose will be highest, in general, to the skin. In fact, for alpha particles or fission product nuclei and for low energy electrons, the range in tissue is so short that the skin is the only body part exposed to external sources. For gamma and X-rays up to approximately 150 keV in energy, again the skin dose is highest and the dose to underlying tissue must be computed carefully, a central problem in radiological practice. As a consequence the risk of immersion in water or air containing only sources of strongly absorbed radiation is rather slight.

The situation changes when radioactive material is introduced into the body by food ingestion, inhalation or injection. The dose to a given tissue now depends on the mode and place of introduction, and the subsequent path of the source material within the body. Certain chemical elements are known to be bone seekers, and hence radioisotopes of barium, strontium or radium are liable to be concentrated in the bone tissue, giving rise to a localized dose to the bone marrow. Iodine is well known for its tendency to concentrate in the thyroid, giving rise to a local dose there and indirectly to any tissue in the vicinity.

Such concentration in specific organs has been measured for all radioisotopes of interest and the organs where preferred concentration occurs are referred to as *critical organs* (6). Knowing the rate of uptake and concentration in a critical organ and allowing for natural elimination, it is possible to estimate the dose to other body portions, such as the gonads and/or the bone marrow. The dose obtained is called the dose equivalent. It is defined by (1)

$$H = D \times Q \times N \qquad (40)$$

where   D = the absorbed dose (in rads)
         Q = the quality factor
         N = the product of other modifying factors such as the distribution factor, a geometry correction to allow for nonuniform distribution of a given radiation source in the body.

The unit of dose equivalent is the *rem.*

In practice the determination of the quality factor is far from simple, particularly in mixed radiation fields. Prêtre (7) has shown that the multitude of steps involved in deriving the quality factor from the LET in many computational methods tends to lead to unduly conservative conclusions.

The calculation of the distribution factor is complicated, particularly when an internal radiation source cannot be treated as a point source. Several model "phantoms" have been devised, such as that of Snyder and Fisher (8), to simulate the distribution of organs and tissue in "standard man" in a manner amenable to computation. Using such a model, whole body exposure from any radionuclide assumed to be concentrated in a critical organ can be calculated and used to determine maximum body burdens for that material. (A critical organ is that organ in the body known to concentrate a particular element most or to have the greatest radiosensitivity of those known to exhibit such concentration.) Since the term *maximum permissible level or concentration* may be open to misinterpretation as setting a maximum safe dose, more recently anticipated internal exposures have been expressed by the *dose commitment,* defined as the dose equivalent, in rem, per microcurie intake calculated for a 50-year period.

The possibility of ingestion or inhalation of radioactive materials tends to extend the possibility of radiation exposure considerably. Unfortunately in assessing environmental impact of nuclear power plants, external and internal exposure conditions are not always specified distinctly. It is evident that to be internally effective, a radioactive substance has to be in a form that can be inhaled and retained in the body or that may enter the food chain. For convenience, Table 20 provides some conversion factors to calculate dose values arising from given concentrations in air of some of the more interesting radioisotopes encountered in nuclear technology.

Table 20. Air Concentration-to-Dose Conversion Factors for
Some Important Radionuclides (9)

Radionuclide	Critical Organ	Conversion Factor[a] $(rem/yr)/(pCi/m^3$ air)
^{85}Kr	Whole body	1.5×10^{-8}
	Gonads (female)	1.5×10^{-8}
	Gonads (male)	2.0×10^{-8}
	Lung	3.0×10^{-8}
	Skin	50.0×10^{-8}
^3H	Whole body	1.7×10^{-6}
^{129}I	Infant thyroid	15
	Adult thyroid	4.6
^{239}Pu	Lung	12

[a]These factors are for continuous exposure to concentrations expressed in pCi/m^3 of air.

THE NATURAL RADIATION BACKGROUND

Although there is a great deal of public agitation against nuclear power because of its feared injection of nuclear radiations into the environment, it does not appear to be common knowledge that everyone is being exposed to such radiations from natural causes all his life. This radiation exposure arises from several sources: cosmic radiation, external terrestrial sources and natural internal sources. The nature and composition of these sources has been discussed extensively in the literature (10-18) and will only be summarized here.

Cosmic Radiation

Primary cosmic radiation enters the earth's atmosphere largely in the form of high-energy protons, which interact with the molecules of the upper atmosphere to produce secondary particles, mainly neutrons and protons, with some pions and kaons as well as electromagnetic radiations. Most of the primary radiation originates in our galaxy, and during periods of solar activity the sun contributes an appreciable number of relativistic protons. This accounts for a significant variation in cosmic-ray intensity throughout a solar cycle. Since the charged particles are subject to the earth's magnetic field, both primary and secondary cosmic radiations vary with latitude (Figure 29) and with altitude. In the middle latitudes at sea level, the neutron flux is of the order of one neutron/m^2-sec, with a neutron dose equivalent of 5.6 mrem/yr (15). The ionizing component produces a dose equivalent of 35.3 mrem/yr at sea level for a total average cosmic-ray dose of 40.9 mrem/year or 4.6 μrem/hr (15). These values are significantly higher than some earlier estimates (19).

The rise in cosmic ray exposure with altitude is of particular interest because many of the world's population live at high altitudes. Figure 30 shows the rise in the cosmic-ray dose with altitude, most of it due to ionizing radiation. In the United States 16.6% of the population lives at elevations above 1000 feet, but in Mexico and the Andean countries, central Europe and central Africa sizable populations live at high altitudes. In addition, aircraft crews and other persons who fly frequently are subject to considerable additional radiation doses from this cause, as illustrated in Figure 30.

Cosmic radiation also produces certain radionuclides in its interaction with the atmosphere and the surface layer of the ground. Of these, by far the most important are tritium (^3H), produced by the (n,t) and (p,t) reactions on nitrogen at a rate of 1-2 tritons/cm^2 sec, and carbon-14, produced by the $^{14}N(n,p)^{14}C$ reaction. This latter reaction is well known as the basis of the carbon-dating method (14). Several other nuclides are produced by cosmic-ray interaction with soil and water components; they are listed in Table 21 and can be seen to form part of the background activity found even in very "pure" air or water.

Terrestrial Sources of External Exposure

Uranium and thorium, which are naturally radioactive, are among the more common elements in the earth's crust. The three longest-lived uranium isotopes (^{234}U, ^{235}U and ^{238}U) have long enough half-lives to be found in most naturally occurring rocks, as does ^{232}Th, which is the most abundant element of the actinide group. Although these nuclides themselves are alpha-emitters, they are at the head of the well-known radioactive decay series, which decays down to lead isotopes via a sequence of alpha and beta decays. Several of the daughter elements in each series emit beta and gamma rays, and it is these radiations that

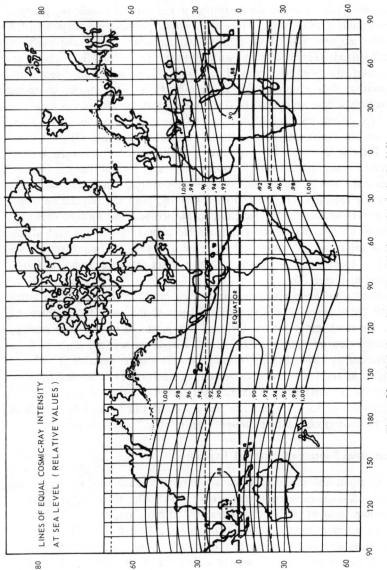

Figure 29. Variation of cosmic radiation with latitude (15).

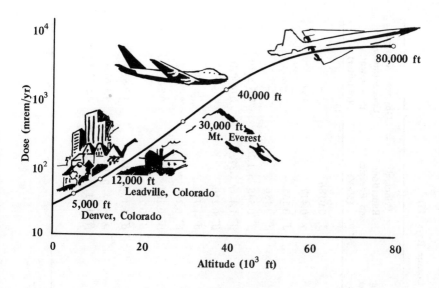

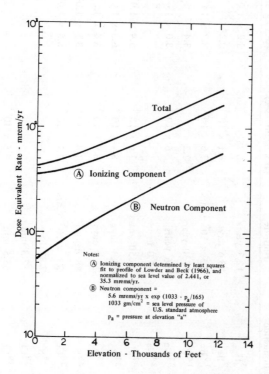

Figure 30. Change in cosmic-ray dose with altitude (15,20).

Table 21.   Natural Radionuclides Produced by Cosmic Rays (11)

Radionuclide	Half-life	Average Atmospheric Production Rate (atoms/cm² sec)	Tropospheric Concentration (pCi/kg air)	Principal Radiations and Energies (MeV)	Observed Average Concentrations in Rainwater (pCi/liter)
^{3}H	12.3 years	0.25	3.2×1.0^{-2}	β^- 0.0186	–
^{7}Be	53.6 days	8.1×10^{-3}	0.28	γ 0.477	18.0
^{10}Be	2.5×10^6 years	3.6×10^{-2}	3.2×1.0^{-8}	β^- 0.555	–
^{14}C	5730 years	2.2	3.4	β^- 0.156	–
^{22}Na	2.6 years	5.6×10^{-5}	3.0×1.0^{-5}	β^+ 0.545, γ 1.28	7.6×10^{-3}
^{24}Na	15.0 hr	–	–	β^- 1.4, γ 1.37; 2.75	0.08-0.16
^{32}Si	~650 years	1.6×10^{-4}	5.4×1.0^{-7}	β^- 0.210	–
^{32}P	14.3 days	8.1×10^{-4}	6.3×1.0^{-3}	β^- 1.71	"a few"
^{33}P	24.4 days	6.8×10^{-4}	3.4×1.0^{-3}	β^- 0.246	"a few"
^{35}S	88 days	1.4×10^{-3}	3.5×1.0^{-3}	β^- 0.167	0.2-2.9
^{36}Cl	3.1×10^5 years	1.1×10^{-3}	6.8×1.0^{-9}	β^- 0.714	–
^{38}S	2.87 hr	–	–	β^- 1.1, γ 1.88	1.8-5.9
^{38}Cl	37.3 min	–	–	β^- 4.91, γ 1.60, 2.17	4.1-67.6
^{39}Cl	55.5 min	1.6×10^{-3}	–	β^- 1.91, γ 0.25, 1.27, 1.52	4.5-22.5

form a significant portion of the external radiation field on the earth's surface. Table 22 lists representative values for radioelement content in common rocks and the corresponding dose values. It also lists potassium-40, the other major radionuclide in soil and rocks, that accompanies all natural potassium with an abundance ratio of 0.0118% and is a beta emitter with a peak energy of 1.32 MeV.

Table 22. Radionuclide Content and Dose-Equivalent Rates from Common Rocks and Soils(15)

Rock	ppm	Uranium (mrem/yr)[a]	ppm	Thorium (mrem/yr)[a]	ppm	Potassium-40 (mrem/yr)[a]	Total (mrem/yr)[a]
Igneous							
Basic	0,9	5.2	2.7	7.3	1.2	14.7	27.2
Silicic (granite)	4.7	26.9	20.0	53.8	5.0	61.3	142.0
Sedimentary							
Shale	3.7	21.1	12.0	32.3	3.2	39.2	92.7
Sandstone	0.45	2.6	1.7	4.6	1.1	13.5	20.7
Limestone	2.2	12.6	1.7	4.6	0.32	3.9	21.1
Upper crustal average	2.8	16.0	10	26.9	2.4	29.4	72.3
U.S. surficial average	1.8	10.3	9.0	24.2	1.8	21.8	56.3

[a]mrem/yr/ppm: uranium, 5.73; thorium, 2.69; potassium-40, 12.3

The widespread occurrence of these radioactive species has several interesting consequences. To start with, most building materials incorporate one or more of the mineral groups listed in Table 22. Consequently, any house built of brick or concrete blocks will contain some uranium, thorium, or potassium in the aggregate material, resulting in a measurable increase in radiation exposure inside over that existing in the open. Similarly, where granite is used in the construction of houses, such as in parts of Scotland, the external radiation dose would be expected to be high. Table 23 shows some measurements on gamma dose rates in buildings (20); note the decrease initially as one goes to higher floors and the soil contribution diminishes and the further increase with height as the cosmic-ray contribution becomes significant.

Table 24, adapted from Spiers *et al.* (21), shows total population doses arising from this background effect, which will be seen to be enormously greater than any dose contribution from power plants.

Table 23.   Gamma-Ray Doses in Some Office Buildings (20)

Building	Year Completed	Construction	Interior Walls	Height, Stories	Floor	Gamma Dose Rate[a] μrads/hr
JFK	1966	Reinforced concrete	Sheetrock partitions	23	Basement	6.7
					5	4.8
					20	4.9
					23	6.5
HC	1962	Reinforced concrete	Sheetrock partitions	10	2	9.0
SO	1917	Steel and concrete	Sheetrock partitions	12	Basement	5.5
					5	7.2
					12	7.3
HSPH[b]	1969	Reinforced concrete	Cinder block	14	Basement	7.3
					1	7.5
					3	7.4
					7	8.9
					9	7.8
					11	4.6
					12	6.7
					13	5.8
					14	6.8

[a]A cosmic-ray contribution of 4.1 μrads/hr has been subtracted from all values.
[b]First four floors, 1962; next 10 floors, 1969.

Some further small internal exposure may arise from radon accumulated in cases of poor ventilation.

Another source of external exposure occurs in locations where unusually high uranium or thorium content in rocks or soil exists, such as the monazite sands in Kerala, India, or the state of Rio de Janeiro, Brazil. Table 25 lists some of these locations and the corresponding population doses. It is evident that these admittedly limited populations receive a significantly higher population dose in comparison with the world's population (22). The γ-ray dose rates in air were measured at a height of 1 meter above the ground; in some of the areas mentioned these natural dose rates are well above the maximum permissible levels laid down for occupational workers.

At this time there appears to be no epidemiological evidence of any increased incidence of radiation-related malignancies or life-shortening in populations living in such areas of high background exposure compared with populations of comparable stock and economic conditions elsewhere.

Table 24. Background Radiation in Buildings (20,21)

Location	Type of Construction	Exposure Rate (mrad/yr) Range	Exposure Rate (mrad/yr) Average	Estimated Exposed Population (10^3)	Population Dose (10^3 man-rem)
Edinburgh	Stone	28-70	48.5	225	545
	Brick		74.5	158	588
	Misc.		48	14	33
Dundee	Stone	45-80	63	104	327
	Brick & concrete		77	58	223
	Misc.		57	15	42
Aberdeen	Granite	70-110	87	155	674
	Other		76	32	121
Aberdeenshire	Granite	55-155	87	89	387
	Other	35-100	66	54	178
Sweden	Concrete & alum. shale	182-228	197	2,500	24,625
	Brick	126-139	130	2,600	16,900
	Wood	75-84	77	2,300	8,855
New York City	Brick & concrete	56-121	87	7,900	34,365
Poland (Warsaw, Lodz)	Concrete	84-106			
Australia, (Darling Range)	Brick	32-193			
(Coastal Plain)	Brick	41-127			

Internal Sources

Potassium-40 is the main naturally occurring source of internal radiation, despite its low isotopic ratio of 0.0118 percent. The activity of ^{40}K in soils is on the average an order of magnitude higher than that of ^{238}U and ^{232}Th. In seawater, ^{40}K constitutes the bulk of the activity with a concentration of about 300 pCi/liter. It is the most common radioactive constituent in many plants; Table 26 shows some measured values for various plants (23).

Potassium enters the body mainly in foodstuffs and is under close homeostatic control; variations in the composition of the diet thus have little effect on the radiation dose received. The potassium content in the body varies considerably from one organ or tissue to another. Some tissues such as muscle, brain, and blood cells contain about 0.3 percent

Table 25. Background Levels in Regions of Higher-than-Average
Radiation Exposure (22).

Areas	Exposure Rate (mrad/yr)		Estimated Exposed Population (10^3)	50 Population Dose $(10^3$ man-rem)
	Range	Average		
Mountain cities: (Denver, Quito, La Paz, Bogota)	150-330	225	4,200	47,250
France (Massif Central)	180-350	200	7,000	70,000
Egypt, All		160	40,000	320,000
Monazite sands	200-475	200	7	70
Brazil Rio de Janeiro Beaches	550-1250	600	50	1,500
Minas Gerais	1700-12000	2,000	0.6	60
India, Kerala	800-8000	1,500	100	7,500
Sri Lanka, All		280	10,000	140,000
Granite areas	3000-7000	3,000	1	150
Niue Island	1000-2000	1,000	6	300
West Central Florida (phosphate area)	150-250	200	1,200	12,000
New England & Georgia granite areas	150-250	200	3,000	30,000

potassium, blood serum has a normal level of 0.01 percent and fat contains none. In bone, a representative value may be taken as 0.05 percent by weight for wet bone without marrow. The average potassium content of the whole body as a percentage of body weight depends upon body build and is smaller in obese persons.

On the basis of a mean potassium content in tissues of 0.2 percent, the dose rate to the soft tissues can be calculated as 19 mrad/yr, the beta and the gamma doses amounting to 17 and 2 mrad/yr, respectively. In bone, the potassium content being about one quarter of that in soft tissues, the annual dose to the osteocytes or to the tissues in the Haversian canals is about 6 mrad/yr. In the active marrow, the potassium content is about 0.2 percent, which leads to an estimated dose rate to the bone marrow and to the cells near the endosteal surfaces in the trabecular cavities of 15 mrad/yr.

The isotopic abundance of ^{87}Rb, which is a naturally occurring pure beta emitter, is 27.85 percent. The average concentration of

Table 26.   Potassium Content in the Ashes of Various Types of Plants (23)

Plant Group	Amount of Ashes %	Potassium Content in the Ashes/13/,%	Specific Radioactivity 10^{-8} curie/kg of the Ashes	of the Living Plant
Angiosperms				
Aquatic plants	16.3	15.5	12	2.0
Gramineae	6.6	23.0	18	1.2
Liliaceae	8.1	30.7	24	1.9
Polygonaceae	9.5	25.0	20	1.9
Chenopodiaceae	20.5	12.4	9.8	2.0
Cruciferae	9.6	23	18	1.7
Legumes	7.9	27.0	22	1.7
Umbelliferae	13.0	28.4	23	3.0
Ericaceae	2.1	16.0	12	2.5
Compositae	13.8	19.7	16	2.2
Cultivated gramineae	7.0	31.8	25	1.7
Legumes	10.4	27.0	22	2.2
Algae				
Green algae	25.3	5.0	4.0	1.0
Brown algae	27.8	18.2	14	3.8
Red algae	20.0	11.0	8.8	1.7
Mushrooms				
Diverse	7.2	28.4	22	1.5
Lichens				
Crustose	8.7	0.98	7.8	0.67
Foliose	4.5	10.0	8.0	0.35
Fruticose	2.6	9.3	7.4	0.19
Bryophyta-Mosses				
Diverse	4.6	8.0	6.4	0.29
Ferns				
Diverse	6.9	35.4	28	1.9
Horsetails				
Diverse	19.0	11.2	9.0	1.7
Gymnosperms				
Coniferous needles	4.5	6.5	5.2	0.23
The whole tree	3.8	15.4	12	0.45

rubidium in the whole body is 17 ppm; in bone, ovaries and testes, the corresponding figures are 10, 4.5, and 12 micrograms per gram of wet tissues, respectively. The dose rates to the gonads and to the bone tissues, resulting from decay of ^{87}Rb in the body, have been estimated from these concentrations to be about 0.3 mrad/yr, while the dose rate to the small tissue inclusions within bone would be approximately 0.4 mrad/yr. Assuming that the concentration of rubidium in the active marrow is the same as that averaged over the whole body, the dose rate to the bone marrow and to the cells near the endosteal surfaces in the trabecular cavities would be 0.6 mrad/yr. Table 27 shows the corresponding doses from the various natural radionuclides.

Because of their low specific activities neither thorium nor uranium contribute significantly to the internal radiation dose to man. Typical daily intake of uranium in food is about 1 μg/day, leading to dose rates to bone ranging from 0.3 to 0.8 mrad/yr (17). Nevertheless, the uranium content in plants constitutes a background level against which any incremental contamination may have to be determined. Table 28 shows examples of the uranium content encountered in some plants (23).

Radium isotopes are normally present in any type of soil at varying levels of radioactive equilibrium with their parents. Since, on the average, ^{232}Th and ^{238}U have about the same activity concentration, the same can be said for ^{226}Ra and ^{228}Ra, the two isotopes under consideration. The normal content of ^{226}Ra may be slightly increased when phosphate fertilizers are used. The uptake of radium by plant shoots is much higher than that of thorium, lead, or polonium, but is low in comparison to calcium.

In air, the natural background concentration is constituted by the airborne particulate matter picked up from the earth. In city air, it can be estimated at around seven attocuries per cubic meter, if it is assumed that uranium and radium have the same activity in the atmosphere. Fossil-fueled power plants release very small amounts of radium, which may be detectable in the air around those installations. The maximum activity of ^{226}Ra dispersed each year by the combustion of coal has been estimated at 150 curies.

In fresh waters, the ^{226}Ra content is highly variable, typical figures ranging from 0.01 to 1 pCi/l. The highest concentrations, ranging up to 100 pCi/l, are found in mineral waters. Drinking-water supplies drawn from surface waters generally do not contain significant amounts of radium. Flocculation and water-softening processes remove the bulk of the radium activity from water. The average daily intake of ^{226}Ra in areas of normal radiation background has been found to be about 1 pCi/g Ca in the United States and many other countries. The average bone dose from ^{226}Ra ranges from 0.6 to 1.6 mrad/yr and from ^{228}Ra from 0.8 to 1.9 mrad/yr, but bone dose rates as high as 16 mrad/yr from radium intake have been observed for the Kerala beach sand region (17).

Finally, radon gas emanated from uranium-bearing rocks and soil is introduced into the body by inhalation. Radon levels vary with uranium

Table 27. Dose Rates Due to Internal and External Irradiation from Natural Sources in Normal Areas (17)

Source of Irradiation	Dose Rates (mrad/yr^{-1})			Cortical Bone		Trabecular Bone	
	Gonads	Bone-Lining Cells	Bone Marrow	Osteocytes	Haversian Canals	Surfaces	Marrow
External irradiation							
Cosmic rays:							
ionizing component	28	28	28				
neutron component	0.35	0.35	0.35				
Terrestrial radiation (including air)	44	44	44				
Internal irradiation							
^{3}H	0.001	0.001	0.001	~0.001	~0.001	~0.001	~0.001
^{14}C	0.7	0.8	0.7	0.8	0.8	0.8	0.7
^{40}K	19	15	15	6	6	15	15
^{87}Rb	0.3	0.6	0.6	0.4	0.4	0.6	0.6
^{210}Po	0.6	1.6	0.3				
^{220}Rn	0.003	0.05	0.05				
^{222}Rn	0.07	0.08	0.08				
^{226}Ra	0.02	0.6	0.1				
^{228}Ra	0.03	0.8	0.1				
^{238}U	0.03	0.3	0.06				
Rounded Total	93	92	89	7.2	7.2	16.4	16.3
Percentage from alpha particles plus neutrons	1.2	4.1	1.2				

Table 28. Uranium Content in Plants (23)

Plant	Uranium Content in the Ashes %	Specific Radioactivity of the Plant curie/kg
Dill		
leaves	2.95×10^{-6}	2×10^{-11}
seeds	9.56×10^{-7}	6.4×10^{-12}
roots	1.54×10^{-6}	1.0×10^{-11}
Celery		
leaves	2.65×10^{-5}	1.8×10^{-10}
roots	7.91×10^{-6}	5.3×10^{-11}
Garlic		
bulb	4.1×10^{-5}	2.7×10^{-10}
Potato	3.2×10^{-8}	2.1×10^{-13}
Wormwood	1.8×10^{-4}	1.2×10^{-9}
Apple tree		
seeds	5.1×10^{-6}	3.4×10^{-11}
Grape		
stem	8.0×10^{-9}	5.4×10^{-14}
fruits	1.6×10^{-6}	1.1×10^{-11}
seeds	2.8×10^{-3}	1.9×10^{-8}
Mistletoe		
branches	3.73×10^{-6}	2.5×10^{-11}
leaves	5.85×10^{-6}	3.9×10^{-11}
berries	2.16×10^{-6}	1.5×10^{-11}
seeds	6.8×10^{-6}	4.5×10^{-11}
Pine tree		
branches	4.5×10^{-7}	3.1×10^{-12}
needles	2.5×10^{-6}	1.7×10^{-11}

soil concentrations. In the lungs, part of the radon will decay and the radon daughters will settle out mainly on the lower bronchi and contribute a local alpha dose as they undergo successive decays (10). Under most conditions radon daughters in the atmosphere contribute a few tenths of a μrem/hr to the dose rate, but low barometric pressure, little wind, atmospheric temperature inversions and low soil moisture may result in increased radiation emanation from the ground and high air concentrations of radon daughters (15). The incorporation of uranium and thorium-bearing minerals in building materials also leads to detectable radon levels in dwellings (Table 29), with indoor radon concentrations

Table 29. Radon-Daughter Concentrations in U.S. Buildings (20)

| Building | Location | Concentrations, pCi/liter | | | Ratio | Number of Air Changes per Hour |
		^{218}Po	^{214}Pb	^{214}Po		
		Single-Family Dwellings[a]				
ASG	Outside	0.04	0.04	0.03	1:1:0.8	
	1st floor	<0.005				
	Basement	~0.1				
MWF	1st floor	0.04	0.04	0.02	1:1:0.5	6
FSH	Outside	0.01	0.01	0.007	1:1:0.7	
	Inside	0.06	0.06	0.06	1:1:1	2
WAB	Outside					
	1st floor	0.23	0.17	0.17	1:0.7:0.7	2
	2nd floor					
	Basement	0.14	0.16	0.05	1:1.2:0.4	3
SP	Outside	0.03	0.02	0.04	1:0.7:1.3	
	1st floor	0.03	0.03	0.02	1:1:0.7	
	2nd floor	0.03	0.02	0.01	1:0.7:0.3	
	Basement	0.30	0.26	0.16	1:0.9:0.3	3
FJV	Outside	<0.01				
	1st floor	0.04	0.04	0.04	1:1:1	3
	Basement	0.94	0.97	0.84	1:1:0.9	1
DWM	Outside					
	1st floor	0.12	0.15	0.13	1:1.2:1.1	2
	2nd floor					
	Basement	0.52	0.46	0.34	1:0.9:0.6	1
		Multiple-Family Dwellings[b]				
MLC	2nd floor	0.01	0.01	0.01	1:1:1	
JS	2nd floor	0.07	0.07	0.03	1:1:0.4	9
OG	Outside	0.15	0.09	0.07	1:0.6:0.5	
	2nd floor	0.19	0.18	0.13		5

[a]All single-family dwellings were wood frame with poured-concrete basements.
[b]All multiple-family dwellings were brick.

three to four times those of outdoor values (17,20-22). Table 30 lists the estimated annual doses to various organs, arising from inhalation of radon and radon daughters from various sources. The special consequences of exceptionally high radon levels, such as may prevail in uranium mines, will be discussed in a later chapter.

Table 30. Estimated Annual Alpha Doses (in mrad) in Human Organs from
Inhalation of Radon Daughters (17)
Mean aspiration rate 13.9 liters/min

	^{222}Rn Concentration 0.3 pCi/l (^{222}Rn and Daughters in the Ratio 1/0.8/0.4/0.3)	^{212}Pb Concentration 3 pCi/l (^{212}Pb-^{212}Bi in Equilibrium)
Lungs (alveolar tissue)	2.2	0.6
Blood	0.2	0.1
Liver	0.1	0.07
Kidneys	0.5	0.2
Adrenal glands	0.1	0.02
Muscle	0.05	0.003
Bone	0.04	0.02
Bone marrow	0.08	0.05
Gonads	0.07	0.003

The Total Radiation Background

In assessing the total radiation background to the general population, one also needs to take into account exposure to two types of man-made radiation sources: fallout from nuclear explosions and medical diagnostic exposures. Since the cessation of atmospheric weapons testing by the U.S. and U.S.S.R., global fallout levels of radioactivity have dropped, but recent French and Chinese tests have been sufficient to maintain a relatively constant annual fallout deposition (24,25). Table 31 lists estimated doses from ingestion of fallout, excluding tritium, and the corresponding whole-body doses for the U.S. population. The aggregated doses are not uniformly distributed, but vary with age group.

The second form of man-made radiation exposure is that due to medical diagnostic radiology. The extent of this exposure is appreciable, though it varies to some extent with the availability of medical services.

Radiation is used to medicine for diagnostic purposes and for the treatment of diseases, particularly cancer. The local doses received by individual patients in the course of diagnostic investigations may vary from being about equal to the average doses received annually from natural sources (~0.1 rem) to 50 times as high. Radiation treatments, on the other hand, may involve individual doses thousands of times

Table 31. Annual Dose from Global Fallout in the U.S. (25)

	Whole Body Doses[a]			Dose from Ingestion (mrem/yr/person)[b]							
				^{14}C		^{137}Cs		^{89}Sr	^{90}Sr		^{131}I
Year	U.S. Population (millions)	Per Capita Dose (mrem)	Man-rem for U.S. Population (millions)	Whole-Body	Bone	Whole-Body	Bone	Bone	Whole-Body	Bone	Thyroid
1963	190	13	2.4	0.3	0.5	4.3		1.3	7.3	73	25
1965	194	6.9	1.3	0.7	1.1	2.3		0.2	7.2	72	4
1969	204	4.0	0.82	0.6	1.0	0.4		0.2	3.2	32	3
1980	237	4.4	1.1	0.6	1.0	0.4		0.2	3.2	32	3
1990	277	4.6	1.3	0.6	1.0	0.4		0.2	3.2	32	3
2000	321	4.9	1.6	0.6	1.0	0.4		0.2	3.2	32	3

aTritium is excluded.

bThe ^{90}Sr doses are delivered over 50 years. The whole-body dose = 0.1 times the dose to bone. Yearly ^{90}Sr whole-body dose rates estimated for the years considered are:

1963 = 0.9 mrem/yr   1980 = 2.5 mrem/yr
1965 = 1.9 mrem/yr   1990 = 2.7 mrem/yr
1969 = 2.1 mrem/yr   2000 = 3.0 mrem/yr

higher than those received for diagnostic purposes and are usually delivered over several weeks to parts of the body only. Both for diagnosis and for therapy, irradiation is mostly external, but an increasing number of radiological procedures now involves the administration of radioactive materials that result in internal irradiation.

The mean doses received by the population are determined by a combination of the doses delivered by the individual medical procedures and the number of cases in which these procedures are applied. The aim of medical radiology is to provide maximum benefit to the population served, and therefore an increase in frequency of examinations may be fully justified, particularly in the developing countries. Since it is probable that a large proportion of the world population does not have easy access to modern X-ray facilities, the number of such facilities must increase greatly if local health standards are to improve substantially. On the other hand, frequent concern has been expressed about the overuse of radiological methods where alternative methods were available or more efficient exposure methods could be applied (26). One of the consequences of this concern has been the discontinuance of population chest X-ray examinations in countries where the incidence of tuberculosis has been reduced to a very low level.

Depending in any particular case on the availability of funds and trained staff, three basic approaches can contribute variously to this improvement: educational programs, surveys of the frequency of examination and of the doses received, and administrative control measures. The provision of educational training programs and the establishment of some administrative control may be much more important than dose surveys, particularly where resources are limited (17). Table 32 lists the estimated doses from medical and dental radiation in the U.S. for the next 25 years. Historically, diagnostic exposures have accounted for

Table 32. Projected Doses—Medical Diagnostic Radiology

Year	Projected Population (millions)	Mean Annual X-Ray Examination Rate (exams/1000 population)	Genetically Significant Dose		Somatic Dose (man-rem)
			(mrem)	(man-rem)	
1960	183	0.57	50	9.1×10^6	13.6×10^6
1970	205	0.70	61	12.5×10^6	18.7×10^6
1980	237	0.83	71	16.8×10^6	25.2×10^6
1990	277	0.97	83	23.0×10^6	34.5×10^6
2000	321	1.10	93	29.8×10^6	44.7×10^6

over 90% of the somatic and genetically significant population dose. Since this is one area of significant radiation exposure that is under human control, it is undoubtedly here that careful review of risks and benefits will always be in order and meaningful reductions can be made. Figure 31 summarizes graphically the various contributions to the population dose (27).

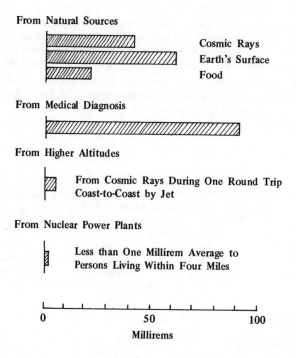

Figure 31. Comparative radiation exposures: approximate annual averages in the U.S. (27).

If we now review all of the sources of background radiation exposure, both natural and man-made, we arrive at the data in Table 33, which distinguish between whole-body exposure and genetically-significant exposure, and which show allocation of the total population dose in the United States to the various sources. In reviewing these figures a number of conclusions can be drawn:

(1)   Unavoidable radiation exposure for much of the world's population is of the order of 90-125 mrem/yr per person.

(2)   This level of radiation exposure is presumably harmless to the extent that no radiation-related illnesses can be clearly associated with it.

(3)   Populations living at higher altitudes and hence in a higher radiation field do not appear to show any health deficiencies; on the contrary, some mountain populations, such as those in the Hunza and Caucasus areas, show incidences of unusual longevity, though again probably not for any radiation-related reason.

(4)   It is reasonable to assume that any increase in radiation dose that is small compared with the natural radiation background and considerably less than the effect of moving from a wooden into a brick house (Table 24) may be considered to be insignificant.

(5)   This consideration does not imply that there is not also an obligation to reduce all man-made radiation levels to values as low as practicable including radiations from nuclear power plants and associated operations.

Table 33.   Estimates of Annual Whole-Body and Genetically-Significant Dose Rates in the United States, 1970 (28)

Source	Average Whole-Body Exposure (mrem/yr)	Genetically Significant Dose (mrem/yr)	Annual Population Dose (10^6 men-rems)
Natural radiation			
Cosmic radiation	44		
Radionuclides in body	18		
External gamma radiation	40		
Subtotal	102	90	20.91
Man-made radiation			
Medical and dental			
Diagnostic	72	30-60	14.8
Radiopharmaceuticals	1		0.2
Global fallout	4		0.82
Occupational exposure	0.8		0.16
Nuclear power (1970)	0.003		0.0007
Miscellaneous	2		0.5
Subtotal	80	30-60	16.48
Total	182	120-150	37.4

RADIATION EFFECTS AT HIGH DOSE LEVELS

The consequences of irradiation at high dose levels are the central concern of radiologists and health physicists. For detailed discussion the reader is referred to the standard texts of Johns and Cunningham (5), Attix and Roesch (2), and Blatz (29), as well as texts on health physics, like those of Morgan and Turner (18), Gloyna and Ledbetter (4), Duhamel (30) or Cember (31). The most complete discussion of the effects of ionizing radiations will be found in the UNSCEAR Report (17), which provides an excellent summary of available data.

The principal effect of high doses of radiation is the destruction of cells in the vicinity of the irradiated volume. For whole-body exposure of small mammals and lesser animals it can be shown that most readily detectable morbidities can be observed with a frequency proportional to the dose received. The effects observed are conventionally divided into *somatic* and *genetic* effects. Somatic effects are those resulting directly from the dose received by the affected cells and show themselves in some radiation-induced malignancy. While there is a great deal of information on irradiation effects in animals, the number of people exposed to substantial doses, other than for therapeutic treatment of cancer, is so small that the relationship between dose and incidence of malignancies in man can only be studied for the most radiosensitive tissues (17). By far the largest and most informative group of irradiated subjects have been the survivors of the atomic bombings of Hiroshima and Nagasaki. To these must be added several groups of patients treated by radiotherapy and followed up for several decades, and a few occupationally exposed radiation workers, especially underground uranium miners.

Leukemia is the best known of the radiation-induced malignancies. All evidence indicates that the incidence of certain types of leukemia increases with dose as a result of postnatal irradiation at high dose rate in the 50-500 rad interval. At higher doses the rise in frequency decreases, possibly because an increasingly large fraction of cells that would otherwise become leukemic are destroyed by radiation. Radiation-induced leukemias tend to occur most frequently within a few years after exposure and after 25 years the frequency tends to return to the levels expected in the absence of irradiation. By that time some 15-40 cases of leukemias per rad per million exposed have been observed (17,32,33).

Lung cancers appear to have been induced at Hiroshima by doses estimated on the basis of crude assumptions to be equivalent to some 30 rads of external gamma radiation delivered at high dose rate, and to have increased with dose up to a dose of about 100 rads. The higher incidence of this type of cancer among irradiated people has been revealed by other surveys also, but it is not yet known whether the increase, which started some 15 years after irradiation, will be sustained for a long time or will eventually subside. Taken at face value, however, the data indicate that from 10 (at 250 rad) to 40 (at 30 rad) cases of cancer per rad per million exposed develop during the first 25 years after exposure to high-dose-rate gamma radiation.

Information is available also in the induction of thyroid and breast cancers. Because those affected by these cancers have long survival times, only in the very long run do mortality data reflect the incidence of these tumors. Thus, while breast cancer mortality at Hiroshima suggests a risk of 6-20 cases per million per rad in the first 25 years after irradiation among women exposed to between 60 and 400 rads, this is probably an underestimate of the total yield. For thyroid cancers, an average figure of about 40 cases per million per rad in the same range of doses over the same period of time is obtained from more reliable morbidity data, but the estimate has large uncertainties due to the small number of cases observed. As for lung tumors, there is no information as to whether the increased annual incidence of tumors in the irradiated populations will subside and when.

Many surveys of externally irradiated people confirm an increase in other types of cancer taken together, although it is not possible at this stage to identify the specific types whose frequency is enhanced. Among the survivors of the atomic bombing at Hiroshima there is a clear trend for mortality from malignancies other than leukemia and lung and breast cancers to increase with increasing dose, but quantitative estimates of the rate of increase are hampered by our ignorance of the doses to the tissues concerned. Only a tentative estimate of 40 cases of cancers (other than leukemias and breast and lung cancer) per rad per million occurring during the first 25 years after exposure to 250 rads can be advanced on the basis of crude assumptions about tissue doses. Here also it is not known how many additional cases may develop at times later than 25 years.

Studies of people exposed to internal irradiation at substantial doses are few. They have concerned workers and patients contaminated with radium isotopes and miners exposed to radon gas. Radium-226 is deposited in bones, irradiates bone-forming cells continuously at a decreasing rate for decades after being absorbed into the body and gives rise to bone tumors. Radium-224 causes similar effects after a shorter period or irradiation.

Miners exposed to high levels of radon and its radioactive daughters show a very high incidence of lung cancers. The frequency appears to rise in proportion to the level and duration of exposure. The range of exposures within which the increased incidence has been reported corresponds to doses of at least a few hundred rads of alpha radiation. However, dosimetry is difficult and the role of other carcinogenic factors such as smoking habits has not yet been fully assessed.

All the evidence on humans and laboratory animals tends to indicate that few, if any, somatic radiation effects can be detected at dose levels below 10 rads. The situation may be slightly different for genetic effects. When the reproductive cells are irradiated, changes may be produced in the genes or in the chromosomes of these cells and subsequently transmitted to the descendants of the irradiated individual. These genetic changes are of different kinds: (a) gene mutations, *i.e.,* alterations in the function of individual genes, (b) chromosome aberrations resulting from breakage and reorganization of the chromosomes, and (c) changes in the number of chromosomes. Some of these changes result in offspring suffering abnormalities that may range from mildly detrimental to severely disabling or lethal.

Since adequate human data are not available, the estimates of genetic risks ensuing from radiation exposure of the human reproductive cells are based on results obtained with other species, notably mice.

The reproductive cell stages most important for the assessment of genetic risks are spermatogonia in the male and oöcytes in the female. At high acute doses of radiation the rate of induction in spermatogonial cells is estimated to range between 100-5,000 recessive mutations per rad per million births. Human populations, however, predominantly receive low doses of radiation under acute (short-term at high dose rate) or chronic (protracted at low dose rate) conditions of exposure. On the basis of experimental studies, it is estimated that under these conditions the rate of mutation induction is about one-third of the above figure. Consequently, for males, a rate of induction of 30-1,500 mutations per million per rad seems a realistic approximation. At high acute doses of radiation the risk of mutation in females conceiving shortly after radiation exposure will be about twice as high as in males, whereas at low doses the risk will be reduced to at least one-third and with chronic exposure to about one-twentieth of that expected after acute exposure to high doses. If the human ovary responds to irradiation as does that of the mouse, which is by no means certain, it can be expected that, if conception occurs after a sufficient interval following irradiation, the resulting frequency of mutations in the descendants of irradiated females might approach zero (17).

Dominant gene mutations are expressed in the first-generation descendants of an irradiated population. There is evidence suggesting that in man about 1,000 genes may contribute to this category. The estimated rate of induction of dominant visible mutations in the human male exposed to low doses or irradiation is two per rad per million descendants.

Spontaneously occurring chromosome aberrations are a source of considerable human hardship, since they are responsible for a large fraction of all spontaneous miscarriages, congenital malformations and mental and physical defects. For instance, the possession of an additional small chromosome (number 21) leads to Down's syndrome, which is associated with severe mental retardation. Extensive data have been collected on the mouse regarding another type of aberration known as translocation. This involves the exchange of parts between two different chromosomes. It is known that in man it may lead to malformations similar to those associated with the presence of additional chromosomes, or may lead to early prenatal death. These effects are associated with the presence of translocations in an unbalanced form, in which there may be loss of one of the exchanged segments and gain of the other. In its balanced form, a translocation usually has no detrimental effect for the person carrying it, but half of his or her offspring are likely to have the translocation in unbalanced form.

In male mice, the yield of balanced translocations in the generation after exposure is more or less proportional to dose for acute X-irradiation at moderate-to-high doses, about 30 being induced per million offspring per rad. It is probably similar in female mice but information is still meagre. The expected yield of unbalanced translocations (causing early death or detectable abnormality) will be 60 per million zygotes per rad after exposure of males and 180 per million per rad after exposure of females.

In attempting to deduce from these figures the probable risk for man under possible exposure conditions, a number of different factors have to be considered. First, there is evidence from work on chromosome aberrations in blood cells that man may be twice as radiosensitive as the mouse in this respect. Second, chronic gamma irradiation is only one-third as effective as acute X-irradiation in inducing translocations in male mice, while acute X-irradiation at very low doses (as used in medical diagnosis) may be only about one-quarter as effective per rad as at higher doses. Therefore the probable yield of balanced translocations in the offspring of exposed males is about 7 per million per rad with chronic gamma irradiation and about 15 per million per rad with low-dose acute X-irradiation. The expected yields of unbalanced translocation products will be about twice these (17).

Information from both man and the mouse suggests that many of these unbalanced zygotes will die at such an early stage in pregnancy that they will lead, at most, to a missed menstrual period. The proportion surviving to produce abnormal newborn babies is difficult to estimate at present, but it is likely to be less than 6%. Therefore one to two additional abnormal babies per million would be expected per rad of a paternal exposure at low doses or dose rates. Although a similar estimate cannot yet be made with confidence for maternal exposure, it seems unlikely that risks will be much higher.

Translocations are only one category of chromosomal aberration. Those occurring spontaneously (in balanced or unbalanced form) represent about one-third of all chromosome aberrations observed in newborn babies. Information from the mouse suggests that very few of the other aberrations (*e.g.,* gains or losses of chromosomes) are likely to be transmitted to the next generation after irradiation of the male because the reproductive cells carrying them will be eliminated before they mature. In the female, however, some are transmitted. Thus evidence from the mouse suggests that irradiation of females at low dose rates results in eight additional zygotes with the XO constitution per million (*i.e.,* with a missing sex chromosome).

Most of these cases will die before birth; those surviving will be sterile and will have certain other symptoms (Turner's syndrome). No doubt loss of other chromosomes will also occur, but these are likely to be associated with such early embryonic death that may not constitute a significant risk to live-born children. Gains of chromosomes form an important component of the human genetic burden. They may be induced by irradiation, especially of females, but positive evidence is lacking so far.

Thus in summary, gene mutations are induced at higher frequencies than chromosome aberrations; furthermore, chromosome aberrations will be eliminated after a few generations, whereas gene mutations may persist through many more generations, thereby affecting a larger number of individuals.

It has been estimated that about four percent of all live-born children suffer from various forms of genetically determined diseases, of which about two percent appear to follow simple rules of inheritance. The other two percent have a more complex mode of transmission. For computational purposes a figure of three percent will be used. Therefore the natural incidence of hereditary diseases maintained by receiving mutation is estimated at 30,000 per million live births (17).

The mutations responsible for that incidence would increase by about 300 per rad under conditions of chronic exposure of males in a parental generation. Up to 20 of these new mutations would contribute to the incidence of hereditary diseases among the immediate descendants of the irradiated males, while the contribution of the remaining new mutations would be distributed over many subsequent generations of descendants.

Genetic effects are essentially statistical phenomena. Although the probability of genetic damage in the offspring of a single irradiated individual is small or may result in a nonviable mutation, the total population dose within a genetic pool should be considered when dealing with a larger population. This subject has been reviewed extensively by Berry (34) and Green (35).

RADIATION EFFECTS AT LOW DOSE LEVELS

Except for cases of acute high-level exposure of reactor personnel due to a major accident, environmental considerations around nuclear facilities are concerned mainly with the impact of low radiation doses due to airborne or liquid effluents from the plant to members of the population in the vicinity of the plant site. For practical and ethical reasons such exposure levels must be kept "as low as practicable" under all conceivable conditions; since 1972 that term has acquired a quasi-legal meaning in the U.S.

In terms that are understandable to the potentially affected lay population it is important to establish beyond any doubt levels that would be "safe" and "acceptable." Neither of these terms can be quantitatively defined but in practice both imply a finite but negligible probability of risk. It is therefore of supreme importance to arrive at a reasonable estimate of the magnitude of any postulated radiation-induced effects.

The dose range of interest is of the order of 2-10 mrem, at dose rates of 0.1-1 mrem/yr. This compares with a dose level of 10 rads, which is the lower limit for observable radiation effects in bomb survivors or animal experiments and potentially much higher local doses from medical exposures.

In general terms, there are two ways at present of arriving at a determination as to what dose levels may be considered acceptable. One of these determines at what dose level the risk of detectable somatic or genetic radiation damage becomes low enough to be negligible in comparison to somatic or genetic damage due to "normal" causes encountered in a person's life time. The other criterion considers the exposure to "background radiation," i.e., to nuclear radiation that forms part of our natural environment, and determines below what level any additional radiation exposure would result in an insignificant increment in the dose received. These considerations have been discussed in detail in the BEIR Report (28), which may be supplemented from the bibliography by Sonnenblick (36).

The difficulties in establishing the effects of very low dose levels are two-fold. One arises from the fact that little statistically valid experimental work is available, mainly because of the long-term nature of such experiments that would need to be conducted under controlled conditions on a large enough population and over several generations in case of genetic effects. To do this for a large enough group of animals is expensive and slow, and it is certainly prohibitive on practical and moral grounds for humans. Even epidemiological studies on comparable populations living exposed to different natural radiation levels have proved difficult and inconclusive. It also becomes imperative to establish a unique cause-and-effect relationship between apparent correlations between morbidity and radiation exposure to rule out other causes such as economic conditions, nutritional factors, or improved diagnostic facilities due to better medical services.

The importance of this aspect is underlined by the repeated attempts by Sternglass to prove such correlations, even though his theories have been widely discredited as unsubstantiated and inconsistent (37-41). Another problem arises from the fact that, although it is quite easy to observe lethalities and major malignancies such as may be observed at dose levels above 50-100 rad, it is obviously more difficult to observe minor discomforts and incidences of illness in a much larger population, such as might result from low-level exposure, and to distinguish them unambiguously from similar troubles that could be caused by a host of other causes.

There are also occasional reports suggesting that low levels of radiation exposure may be beneficial, such as a study on the mortality experience of uranium and nonuranium workers by Scott et al. (42). Other papers have attempted to show beneficial effects of radiation (43,44) when in fact they merely show the existence of coincident radiation effects, some of which predominate at different dose levels. Some effects may well be judged beneficial by some subjective criterion; either way, however, any *detectable* radiation effects should be considered to represent an unacceptable environmental impact.

In the fact of these difficulties, a common approach is to take the dose-effect values obtained at high dose levels and to extrapolate them down to low dose levels. This assumes a linear relationship between dose and effect independent of dose rate, a fact that appears to be well established for acute high dose levels (no chronic high dose effects seem to be on record, for obvious reasons).

Thus, one can express an estimate of risk in absolute or relative terms (28). The *absolute risk* is the excess risk due to irradiation. In practice, this is the difference between the risk of suffering a certain malignancy in the irradiated population and the nonirradiated population. For example, under the linear dose-incidence model, the absolute risk may be expressed as the number of excess, radiation-related cases of cancer per unit of time in an exposed population of given size per unit dose (*e.g.,* 1 case per 10^6 exposed persons per year per rad). The *relative risk* is the ratio between the risk in the irradiated population and the risk in the nonirradiated population. The doubling dose, *i.e.,* the dose that will double the standard (natural) risk, is a special example of relative risk.

As suggested by the ICRP (19), the expression of risk estimates in absolute terms, *i.e., x* cases per million exposed people per year per rem, might be interpreted as implying considerably greater accuracy than is warranted.

In order to minimize their misinterpretation and misuse, numerical risk estimates should be accompanied regularly by the following qualifying information:

(1)   Range of doses, dose rates, and exposure conditions on which the risk estimate is based and for which it is, therefore, scientifically valid.

(2)   Any biomedical conditions or indications for irradiation affecting the population for which the risk estimate is scientifically valid.

(3)   Age range (at irradiation) for which the risk estimate is scientifically valid.

(4)   Sex for which the risk estimate is scientifically valid.

(5)   Years of observation or person-years at risk for which the estimated average annual risk or total risk are valid. The expression of risk on an annual basis averaged over a short period may give an underestimation of the lifetime risk for cancers with a longer modal latent period, or may give an overestimation of lifetime risk for cancers with a short modal latent period. One of the most serious problems is the fact that existing knowledge of cancer induction in man is based on a limited number of years after exposure, and information is lacking about risk during later years when the natural cancer incidence increases greatly.

On this basis, incidence figures have been developed for leukemia ranging from 2 to 30 x 10^{-6} cases per person per rad, and for thyroid carcinoma of the order of 3.5 x 10^{-6} per person per rad from ^{131}I, with a mortality risk of 5 x 10^{-6} per person per rad. These figures should be compared with the normal incidence of leukemia, approximately 9 cases per 100,000 people for Japan, and with an incidence of thyroid cancer estimated at 20-25 cases per million people per year, with a mortality in the U.S. in 1955 of 5 per million men and 8 per million women.

Invariably the above risk figures have been obtained for dose levels of the order of 100-1000 rads. Extrapolating them down to dose levels a thousand times lower presents a number of serious problems. This is usually expressed in terms of the conflict of the "linear" versus the "threshold hypothesis," as shown in Figure 32. The linear hypothesis

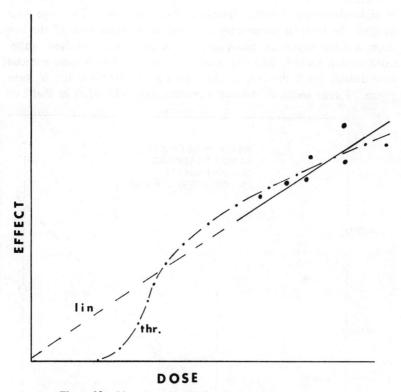

Figure 32. Linear *vs.* threshold radiation-effect hypotheses.

assumes that the proportionality between radiation dose and observed effects, that has been well established at high individual dose rates, can be projected to much lower dose levels for large populations. The threshold hypothesis takes into account the fact that few if any radiation-induced effects have in fact been observed in individuals at very low dose levels, that there appears to be evidence of recovery of cell damage by replacement in most tissues, and that many radiation damage phenomena clearly require some synergistic effects that can occur only if a certain minimum amount of energy is deposited in a given cell volume (17). In the absence of clear experimental evidence supporting either hypothesis at very low dose levels at low dose rates it is usually considered *prudent* to adopt the linear hypothesis simply as offering a more conservative answer, even though it is evident that it leads to unduly pessimistic predictions. Estimates based on them can be considered as upper limits of effect, not precise estimates of the effects that might actually occur.

Figures 33 and 34 from the BEIR Report (28) illustrate the difficulty of making meaningful extrapolations to low dose levels. The large error margins, the inherent uncertainty as to the appropriate value of the "true" slope, and the enormous discrepancy of the dose ranges involved, make extrapolation through four orders of magnitude of dose a futile exercise; nevertheless, this is the only available data at this time and has to serve. Figure 34 even seems to indicate a possible threshold effect in that case.

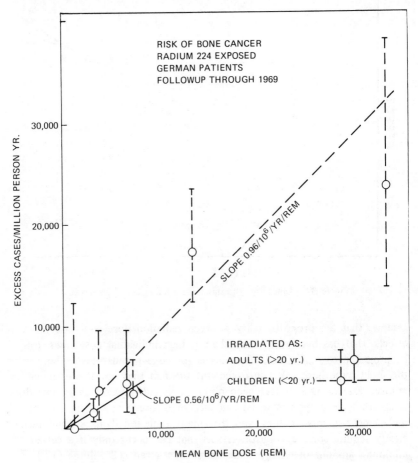

Figure 33. Dose-response data for bone cancer in German patients given radium-224 therapeutically (28).

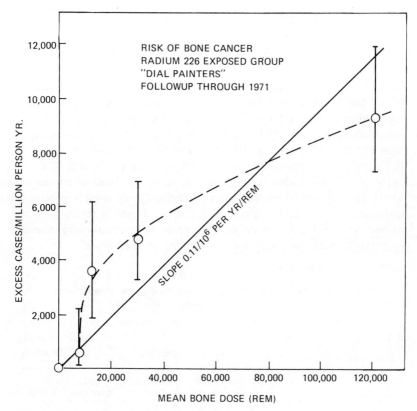

Figure 34. Dose-response data for bone cancer in Argonne National Laboratory series of subjects exposed to radium-226 from 1915 to 1935 (28).

It must also be remembered that most estimates of somatic damage are expressed in terms of "extra incidences" of disease or "additional deaths." Many health physicists have pointed out that such statements tend to be meaningless or misleading, unless they are related to the normal incidence rate for the specific disease, *i.e.,* as a percentage increment or, in the case of mortality figures, as percentage life shortening. Although absolute incidence figures tend to have a more dramatic or emotional impact for those anxious to indict the nuclear industry, in many cases the predicted life shortening or cancer incidence is small compared with annual statistical fluctuations. Even so, it is obviously desirable to reduce radiation exposure to the fullest extent possible.

This applies especially when considering genetic risks. These are currently expressed in four ways:

(1)   risk relative to natural background radiation
(2)   risk estimates for specific genetic conditions
(3)   risk relative to current prevalence of serious disabilities
(4)   risk in terms of overall ill health.

Table 34 presents estimated radiation effects for specific genetic damage
(28). The 5 rem dose is listed because that dose value is used in many
regulations as the maximum radiation dose that may be received by a
nonradiation worker over his reproductive lifetime from sources other
than natural background or medical treatment. Assuming this period to
be 30 years, the corresponding maximum exposure per year is of the
order of 170 mR. This exposure level is vastly in excess of any exposure
to the general population that can be visualized from nuclear power plants
under construction at this time. The situation is not as satisfactory at
present with regard to the rather small group of occupationally exposed
radiation workers, who may well be exposed to such doses.

Table 34. Estimated Radiation Effects for Specific Genetic or
Cytogenetic Damage (28)[a]

	Current Incidence per Million Live Births	Number that are New Mutants	Effect of 5 rem per Generation	
			First Generation	Equilibrium
Autosomal dominant traits	10,000	2,000	50-500	250-2500
X-Chromosome-linked traits	400	65	0-15	10-100
Recessive traits	1,500	?	very few	very slow increase
Congenital anomalies				
Unbalanced rearrangements[b]	1,000		60	75
Aneuploidy	4,000		5	5
Recognized abortions				
Aneuploidy & polyploidy	35,000		55	55
XO	9,000		15	15
Unbalanced rearrangements	11,000		360	450

[a]Values of expected numbers per million live births.
[b]Unbalanced rearrangements based on male radiation only.

In summary, extrapolation down to low dose rate values tends to give quantitative but probably overconservative estimates of the probable effects of any radiation exposure to the general population. This approach leads to a possible means of estimating the environmental impact of nuclear installations if a reasonable estimate can be made of the total radiation exposure. According to the UNSCEAR Report (17) it has been estimated that about four percent of all live-born children suffer from various forms of genetically determined diseases, of which about two percent have a more complex mode of transmission. For computational purposes a figure of three percent will be used. Therefore the natural incidence of hereditary diseases maintained by receiving mutation is estimated at 30,000 per million live births. The mutations responsible for that incidence would increase by about 300 per rad under conditions of chronic exposure of males in a parental generation. Up to 20 of these new mutations would contribute to the incidence of hereditary diseases among the immediate descendants of the irradiated males, while the contribution of the remaining new mutations would be distributed over many subsequent generations of descendants. Again, it must be emphasized that observations of the effects of comparable increases in the natural radiation background dose levels do not support the above estimates.

RADIATION STANDARDS

In attempting to define "safe" or "acceptable" radiation exposure levels several approaches have been taken. A number of recommended guidelines have been issued, mainly under the auspices of the International Commission on Radiological Protection (ICRP), which set out certain principles and procedures of good practice (6,45). The usual method has been to define upper limits of permissible exposure, based on previous experience with radiation injuries and atomic weapons effects, with a reasonable safety factor thrown in. This approach has been attacked as giving the public the illusion of a sharp demarkation between "safe" and "unsafe" conditions, but careful reading of the documents does not bear out this contention. Nevertheless in recent years great pains have been taken by the nuclear industry to show that any projected radiation levels associated with their operations can be maintained well below those maximum values. The guidelines distinguish between doses received by a radiation worker —a person knowingly and occupationally exposed to identified radiation sources in the course of his normal employment—and a member of the general public. A radiation worker is supposed to be aware of the existence of radiation fields, to carry radiation-monitoring devices, and to be subject to stringent operational rules and regulations.

As such it is considered reasonable to permit him to be exposed, if necessary, to somewhat higher radiation levels than the general public.

One of the fundamental criteria for the upper limit of permissible occupational exposure is that an employee should not accumulate more than a whole-body dose of 5(N - 18) rem, where N is the employee's age in years. Stated another way, the employee should not work with ionizing radiation until he or she is at least 18 years old and then should not be exposed to more than an average dose of 5 rem/year. The guidelines then direct that certain organs of the body should receive no more than is permissible for whole-body exposure, but that other organs or parts of the body may be more resistant to the effects of radiation exposure and, therefore, can be permitted to receive a higher dose. Thus, the skin and portions of the extremities such as the hands and feet can be permitted more than 5 rem/year. The limiting dose equivalents for various types of exposure are given in Table 35 (11).

When internal radiation exposure is involved, the ICRP methodology introduces the concept of the critical organ, already defined. For occupational exposure and but for the exceptions noted in Table 35, the permissible dose to the individual organs of the body is limited to 15 rem/year.

Table 35.   Dose-Limiting Recommendations, Proposed by NCRP, 1971 (11)

Combined whole-body occupational exposure	
Prospective annual limit	5 rem in any one year
Retrospective annual limit	10-15 rem in any one year
Long term accumulation to age N years	$(N-18)$ x 5 rem
Skin	15 rem in any one year
Hands	75 rem in any one year (25/qtr)
Forearms	30 rem in any one year (10/qtr)
Other organs, tissues, and organ systems	15 rem in any one year (5/qtr)
Fertile women (with respect to fetus)	0.5 rem in gestation period
Dose limits for the public or occasionally	
exposed individuals	0.5 rem in any one year
Population dose limits	
Genetic	0.17 rem average per year
Somatic	0.17 rem average per yer
Emergency dose limits—life saving	
Individual (older than 45 years if possible)	100 rem
Hands and forearms	200 rem, additional (300 rem, total)
Emergency dose limits—less urgent	
Individual	25 rem
Hands and Forearms	100 rem, total

For occupational external exposure the maximum permissible dose is assumed to be administered over a 13-week period and, although the permissible annual dose is limited to an average of 5 rem, the quarterly (13-week) dose is limited to 3 rem. For occupational exposure, it is seen from Table 35 that the NCRP has recommended a limit of 100 rem for persons older than 45 years who become involved in lifesaving emergencies such as might be necessary in the event of an industrial accident. The hands and forearms can receive an additional 200 rem.

The NCRP recommendations recognize that, although one may plan to limit the exposure to 3 rem/quarter or 5 rem per year, this may be exceeded for various unanticipated reasons. The NCRP position is that no deviation from sound protection practice is implied if the retrospective dose does not exceed 10 to 12 rem for dose increments well distributed over time or even 15 rem for exceptionally well-distributed increments. Of course, the age-dependent cumulative dose limitation, 5 (N - 18), is overriding.

The standards for human exposure to ingested or inhaled radioactive substances are based on the assumption that the dose from any radioactive material in the body or in the critical organ should not exceed the applicable annual permissible dose. These figures are then translated into maximum permissible concentrations of each radionuclide in air or water using a set of physiological parameters that describe the movement of each element in and out of the critical organ, the mass of the organ, and the daily rate at which the contaminants are inhaled or ingested. For purposes of uniformity, the ICRP has defined the characteristics of the "standard man," and from these calculations the ICRP and NCRP have prepared tabulations of the maximum permissible concentrations of each radionuclide in air or water which if inhaled or ingested continuously would, in a lifetime exposure of 50 years, result in a body burden that would deliver the maximum permissible dose to one or more organs of the body. The values for some of the principal parameters of the ICRP standard man are given in Table 36.

Permissible Exposure to Internal Emitters

From time to time the NCRP and ICRP publish tabulations of maximum permissible values for more than 130 radionuclides (6,46). The American values are now essentially the basis for the NRC regulations as expressed in 10 CFR 20 (47).

The permissible dose due to internal emitters is calculated on the assumption that there is no exposure to external radiation. Where such exposure exists, the permissible concentration of radionuclides in air,

Table 36. Physiological Parameters of "Standard Man" (11)

Intake and Excretion of the Standard Man

Water Balance			
Intake (cm^3/day)		Excretion (cm^3/day)	
Food	1000	Urine	1400
Fluids	1200	Sweat	600
Oxidation	300	From lungs	300
		Feces	200
Total	2500	Total	2500

	Air Balance		
	O_2 (vol. %)	CO_2 (vol. %)	N_2 + Others (vol. %)
Inspired air	20.94	0.03	79.03
Expired air	16	4.0	80
Alveolar air (inspired)	15	5.6	–
Alveolar air (expired)	14	6.0	–

Vital capacity of lungs	3-4 liters (men)
	2-3 liters (women)
Air inhaled during 8-hr workday	10^7 cm^3/day
Air inhaled during 16 hr not at work	10^7 cm^3/day
Total air inhaled	2×10^7 cm^3/day
Interchange area of lungs	50 m^2
Area of upper respiratory tract, trachea, bronchi	20 m^2
Total surface area of respiratory tract	70 m^2
Total water in body	4.3×10^4 g
Average life span of man	70 years
Occupational exposure time of man	8 hr/day; 40 hr/week;
	50 weeks/year; 50 years total

Gastrointestinal Tract of the Standard Man

Portion of GI Tract that is the Critical Tissue	Mass of Contents (g)	Remains (day)	Fraction from Lung to GI Tract (fa)	
			Soluble	Insoluble
Stomach	250	1/24	0.50	0.625
Small intestine	1100	4/24	0.50	0.625
Upper large intestine	135	8/24	0.50	0.625
Lower large intestine	150	18/24	0.50	0.625

Table 36. continued

Particulates in Respiratory Tract of the Standard Man[a]

Distribution	Readily Soluble Compounds (%)	Other Compounds (%)
Exhaled	25	25
Deposited in upper respiratory passages and subsequently swallowed	50	50
Deposited in the lungs (lower respiratory passages)	25	25

[a]Retention of particulate matter in the lungs depends on many factors, such as the size, shape, and density of the particles, the chemical form, and whether or not the person is a mouth breather.

water, or food must be reduced so that the total dose to the organ from both internal and external sources does not exceed the basic guide. To accomplish this, the MPC values must be reduced by the factor $(D - E)/D$, where D is the maximum permissible dose permitted to an organ and E is the dose received from external radiation.

The published tabulations of the permissible concentrations in air and drinking water for exposure of the general public are a convenient method of presenting the individual calculated values. (To avoid misinterpretation of the intent of the "Maximum Permissible Concentration" (MPC) values, the term "Radiation Concentration Guide" (RGG) value is used to an increasing extent.) However, since human exposure may result from contamination of foods, allowance must be made for varying biological concentration processes that may require that in some situations the "permissible concentration value" in air or water should be lower than the concentrations that are acceptable for direct inhalation or ingestion.

Again, it is important to distinguish between external and internal radiation levels and the different methods of applying them. External radiation fields can be readily monitored and recorded by means of personnel dosimeters. Internal exposures involve a measurement of radioisotope concentrations in air, food and water, estimates of the mode of ingestion and retention in the body, and consequent calculation of whole-body and gonadal doses from accumulated body burden concentrated in critical organs (4,6,11,26).

Obviously, it would not be proper to permit an employee to continue to receive the maximum permissible annual dose equivalent for many years as a result of the intake of a radionuclide during a single year, for such a commitment essentially would permit him no further occupational exposure. As a consequence, the ICRP has developed the concept of "dose commitment." The maximum permissible annual dose commitment is defined as the dose resulting from a body intake of a radionuclide by a person occupationally exposed for one year at the maximum permissible concentration (MPC) of a radionuclide. It can be shown that from a maximum permissible annual dose commitment the resulting critical organ dose-equivalent integrated over $50 + t/2$ years (time beginning when the intake of the radionuclide begins) would be equal numerically to the annual permissible dose equivalent, R, (where R equals 5 rem to total body, gonads or red bone marrow, 30 rem to thyroid, skin or bone and 15 rem to most body organs) and t is the period of intake of the radionuclide at the MPC (26). This is subject to the limitation that the sum of all planned special exposures in a lifetime should not exceed a $5R$ dose commitment (45).

Of course, in case of many unplanned situations, accidents or serious emergencies, *e.g.,* to save a life, there can be set no fixed levels of maximum permissible exposure and each responsible person must be ready to take whatever action seems wise in view of the circumstances. In such a case, useful and intelligent decisions can be made only by those who have a considerable knowledge of the local situation, a relatively complete picture of the potential radiation exposures with an appreciation of the probable errors in the estimations and a broad case of understanding of what damage might be expected from a given dose. This is one reason why the health physicist is an essential person to have available for advice and quick decisions during a time of radiation emergency.

It is evident that most of the regulating agencies have been concerned essentially with setting guidelines for maximum exposure under chronic or acute conditions. In practice they stand at 500 mrem per year to any individual in the general public from all sources except background radiation and medical devices; the exposure to the whole population is limited to an average of 170 mrem/year per person. It is widely recognized that such maximum exposures merely define, principally on legal grounds, an upper limit to radiation levels and radioisotope concentrations in air and water that would be acceptable in various circumstances. If it is prudently assumed that there is no threshold dose for the various radiation effects, it is obviously incumbent on all concerned to maintain the lowest possible radiation levels at all times. The U.S. Atomic Energy Commission has required in recent years that all nuclear facilities design for *lowest practicable levels* in the effluents, both under routine and emergency conditions. These levels are not defined numerically, but are interpreted as being attainable with existing technology and at present are considered to be equivalent to 5 mrem/yr to an individual at the fence line from airborne effluents and the same from liquid effluents. As technology improves, these levels can undoubtedly be reduced further, given unlimited financial resources, but at some point the law of diminishing returns will require an evaluation of the effects of residual exposures in terms of any additional hazards imposed by them on the surrounding population, especially in relation to natural background and medical exposures. This requires an application of sound judgment and common sense, with a realistic estimate of any risks involved. This assessment will be discussed in Chapter 13.

REFERENCES

1. International Commission on Radiation Units and Measurements. "Radiation Quantities and Units," ICRU Report 19, Washington, D.C. (1971).
2. Attix, F. H. and W. C. Roesch, Eds. *Radiation Dosimetry*, Vol. 1, (New York: Academic Press, 1968).
3. Grodstein, G. W. "X-Ray Attenuation Coefficients from 10 keV to 100 MeV," NBS Circular 583 (Washington, D.C.: National Bureau of Standards, 1957).
4. Gloyna, E. F. and J. O. Ledbetter. *Principles of Radiological Health*. (New York: Marcel Dekker, 1969).
5. Johns, H. E. and J. R. Cunningham. *The Physics of Radiology*, 3rd ed. (Springfield, Illinois: C. C. Thomas, 1969).
6. International Commission on Radiological Protection (ICRP). "Report of Committee II on Permissible Dose for Internal Radiation (1959), ICRP Publication 2 (Oxford: Pergamon Press, 1960), Supplement ICRP Publication 6 (New York: Macmillan Co., 1964).
7. Prêtre, S. "Quantities for the Description of Neutron Irradiations of Man—A Look at the Mechanism of Over-Conservatism," *Health Phys.* **23**, 169 (1972).
8. Snyder, W. S., M. R. Ford, G. G. Warner and H. L. Fisher. "Estimates of Absorbed Fractions for Monoenergetic Photon Sources Uniformly Distributed in Various Organs of a Heterogeneous Photon Sources Uniformly Distributed in Various Organs of a Heterogeneous Phantom," *J. Nucl. Medicine* **10**, Suppl. No. 3, 5 (1969).
9. "Environmental Radiation Dose Commitment: An Application to the Nuclear Power Industry," Rept. EPA-520/4-73-002. (Washington, D.C.: U.S. Environmental Protection Agency, 1974).
10. Adams, J. A. S. and W. M. Lowder, Eds. *The Natural Radiation Environment* (Chicago: University of Chicago Press, 1964).

11. Eisenbud, M. *Environmental Radioactivity*, 2nd ed. (New York: McGraw-Hill, 1973).

12. Freiling, E. C. Ed. *Radionuclides in the Environment*, Advances in Chemistry Series 93 (Washington, D.C.: American Chemical Society, 1970).

13. Junge, C. E. *Air Chemistry and Radioactivity* (New York: Academic Press, 1963).

14. Lal, D. and H. E. Suess. "The Radioactivity of the Atmosphere and Hydrosphere," *Ann. Rev. Nucl. Sci.* **18**, 407 (1968).

15. Oakley, D. T. "Natural Radiation Exposure in the United States," Report ORP/SID 72-1 (Washington, D.C.: U.S. Environmental Protection Agency, 1972).

16. Otway, H. J. "The Application of Risk Allocation to Reactor Siting and Design," U.S. Atomic Energy Commission Report LA-4316, Los Alamos Scientific Lab (1969).

17. UNSCEAR. "Ionizing Radiations: Levels and Effects," (New York: United Nations, 1972).

18. "Natural Background Radiation in the United States," NCRP Report 45. (Washington, D.C.: National Council on Radiation Protection and Measurements, 1975).

19. International Commission on Radiological Protection (ICRP). "The Evaluation of Risks from Radiation," *Health Phys.* **12**, 289 (1966).

20. Yeates, D. B., A. S. Goldin and D. W. Moeller. "Natural Radiation in the Urban Environment," *Nucl. Safety* **13**, 275 (1972).

21. Spiers, F. W., M. J. McHugh and D. B. Appleby. in *Environmental X-Ray Dose to Populations in the Natural Radiation Environment* J. A. S. Adams and W. M. Lowder, Eds. (Chicago: University of Chicago Press, 1964).

22. Auxier, J. A. "Contribution of Natural Terrestrial Sources to the Total Radiation Dose to Man," Thesis, Georgia Institute of Technology, Atlanta (1972).

23. Pertsov, L. A. "The Natural Radioactivity of the Biosphere," (trans.), U.S. Atomic Energy Commission Report AEC-tr-6714 (1967).

24. Klement, A. W., Ed. *Radioactive Fallout from Nuclear Weapons Tests*, AEC Symposium Series No. 5. (Washington, D. C.: U.S. Atomic Energy Commission, 1965).

25. Klement, A. F., C. R. Miller, R. P. Minx, B. Shlein, Eds. "Estimates of Ionizing Radiation Doses in the United States 1960-2000," Rept. ORP/CSD72-1 (Rockville, Maryland: U.S. Environmental Protection Agency, 1972).

26. Morgan, K. Z. "Present Status of Recommendations of the ICRP, NCRP and FRC," in A. M. F. Duhamel, Ed. *Health Physics*, Progress in Nuclear Energy, Series XII (Oxford: Pergamon Press, 1969).

27. "The Nuclear Industry—1971," U.S. Atomic Energy Commission Report WASH 1174-71 (Washington, D.C.: U.S. Atomic Energy Commission, 1972).

28. BEIR Committee. "The Effects on Populations of Exposure to Low Levels of Ionizing Radiation," (Washington, D.C.: National Academy of Sciences/National Research Council, 1972).

29. Blatz, H. *Radiological Health.* (New York: McGraw-Hill, 1964).
30. Duhamel, A. M. F., Ed. *Health Physics,* Vol. 2, Pt. I, Progress in Nuclear Energy, Series XII (Oxford: Pergamon Press, 1969).
31. Cember, H. *Introduction to Health Physics.* (London: Pergamon Press, 1969).
32. Brill, A. B., M. Tomonaga and R. M. Heyssel. "Leukemia in Humans Following Exposure to Ionizing Radiation," Tech. Rept. 15-59 (Hiroshima-Nagasaki, Japan: Atomic Bomb Casualty Commission, 1959).
33. Ishimaru, T., T. Hoshino, M. Ichimaru, H. Okada, T. Tomiyasu, T, Tsuchimoto, and T. Yamamoto. "Leukemia in Atomic Bomb Survivors: Hiroshima and Nagasaki," 1 Oct. 1950-30 Sept. 1966, Technical Report 25-69 (Japan: Atomic Bomb Casualty Commission, 1969).
34. Berry, R. J. "Genetical Effects of Radiation on Populations," *Atomic Energy Review* **10**, 67 (1972).
35. Green, E. L. "Genetic Effects of Radiation on Mammalian Populations," *Ann. Rev. of Genetics* **2**, 87 (1968).
36. Sonnenblick, B. P. "Bibliography—Low and Very Low Dose Influences of Ionizing Radiations on Cells and Organisms, Including Man," (FDA) 72-80029; BRH/DBE 72-1 (Washington, D.C.: Department of Health, Education and Welfare, 1972).
37. Graham, S. and E. Thro. "The Sternglass Phenomenon," *Nuclear News* **12**(10), 37 (1969).
38. Shaw, R. F. and A. P. Smith. "^{90}Sr and Infant Mortality in Canada," *Nature* **228**, 667 (1970).
39. Sternglass, E. J. *Low Level Radiation* (New York: Ballantine Books, 1972).
40. Stewart, A. "The Carcinogenic Effects of Low Level Radiation. A Reappraisal of Epidemiologists Methods and Observations," *Health Phys.* **24**, 223 (1973).
41. Tompkins, E. and M. L. Brown. "Evaluation of a Possible Causal Relationship between Fallout Deposition of Sr-90 and Infant and Fetal Mortality Trends," Public Health Service Report DBE 69-2 U.S. Department of Health, Education and Welfare (1969).
42. Scott, L. M., K. W. Bahler, A. de la Garza and T. A. Lincoln. "Mortality Experienced of Uranium and Nonuranium Workers," *Health Phys.* **23**, 555 (1972).
43. McGregor, J. F. and H. B. Newcombe. "Decreased Risk of Embryo Mortality Following Low Doses of Radiation to Trout Sperm," *Radn. Res.* **52**, 536 (1972).
44. Newcombe, H. B. " 'Benefit' and 'Harm' from Exposure of Vertebrate Sperm to Low Doses of Ionizing Radiation," *Health Phys.* **25**, 105 (1973).
45. International Commission on Radiological Protection (ICRP). "The "Recommendations of the International Commission on Radiological Protection," ICRP Publication 9 (London: Pergamon Press, 1966).
46. *Radiological Health Handbook* (revised ed.) (Rockville, Maryland: U.S. Dept. of Health, 1970).
47. U.S. Atomic Energy Commission. "Standards for Protection Against Radiation," Code of Federal Regulations, Title 10, Part 20, as amended.

ADDITIONAL REFERENCES

Bair, W. J., Ed. "Inhaled Radioactive Particles and Gases," *Proceedings Hanford Symposium* (London: Pergamon Press, 1964).

Brode, H. L. "Review of Nuclear Weapons Effects," *Ann. Rev. Nucl. Sci.* **18**, 153 (1968).

Fitzgerald, J. J. *Applied Radiation Protection and Control.* (New York: Gordon & Breach, 1969).

International Commission on Radiological Protection (ICRP). "The Evaluation of Risks from Radiation," ICRP Publication 8 (Oxford: Pergamon Press, 1966).

Jablon, S., S. Fujita, K. Fukushima, T. Ishimaru and J. A. Auxier. "RBE of Neutrons in Atomic Bomb Survivors, Hiroshima-Nagasaki," Technical Report 12-70 (Japan: Atomic Bomb Casualty Commission, 1970).

Mays, C. W. "Cancer Induction in Man from Internal Radioactivity," *Health Phys.* **25**, 585 (1973).

Schildt, E. *Nuclear Explosion Casualties.* (Stockholm: Almqvist & Wiksell, 1967).

Snyder, W. S., M. J. Cook, E. S. Nasset, L. R. Karhausen, G. P. Howells and I. H. Tipton. "Reference Man," ICRP Report No. 23 (Oxford: Pergamon Press, 1975).

Spaulding, J. F., R. F. Archuleta, O. S. Johnson and J. E. London. "Comparative Effects of Exposure on the Mouse to Radiation at Constant and Changing Dose Rates," *Health Phys.* **25**, 381 (1973).

Starr, C., J. H. Sterner, C. R. McCullough, F. J. Jankowski, M. Eisenbud. "Radiation in Perspective," *Nuclear Safety* **5**, 226; **5**, 325; **6**, 31; **6**, 142; **6**, 380; **7**, 12 (1964, 1965).

Upton, A. C. "Effects of Radiation on Man," *Ann. Rev. Nucl. Sci.* **18**, 496 (1968).

Volchok, H. L., R. H. Knuth and M. T. Kleinman. "The Respirable Fraction of Plutonium at Rocky Flats," *Health Phys.* **23**, 395 (1972).

CHAPTER 5

NUCLEAR POWER PLANTS

INTRODUCTION

At the heart of the nuclear industry lies the nuclear reactor as the prime source of energy for the generation of electricity. Although details of reactor design are beyond the scope of this book, certain features of the construction and operation require discussion to explain their potential impact on their environment.

The basic feature of nuclear energy is the fission chain reaction: a neutron impinging on a uranium nucleus causes it to split into two unequal halves called fission products, which represent isotopes of two elements near the middle of the periodic table, and several neutrons and gamma-ray photons. If at least one emitted neutron from each fission reaction interacts with another fissile uranium atom so as to initiate another fission, this fission chain reaction can go on as long as fissile fuel is available and other conditions are satisfied to prevent excessive loss of fission-initiating neutrons. The energy released is of the order of 200 MeV per fission, initially in the form of kinetic energy of the neutrons and fission products, but ultimately converted into heat as these particles are brought to rest by successive collisions within the reactor core. If this heat is carried off by a suitable coolant and is used to drive a steam turbine, some of this energy can be converted into useful form, electrical power.

Various engineering designs have been developed over the past few decades to provide safe and efficient power generation. Reactors are classified according to the type of fissile fuel used, the type of coolant, the neutron energy, and any other special feature employed. These characteristics are listed in Table 37 for reference.

Since uranium-235 is the only natural uranium isotope that is readily fissioned by slow, "thermal" neutrons, thermal reactors require for fuel

Table 37. Characteristics of Nuclear Reactor Systems

Reactor Type	Abbrev.	Fuel	Cladding	Coolant	Moderator	Primary Coolant Conditions	Steam Conditions	Cycle Efficiency	Av. Fuel Exposure (MWd/Tonne)
Pressurized water	PWR	UO_2	Zircaloy	H_2O	H_2O	2300 psi (161.8 kg/cm^2) 610°F (320°C)	780 psi (54.9 kg/cm^2) 514°F (268°C)	33%	33,000
Boiling water	BWR	UO_2	Zircaloy	H_2O	H_2O	1000 psi (70.3 kg/cm^2) satur. temperature	1000 psi (70.3 kg/cm^2) 545°F (285°C)	33%	27,500
Gas-cooled	GCR	UO_2	Mg alloy	CO_2	Graphite				
High-temperature gas-cooled	HTGR	$(Th-U^{235})C_2$	Pyrolytic carbon	He	Graphite	700 psi (49.2 kg/cm^2) 1430°F (777°C)	2400 psi (169 kg/cm^2) 950°F (510°C)	39.4%	100,000
Organic-cooled	OCR	U-Mo	Zr alloy	HB-40 Terphenyl	D_2O	7 kg/cm^2 400°C		35.4%	6,500 20,000 Pu recycle
Heavy water	HWR	UO_2	Zr-Nb	D_2O	D_2O	88.7 kg/cm^2 300°C		31%	10,000 20,000 Pu recycle
Fast breeder	LMFBR	(U-Pu)C	304 Stainless steel	Na-K	–	100 psi (7 kg/cm^2) 1025°F (552°C)	2415 psi (170 kg/cm^2) 950°F (510°C)	28.1%	75,000

either a large volume of natural uranium, which contains 0.71% U-235 by weight, or a more compact core of "enriched uranium," where the U-235 content has been increased, at considerable cost, to a higher concentration, typically 2-3 wt %. In order to slow down fission neutrons, which are generated with kinetic energies of 2-3 MeV, to the thermal energies (about 0.025 eV) needed to sustain a thermal chain reaction, natural uranium reactors need a highly efficient moderator, such as heavy water, D_2O. With the higher interaction probability for fission provided by enriched fuel, light water, H_2O, is an adequate and economic moderator. Graphite also is employed as a moderator, especially in gas-cooled reactors. The cooling fluid used to carry off the heat from the reactor to the turbine, either directly or through an intermediate heat exchanger, further characterizes the reactor type. Thus we have the following classification:

1. Light-water moderated reactors: This category includes both pressurized-water and boiling-water reactors, as well as modifications utilizing nuclear superheating.

2. Heavy-water moderated reactors: These reactors are generally also cooled with heavy water, but this is not essential.

3. Organic-cooled reactors: The organic liquid may serve as moderator as well as coolant, or a different moderator can be employed.

4. Sodium-graphite reactors: In these reactors, graphite is the moderator and sodium the coolant.

5. Gas-cooled reactors: These may utilize either natural or enriched uranium as fuel; carbon dioxide and helium have been used as coolants.

6. Fast breeder reactors: The coolant is generally sodium, although sodium-potassium alloy has been used in experimental systems.

7. Fluid-fuel reactors: This category includes thermal breeders (aqueous homogeneous reactors), liquid-metal and fused-salt-fueled thermal reactors, as well as some fast-reactor concepts.

Historically, light-water reactors have been developed most actively in the United States and the USSR, heavy-water reactors in Canada, and gas-cooled reactors in Great Britain and the U.S. Organic-cooled reactors and fluid-fuel reactors have not passed beyond pilot operation and are less actively pursued at present.

Fast breeder reactors are under development in several countries, but are still many years from constituting accepted commercial power sources. In these reactors, fission neutrons are not slowed down but interact at high energy with uranium-238 to cause fission. This utilizes the much more plentiful U-238, but requires a nonmoderating heavy coolant, such as sodium, which imposes novel technological problems. However, the main benefit of breeder reactors is the production of additional fissile material, plutonium-239, by neutron capture in U-238 in sufficient quantity to supplement appreciably the available supply of fissile fuel resources.

WATER-COOLED REACTORS

In terms of basic operating principles the various reactor types do not differ much, as is shown in Figure 35, but the differences in engineering detail are sufficient to affect significantly their potential environmental effects. Although the reactor systems superficially resemble a fossil-fueled generating plant as Figure 35 indicates, there are several characteristic differences in the heat utilization that determine both their conversion efficiency and the choice of steam cycle operation.

Figure 36 shows diagrammatically the general internal arrangements of a typical pressurized water reactor (PWR). The fuel consists of uranium oxide contained by a metal sheath (cladding) that has the functions of providing mechanical support, facilitating uniform heat transfer from the fuel to the coolant, protecting the fuel from corrosive action by the coolant, and retaining all fission products within the fuel element. A number of fuel elements are arranged in bundles (fuel assemblies), one of which is illustrated in Figure 37. The assembly also contains the control rods, made up of strong neutron absorbers such as cadmium, boron-containing steel, or other materials, that serve to adjust and control the total number of neutrons in the system and hence the reactor power level. The combined fuel assemblies form the reactor core. The coolant circulates between the fuel elements, carrying off the excess heat to the steam generator. The main differences between the various reactor types lie in the method of coolant circulation, the location of the steam generator and methods of assuring uniform burn-up, and heat distribution; many of these parameters are being modified in new reactors in the light of accumulated operating experience.

In the boiling water reactor (BWR), steam is generated within the core and passed through a series of separators and driers to the turbine generator. The pressure in a typical BWR is maintained at about 1000 psi (6.89 M Pascal). The coolant recirculation pumps are generally jet pumps and increase the amount of heat that can be taken from the core. In the turbine generator, the steam pressure is converted to electrical energy. Steam exiting the turbine generator passes into the condenser and is pumped back to the annulus of the reactor (3).

Two special auxiliary systems important as potential sources of effluents in a boiling water reactor are the air-ejectors and the turbine gland seal system. An air ejector passes steam through a series of nozzles and creates a vacuum that removes air from the condenser. Air ejector exhaust is collected and eventually vented from the plant. Gland seal steam is used to seal the main turbine by passing high pressure steam over a series of ridges and evacuating the steam when it reaches a low pressure. This is also vented from the plant.

The pressurized water reactor (PWR) differs from the boiling water reactor primarily by the mechanism of generating steam that runs the turbine generator. The PWR may be divided into two separate systems—the primary and the secondary systems. The former consists of a closed loop that is completely filled with coolant (water) except for the pressurizer. It operates under a typical pressure of 2000-2500 psi (13.8-20.7 MPa), which implies that the water can be heated to temperatures of approximately 500°F (260°C) without boiling. Water is continuously circulated by means of a large coolant pump through a steam generator. It is here that hot primary water passing through closed tubes causes boiling of water on the secondary side.

The pressurizer maintains primary loop pressure using a heater bank and cooldown spray. It also absorbs and injects primary coolant when required by minute changes in operating temperature (T_{ave}). The upper portion of the pressurizer is a saturated vapor while the lower portion contains coolant. Noncondensible gases collect in the top of the pressurizer and must periodically be vented off to the atmosphere.

Most PWR's today remove ionic and particulate activities in the chemical and volume control system (CVCS). This system acts as a surge reservoir during normal operations. The water is demineralized and stripped of noncondensible gases in the volume control tank. This is also the system that removes dissolved boron for recycling back into the plant as boric acid (4).

The secondary system is separate from the primary. Steam produced in the steam generators is passed to the turbines where electricity is generated. The steam is then condensed and returned to the steam generator. Like the BWR, the PWR contains an air-ejector to control vacuum and a gland steam system. However, in the PWR the steam is supplied from the secondary, and the probability of radioactive materials being exhausted from the plant from these sources is minimized.

HIGH TEMPERATURE GAS-COOLED REACTORS (HTGR)

Reactors that use gas as a coolant may be classified into two types—those that operate at temperatures below about 1000°F (540°C) and those that operate above this temperature. The main features of the high-temperature gas-cooled reactor (HTGR) are high-temperature operation,

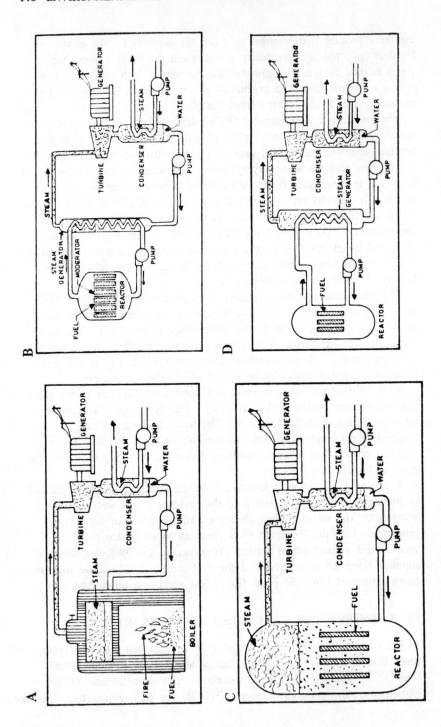

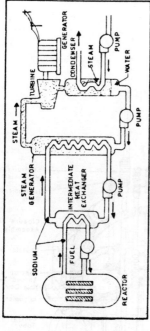

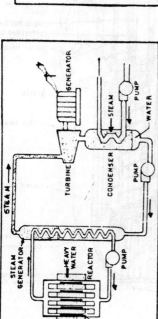

Figure 35. Schematic diagrams of the principal steam-generating plants (1).
A: fossil-fueled plant; B: gas-cooled reactor; C: boiling water reactor;
D: pressurized-water reactor; E: heavy-water reactor; F: LMFBR.

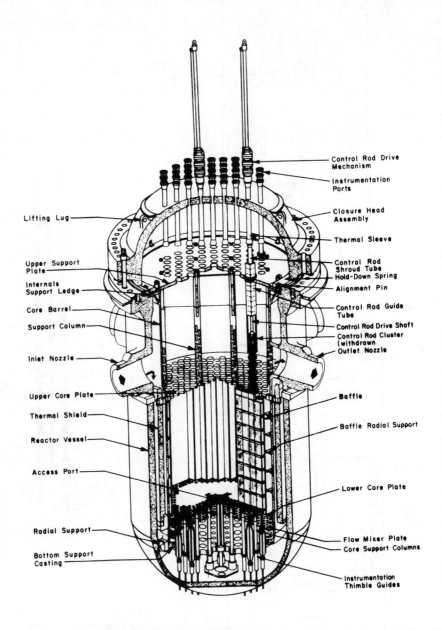

Figure 36. PWR: reactor vessel and internals (2).

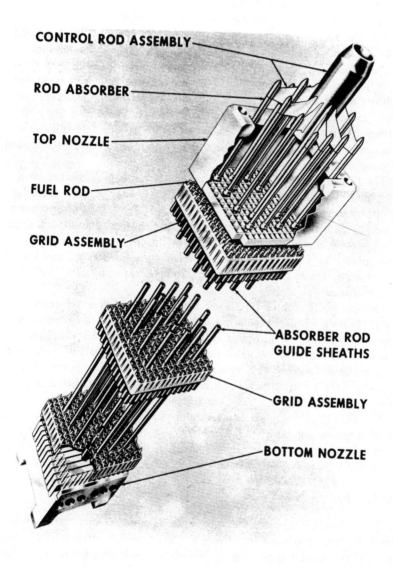

CONTROL ROD ASSEMBLY

ROD ABSORBER

TOP NOZZLE

FUEL ROD

GRID ASSEMBLY

ABSORBER ROD GUIDE SHEATHS

GRID ASSEMBLY

BOTTOM NOZZLE

Figure 37. View of a nuclear fuel assembly.

gas coolant, graphite moderator and reflector, a highly enriched uranium-thorium fuel, fission product control, and homogeneous fuel with graphite structure, cladding and fuel diluent.

Helium appears to be the best coolant for HTGR's, and the temperature at the reactor outlet is typically 1300-1500°F (700-820°C), providing 1000°F (540°C) steam at 1450 psia (9.65 MPa). These temperatures and pressures result in optimum plant efficiency (5).

Helium is chemically inert with graphite-clad fuel and with other components of the primary system. Thus there is no corrosion hazard that might eventually lead to loose activation products. Although the neutron activation cross section of ^{4}He is zero, the rare isotope ^{3}He will capture neutrons. The product of this reaction is tritium. This will be discussed in more detail later.

The HTGR has a closed-loop primary system in which the helium is circulated by means of turbine pumps. The secondary system is similar in operation to the PWR. As in the PWR, an air ejector uses secondary steam to maintain condenser vacuum, and a gland seal prevents inleakage of air around the turbine. However, since this steam does not come into contact with primary coolant, it remains uncontaminated and should not be a source of gaseous effluents. Table 38 summarizes HTGR plant characteristics.

One unique feature of HTGR's is their prestressed concrete reactor vessel (PCRV). This structure has the function of containing the coolant operating pressure and providing radiological shielding. Since helium could readily escape the type of containment used in the BWR and PWR, all penetrations through the reactor vessel have a double closure. The concrete walls are constructed around a carbon steel liner that is about 3/4-in. (1.8 cm) thick and provides a helium-tight membrane.

The Fort St. Vrain plant has a helium purification system that provides purified helium for purging seals, control rod drives, and PCRV penetrations. This system removes particulates, CO_2, and noble gases and helps to reduce the total amount of radioactivity that escapes from the reactor.

Heavy-Water Reactors

The natural-uranium fueled, heavy-water cooled and moderated reactor has been developed most intensively in Canada and is often referred to as the CANDU reactor type. Figure 38 shows a flow diagram for a CANDU reactor; its most notable feature is the horizontal fuel arrangement that permits on-power refueling. Since the fuel is natural-uranium oxide, no enrichment is required and fuel reprocessing for the recovery of fissile components is not considered worthwhile. Consequently, the fuel

Table 38. Operating Parameters for a Large HTGR

General

Thermal power, MW(th)	3000
Electric power, MW(e)	1160
Plant lifetime, years	40
Conversion ratio	0.66

Reactor

Fuel, startup	Th/U^{235} (93% enriched)
recycle	Th/U^{235} (93% enriched)/U^{233} (recycle)
Fuel form	Coated particles in cylindrical bonded rods
Moderator	Graphite
Avg. power density, kW/liter	8.4
Outlet temperature	1366°F (741°C)
Temperature rise across core	760°F (404°C)
Fuel temperature, Avg/Max.	1634/2467°F (890/1353°C)
Reactor vessel, height	20.8 ft (6.34 m)
Reactor vessel, diameter	27.8 ft (8.47 m)
Coolant inlet pressure, psi	710
Vessel material	prestressed concrete

Other Components

Number of circulators	6
Circulator speed, rpm	7050
Number of steam generators	6
Steam conditions	
pressure, psi	2400
temperature	950°F (510°C)

cycle for a CANDU reactor is much simpler than that of a LWR; however, the larger reactor core leads to a higher capital cost for initial construction. Tritium generation and the production of delayed fast neutrons in the heavy water coolant may pose some problems that can, however, be overcome fairly easily. In other respects the HWR resembles the LWR and most environmental factors will be similar.

THE FISSION PRODUCT INVENTORY

Both in terms of engineering complexity and overall costs the reactor by no means represents the largest component in the complete power plant. However, it obviously constitutes the portion requiring the closest attention for quality assurance and care in design in view of the limited operational experience at present available for specific large reactors and the practical difficulties that would arise if subsequent major

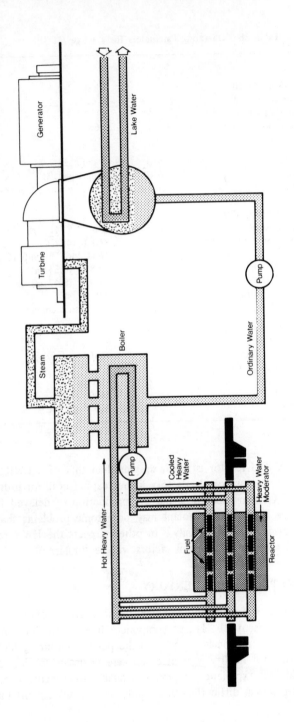

Figure 38. Flow diagram for the CANDU reactor.

maintenance or back-fitting work were needed. For reference, Table 39 lists design parameters for typical thermal reactors of current U.S. design (6).

As the fuel is burned, that is as uranium-235 undergoes fission, each of the fissioned atoms is replaced by two fission products and several neutrons. Since uranium-235 contains 143 neutrons but only 92 protons, the fission product nuclides have an excess of neutrons over the near balance between protons and neutrons required for nuclide stability for elements in the middle of the periodic table. Consequently the fission products undergo successive beta decays along their isobaric chain, through nuclides of increasingly longer half-lives, until they reach a stable isobar. Some of the intermediate isobars may already be very long-lived and the decay chain may be virtually stopped in that case. This fission product decay has several important consequences. The core becomes highly radioactive as delayed gamma rays are emitted by the decaying fission products, leading to substantial self-heating of the core even when the reactor is shut down. The relative composition of the fission products in the fuel changes with burn-up time as the longer-lived fission products slowly build up to secular equilibrium concentration between their rate of production and their rate of decay. The composition of the fission product inventory in the core thus varies with operating conditions and burn-up time. Table 40 shows the more important fission products with half-lives over a few hours that will accumulate in the reactor. Empirically, the decay heat power produced by the fission products follows a law of the form

$$\frac{P}{P_o} = 6.1 \times 10^{-3} \ [(t - t_o)^{-0.2} - t^{-0.2}] \tag{41}$$

where P and P_o are the decay heat power and the reactor power, t_o = reactor operating time to shutdown, and $(t - t_o)$ is the cooling period (7). Table 41 lists the quantity of some of the more important fission products that build up during reactor operation, and the associated decay.

In addition to uranium-235 fission, some of the neutrons will undergo capture in the more abundant uranium-238 present in the fuel. This capture reaction transforms U-238 to U-239, which beta-decays successively to Np-239 and then to Pu-239, which has a half-life of 24,360 years and is also fissile. This plutonium "breeding" is the basic process in the breeder reactor, though most breeders are designed to operate with fast neutrons for better conversion efficienty. In high-flux reactors and under long burn-up conditions further neutron captures will produce heavier nuclides such as Pu-240, Pu-241 which is also fissile, and americium and curium isotopes, some of which have long half-lives.

Table 39. Thermal Reactor Characteristics[a] (6)

	BWR		PWR		HTGR	AGR	
	Thru 1980	After 1980	Thru 1980	After 1980		Inner Core	Outer Core
Thermal Efficiency (%)	34	34	33	33	39	42	
Specific Power (MWth/MT)	26	28	38	41	82	13	
Initial Core (Average)							
Irradiation level (MWD/MT)	17000	17000	24000	24000	54500	13000	
Fresh fuel assay (wt % U-235)	2.03	2.03	2.63	2.63	93.15	1.49	1.78
Spent fuel assay (wt % U-235)	0.86	0.86	0.85	0.85	d	0.75	1.00
Fissile Pu recovered (kg/MT)[b]	4.8	4.8	5.8	5.8	d	2.5	
Feed required (ST U_3O_8/MWe)[c]	0.625	0.580	0.591	0.548	0.456	0.737	
Separative work required (SWU/MWe)[c]	200	185	224	208	311	188	
Replacement Loadings (Annual rate at steady state and 80% Plant Factor)							
Irradiation level (MWD/MT)	27500	27500	33000	33000	95000	20000	
Fresh fuel assay (wt % U-235)	2.73	2.73	3.19	3.19	93.15	2.10	2.54
Spent fuel assay (wt % U-235)	0.84	0.84	0.82	0.82	d	0.59	0.87
Fissile Pu recovered (kg/MT)[b]	5.9	5.9	6.6	6.6	d	4.0	
Feed required (ST U_3O_8/MWe)[c]	0.191	0.191	0.205	0.205	0.113	0.176	
Separative work required (SWU/MWe)[c]	89	89	99	99	77	73	
Replacement Loadings (Annual rate at steady state, 80% plant factor, and plutonium recycle)							
Fissile Pu recycled (kg/MWe)	0.174	0.174	0.167	0.167			
Fissile Pu recovered (kg/MT)[b]	10.3	10.3	11.4	11.4			
Feed required (ST U_3O_8/MWe)[c,e]	0.168	0.168	0.179	0.179			
Separative work required (SWU/MWe)[c]	70	70	80	80			

aMWth is thermal megawatts, MWe is electrical megawatts, MWDt is thermal megawatt days, MTU is metric tons (thousands of kilograms) of uranium, and ST U_3O_8 is short tons of U_3O_8 yellowcake from an ore processing mill. One SWU is equivalent to one kg of separative work.

bAfter losses.

cBased on operation of enriching facilities at a tails assay of 0.3%. For replacement loadings, the required feed and separative work are net, in that they allow for the use of uranium recovered from spent fuel. Allowance is made for fabrication and reprocessing losses.

dAll spent fuel and fissile production (primarily U-233) are recycled on a self-generated basis. Only one recycle of U-235 is assumed.

e... plutonium: 0.0087 ST U_3O_8/MWe for BWR and 0.0067 for PWR.

Table 40. Decay of Individual Fission Products (8)

Isotope	Isotope Half-Life	Reactor (1095 days operation) Decay, curies/MW		Excursion (4200 MW-sec) Decay, curies	
		1 day	90 days	1 day	100 days
^{85m}Kr	4.36 hr	140		38	
^{85}Kr	10.57 years	360	350	0.03	0.03
^{89}Sr	53 days	23,000	6,600	21	6.1
^{90}Sr	28 years	2,600	2,600	0.2	0.2
^{91}Sr	9.7 hr	5,300		700	
^{90}Y	64.8 hr	2,700	2,600	0.05	0.2
^{91m}Y	51 min	3,500		300	
^{91}Y	58.3 days	30,000	11,000	21	8.3
^{92}Y	3.5 hr	1,100		330	
^{93}Y	10 hr	7,800		790	
^{95}Zr	65 days	50,000	19,000	270	97
^{97}Zr	17 hr	18,000		950	
^{95m}Nb	90 hr	1,000	410	0.09	0.2
^{95}Nb	35 days	51,000	32,000	0.5	12
^{97}Nb	60 sec	17,000		950	
^{97}Nb	74 min	20,000		990	
^{99}Mo	68 hr	40,000		470	
^{99m}Tc	6.04 hr	40,000			
^{103}Ru	39.8 days	40,000	8,600	24	4.4
^{105}Ru	4.5 hr	610		38	
^{106}Ru	1 year	19,000	16,000	0.4	0.3
^{103}Rh	57 min	40,000	8,600	24	4.4
^{105m}Rh	45 sec	170		38	
^{105}Rh	36.5 hr	16,000		130	
^{106}Rh	30 sec	19,000	16,000	0.4	0.3
^{109}Pd	13.6 hr	2,600		4.1	
^{112}Pd	21 hr	330		2.1	
^{109}Ag	40 sec	2,600		4.1	
^{111}Ag	7.6 days	1,300			
^{112}Ag	3.2 hr	390		2.4	
^{121}Sn	27.5 hr	150		1.8	
^{125}Sn	9.4 days	420	0.6	0.7	
^{125}Sb	2.0 years	200	190	0.0005	0.007
^{127}Sb	93 hr	2,600		10	
^{128}Sb		820			
^{129}Sb	4.6 hr	180		30	
^{127m}Te	105 days	580	340	0.015	0.05
^{127}Te	9.3 hr	2,800	330	8.7	5.1
^{129}Te	33.5 days	3,000	460	1.6	0.2
^{129}Te	74 min	3,100	460	31	0.02

Table 40, continued

Isotope	Isotope Half-Life	Reactor (1095 days operation) Decay, curies/MW		Excursion (4200 MW-sec) Decay, curies	
		1 day	90 days	1 day	100 days
^{131m}Te	30 hr	2,600		45	
^{131}Te	25 min	580		46	
^{132}Te	77 hr	38,000		280	
^{130}I	12.6 hr	235			
^{131}I	8 days	28,000	13		0.02
^{132}I	2.4 hr	35,000		285	
^{133}I	20.5 hr	22,000		930	
^{135}I	6.68 hr	4,700		490	
^{131}Xe	≈ 12 days	270	4	0.05	
^{133m}Xe	2.3 days	1,500		8.2	
^{133}Xe	5.27 days	55,000		160	
^{135}Xe	15 min	1,300		160	
^{135}Xe	9.2 hr	14,000		1,300	
^{134}Cs	2.1 years	2,000	1,800		
^{136}Cs	12.9 days	910	8		
^{137}Cs	30.2 years	3,600	3,600	0.1	0.1
^{137}Ba	2.63 min	3,300	3,300	0.1	0.1
^{140}Ba	12.8 days	45,000	370	130	0.7
^{140}La	40.5 hr	49,000	420	44	0.8
^{141}La	3.7 hr	750		140	
^{141}Ce	32.8 days	48,000	7,200	51	6.7
^{143}Ce	33 hr	29,000		700	
^{144}Ce	286 days	38,000	30,000	4.9	3.9
^{142}Pr	119 hr	900			
^{143}Pr	13.7 days	46,000	550	44	1.0
^{144}Pr	17.5 min	38,000	30,000	4.9	3.9
^{145}Pr	4.5 hr	1,700		280	
^{147}Nd	11.3 days	19,000	72	67	0.2
^{147}Pm	2.6 years	7,600	7,400	0.04	0.6
^{148m}Pm	42 days	620	140		
^{148}Pm	5.3 days	4,100	160		
^{149}Pm	54 hr	12,000		150	
^{151}Pm	27.5 hr	3,400		70	
^{153}Sm	47 hr	6,300		16	
^{156}Eu	14 days	4,800	78	0.2	0.04

Table 41. Calculated Radioactivity of 1100 MWe PWR Core After 293 Days at Power and After Various Decay Times (9)[a]

Decay Time (days)	Radioactivity (megacuries)						Total Thermal Power (kW)
	Iodine and Bromine Isotopes	Noble Gases	All Fission Products	Actinides	Activation Products	Total	
0	1,435	1,240	13,800	3,450	10.6	17,250	225,000
1	265	221	2,890	1,330	9.19	4,230	17,400
5	101	105	1,870	432	8.42	2,310	9,720
15	28.7	29.0	1,280	39.7	7.50	1,330	5,600
30	6.74	4.77	947	9.35	6.40	963	4,060
60	0.494	0.784	656	6.32	4.76	666	2,350
120	0.00282	0.659	401	5.90	2.76	410	1,740
210	0.00000309	0.648	244	5.56	1.36	250	1,100
365	0.00000218	0.630	146	5.17	0.614	152	659
1,097	0.00000218	0.553	47.3	4.45	0.324	52.0	204
3,653	0.00000218	0.353	17.9	3.27	0.132	21.3	67

[a]Reactor is assumed to be shut down just before refueling after a sustained (293 days) period at a specific power of 37.5 MW/metric ton. The time-average specific power over the previous 1100 days is 30 MW/metric ton. The reactor is fueled with 3.3% enriched uranium totaling 82 metric tons of enriched uranium fuel.

The gradual build-up of fission products in the reactor core limits the extent of burn-up that is practical. Many of the fission products have fairly high neutron capture cross sections and hence poison the core by absorbing neutrons unproductively, thus removing them from the fission chain reaction. In addition, the insertion of two fission product atoms into the core for only one initial uranium atom gradually damages the fuel structure, causing it to swell and distort, endangering the integrity of the cladding. Since some of the fission products are gaseous, they can readily diffuse through the fuel, forming vapor bubbles and building up gas pressure behind the thin cladding sheath. To maintain safe and efficient operation, it is necessary, therefore, to limit burn-up and remove the fuel, typically after burn-up of the order of 30,000 megawatt-days per ton of uranium (MWd/T), at which time about 75% of the uranium-235 contained in the original fuel has been consumed. To recover and recycle the unused U-235 and the fissile plutonium now present in the fuel, the spent fuel is shipped to a reprocessing plant after being stored for three to six months, during which time the larger part of the fission product activity will have a chance to decay, making subsequent handling and shipping a little easier and less hazardous.

If we inspect the list of fission products in Table 40, we note that there are a few fission products with half-lives in excess of about two months. These include Zr-95 (65d), Ru-106 (1 yr), Ce-144 (286d), Pm-147 (2.6 yr), Cs-134 (2.1 yr), Kr-85 (10.57 yr), Sr-90 (28 yr) and Cs-137 (30.2 yr), as well as I-129 (1.6×10^7 yr) which is not listed. It is these fission products, together with activation and corrosion product activities such as Co-60 (5.2 yr), Fe-59 (45 days) and H-3 (12.33 yr), and the transuranium isotopes bred in the fuel, that account for most of the reprocessing problems and waste disposal requirements. They also represent the bulk of the source term for reactor accident calculations.

The environmental impact of such a reactor arises from its size and location, from the waste heat rejected, from the possible radioactive effluents associated with its operation, the need to transport fuel to and from the plant and the need to dispose of small but finite amounts of radioactive waste. In addition, it is still necessary at this time to reassure the public that such a reactor is safe and does not constitute a potential bomb. The question of thermal releases and the impact of transportation will be discussed in later chapters. Here we will deal with the question of reactor safety, as far as it is related to environmental considerations, and the nature and origin of radioactive materials in a reactor.

REACTOR SAFETY

Whenever a nuclear power plant is being built, one of the prime considerations is whether such a plant is "safe." This concern is a natural one considering the history of nuclear energy and the frequent exposure of the public to various books and publications that proclaim that nuclear energy is not safe. In the face of this scepticism, the utilities and governmental agencies have had a hard struggle to convince the public that nuclear power plants are, in fact, safe and to demonstrate the excellent safety record of the industry (10-13). The long period of operation of a large number of reactors of various types with a very small number of plant accidents is a tribute to the skill of the designers, who in many cases had to scale up from relatively limited pilot operations, and to the extreme care and quality control that went into the construction and assembly of reactors all over the world. Though no serious accidents have occurred so far in commercial nuclear power plants, such lesser accidents that have occurred could be attributed in almost every case to almost willful disregard of established safety and operating procedures.

Safety is usually defined as an absence or low probability of danger or risk. For a nuclear power plant this definition can be further refined to mean:

(1) negligible risk of damage to the integrity of the generating system from any internal or external cause

(2) negligible risk of harm to operating personnel from normal or abnormal operation of the plant

(3) negligible risk or hazard to the surrounding population from any cause related to operation of the plant.

It is the last definition that is normally applied in examining the environmental impact of a nuclear plant, but in practice it is impossible to divorce it from the others. Clearly, no hazard from accidental events would exist if from the start the plant is designed to rule out the possibility of a major accident from any cause other than sabotage or a cataclysmic act of God. While this aim has not been achieved fully, as yet, it is an underlying design philosophy that goes far beyond the safety principles practiced in industry hitherto. Such safety consciousness leads to multiple barriers, engineered safeguards, and a planned redundancy of controls and devices that exact their price in cost and complexity and require periodic review as operational experience is gained. At the same time, there can be no moral or economic justification for the construction of a power plant that cannot be regarded as safe by any reasonable standard.

The primary aim is to keep the reactor readily controllable under all conditions of burn-up, power level and transient response that can be anticipated, and to ensure immediate and automatic shutdown or "scram" of the reactor whenever the neutron flux or power level exceed preset conditions. This is accomplished by maintaining a negative temperature coefficient in the reactor core so that any unplanned rise in power, and hence in core temperature, immediately causes the reactor to lose its critical condition, closing down the reactor. Any other type of abnormal behavior or external problems, such as loss of instrument power or turbine malfunction, also promptly shut down the reactor by causing rapid insertion of the control rods. Furthermore it is important to eliminate or minimize possible hazardous conditions that might arise from any human error.

Considerable thought has been devoted to means of ensuring operational safety and reliability of the reactor systems under all conditions; these methods are often referred to as engineered safeguards. They include the following major provisions:

(1)   quality control during installation and maintenance of all reactor components

(2)   redundancy of all monitoring and control equipment

(3)   standby power sources for all vital reactor components (pumps, fans, instruments)

(4)   successive barriers to the accidental release of fission products presented by the cladding, the reactor vessel, and the containment building

(5)   careful operator training programs and development of operating procedures covering all routine and emergency operations around the plant

(6)   provision for remedial action in case of all accidents, particularly the most severe accident conceivable, often referred to as the "design-basis accident" (DBA).

The engineered safeguards thus provide additional assurance beyond the design of an inherently safe reactor. For all thermal reactors, design of reactors with negative temperature coefficients and rapid control rod action has been standard and relatively straightforward. For fast reactors now being designed and for fusion-power plants, this requirement is a little harder to attain, but the existing fast demonstration reactors have shown that these objectives can be achieved there, too.

The safety analysis of reactors has involved the identification of all possible accidents and an assessment of their severity, their postulated consequences as well as their probability of occurrence. Table 42 shows a listing of postulated accidents by classes of increasing severity. Since the more severe incidents trigger responses from a chain of control instruments and devices, and become severe only if all or a sequence of them fail in turn, it is possible to analyze such a system by means of a "fault tree," a computational

Table 42. Classification of Postulated Accidents and Occurrences
at Reactor Facilities (9)

No. of Class	Description	Example(s)
1	Trivial incidents	Small spills Small leaks inside containment
2	Misc. small releases outside containment	Spills Leaks and pipe breaks
3	Radwaste system failures	Equipment failure Serious malfunction or human error
4	Events that release radioactivity into the primary system	Fuel defects during normal operation Transients outside expected range of variables
5	Events that release radioactivity into secondary system	Class 4 and heat exchanger leak
6	Refueling accidents inside containment	Drop fuel element Drop heavy object onto fuel Mechanical malfunction or loss of cooling in transfer tube
7	Accidents to spent fuel outside containment	Drop fuel element Drop heavy object onto fuel Drop shielding cask—loss of cooling to cask, transportation incident on site
8	Accident initiation events considered in design–basis evaluation in the safety analysis report	Reactivity transient Rupture of primary piping Flow decrease—steamline break
9	Hypothetical sequences of failures more severe than Class 8	Successive failures of multiple barriers normally provided and maintained

model that establishes interrelationships between various components and their functions and attempts to predict, on the basis of known individual failure probabilities for each component, what the probability would be for an extended failure or a system malfunction. The application of this fault tree approach is limited at this time mainly by a lack of precise failure statistics for many components and by the size of the computational program needed to encompass anything but the simplest systems. Table 43 lists postulated reactor accidents in terms of probability of occurrence. Obviously the aim of safety design must be to make severe

Table 43. Probability Rating of Postulated Reactor Plant Accidents (9)

A. **Moderate Frequency Events** (no abnormal radioactive release from the facility)
1. Withdrawal of control rod at maximum speed due to malfunction or error.
2. Failure of one safety rod to scram.
3. Partial loss of normal forced reactor coolant flow
4. Unintentional startup of an inactive reactor coolant loop
5. Loss of external electrical load and/or turbine trip
6. Loss of off-site electrical power.
7. Excessive load increase.
8. Loss of normal feedwater flow.
9. Inadvertent depressurization of the primary coolant system.

B. **Infrequent Accidents of Small Probability** (abnormal radioactive release possible, but not expected)
1. Small leaks and breaks in pipes (or minor leaks in large primary or secondary system pipes).
2. Inadvertent loading of a fuel assembly into an improper position.
3. Complete loss of normal forced reactor coolant flow.
4. Complete loss of all A-C power (station blackout).
5. Major leakage in radioactive waste decay tank.

C. **Highly Unlikely Accidents** (postulated for evaluating site acceptability)
1. Major rupture of pipes containing reactor coolant up to and including double-ended rupture of largest pipe in the primary coolant system (loss-of-coolant accident)
2. Major secondary or steam system pipe rupture up to and including double-ended rupture of a main steam pipe.
3. Control rod ejection.
4. Severe fuel handling accident.
5. Tornadoes, flooding and earthquakes.

accidents exceedingly improbable. The danger of focusing attention excessively on the "design basis accident," defined as the most serious postulated accident, is to divert attention from other far more plausible ones.

The benefits of a meticulous system of inspection and quality control, including 100% inspection of all welded joints and separate and assembled testing of all pressure system components, have been clearly demonstrated in the American space and naval programs. In the nuclear program they have paid off in terms of reliability and low accident rates, in spite of the considerably higher construction costs entailed, and there can be no serious argument for relaxing them at this time.

A severe contamination problem may arise if the reactor core is allowed to overheat due to a sudden loss of coolant, resulting perhaps from a massive fracture in the coolant feed pipe, so that the decay heat in the fuel causes the fuel temperature to rise above the softening temperature of the cladding, about 2250°F for zircaloy-4. Under these conditions the fuel might split or melt down, giving rise to what is considered the design basis accident for a water-cooled reactor. In this case a large fraction of the fission products might conceivably escape from the fuel, and their future fate then becomes a central concern of the formal reactor safety analysis (14).

Engineered Safeguards

Engineered safeguards to minimize escape to the environment of accidentally released radionuclides may take the form of active or passive systems. Examples of passive systems are large pressure containment structures housing all primary system components, and vapor suppression pools and their associated pressure vessels. Active systems may include: internal air recirculation cleanup systems; recirculating liquid cleanup systems; spray systems to reduce the internal driving pressure causing leakage from the containment; pressurization of all containment penetrations in such a way as to prevent leakage; and double containment schemes that tend to prevent any leakage, assure that leakage can be filtered before release, and enable gases to be directed up a tall stack.

The adequacy of safeguards is judged after consideration of:

- the objective of the safeguard
- the design criteria
- the performance specifications
- the design derived from the specifications
- the extent to which tests of the system are practical and necessary to demonstrate continued operability and reliability of the system
- safety analysis of the safeguard itself (implications of malfunctions)
- the margin of safety offered by the system.

The effectiveness of engineered safeguards is illustrated by examination of some actual installations. The San Onofre plant uses an automatic spray injection system of borated water to limit the fission release to containment in the loss-of-coolant accident to 6% of a full core meltdown. This reduces the low population and population center distances specified in TID-14844 by a factor of 10 and reduces the exclusion area radius by a factor of 7.5 (15).

The Connecticut Yankee plant uses both the automatic safety injection system, a spray system that washes out fission products from the containment and serves to reduce pressure during the design basis accident, and recirculation and filtering which serve as cooling during normal operation and can remove fission products from the air during an accident. These safeguards can reduce the iodine available for leakage in the first two hours by a factor of eight (equivalent to a factor of four in the exclusion radius) though a reduction of only five was necessary to meet the 10 CFR 100 guidelines (15).

The Malibu plant employs automatic safety injection, a spray system, air recirculation, and double containment. Double containment alone justifies a reduction in site dimensions by a factor of 20 as compared to standard single containment (15).

The above installations are of the light-water type in wide use today. Because these reactors are cooled by water, which is at high temperature and pressure, a great deal of energy is stored in the coolant. This means that any break in the primary system will release this energy to the containment system. Thus, containment and other safeguards for water-cooled reactors are designed largely to reduce the pressure in the containment. Reactors using coolants other than water do not necessitate the use of safeguards capable of withstanding such large pressures.

Among the different engineered safeguards employed is the emergency core cooling system (ECCS), shown in Figure 39, which rapidly floods the reactor core to prevent any undue rise in temperature. The effectiveness of such systems is currently under investigation and constitutes a significant problem in heat transfer analysis (16-18).

Another major safeguard is provided by the physical barriers to dispersion of the escaping fission products presented by the reactor vessel and the containment building. Both the shape and configuration of these structures vary widely, as can be seen diagrammatically in Figure 40. The reactor vessel proper may house the reactor core and the primary cooling system; in the BWR it is surrounded by a containment *vessel,* which is distinct from the containment *building.* In French and British reactors, the concrete shield surrounding the reactor vessel provides the primary safety barrier and containment (19). Figure 41 shows the type of containment building representative of U.S. water-cooled reactors (2,9).

A spherical containment is favored by German designers (42), and a horizontal fuel arrangement inside a reactor vessel, the calandria, is used in Canadian reactors. The function of the reactor vessel is to contain the fuel and coolant, even in case of fuel failure. A biological shield, usually heavy concrete, surrounding the reactor provides gamma-ray and neutron shielding to protect the operating personnel. The containment building encloses not only the reactor but also the primary steam generator and usually any storage pools needed for spent fuel storage. It

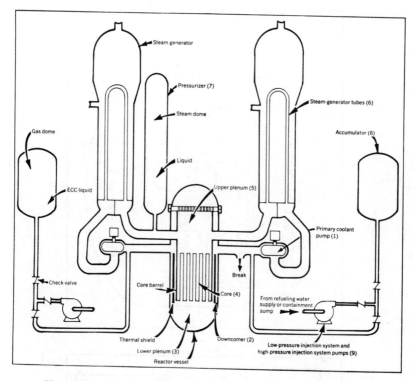

Figure 39. Primary cooling loops of a PWR, showing emergency core cooling system (17).

provides the principal barrier to dispersion of radioactive contaminants in case of a fuel spill, failure of fuel elements or the coolant pipes, or a major core rupture like the loss-of-coolant accident. Interlocks have to be installed to prevent immediately any venting up the stack to ensure complete containment whenever the radiation level in the building air rises above a preset level. The containment building should also be maintained at a slightly negative pressure compared with the surrounding atmosphere, should be capable of withstanding any shock waves associated with the design basis accident, should resist penetration by flying objects, such as torn pipe fragments or valve handles, and should maintain its integrity in case of floods or earthquakes. Current U.S. practice uses 4-inch thick prestressed concrete shells for this purpose. Cost comparisons of different containment shells have been made by Verstraete (16).

Specifications for containment buildings call for regular periodic leak tests and careful attention to the design and installation of all pressure

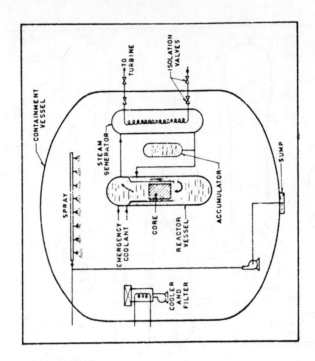

B

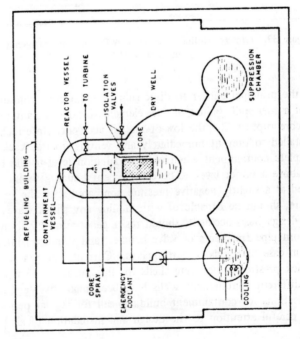

A

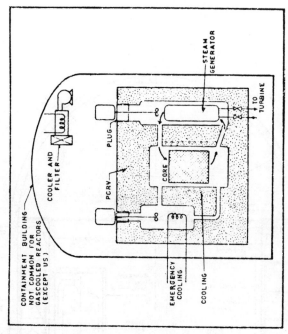

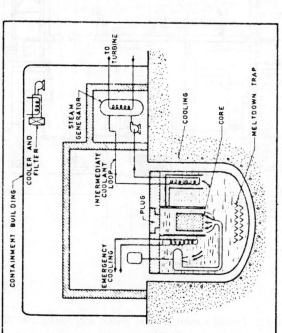

Figure 40. Safety features of typical nuclear power plants. A: BWR; B: PWR; C: LMFBR; D: GCR (1).

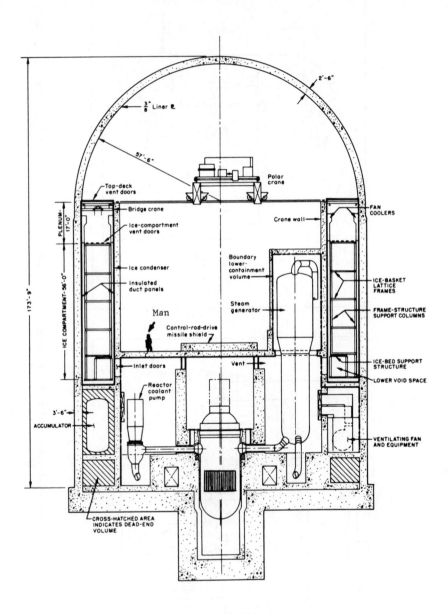

Figure 41. Typical primary coolant system and containment for a large PWR system (9).

vessel penetrations (16). The combination of concrete shielding and containment building also provides some reassurance against relatively improbable accidents such as airplane crashes or acts of sabotage or civil disturbance, since the reactor portion of the generating plant is relatively less vulnerable to such acts than conventional portions of the system. The provision for containment vessels, spray systems and control of effluent release permits a significant reduction in the controlled-access exclusion zone specified to surround a nuclear power plant (16).

Thus, safety is not an accidental feature of a large engineered system, but requires a definite design philosophy. This is often referred to as defense in depth, involving three levels of safety (9):

1. Design for maximum safety in normal operation and maximum tolerance for system malfunctions. Use design features inherently favorable to safe operation; emphasize quality, redundancy, inspectability and testability prior to acceptance for sustained commercial operation and over the plant lifetime.

2. Assume incidents *will* occur in spite of care in design, construction and operation. Provide safety systems to protect operators and the public, and to prevent or minimize damage when such incidents occur.

3. Provide additional safety systems as appropriate, based on evaluation of effects of hypothetical accidents, where some protective systems are assumed to fail simultaneously with the accident they are intended to control.

To this might be added the need for a well-trained and responsible staff that will follow procedures and respond coolly to unexpected situations.

The cost of items installed for accident prevention and limitation has been estimated to be $8/kWe for a 600 MWe plant. The cost of a standard containment building has been estimated at $4.5 to $5 million for a similarly sized plant, which corresponds to an additional $8/kWe. Multiple containment and the associated equipment for filtering fission products cost about $4/kWe. Thus, the total cost of these engineered safeguards is approximately $19/kWe in a typical 600 MWe plant in 1967 dollars (20). It is interesting to note that the estimated costs for the prevention of atmospheric pollution in a coal or oil-fired plant have been given as $12 to $24/kWe.

For advanced reactors that use prestressed-concrete containment, capital-cost savings of $10 to $15/kWe are believed possible. Since they are in limited use at present, it is difficult to ascertain the relative cost position of these reactor containments (16).

Nuclear power plants must be designed so that the plants remain in a safe condition in the event of the most severe tornado that can reasonably be predicted to occur at a site as a result of severe meteorological conditions. For the last several years, many investigators have studied the meteorological circumstances preceding and during tornado occurrences and the destruction following tornado strikes. Because of the inherent difficulty in directly observing a tornado and the great uncertainty in predicting the time and location of its occurrence, the possibility of directly sensing tornado characteristics such as a maximum wind speed is remote. Essentially all conclusions concerning tornado properties have been based on indirect observations of subsequent destruction, structural failure, generated missiles, or tornado markings rather than direct measurement of the tornado. Determinations of maximum speeds in tornadoes have been rough approximations. As a result of these studies, however, tornadoes have been characterized by a set of properties that are significant for purposes of structural design and siting. These parameters are: (1) geographical distribution of frequency of occurrence, (2) rotational wind speed, (3) translational wind speed, (4) pressure drop across the tornado, (5) rate of this pressure drop, and (6) radius of maximum rotational wind speed. Using these properties, it is possible to develop a definition for a design basis tornado in terms of the six tornado parameters and to use analytical techniques for estimating values of these parameters for purposes of design with an adequate level of conservation (21).

The design basis tornado is defined in terms of values for the six descriptive parameters in Table 44. These values are listed for each of the three tornado intensity regions within the contiguous United States.

COOLANT ACTIVITIES

As we have seen, a sizable inventory of fission products accumulates in the reactor fuel during burn-up, representing an enormous source of radioactivity. Since the cladding is only a fairly thin metal sheath around the fuel, which is subjected to thermal stresses, mechanical forces, internal gas pressures, and corrosive action by the coolant, almost inevitably small cracks will develop in it during operation that will permit a small but finite fraction of the fission products to leak into the coolant. While the existence of such leaked activity in the coolant is readily detected, it is almost impossible to pinpoint a small leak with any certainty while the reactor is operating and it is not practical, nor economically feasible, to shut down the reactor and change fuel assemblies every time such a small leak occurs. Instead, a reactor is designed to continue to operate as long as not more than 0.25 to 1.0% of the fuel elements are affected. During this time, part of the coolant is

Table 44. Design Basis Tornado Characteristics (21)

Region	Maximum Wind Speed[a] (mph)	Rotational Speed (mph)	Translational Speed (mph) Max.	Translational Speed (mph) Min.[b]	Radius of Maximum Rotational Speed (feet)	Pressure Drop (psi)	Rate of Pressure Drop (psi/sec)
East of Rockies	360	290	70	5	150	3.0	2.0
Pacific Coast	300	240	60	5	150	2.25	1.2
Rocky Mt. States	240	190	50	5	150	1.5	0.6

[a]The maximum wind speed is the sum of the rotational speed component and the maximum translational speed component.

[b]The minimum translational speed, which allows maximum transit time of the tornado across exposed plant features, is to be used whenever low travel speeds (maximum transit time) are a limiting factor in design of the ultimate heat sink. The ultimate heat sink is that complex of water sources, including associated retaining structures, and any canals or conduits connecting the sources with, but not including, the intake structures of nuclear reactor units.

cycled continuously through a demineralizer column that decontaminates the coolant by removing the dissolved tracer activity by ion exchange.

In addition to fission products and transuranium elements, hydrogen and helium will build up in the fuel. These are produced in some of the fission reactions as well as in (n,p), (n,d), (n,t) and (n,α) reactions, which can occur for fast neutrons interacting with both fuel and fission products. The helium and hydrogen thus generated may contribute to the gas pressure building up inside the fuel element and they will diffuse through the cladding. Stainless steel appears to be more permeable to hydrogen diffusion than zircaloy, where zirconium hydride forms readily, and this affects the amount of tritium from fuel that appears in the coolant. Otherwise helium and hydrogen are of importance mainly as leading to embrittlement of the cladding and affecting the conductivity and pH of the coolant.

In both BWR's and PWR's about one fission in 10,000 is a "ternary fission" event, that is, three fission fragments arise, one of which is tritium. The annual production of such fission-product tritium is estimated to be in the range of 15-25 kCi or about 1.5-2.5 g of tritium for a 1000 MWe LWR. Ninety-nine percent or more of this tritium remains with the fuel rods, which are clad with zirconium alloy, until they are reprocessed (9,22).

The rate of leakage into the coolant varies with the fuel temperature, the size of cracks in the cladding, the chemical nature of each individual fission product, and the pressure and concentration gradients across the cladding. Many elements will remain chemically bound to the uranium oxide fuel and the cladding alloy; however, as one might expect, the noble gases (krypton, helium and xenon) will diffuse out fairly readily and to a lesser extent so do hydrogen and the volatile halogens. Most of the other fission products diffuse or leach out into the coolant at a very slow rate. Experimental results on the release of various fission products from UO_2 fuel under specific experimental conditions show wide variations of inert gas and iodine escape from ruptured fuel elements (23). The actual amounts escaping obviously depend on the particular conditions of the fuel at the time, and vary along the length of the fuel element and with the extent of pressure distribution within the fuel element.

Finally, radioactivity will occur in the coolant due to tramp uranium and corrosion products. The term *tramp uranium* refers to uranium dust left on the outside of the fuel elements after fabrication and reflects insufficient cleaning of the fuel assemblies before shipment. Any such uranium will, of course, undergo fission and the fission products will enter the coolant. Recent advances in processing and good quality control have helped to minimize this source of contamination. The corrosion products can arise anywhere in the coolant circuit which, typically, is composed of zircaloy cladding within the core and stainless steel in the plenum and heat exchanger or steam generator loops. As a result of radiolytic decomposition of the cooling water in the core, free H and OH radicals as well as some H_2O_2 and free oxygen are formed which will change the pH of the coolant and may promote corrosion (24-26). In some plants small amounts of lithium hydroxide (60 lb/yr) are added to the reactor coolant system for control of acidity. During shutdowns 100 lb/yr of hydrazine are added for removal of radiolytic oxygen.

Most of the activity buildup, in practice, is due to neutron activation of the corrosion and wear products of the materials composing the coolant loop. These materials include stainless steel, zircaloy, Inconel, carbon steel and other steel and copper alloys depending on the type of reactor. These alloys are rich in nickel, chromium and, in the case of stellite, cobalt. Though corrosion may be slow, activation activities, such as Fe-59, Cr-51, Mn-54, Co-60 and Zr-95, will build up in the coolant and their presence may give rise to significant radiation doses to plant personnel during maintenance unless layout and access for maintenance are well planned. Cobalt-58 from the reaction $^{58}Ni(n,p)^{58}Co$ in Inconel piping has been found to be the principal "crud" activity in PWR's. In

BWR's cobalt-60 from the activation of ^{59}Co in the stellite in flow valves represents the chief crud activity. Most of this activity will appear ultimately in the radwaste treatment system, some of it in soluble form, some of it as particulates.

Even though most of the cores today use Zircaloy-4 as a fuel clad, the zirconium content in the crud is actually very low because of the adherent nature of the oxide film of Zircaloy. Proper pH control of the primary loop helps minimize corrosion products.

Activation of primary coolant impurities is also a source of activation products. This source is minimized by using high purity water as makeup for the primary. One example of this source is the activation of argon-40 dissolved in primary makeup water by the reaction ^{40}Ar$(n,\gamma)^{41}$Ar, which has a half-life of 1.83 hours. Since argon is about 1% by volume in standard atmosphere, exposure of makeup water to air should be minimized. However, the short half-life of the product ^{41}Ar essentially eliminates this activation product as a major radioactive source in plant effluents when decay tanks are used.

For an 1100 MWe BWR reactor the principal activation gases in reactor steam are shown in Table 45 together with the concentration and release rates. Of these gases, N-16 from the ^{16}O$(n,p)^{16}$N reaction is important because of the high-energy gamma-ray emitted. ^{13}N produced by ^{16}O$(p,\alpha)^{13}$N may be responsible for potential off-site doses due to its 10-minute half-life. ^{19}O is also included in off-site dose estimates. In fast reactors, the ^{16}O$(n,\alpha)^{13}$C reaction is of importance, too, though mainly as a source of CO_2. In heavy-water reactors the N-16 gamma ray gives rise to delayed fast-neutron emission from the coolant, which may pose an external hazard.

Table 45. Principal Activation Gases in a BWR

Isotope	$\tau_{1/2}$	Concentration	Release from Reactor (μCi/sec)
N-13	9.9 min	6.5×10^{-3}	1.2×10^{4}
N-16	7.13 sec	1.0×10^{2}	1.7×10^{8}
N-17	4.14 sec	1.6×10^{-2}	2.6×10^{4}
O-19	26.8 sec	8.0×10^{-1}	1.4×10^{6}
F-18	109.8 min	4.0×10^{-3}	7.2×10^{3}

Tritium Generation

In PWR's boric acid is the chemical added to the reactor coolant in greatest quantities, about 20,000 lb/yr. It is used as the "chemical shim," a burnable poison used for obtaining better power distribution in the core and as a convenient means for fine-controlling the reactor power level (24,26). Its concentration is controlled either by special demineralizers or by "feed and bleed" operations, i.e., simultaneous injection into and withdrawal from the reactor coolant system. As a result of neutron capture reactions in B-10 and B-11, tritium is formed. Whereas the boron can be removed and recycled, the tritium cannot be readily extracted and will build up in the coolant loop. Boron-10 may undergo a $^{10}B(n,t)2\alpha$ reaction if the neutron energy is greater than about 1.5 MeV. Lithium-6 may undergo a $^6Li(n,\alpha)T$ reaction; this reaction does not have a neutron energy threshold and thus occurs with very low- as well as high-energy neutrons. Other tritium-producing neutron-capture reactions, i.e., $D(n,\gamma)T$, are normally insignificant when compared with the tritium production from ^{10}B and/or ^6Li.

Lithium enters the primary loop through an ion exchanger or demineralizer. It is used to help maintain primary pH at a value of approximately 9.5. An alternative to this method is the use of an NH_4OH or KOH demineralizer instead of LiOH. A change from LiOH resin to KOH in the Bettis pressurized water reactor reduced the tritium concentration by a factor of 100 (27).

It may be assumed that 1% of the tritium produced by ternary fission escapes when zircaloy-clad fuel is used and 80% escapes when stainless steel is used (28).

Tritium generated by these processes becomes an integral part of the coolant water and accounts for more activity in reactor plant effluents than any other source. In BWR's neither soluble boron nor lithium compounds are used and this is the major reason for the smaller amount of tritium in BWR liquid effluents. Furthermore, the lithium content has been responsible for appreciable production of beryllium-7 at some plants.

An additional 400 curies/year are added to the above if boron control rods are used in the plant (29). Some nuclear plants are being designed using Ag-In-Cd control rods instead of B_4C to reduce tritium generation.

Table 46 shows the estimated quantities of tritium expected in the coolants of LWR's and HTGR's, as currently designed. Actual measurements have shown much higher tritium concentrations for reasons that are not quite clear at this time. It should be noted, however, that most of the tritium not released to the environment at the LWR plant may be

Table 46.   Tritium Disposition in 1000 MWe BWR's and 120 MWe HTGR (9)

Tritium Source	Maximum Expected Releases to the Coolant (Ci/yr)		HTGR Production Rate (Ci/yr)
	PWR	BWR	HTGR
Ternary fission (assuming 0.1% enters coolant)	40	40	605
Soluble boron	560	—	
Lithium−reactions	17	—	3
Helium 3−capture	−	−	91
Deuterium reaction	10	10	−
Total	627	50	699

released at the fuel reprocessing plant when the fuel is eventually reprocessed. The fate of the tritium present in the cladding during reprocessing is uncertain.

Tritium in contact with water or water vapor quickly becomes a part of the water. Tritium gas, either as T_2 or HT, exchanges with ordinary hydrogen from water (H_2O) to form HTO; for example:

$$HT + H_2O \rightarrow HTO + H_2$$

Thus, when tritium enters or is formed in the coolant of an LWR, or enters the aqueous dissolver solution in the fuel reprocessing plant, it becomes an integral part of the water, from which it is difficult to separate. Since the condenser water of most LWR's and the water used in a reprocessing plant are currently discharged to the environment more or less continuously, all its tritium goes into the environment with it (9,29,30). Some additional tritium is also formed by neutron capture in the deuterium normally present at 0.15% abundance in fresh water. This process also accounts for massive tritium inventories in heavy-water reactors.

In high-temperature gas-cooled reactors production of tritium will occur by ternary fission and by activation of helium-3 found in trace amounts in the helium coolant (31). Table 46 also shows the tritium reactions and estimated production rates in a 120 MWt gas-cooled reactor. Tritium can be removed from the helium coolant by absorption on titanium sponge. With two loops, the titanium needs to be regenerated every six months.

COOLANT PURIFICATION AND RELEASE PATHS

From the operational point of view, the various activities contained in the coolant, and listed in Table 47 (29,32), are relatively small compared with the fission-product inventory in the reactor core, yet in routine operation this activity forms the principal source of radioactive effluents from the nuclear power plant. It should be noted that the activities listed are maximum estimates, corresponding to 1% failed fuel, and are much higher than measured values (29). Because of the different system designs the escape paths for the activities differ between reactor types. In the boiling water reactors steam is raised immediately above the reactor core, within the reactor containment. This steam is then passed to the turbine without an intervening heat exchanger and subsequently condensed and cleaned. Any escape of steam via the turbine, gland seals or condensate phase separators will introduce some radioactivity in gaseous form into the building atmosphere. Consequently any release of radioactivity in a BWR is most likely by way of the airborne effluents traveling up the stack. In contrast, in a pressurized water reactor the primary coolant passes through a heat exchanger still in the liquid state. In such a system, escape of activity is less probable and would proceed mainly through cracks and defects in the primary loop piping, boiler tube defects, and pump shaft and valve stem seal leaks. The treatment of this activity in the liquid effluent stream will be discussed in a subsequent chapter.

In both cases it is evident that the radioactivity released can be greatly reduced by continuous decontamination of the primary coolant stream, at least by removing most of the particulate and soluble-ion material. This is done by an on-line filter and demineralizer ion exchange system in the coolant loop, and this is where the bulk of the activity will accumulate. The demineralizers consist of steel tanks filled with a bed of ion exchange resins, which are organic particles having the appearance of fine sand. The resins have been chemically treated so that they can adsorb and retain dissolved materials, both radioactive and nonradioactive. The removal efficiency of ion exchange resins is very high for such dissolved materials when a fresh bed of resin is used, but the removal efficiency decreases as the resin becomes saturated with adsorbed materials. In PWR's boric acid recovery from the primary coolant requires further ion exchange treatment; currently temperature-controlled exchange resins are being introduced for this purpose. The demineralizer beds used to purify radioactive waste waters generally are used only once; thus, the resins containing the absorbed radionuclides constitute a major source of solid waste from the plant. Of course, the gaseous components in the

Table 47. Predicted Reactor Coolant Inventory of Fission Products and Corrosion Products (1000 MWe PWR at 578°F) (32)[a]

| Noble Gas Fission Products | | Fission Products | |
Isotope	μCi/ml	Isotope	μCi/ml
Kr-85	1.11	Br-84	3.0×10^{-2}
Kr-85m	1.46	Rb-88	2.56
Kr-87	0.87	Rb-89	6.7×10^{-2}
Kr-88	2.58	Sr-89	2.52×10^{-3}
Xe-133	1.74×10^2	Sr-90	4.42×10^{-5}
Xe-133m	1.97	Y-90	5.37×10^{-5}
Xe-135m	0.14	Y-91	4.77×10^{-4}
Xe-138	0.36	Sr-92	5.63×10^{-4}
Total		Y-92	5.54×10^{-4}
Noble Gases	187.3	Zr-95	5.04×10^{-4}
		Nb-95	4.70×10^{-4}
		Mo-99	2.11
		I-131	1.55
Corrosion Products		Te-132	0.17
		I-132	0.62
Mn-54	4.2×10^{-3}	I-133	2.55
Mn-56	2.2×10^{-2}	Te-134	2.2×10^{-2}
Co-58	8.1×10^{-3}	I-134	0.39
Fe-59	1.8×10^{-3}	Cs-134	7.0×10^{-2}
Co-60	1.4×10^{-3}	I-135	1.4
Total Corrosion		Cs-136	0.33
Products	3.7×10^{-2}	Cs-137	0.43
		Cs-138	0.48
		Ce-144	2.3×10^{-4}
		Pr-144	2.3×10^{-4}
		Total Fission	
		Products	12.8

[a]Contamination concentrations corresponding to 1% failed fuel near end of fuel life.

coolant shown in Table 48 will not be retained and may ultimately escape to the plant environment, where they will appear in the gaseous effluent (Chapter 7).

In gas-cooled reactors the coolant is similarly decontaminated continually by passing it over charcoal filters to remove iodine and other volatile materials. In this case the accumulation of carbon-14, from activation of the graphite moderator and the reaction $^{14}N(n,p)^{14}C$ on traces of air, in the form of CO or CO_2 may become significant. Molecular sieves are used to remove CO_2 and water vapor from the helium stream and the noble gases are removed by a low-temperature

Table 48. Concentrations of Gaseous Fission Products in PWR and BWR Primary Coolants

Nuclide	Half-Life	PWR $(\mu Ci/ml)^a$	PWR $(\mu Ci/ml)^b$	BWR $(\mu Ci/ml)$
Kr-83m	1.86 h	5.5×10^{-2}	5.3×10^{-2}	1.2×10^{-3}
Kr-85m	4.4 h	2.9×10^{-1}	2.7×10^{-1}	1.9×10^{-3}
Kr-85	10.74 y	2.0×10^{-1}	3.1×10^{-2}	9.7×10^{-6}
Kr-87	76 m	1.6×10^{-1}	1.6×10^{-1}	5.7×10^{-3}
Kr-88	2.79 h	5.1×10^{-1}	4.9×10^{-1}	6.1×10^{-3}
Kr-89	3.18 m	1.2×10^{-2}	1.2×10^{-2}	2.4×10^{-2}
Xe-131m	11.96 d	2.3×10^{-1}	6.6×10^{-2}	8.4×10^{-6}
Xe-133m	2.26 d	5.4×10^{-1}	3.1×10^{-1}	1.2×10^{-4}
Xe-133	5.27 d	4.1	1.6	3.3×10^{-3}
Xe-135m	15.7 m	3.4×10^{-2}	3.4×10^{-2}	1.0×10^{-2}
Xe-135	9.16 h	8.5×10^{-1}	7.5×10^{-1}	9.6×10^{-3}
Xe-137	3.82 m	2.5×10^{-2}	2.5×10^{-2}	4.1×10^{-2}
Xe-138	14.2 m	1.2×10^{-1}	1.2×10^{-1}	3.2×10^{-2}
Total Noble Gases		7.1	3.9	1.3×10^{-1}
I-131	8.06 d	7.1×10^{-1}	5.6×10^{-1}	5.4×10^{-3}
I-133	20.8 h	8.6×10^{-1}	7.4×10^{-1}	3.1×10^{-2}

aWithout volume control tank purge.
bWith volume control tank purge.

delay bed system (33) or by chromatographic columns. These may need regenerating every 90 days, when the partially decayed noble gases may be released or recovered for long-storage by cryogenic methods.

The release of radioactive gases from water in the reactor system and from water that leaks from the reactor is an important consideration in estimating the release of radioactive gases to the environment coefficients. The *partition factor* is defined as the fraction of iodine or noble gas originating in a particular liquid stream that transfers to the gas stream. For example, the partition factor of iodine for the blowdown tank vent is the fraction of iodine in the liquid stream blowdown that goes out with the gas phase through the blowdown tank vent. The *partition coefficient* is defined as the ratio of the concentrations in a liquid phase and a gas phase at equilibrium. For example, the partition coefficient of iodine is the sum of the concentrations of all the iodine species (I^-, I_2, I_3^-, etc.) in the liquid phase divided by the concentration of iodine in the gas phase. Thus, the partition factor is the gas volume (or flow rate) divided by the liquid volume (or flow rate) and that ratio divided by the partition coefficient. Where the partition factor approaches unity, a more precise definition is:

$$\text{Partition factor} = F = \frac{(G/L)/P}{1 + (G/L)/P}$$

where the gas-to-liquid volume ratio is designated as G/L and the partition coefficient as P. The volume or flow rate ratios are determined by the assumed reactor operating conditions. The partition coefficients are calculated from theoretical conditions and data from the literature. For the noble gases, Kr and Xe, the partition coefficients are about 0.1 (liquid to gas concentration ratio) at room temperature and they decrease with increasing temperature to about 0.03 at 100°C and increase above 100°C. Since Kr and Xe do not chemically react with water, the partition coefficients do not change significantly with pH (relative acidity or alkalinity) and concentration of the gases. Since the gas phase is about ten times more concentrated than the liquid, entrainment (the presence of liquid droplets in the gas phase) will not significantly affect the partition coefficient. A partition coefficient of 0.1 may be considered as an appropriate approximate value.

The partition of iodine between gaseous and liquid phases is much more complex. Iodine is quite soluble and at room temperature the liquid phase concentration of I_2 is about 10^2 times higher than the gas phase concentration of I_2. Thus, entrainment can be important. One percent carryover would result in as much I_2 in airborne droplets as is in the gas phase at equilibrium. Another complication arises because I_2 hydrolyzes in water to form I^- and HOI, and additional I_2 reacts with I^- to form I_3^-. Other dissolved species have also been postulated. The partition coefficient is affected by this system of equilibria because some of the volatile I_2 is transformed to nonvolatile species. Increases in pH, decreases in concentration, and increases in temperature each increase the extent of hydrolysis and the fraction of iodine which is nonvolatile and hence the partition coefficient. The partition coefficient (due to I_2 volatility) can be calculated from reported values of the equilibrium constants, and the agreement between calculated and experimentally observed values is reasonably good over reasonably wide ranges of parameters. The very high partition coefficients that can be calculated for solutions at high pH have however, not been observed. Observations of partition coefficients greater than 10^4 are rare. The probable explanation is that there are mechanisms other than I_2 volatility by which iodine can become gasborne. For example, (1) HOI as well as I_2 may be volatile; (2) in basic solutions, iodination of organic compounds occurs rapidly and may form volatile species such as methyl iodide (CH_3I); and

(3) entrainment of droplets may account for observations of partition coefficient values that are smaller than predicted. The partition coefficient of CH_3I is about 3.5 at 25°C and about a factor of ten smaller at 100°C. In addition to hydrolysis and all of its complications, iodine can also react with oxidizing or reducing agents. Such reactions (unless HOI is volatile) will convert more I_2 to nonvolatile species, which will increase the partition coefficient.

In practice, one is concerned with extremely small concentrations of iodine: in the PWR $\sim 5 \times 10^{-11}$ molar in the primary coolant, $\sim 10^{-13}$ in the secondary liquid and $\sim 10^{-15}$ in the secondary condensate; in the BWR $\sim 10^{-13}$ molar in the coolant liquid and $\sim 10^{-15}$ molar in the condensate. Experimental values of partition coefficients are not available for such small concentrations. The calculated values for these low concentrations are greater than 10^4. Since very few experimental values under any conditions of pH, concentration, basicity, and temperature have exceeded 10^4, a partition coefficient value of a maximum of 10^4 can be used for the theoretical calculations. Thus, any variation in partition coefficients due to acidity differences in PWR and BWR systems will not affect these calculations.

For example, for a PWR system the amount of Xe, Kr and I volatilized from the shim bleed in the vacuum degasifier has been calculated by assuming that the gas-liquid volume ratio was equivalent to the average flow of gases removed from the shim bleed, 0.1 cfm, divided by the shim bleed flow rate of 1 gpm, which (with appropriate change of units) is 0.5. (Note that this volume ratio could be a factor of approximately 20 higher for a system using a nitrogen degasifier rather than a vacuum degasifier.) For the noble gases with a partition coefficient of 0.1, the partition factor becomes (0.5/0.1)[(0.5/0.1) + 1] or 0.8. For iodine, with a partition coefficient of 10^4, the partition factor becomes about $0.5/10^4$ or 5×10^{-5}. The use of a partition coefficient of 1 rather than 0.8 is valid, since the water and any retained noble gases would be retained in nonaerated tanks. These gases would ultimately be combined with the gases removed in the degasifier as the water is recycled for reuse.

Kr, Xe and I can escape the secondary system with the blowdown, through the steam-jet air ejector, with the gland seal steam condenser effluent, and with leaks into the turbine building. The leaks into the turbine building contribute to building ventilation releases and the others are released with secondary coolant gaseous effluents. A liquid stream is blown down from the steam generators and flashed into the blowdown tank. The vent for this tank is one of the gaseous radwaste streams. The condensate is sent to the liquid radwaste system. For a typical case, from a consideration of the heat balance in the blowdown tank where a 4800-lb/hr stream of 580°F liquid water becomes a mixture of 212°F steam and water, it is estimated that about 40% of the water exits as steam and that the ratio of the volumes of the steam and liquid is about 600. The partition factor for noble gases through the blowdown vent is therefore $(600/0.1)[(600/0.1) + 1]$ or nearly unity. However, the blowdown liquid does not contain much of the noble gases because most of these gases are evolved with the steam in the steam generator.

Emergency Decontamination

In considering the consequences of a loss-of-coolant accident, separate consideration must be given to the fate of the activity in the fuel core and that contained in the coolant loop. At high temperatures, most of the fission products may remain in the molten fuel, except for the volatile and gaseous elements. Considerations of heat loss rates to surrounding surfaces and distortion or flattening of the molten core tend to militate against a "China syndrome effect," where the core is assumed to melt its way into the ground. A more likely event is the spalling off of some of the fuel material under the shock wave of any sudden pressure release or under the impact of a jet of high pressure steam. Figure 42 shows the growth and decay of the main fission product categories in the reactor and refueling area following an accident (23).

Under these conditions the activity of escaping fission products from the core would greatly exceed that in the coolant loop. Investigations have been conducted on means of precipitating and fixing the airborne activity to ensure its retention in the building and minimize inhalation hazards. Such methods involve release of aerosol sprays, use of special coatings, sprinkler systems and use of adsorptive filter systems (23). Figure 43 shows the life history of fission products in the major groups as they move within the reactor and containment structures following a postulated accident. It is evident that any form of holdup will lead to a rapid decay in a few days, though with a rapid initial build-up in gases and volatiles. Experimental data have been obtained for a simulated fuel meltdown on the release and filter decontamination of some major fission products (23). Table 49 lists some data on the proportionate release of the major groups under design-basis accident conditions.

Iodine release has been of prime concern and can be reduced by use of some wall coating materials tested for iodine removal for the LOFT tests (34). The form and compound formation of released iodine will affect adsorption from steam and aerosols (35). Entrainment of iodine and other volatile materials in water droplets presents a simple method

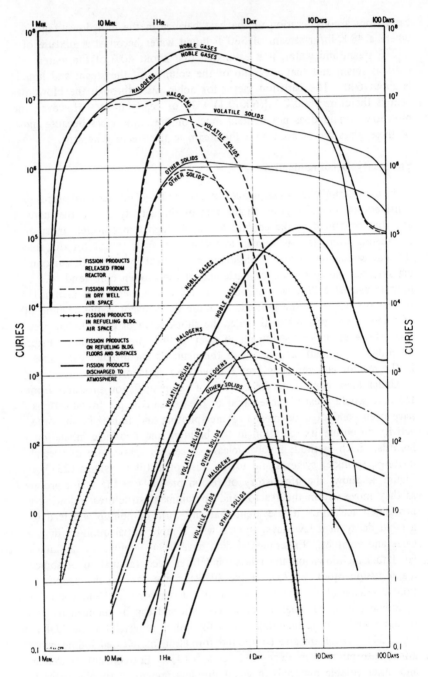

Figure 42. Fate of fission product inventory following loss of coolant accident (23).

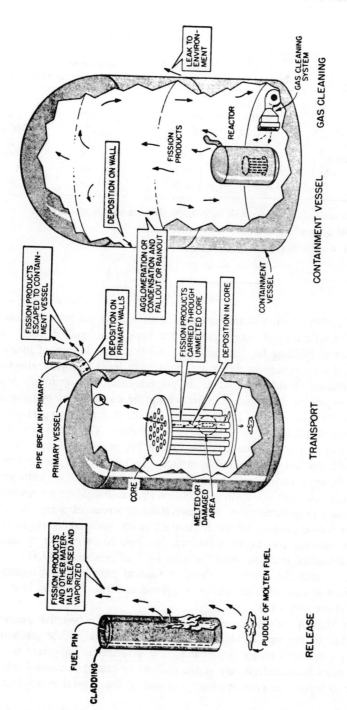

Figure 43. Life history of fission products following a reactor accident.

Table 49. Fission Product Release for Core-Melt Accident Conditions (23)

	Core[a] Inventory (g)	Fractional[b] Release	Release to Containment (g)	Containment[c] Conc. (mg/m^3)
UO_2	1.2×10^6	10^{-4}	120	140
Cladding	2×10^5	10^{-3}	200	250
Rare gases	1.7×10^3	10^0	1,700	2,000
Halogens	9×10^1	5×10^{-1}	45	50
Alkali metals	1.3×10^3	3×10^{-1}	390	450
Alkaline earths	9.6×10^2	1×10^{-1}	96	110
Other	1.0×10^4	1×10^{-3}	10	12

[a]Assumed basis of 1/5 linear scale of a 3,000 MWt reactor, 10,000 MWD/T.
[b]Estimated on basis of available experimental studies.
[c]Based on a uniform release into 850 m^3.

of air decontamination following an accident; much has been published on the effectiveness of spray systems that have been installed inside the containment building for this purpose (36-39). The addition of hydrazine to the emergency cooling water apparently leads to highly effective spray removal of airborne iodine. In practice much will obviously depend on the degree of turbulence existing within the containment building and the concentration of fuel particulates in the air that may serve as condensation nuclei for some of the volatile elements. Of special concern would also be the plutonium in the fuel, which one would expect to precipitate out on any oxidizable surface. To minimize explosion hazards and to improve tritium retention, fixed or mobile recombiner units are frequently specified for hydrogen control after a loss-of-coolant accident.

The safe decontamination and restoration to service of a reactor after a major fission product release is clearly a major undertaking requiring detailed planning and many operations that have to be broken up into small, individual tasks to avoid "burning out" of personnel (40,41). Procedures must be developed ahead of time to provide for emergency operations in case of both on-site or off-site accidents involving high-level radioactivity (42).

Even though the release of a massive quantity of radioactive material from the containment building in the case of a severe reactor accident may be considered highly improbable, it is nevertheless important to devise procedures and allocate responsibilities in case it is deemed advisable to inform or even evacuate members of the general population in the vicinity of the plant site, working in conjunction with appropriate local authorities. A recent report by Hans and Sell (43) has pointed out the problems associated with large-scale evacuation; though it is rarely seriously contemplated, orderly evacuation plans must be carefully worked out in advance to be of any use at all.

Spent Fuel Storage

In water-cooled reactors, unloading spent fuel from the reactor and storing it for a cooling-off period before shipment is an operation requiring particularly careful planning and adequate facilities. Fuel storage and handling systems must be designed to assure adequate safety under normal and postulated-accident conditions, with appropriate containment and filtering systems, and to prevent significant reduction in the coolant inventory of the storage facility under accident conditions. Since the fuel normally requires forced cooling to remove the fission-product heat, it is usually stored, in current American power reactors, in a pool adjoining the reactor vessel. During refueling the reactor vessel is flooded so that all transfer operations take place under water to minimize operator exposure. Obviously the fuel handling and storage facilities must be designed with the following objectives:

(1) to prevent loss of water from the fuel pool that would uncover the fuel

(2) to protect the fuel from mechanical damage

(3) to provide the capability for limiting the potential offsite exposures in the event of significant release of radioactivity from the fuel.

If spent fuel storage facilities are not located within the primary reactor containment nor provided with adequate protective features, radioactive materials could be released to the environs as a result of either loss of water from the storage pool or mechanical damage to fuel within the pool.

Unless protective measures are taken, loss of water from a fuel storage pool could cause overheating of the spent fuel with resultant damage to fuel cladding integrity and result in release of radioactive materials to the environment. Natural events, such as earthquakes or high winds, could damage the fuel pool either directly or by the generation of missiles. Structures and cranes also could fall into the pool during an earthquake or high winds. Design of the facility to withstand these occurrences without significant loss of watertight integrity would alleviate these concerns. Dropping of heavy loads, such as a 100-ton fuel cask, although of low probability, cannot be ruled out in plant arrangements where such loads are deliberately placed in or over the fuel pool. Two possible solutions to this problem are: (1) design the pool to withstand dropping of the load without significant leakage occurring from the pool area in which fuel is stored, or (2) prevent by design rather than interlocks heavy loads from being lifted over the pool. Water could be lost from the pool as a result of equipment failures in systems connected to the pool if such loss is not prevented by design.

Even if the steps described above to prevent leakage are taken, small leaks may still occur from structural failure or other unforeseen events. A permanent fuel pool coolant makeup system with a moderate capability, and with suitable redundancy or backup, could prevent uncovering of the fuel if such leaks should occur. Early detection of pool leakage and fuel damage should be provided by pool water level monitors and radiation monitors designed to alarm both locally and in a continuously-manned location. Timely operation of building filtration systems can be assured by actuation of these systems by a signal from local radiation monitors.

The release of radioactive material from fuel may occur during the refueling process, and at other times, as a result of fuel cladding failures or mechanical damage resulting from the dropping of fuel elements, or the dropping of objects onto fuel elements. A relatively small amount of mechanical damage to the fuel might cause significant offsite doses if no dose reduction features, such as a controlled leakage building, are provided (44).

Whether or not damaged fuel elements are present, the unloading of fuel from a reactor in many ways represents the most difficult portion of reactor operations and one that can entail the highest radiation exposure levels to the operators involved. Where the fuel transfer is not done under water, such as in gas-cooled reactors or heavy-water reactors, a heavily shielded fuel handling machine is placed over the fuel position and the elements are cooled by circulating helium gas (45). Great care is obviously required to remove and contain any trace fission products that may enter the building atmosphere from damaged fuel elements, and to prevent their release to the environment.

ADVANCED REACTORS

Although most experience on the safety and probable environmental effects of nuclear reactor operation has been gained on the current generation of commercial power reactors, whether gas-cooled or water-cooled, it is appropriate to consider also the consequences of the technological developments attendant on the adoption of more advanced reactor types.

The term *advanced reactor* tends to be used rather loosely to describe the next engineering extrapolation of current reactor types, such as the advanced gas-cooled reactor (AGR), where the anticipated environmental effects relate mainly to an extension in size or power rating, without introducing any novel problem in principle (46). For the present purpose the term will be used to denote reactor types at an advanced stage of research and prototype development, namely fast breeder reactors and fusion reactors.

Fast Breeder Reactors

The reasons for developing breeder reactors have been reviewed in many places (47-51). Obviously, any system capable of conserving fuel resources by generating additional fuel material during operation can make a significant contribution towards conserving increasingly scarce and costly raw materials. In fact, without breeder reactors the demand for readily accessible uranium ore of high and intermediate U_3O_8 content may well exhaust most known and anticipated raw material resources by the end of the century. In a fast breeder reactor fast neutrons are allowed to be captured by uranium-238, the most abundant uranium isotope, and will produce plutonium-239, which is fissile, by the reaction chain:

$$^{238}U + n \rightarrow {}^{239}U \quad \tau_{1/2} \text{ 23.5 min}$$
$$\downarrow \beta$$
$$^{239}Np \quad \tau_{1/2} \text{ 2.35 d}$$
$$\downarrow \beta$$
$$^{239}Pu \quad (+ n \rightarrow {}^{240}Pu, \text{ etc.})$$
$$\downarrow \text{ fission}$$

Consequently, a better utilization of available uranium resources can be assured and a greater degree of fuel conservation can be obtained. By 1986 it is hoped that increasing use of breeder reactors will slow down the demand curve for uranium ore and for enrichment plant capacity; that forecast may turn out to be over-optimistic. Since uranium-238 undergoes fission with fast neutrons it is both a fissile and a fertile material. The fertile material can be introduced by mixing fissile and fertile material in the core or by surrounding the core with a fertile blanket to utilize any neutrons that would otherwise escape from the core

To optimize breeding it is important to minimize parasitic absorption of neutrons in structural materials and the coolant; this is achieved by operating the reactor without slowing down the neutrons in the core.

For this reason, water is no longer a suitable coolant and it is necessary to use a nonmoderating coolant medium. One such coolant, widely used in fast reactors, is metallic sodium. Prototype breeder reactors have been operated in several countries, such as the Dounreay reactor in Great Britain, RAPSODIE and Phenix in France, EBR II and Enrico Fermi in the United States, and BR-2 and BR-5 in the USSR, and their general feasibility seems to have been established (47,51). On the face of it, the use of a highly reactive metal, such as sodium, would seem to carry with it some real fire hazards, but by careful design and handling it has been amply demonstrated that such cooling systems can be operated safely over very long periods (52).

Sodium is an alkali metal that melts at about 210°F (97.5°C). Its cross section for absorbing and thermalizing neutrons is very low (*i.e.,* it does not capture or slow large amounts of them), and its ability to transfer heat is excellent. It has a high boiling point (1640°F, 883°C) and a low vapor pressure over a wide range of temperatures. These properties make it nearly ideal as a reactor coolant. Unlike water, it can be heated to very high temperatures without generating pressure, and its excellent ability to transfer heat makes it much less sensitive to short term disturbances at the surfaces from which heat is being transferred. Because the coolant systems operate at low pressure, in the event of a failure of a pipe or other piece of equipment containing sodium the liquid does not escape as rapidly as it would with high-pressure systems.

The chemical reactivity of sodium is a safety asset in one respect: its ability to combine with or retain other elements. For example, during irradiation of fuel, many radioactive isotopes are formed as fission products: iodine-131, cesium-137, and niobium-95. In some types of fast reactors, these fission products are deliberately vented or discharged from fuel in the reactor into the sodium coolant. In other fast reactors, failure of the fuel outer cladding or covering can release these fission products to the sodium. Fortunately, sodium tends to retain many of these fission products, so they are not readily released to the inert gases (such as helium and argon) that normally are used to blanket the sodium. Radioactive iodine, for example, combines with the sodium to form sodium-iodide, and cesium is retained in solution. Niobium and certain other solid fission products tend also to be retained in the sodium. This does not mean that sodium retains all fission products; nearly all of the radioactive xenon and krypton gases bubble up through the sodium and are released into the inert cover gas.

This property of sodium to retain these materials, although complicating maintenance, acts as a safety advantage since accidental spillage or slow combustion of sodium does not free large quantities of the fission products. The chemical reactiveness of sodium also results in certain undesirable aspects. When exposed to air, sodium oxidizes rapidly if it is in the solid state and burns if liquid. This burning is at a controlled rate (2-14 $lb/hr/ft^2$ of exposed surface), and is easily extinguished by eliminating oxygen. When exposed to water, sodium reacts violently and forms hydrogen. The hydrogen in turn can combine with oxygen and increase the reaction energy.

Sodium has certain other features that are undesirable for a reactor coolant. One of these is that under irradiation it forms radioactive isotopes (Na-22 and Na-24) which emit gamma radiation. Fortunately, most of this radiation decays within a few days. However, this characteristic of sodium to become radioactive and to contain radioactive products from other sources makes it potentially hazardous. Table 50 presents an estimate of radioactive coolant concentration for a breeder reactor. This does not mean that under properly controlled conditions sodium is dangerous, but it does mean that care must be taken to assure that the sodium is properly controlled (51).

Table 50. Radioactive Inventory in Sodium Coolant of a Fast Breeder Reactor (49)

Isotope	Na Concentration[a] ($\mu Ci/cc$)	MPC (A) ($\mu Ci/cc$)	Na Concentration/ MPC (A)
^{24}Na	4×10^4	1×10^{-6}	4×10^{10}
^{131}I	5×10^0	9×10^{-9}	0.05×10^{10}
Pu	6×10^{-4}	4×10^{-11} (α)	0.0015×10^{10}
^{22}Na	1×10^0	2×10^{-7}	0.0005×10^{10}
^{90}Sr	2×10^{-3}	1×10^{-9}	0.0002×10^{10}

[a]Assumes equilibrium ^{24}Na and ^{22}Na activation; 0.1% defective fuel claddings with 10% halogen and 0.1% release of other fission products and plutonium to sodium. All fission product and plutonium contamination is assumed to remain in sodium (e.g., no plating on surfaces or removal by sodium cold traps).

The practical effect of the radioactivity in sodium is to limit direct human access during plant operation. To minimize this problem a fast breeder reactor includes two separate cooling circuits containing sodium and one containing water. The first circuit circulates sodium through the reactor core and becomes highly radioactive. This circuit is shielded from access; any maintenance is accomplished with remote mechanisms. The second circuit picks up heat from the first without becoming radioactive. Because sodium has such excellent heat transfer characteristics, it is possible to use two loops and still have an economically attractive system (50).

In the fast reactor, the sodium is normally safely contained within the piping and equipment. In order to protect against the possibility of leaks in equipment or piping, all equipment with radioactive sodium is contained within shielded vaults that exclude oxygen by substituting an inert gas like nitrogen so that any escaping sodium merely spills harmlessly into these vaults. Nevertheless the possibility of a leak and a consequent sodium fire is an ever-present possibility that must be taken into account in safety design and emergency procedures.

Current reactor designs require sealing of the part of the structure that is in direct contact with the radioactive core and blanket. In routine operation there would be no release of radioactive fission products to the environment. Because of the inherently low pressure of the sodium coolant, the reactor vessel and its associated piping must be designed to withstand only moderate operating stresses, in marked distinction to the pressure vessels and other primary-system components of a pressurized-water reactor, a boiling-water reactor or a gas-cooled reactor. Table 51 presents some of the characteristics of an LMFBR as currently envisaged; reactors operating or planned at present have been essentially only demonstration systems that were not intended to produce commercial power. To obtain good resistance to radiation damage and compatibility with the sodium coolant, stainless steel is usually chosen as the fuel cladding material.

All materials have markedly lower neutron-absorption probabilities for fast neutrons than for thermal ones. The lower cross sections reduce the effectiveness of fast-reactor control rods of sizes comparable to those in thermal reactors. On the other hand, a large amount of excess fuel is present in the core of a thermal reactor to compensate for the fuel that will be consumed by fission and to counteract the poisoning effect of the fission products. (The fission products capture neutrons without yielding significant amounts of energy.) With extra fuel there must be extra controls. Fast breeder reactors require fewer control rods because their greater effectiveness in converting uranium-238 to fissionable plutonium-239 compensates for depletion of the initial fuel charge and because fast neutrons are not absorbed by fission products as much as thermal neutrons are.

During a fission reaction not all the neutrons are released at the precise instant that each nucleus disintegrates. A small proportion of the neutron population is created by the decay of fission products. One thus distinguishes delayed neutrons from the "prompt" neutrons emitted directly by the fissioning nuclei. It is the delayed neutrons that keep

Table 51.   Fast Reactor Characteristics (6)

	Reference	Advanced
Net electrical output (MWe)	1000	1000
Total thermal power[a] (MWth)	2400	2417
Core	2219	2081
Axial blanket	107	195
Radial blanket	74	141
Average specific power in core (MWth/MT)[b]	116	155
Reactor inventory[a]		
Core uranium, kg	15460	10590
Core fissile plutonium, kg	2360	1690
Blanket uranium, kg	28270	31110
Blanket fissile plutonium, kg	305	362
Burnup, MWD/MT average[c]		
Core	67600	104059
Axial blanket	4740	8725
Radial blanket	7970	9051
Breeding ratio	1.30	1.42
Compound doubling time (years)	13	7.8

[a]Mid-cycle equilibrium.

[b]Mid-cycle equilibrium based on mass charged initially $(^{238}U + {}^{235}U + {}^{239}U + {}^{240}Pu + {}^{241}Pu + {}^{242}Pu)$.

[c]End-of-cycle equilibrium based on mass charged initially.

the chain reaction from escalating into an essentially instantaneous propagation of one generation of neutrons to the next.

The fraction of delayed neutrons depends appreciably on which nucleus is fissioning. Most thermal reactors are fueled with uranium-235, whereas the fast breeder will be fueled with plutonium-239. The fraction of delayed neutrons produced by the fission of uranium-235 is about 0.0065 and by plutonium-239 fission about 0.003. The smaller fraction of delayed neutrons present in a fast reactor is not of major concern under normal operation. It does increase the sensitivity of the reactor to adjustments of the control rods and to other inputs that affect reactivity, such as temperature variations in the core.

An alternative to the sodium-cooled breeder reactors that avoids some of the problems inherent in the use of reactive liquid metal coolants is the gas-cooled breeder reactor, using helium gas as the coolant (50). The Gas-Cooled Fast-Breeder Reactor System (GCFR) is a potential means of achieving high breeding ratios, high thermal efficiency, greater systems simplicity, and lower power costs in fast breeder reactors. The system offers the potential for breeding ratios of 1.4 or greater with stainless steel-clad oxide fuel, fuel doubling times of about 10 years, and system simplicity, which does not require an intermediate heat exchange loop between the core and steam system or an inert atmosphere in the secondary containment. This simplicity should be reflected in lower capital costs, greater reliability, and easier maintenance. However, a substantial program of development, proof testing and subsequent demonstration projects will be required before this promising concept can be considered a viable alternative..

For environmental assessment, one may summarize the advantages and disadvantages of breeder reactors as follows: the benefits include conservation of energy resources, minimum ore requirements per unit of energy utilized, minimum waste disposal, and high thermal efficiency. Drawbacks at this time relate to the lower level of demonstrated reliability, the inherent problems of utilizing large quantities of highly radioactive sodium metal, and the high plutonium inventory both in the reactor and in any spent fuel shipped for reprocessing. None of these problem areas can be regarded as insuperable, and considering the long-range need for conservation of nuclear fuel resources, the need for a breeder-type of reactor, though not necessarily a sodium-cooled one, would seem to justify the expense and some of the risks entailed in developing such power sources. There are, undoubtedly, some reasons for caution (53), but with careful procedures and good use of existing prototype breeders one may hope for a successful demonstration of the feasibility of fast breeder reactors by the mid-1980's.

Fusion Reactors

Nuclear energy can be produced in two ways: by the neutron-induced splitting of the heaviest nuclei or by the forcible fusion of some of the lightest. The reality of both concepts has been demonstrated, overdramatically, in the construction and testing of explosive devices of both types. However, while the design and operation of sustained fission reactions for power generation may be considered well-established by now and forms the basis of a sizeable industry, the production and control of sustained and controlled fusion reactions suitable for power generation still eludes us at this time, in spite of the vast and expensive research effort that has been devoted to it in many countries over the past two decades.

Fusion reactions are based on one of the following reactions:

(a) D-T $^2H + {}^3H \longrightarrow {}^4He(3.5 \text{ MeV}) + n(14.1 \text{ MeV})$

(b) D-D ^{2}H + ^{2}H $\begin{cases} \longrightarrow ^3\text{He}(0.82 \text{ MeV}) + \text{n}(2.45 \text{ MeV}) \\ \longrightarrow ^3\text{H}(1.01 \text{ MeV}) + ^1\text{H}(3.02 \text{ MeV}) \end{cases}$

(c) D-^{3}He ^{2}H + ^{3}He $\longrightarrow$ ^{4}He(3.6 MeV) + ^{1}H(14.7 MeV)

The two D-D reactions occur with equal probability. Relative to the other fuel cycles, the D-T cycle exhibits the least-demanding ignition requirements with a reaction energy of 6 keV and is usually considered the most probable candidate for the first fusion reactors. It does, however, involve the largest inventories of radioactive materials and hence poses the most demanding requirements for minimum radiological impact. In addition, questions have been raised as to the availability of the tritium in the quantities required to permit the operation of large-scale fusion reactors. Practical feasibility depends on four factors: (1) confinement of the high-temperature plasma in which the reaction takes place, (2) engineering of the complex materials and heat transfer systems required, (3) environmental safety, related primarily to the enormous tritium inventory and the intense fast neutron output, and (4) the economics of such reactors in competition with other energy sources (54-57).

In plasma fusion devices, the crucial problem is to maintain a mixture of deuterium and tritium at a temperature between 10^8 and 10^9 °K, equivalent to 10^4-10^5 eV, at a high enough number density ρ for a long enough time t to meet the Lawson criterion that ρt must exceed 3×10^{14} sec-cm^{-3}. Current research systems have achieved values of ρt some two orders of magnitude below the Lawson criterion and little engineering design work can be done until demonstration of the scientific feasibility of such a system has been accomplished.

Since hydrogen is the most abundant naturally occurring element, there is a clear incentive to the development of fusion devices particularly when considering the limited availability of high-grade uranium ores. Two types of systems are currently under active investigation. In one, a high-temperature plasma is contained by means of strong magnetic fields that constrain the ionized atoms in a large volume within a low-pressure enclosure. This enclosure is surrounded by a breeding "blanket" containing lithium, which serves both as a primary coolant and as a source of tritium regeneration, via the reactions ^{6}Li + n $\rightarrow$ ^{3}H + ^{4}He and ^{7}Li + n $\rightarrow$ ^{3}H + ^{4}He + n. The lithium in the blanket cooling system heats banks of potassium boiler tubes installed in the blanket, which are proposed to be made of niobium to minimize corrosion and radiation damage (57).

The second approach being investigated involves the heating of DT pellets by intense pulsed laser beams. This concept has not yet been developed sufficiently to outline specific engineered designs, but, if successful, it would avoid some of the complications introduced otherwise by the need for intense superconducting magnetic fields. Some of the design problems associated with laser-induced fusion have been described by Lubin (58) and Williams (54).

From the environmental point of view, attention is focused on the question of system reliability and on the radioactive inventory, some of which might conceivably enter the environment in case of an accident. Table 52 compares the hazard of tritium inventories in fusion reactors with that of postulated iodine releases from fission reactors (56,57). It

Table 52. Principal Radioactive Inventories of the Reference Fusion Reactor (RFR) and Advanced Fission Reactors (57)

Inventory	Activity, Ci/kW(t)	Maximum Permissible Airborne Concentration, $\mu Ci/cm^3$	Biological-Hazard Potential, Activity ÷ MPC, km^3 of air/kW(t)[d]
RFR			
^3H	60^a	2×10^{-7} (HTO)	0.30
Niobium as the Blanket Structure			
^{95}Nb	155	3×10^{-9}	52
Total niobium structure	714		240
Vanadium as the Blanket Structure			
^{48}Sc	4.20	5×10^{-9}	0.84
Total vanadium structure	55.1		0.86^b
Advanced Fission Reactors			
^{131}I (air)	31.6	1.0×10^{-10}	330
^{131}I (milk pathway)	31.6	1.4×10^{-13}	230,000
$^{239}Pu^c$	0.06	6×10^{-14}	1,000
Total plutonium isotopes	18.2		8,300

[a]The specific activity of tritium is approximately 10^4 Ci/g.

[b]Impurities within the vanadium might increase this number by a factor of 2.

[c]The plutonium inventory is characteristic of a liquid-metal-cooled fast breeder reactor.

[d]These figures assume complete loss of inventory to the environment and dispersion along indicated paths.

should be noted, however, that decontamination and retention of iodine is probably easier with current technology than for tritium; hence the actual iodine release levels would be many orders of magnitude lower than indicated. Since tritium will be contained both in the fusion vessel and in the breeder blanket, a variety of escape paths can be postulated. In addition, the intense neutron flux will produce activation products in the structural materials, blanket and coolant surrounding the fusion volume.

The chemical effects of the plasma interacting with the reactor surface, the diffusion of the hydrogen isotopes, particularly tritium, through blanket materials and the molten salt or metal coolant have been discussed by Gruen and Stickney (58), Hickman (59), and Hansborough (60). Ultimately, a certain fraction of the tritium will escape to the plant atmosphere in routine operations to be dealt with in the same way as in other reactors. Such leakage may be accentuated in molten salt coolants by their radiolytic dissociation and the effects of fluorine compounds on the pipe walls (61). On the other hand, the absence of long-lived fission products greatly simplifies the waste disposal problem. There is no fuel reprocessing step required as such, and the principal chemical processes are those associated with the recovery of tritium bred in the blanket and the removal of impurity activation products from the reactor coolant and the secondary heat transfer loop (58,62-64).

Draley and Greenberg (65) have reviewed some features of a conceptual fusion reactor plant. Assuming a concentration of 1 ppm of tritium in the molten lithium blanket, they calculated tritium permeation through stainless steel or other alloys by the equation

$$\text{Permeation rate} = 10^3 - \frac{3571}{T} \left(\frac{P_T}{3 \times 760} \right)^{1/2} \cdot \frac{A}{d} \text{ cm}^3/\text{hr} \qquad (42)$$

where: P_T = pressure, in torr
  A = area, in cm^2
  d = thickness, in mm

Table 53 lists some of the tritium release rates obtained. The tritium inventory in the lithium amounts to 460 g (4.4 MCi) and in the potassium secondary coolant to 0.06 g ($\sim$540 Ci). The rate of entry into the steam system is estimated as 22 mg/year or 210.8 Ci/yr. Most of the tritium losses to the environment would be by way of blowdown releases.

In regard to system reliability, the choice of containment materials is difficult and crucial. Such materials will be subjected to radiation damage from intense fast-neutron and gamma ray fluxes, hydrogen embrittlement, severe temperature gradients and corrosion from alkali metal coolants. Both refractory metals and ceramics have been proposed for this purpose and a great deal of research still needs to be done (66).

It is evident that these problems require a great deal of further study before a realistic engineering design can be attained and the reactor

Table 53. Predicted Tritium Release in Fusion Reactor Blanket (65)

Source	Temp.[a] °C	P_T, torr	Thickness mm	Area[b] cm^2	Release cm^3/hr	Ci/day
Potassium piping	788	6.3×10^{-8}	10.3	3.1×10^5	0.066	2.0
Potassium turbine	788	6.3×10^{-8}	10.3	3×10^3	6.6×10^{-4}	0.02
Steam generator						
Stainless steel	593	1.2×10^{-8}	2.4	3×10^7	2.16	66.7
Tungsten clad	593	1.2×10^{-8}	0.3(W)	3×10^7	0.0187	0.58
Potassium cold trap	532	7.8×10^{-9}	7.1	3.1×10^4	2.9×10^{-4}	0.003
Reactor cell	27	5×10^{-4} c	6.35	7.54×10^8	7.0×10^{-5}	0.002

aWhere a temperature gradient exists, the arithmetic average is used.

bAreas based on following assumptions: potassium piping, 100 ft of 12-in. pipe; potassium turbine, 1% of potassium piping; steam generator, Fraas design; potassium cold trap, 10% of potassium piping; cell cylinder, 25 m high by 60 m in diameter.

cIt is assumed that the cell is evacuated to a total pressure of 10^{-3} torr and that half of this is tritium.

concept can be reviewed for minimum environmental impact. The environmental assets of fusion include: volatile radioactivity is 10^4 to 10^6 lower than the fission breeder; maximum internal energy release can be contained; long-lived waste is 10^2 to 10^5 times lower than the breeder; there are no bomb-grade materials to steal, hence there is no need for elaborate safeguards; potential for waste heat reduction is good because of higher efficiency. The problems that fusion does not solve, or may create, are severe thermal pollution; demand for rare materials, which may impose a practical barrier to the large-scale introduction of fusion reactors; and the significant buildup of tritium.

DECOMMISSIONING

Due to technical obsolescence, deterioration of mechanical and electrical equipment, and gradual build-up of stray radioactive contaminants in the reactor system, the useful life of a nuclear steam supply system will be limited. At the time the reactor system is taken out of service a careful assessment must be made of the disposition of the reactor system and all its components. In general it may be assumed that there will be no further use for the reactor itself and its containment, and direct replacement of the reactor by a new one of different design will be impractical.

Up to this time only a small number of reactors have been decommissioned and hence in each case procedures had to be tailor-made. None of the reactors were of a size comparable with the current and projected power reactors. However, in every case the decontamination procedures developed served as a useful training ground in ensuring full containment of all radioactive material and minimum environmental disturbance (40,67). Among those reactors, the shut-down procedures for the Hallam reactor (68), Elk River reactor (69,70), BONUS (70), and MERLIN and JASON reactors (67) have been well-documented, as well as the more recent deactivation and dismantling of the SEFOR reactor.

In every case, the decommissioning procedure must meet four objectives: (1) thorough decontamination of site and equipment, (2) long-term security of the site to prevent unauthorized access and exposure hazards, (3) recovery of all usable components and equipment, and (4) minimum capital and surveillance expenses (71-73).

The dismantling program will typically comprise the following steps:

1. Removal of reactor fuel and all associate core structures.
2. Removal of reactor pressure vessel and internals, biological shielding, and all equipment and materials located within the containment building.
3. Removal and possible salvage of all external piping, electrical equipment and control room facilities.
4. Removal of all pipes, cables, switches and airlines if they could potentially contain reactor-originated radioactivity.
5. Weather-proof sealing and finishing of all openings in any remain - ing structures.
6. Filling of all cavities remaining after removal of structures and equipment with clean rubble topped with earth to approximate grade level.
7. Shipping for appropriate disposal of all items that contain detectable reactor-originated radioactivity.
8. Thorough radiation survey of the site prior to backfill, on completion of the work and periodically after completion, consistent with subsequent use of the site which should not necessitate any radiological limitations.

During dismantling, the immediate work area should be enclosed within a contamination control envelope (69). Prevention of excessive exposure of personnel to fixed radioactivity or minute dust-borne radioactivity must be accomplished by use of written procedures, mock-up testing and careful radiological control. Initially, the flow of airborne toxic and radioactive materials will be controlled primarily by the reactor building ventilation system. Localized exhaust and protective clothing must be provided where toxic vapors or radioactive contamination may be produced.

All liquid wastes and contaminated water generated during dismantling operations should be collected and decontaminated in the radwaste system before being released to the environment. The waste should be sampled and analyzed for gross beta and gamma activity to determine the need for additional purification. The usual precautions need to be adopted for shipping out consolidated solid and liquid low-level radioactive wastes.

The time required for the complete deactivation of a reactor site may well exceed one year. The cost of this operation can be very high. Manion (74) has estimated these costs for an 1100 MWe light-water reactor: for mothballing, $1.53M plus $155,000 per year for maintenance and surveillance; for complete dismantlement, $19.85M, in 1973 dollars. Allowing for 7% compound inflation rate, he estimated the cost 40 years hence as $299M. To meet this enormous cost the utility might establish a sinking fund or set aside annual installments out of earnings to cover the decommissioning commitment. For another plant the decommissioning costs without removal of major pieces of equipment, such as the nuclear steam supply system components or turbine building equipment, have been estimated as $3,240,000, based on 1972 costs. This work would involve removal of spent fuel elements, decontamination of auxiliary systems, disposal of resins, radioactive material and about three million gallons of contaminated water, sealing the containment, radiation surveys and surveillance.

With all these precautions there is no reason to assume that any significant adverse environmental impact would result during dismantling operations or that the site could not be converted to alternative uses without long-term radiological limitations.

REFERENCES

1. *Nuclear Power and the Environment.* (Vienna: International Atomic Energy Agency, 1973).
2. *Steam—Its Generation and Use*, 38th ed. (New York: Babcock and Wilcox, Co., 1972).
3. "Status and Future Technical and Economic Potential of Light Water Reactors," Report WASH-1082 (Washington, D.C.: U.S. Atomic Energy Commission, 1968).
4. Van Hollen, J. and J. A. Webb. "PWR Nuclear Power Plant Systems for Reducing Radioactive Releases," American Power Conference (1971).
5. Landis, J. W., E. W. O'Rorke, C. L. Rickard, P. Fortescue, A. J. Goodjohn, J. L. Everett, R. F. Walker and D. B. Trauger. "Gas-Cooled Reactor Development in the USA," *Proceedings 4th Geneva Conference on Peaceful Uses of Atomic Energy* 5, 345 (1972).

6. "Nuclear Power 1973-2000," Report WASH-1139 (Washington, D.C.: U.S Atomic Energy Commission, 1972).
7. Glasstone, S. and A. Sesonske. *Nuclear Reactor Engineering* (Princeton, N.J.: D. Van Nostrand Co., 1955).
8. Slade, D. H., Ed. *Meteorology and Atomic Energy—1968* (Washington, D.C.: U.S. Atomic Energy Commission, 1968).
9. "The Safety of Nuclear Power Reactors (Light-Water Cooled) and Related Facilities," Report WASH-1250 (Washington, D.C.: U.S. Atomic Energy Commission, 1973).
10. Hess, D. N. "Nuclear Power in Perspective: The Plight of the Benign Giant," *Nuclear Safety* 12, 283 (1971).
11. Salvatori, R. *The Environmental Impact of Nuclear Power Plants in the United States* (Pittsburgh: Westinghouse Electric Corp., 1973).
12. Smith, L. D. "Evolution of Opposition to the Peaceful Uses of Nuclear Energy," *Nucl. Eng. Internat.* 17, 461 (1972).
13. Starr, C., M. A. Greenfield, and D. F. Hausknecht. "A Comparison of Public Health Risks: Nuclear vs. Oil-Fired Power Plants," *Nuclear News* 15 (10), 37 (1972).
14. Ritzman, R. L., P. C. Owzarski, A. K. Postma, D. L. Lessor, D. L. Morrison, *et al.* "Release of Radioactivity in Reactor Accidents," Appendix VII, Report WASH-1400 (Washington, D.C.: U.S. Atomic Energy Commission, 1974).
15. Koffman, E. "Siting of Nuclear Plants near Population Centers," CONF-650201 (1965).
16. "Containment and Siting of Nuclear Power Plants," *Proceedings Vienna Symposium* (Vienna: International Atomic Energy Agency, 1967).
17. Leeper, C. K. "How Safe are Reactor Emergency Cooling Systems?" *Physics Today* 26(8), 30 (1973).
18. Sesonske, A. "Nuclear Power Plant Design Analysis," TID-26241 (Washington, D.C.: U.S. Atomic Energy Commission, 1973).
19. Coxon, T. "The Design of Nuclear Plant with Integral Coolant Circuits as Related to the Maximum Credible Accident Approach," *Proceedings 4th Geneva Conference on Peaceful Uses of Atomic Energy* 3, 227 (1972).
20. Lane, J. A. "Nuclear Safety Economics," *Nuclear Safety* 8, 316 (1967).
21. "Technical Basis for Interim Regional Tornado Criteria," Report WASH-1300 (Washington, D.C.: Supt. of Documents, 1973).
22. Jacobs, D. G. "Sources of Tritium and Its Behavior upon Release to the Environment," TID-24635 (Washington, D.C.: U.S. Atomic Energy Commission, 1968).
23. "Fission Product Release and Transport under Accident Conditions," *Proceedings Oak Ridge Symposium 1965*, U.S. AEC Report CONF-650407 (1965).
24. Cohen, P. *Water Coolant Technology of Power Reactors* (New York: Gordon & Breach, 1969).
25. Martucci, J. A. and B. J. Reckman. "Water Technology for PWR Nuclear Power Plants," *Proceedings 6th Meeting of Chemical Engineers for Electrical Power Generation*, Mexico, DF (1970).

26. Miller, D. A. and P. E. C. Bryant. "Corrosion and Coolant Chemistry Interactions in Pressurized Water Reactors," 25th American National Association of Corrosion Engineers Conference, Philadelphia, Pa. (1970).

27. Lechnick, W. "Tritium Concentration in PWR's Using Lithium Resin," Report WAPD-PWR-TE-103, Westinghouse Electric Corp. (1962).

28. Kouts, H. and J. Long. "Tritium Production in Nuclear Reactors," in *Tritium*, A. A. Moghissi and M. W. Carter, Eds. (Phoenix, Arizona: Messenger Graphics, 1973).

29. Denton, H. R. *Statement on the Sources of Radioactive Material in Effluents from Light-Water Cooled Nuclear Power Reactors and State of Technology of Waste Treatment Equipment to Minimize Releases* (Washington, D.C.: U.S. Atomic Energy Commission, 1972).

30. Hornyik, K. "Tritium Generation in the Coolant-Moderator of Pressurized Water Reactors," *Nucl. Sci. Eng.* **49**, 247 (1972).

31. Weaver, C. L., E. D. Harward, and H. T. Peterson. "Tritium in the Environment from Nuclear Power Plants," *Public Health Reports* **84**, 363 (1969).

32. "Environmental Analysis of the Uranium Fuel Cycle—Pt. II. Nuclear Power Reactors," Report EPA-520/9-73-003-C. (Washington, D.C.: U.S. Environmental Protection Agency, 1973).

33. Knapp, W. J. and P. E. Lutz. "Performance of Coolant Clean-up System in the Peach Bottom Reactor," *AIChE Chemical Engineering Progress Symposium Series* **66** (104), 41 (1970).

34. Newby, B. J. and K. L. Rohde. "Fission Product Sorption by Protective Coatings," in *Proceedings, Symposium Fission Product Release and Transport under Accident Conditions*, CONF-650407 (1965).

35. Keller, J. H., F. A. Duce, D. T. Pence and W. J. Maeck. "Iodine Chemistry in Steam Air Atmospheres," *Proceedings Health Physics Society, Symposium on Health Physics Aspects of Nuclear Facility Siting*, Idaho Falls, Idaho (1970).

36. Hilliard, R. K., A. K. Postma, J. D. McCormack and L. F. Coleman. "Removal of Iodine and Particles by Sprays in the Containment Systems Experiment," *Nucl. Technol.* **10**, 499 (1971).

37. Milioti, S. J., A. Sherman, R. L. Ritzman and J. A. Giesecke. "Analytical Studies of Elemental Iodine Removal by Sprays in the Donald C. Cook Nuclear Plant," *Nucl. Technol.* **16**, 497 (1972).

38. Parsly, L. F. "Pilot Plant Studies of Methyl Iodide Cleanup by Sprays," *Nucl. Appl. Technol.* **8**, 13 (1970).

39. Parsly, L. F. "Spray Program at the Nuclear Safety Pilot Plant," *Nucl. Technol.* **10**, 472 (1971).

40. Ayres, J. A., Ed. *Decontamination of Nuclear Reactors and Equipment* (New York: Ronald Press, 1970).

41. *Radiation Protection Procedures*, Safety Series No. 38. (Vienna: International Atomic Energy Agency, 1973).

42. Braun, H. and K. Traube. "Safety of Light-Water Reactors Designed in the Federal Republic of Germany," *Proceedings, 4th Geneva Conference* **3**, 51 (1972).

43. Hans, J. M. and T. C. Sell. "Evacuation Risks–An Evaluation," Report EPA-520/6-74-002 (Las Vegas, Nevada: U.S. Environmental Protection Agency, 1974).

44. "Fuel Storage Facility Design Basis," Safety Guide 13 (Washington, D.C.: U.S. Atomic Energy Commission, 1971).

45. Davis, D. E. and J. M. Krase. "Nuclear Components (of Fort St. Vrain Nuclear Power Station)," *Nuclear Eng. Internat.* **14**, 1085 (1969).

46. Cave, L. and R. E. Holmes. "Suitability of the Advanced Gas-Cooled Reactor for Urban Siting," in *Containment and Siting of Nuclear Power Plants* (Vienna: International Atomic Energy Agency, 1967).

47. Buhl, A. R. and J. Graham. "Safety and Environmental Impact of Liquid Metal Fast Breeder Reactors," Report ORBRP-PMC-76-01 (Oak Ridge, Tennessee: Breeder Reactor Corp., 1976).

48. Gibson, A. S. "The Fast Breeder Reactor," (San Jose, California: General Electric Co., 1970).

49. Heine, W. F. and R. S. Hart. "Health Physics Aspects of Fast Breeder Reactors," *Health Phys.* **21**, 449 (1971).

50. Seaborg, G. T. and J. T. Bloom. "Fast Breeder Reactors," *Scient. Amer.* **223** (5), 13 (1970).

51. Argous, J. P., H. Chantot, J. Petit and S. J. Stachura. "Safety Experience Gained from Rapsodie Operation," *Nucl. Safety* **14**, 27 (1973).

52. Gibson, A. S. "Ecological Considerations and the Fast Breeder Reactor," IEEE Region Six-Conference, Sacramento (1971).

53. Lovins, A. B. "The Case Against the Fast Breeder Reactor," *Bull. Atomic Scientists* **29** (3), 29 (1973).

54. Draper, E. L., Ed. "Technology of Controlled Thermonuclear Fusion Experiments and the Engineering Aspects of Fusion Reactors," *Proceedings 1972 Austin Symposium* (Washington, D.C.: U.S. Atomic Energy Commission, 1974).

55. Gough, W. C. and B. C. Eastlund. "The Prospects of Fusion Power," *Scient. Amer.* **224** (2), 50 (1971).

56. Postma, H. "Engineering and Environmental Aspects of Fusion Power Reactors," *Nuclear News* **14** (4), 57 (1971).

57. Steiner, D. and A. P. Fraas. "Preliminary Observations on the Radiological Implications of Fusion Power," *Nucl. Safety* **13**, 353 (1972).

58. Gruen, D. M., Ed. *The Chemistry of Fusion Technology* (New York: Plenum Press, 1972).

59. Hickman, R. G. "Tritium in Fusion Power Reactor Blankets," *Nucl. Technol.* **21**, 39 (1974).

60. Hansborough, L. D. "Tritium Inventories and Leakage: A Review and Some Additional Considerations," in *Technology of Controlled Thermonuclear Fusion Experiments and the Engineering Aspects of Fusion Reactors*, E. L. Draper, Ed. (Washington, D.C.: U.S. Atomic Energy Commission, 1974).

61. Cantor, S. and W. R. Grimes. "Fused-Salt Corrosion and its Control in Fusion Reactors," *Nucl. Technol.* **22**, 120 (1974).

62. Natesan, K. and D. L. Smith. "Effectiveness of Tritium Removal from a CTR Lithium Blanket by Cold Trapping Secondary Liquid Metals," *Nucl. Technol.* **22**, 138 (1974).
63. Strehlow, R. A. and H. C. Savage. "The Permeation of Hydrogen Isotopes Through Structural Metals at Low Temperatures and Through Metals with Oxide Film Barriers," *Nucl. Technol.* **22**, 127 (1974).
64. Vogelsang, W. F., G. L. Kulcinsky, R. G. Lott and T. Y. Sung. "Transmutations, Radioactivity and Afterheat in D-T Tokamak Fusion Reactor," *Nucl. Technol.* **22**, 379 (1974).
65. Draley, J. E. and S. Greenberg. "Some Features of the Environmental Impact of a Fusion Reactor Power Plant," in *Proceedings 1972 Austin Symposium,* E. L. Draper, Ed. (Washington, D.C.: U.S. Atomic Energy Commission, 1974).
66. Coffman, F. E. and T. M. Williams. "Environmental Aspects of Fusion Reactors," *Nucl. Technol.* **27**, 174 (1975).
67. Blythe, H. J., Ed. *Proceedings, 1st International Symposium on the Decontamination of Nuclear Installations* (Cambridge: Cambridge University Press, 1967).
68. Strauss, S. D. "Hallam: Death of a Concept," *Nucleonics* **24** (10), 63 (1966).
69. "Elk River Reactor Dismantling," Environmental Statement, U.S. Atomic Energy Commission Report WASH-1516, 1972, in *AEC Authorizing Legislation FY 1973*, Hearings, Joint Committee on Atomic Energy, 92nd U.S. Congress, Part 5: vol. 2 (1972).
70. Giambusso, A., B. Ureda, W. Willoughby, H. T. Babb, M. Iriarte, J. Hernandez-Fragoso and M. N. Bjeldanes. "Four Decommissioning Case Histories—Hallam, CVTR, Bonus and Pathfinder," *Nuclear News* **13** (6), 39 (1970).
71. European Nuclear Energy Agency. "Radioactive Waste Management Practices in Western Europe," (Paris: Organization for Economic Corporation and Development, 1971).
72. Pittman, F. K. "Plan for the Management of AEC-Generated Radioactive Wastes," Report WASH-1202 (73), U.S. Atomic Energy Commission (1973).
73. "Termination of Operating Licenses of Nuclear Reactors," Regulatory Guide 10.XX (Draft 3) (Washington, D.C.: U.S. Atomic Energy Commission, 1974).
74. Manion, W. J. Quoted in *Nucleonics Week* (September 27, 1973).

ADDITIONAL REFERENCES

Burchsted, C. A. and A. B. Fuller. "Design, Construction and Testing of High-Efficiency Air Filtration Systems for Nuclear Application," Report ORNL-NSIC-65, Oak Ridge National Laboratory (1970).
Couez, H. M. and L. F. Picone. *Decontamination of a PWR Primary System: SENA Plant* (Pittsburgh, Pa.: Westinghouse Electric Corp., 1971).

Eisenbud, M. *Environmental Radioactivity*, 2nd ed. (New York: McGraw-Hill, 1973).

"Environmental Aspects of Nuclear Power Stations," *Proceedings New York Symposium* (Vienna: International Atomic Energy Agency, 1971).

"Environmental Effects of Producing Electric Power," Hearings before the Joint Committee on Atomic Energy, 91st Congress of the United States (Washington, D.C.: U.S. Government Printing Office, 1969).

Etherington, H., Ed. *Nuclear Engineering Handbook* (New York: McGraw-Hill, 1958).

Gitterman, M. "Helium Purification for High-Temperature Gas-Cooled Reactors," *AIChE Chemical Engineering Progress Symposium Series* 66 (104), 31 (1970).

Hart, E. J., W. R. McDonald, and S. Gordon. "The Decomposition of Light and Heavy Water Boric Acid Solutions by Nuclear Reactor Radiations," *Proceedings U.S. Conference on Peaceful Uses of Atomic Energy, Geneva 1955* 7, 593 (1956).

Hart, R. G., L. R. Haywood and G. A. Pon. "The CANDU Nuclear Power System: Competitive for the Foreseeable Future," *Proceedings 4th Geneva Conference on Peaceful Uses of Atomic Energy* 5, 239 (1972).

Hickman, R. G. "Some Problems Associated with Tritium in Fusion Reactors," in *Proceedings 1972 Austin Symposium*, E. L. Draper, Ed. (Washington, D.C.: U.S. Atomic Energy Commission, 1974).

Hull, A. P. "Comparing Effluent Releases from Nuclear and Fossil-Fueled Power Plants," *Nuclear News* 17, 51 (1974).

Keilholtz, G. W., C. E. Guthrie and G. C. Battle. "Air Cleaning as an Engineered Safety Feature in Light-Water-Cooled Power Reactors," Report ORNL-NSIC-25, Oak Ridge National Laboratory (1968).

Kent, C. E., S. Levy, and J. M. Smith. "Effluent Control for Boiling Water Reactors," in *Environmental Aspects of Nuclear Power Stations* (Vienna: International Atomic Energy Agency, 1971).

Mackintosh, W. D., J. A. Atherly and P. J. Dyne. "Radiolytic Gas Production in Liquid Control Elements for Nuclear Reactors Using Boron as the Absorber," AECL Report AECL-2670, Chalk River (1967).

Martin, L. E., E. D. Harward, D. T. Oakley, J. M. Smith and P. H. Bedrosian. "Radioactivity from Fossil-Fuel and Nuclear Power Plants," in *Environmental Aspects of Nuclear Power Stations* (Vienna: International Atomic Energy Agency, 1971).

"The Nuclear Industry 1974," Report WASH-1174-74 (Washington, D.C.: U.S. Government Printing Office, 1974).

Schikarski, W., P. Jansen and S. Jordan. "An Approach to Comparing Air Pollution from Fossil-Fuel and Nuclear Power Plants," in *Environmental Aspects of Nuclear Power Stations* (Vienna: International Atomic Energy Agency, 1971).

Wright, J. H. "Environmental Aspects of the Clinch River Breeder Reactor Project," *Nuclear Eng. Internat.* 19, 221 (1974).

CHAPTER 6

WASTE HEAT DISSIPATION

INTRODUCTION

The energy released in the process of nuclear fission is converted to heat within the reactor core by the slowing down of the fission products and must be extracted by means of a suitable fluid that either directly or indirectly converts water to steam, which powers a turbine to generate electricity. At the exhaust of the turbine the steam is condensed to water to maximize the energy conversion and then is returned to the boiler or reactor to repeat the cycle. The thermodynamic efficiency of the system depends on the temperature difference between the coolant inlet and outlet temperatures.

A large amount of heat is rejected in the condensing process, and even in the most efficient plants the rejected heat is substantially greater than the heat equivalent of the electric energy generated. As Table 37 showed, the thermal efficiency of light-water reactors at present is approximately 33%, that of gas-cooled reactors about 39-41%. This means that almost two-thirds of the heat energy generated in a reactor core has to be rejected to the environment. Although there have been considerable improvements in the thermal efficiency of fossil-fueled plants over the past four decades, the most efficient fossil-fueled plants at present operate at about 40% efficiency, although the national average for the United States is about 32.8%, requiring transfer of waste heat to the cooling water in the amount of 5400 Btu/kWh (1). The thermal efficiencies quoted include an allowance for in-plant and stack losses. Large requirements for power to run the plant and its associated cooling facilities would further reduce its overall efficiency (2); this may amount to 4% of the power generated for an LWR, and more for a coal-fired station.

Research is under way to explore direct conversion of heat from gas-cooled reactors to electricity by means of gas turbines and the use of air-cooled condensers to improve plant efficiency. All waste heat from the electric power plant must eventually be dissipated to the environment. In practice, most of it ultimately is transferred to the atmosphere, which acts as a major heat sink. Some heat is transferred directly to the ambient air and, in the case of fossil-fueled plants, some is discharged up the stacks. The bulk of the waste heat, however, is transferred from the steam to the cooling water in the condensers. Water is commonly used as the absorbent because of its general abundance, low cost, high specific heat, and ability to dissipate heat in the evaporation process. Figure 44 shows diagrammatically the cooling scheme for a power plant.

NUCLEAR PWR

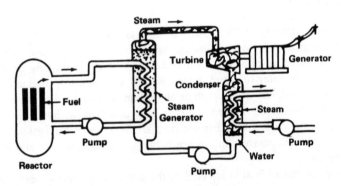

Figure 44. Cooling water flow in a nuclear power plant.

The real problem in accommodating this waste heat in the environment arises from the ever-increasing size and concentration of power plants, which strain available water resources and dump large quantities of heat into limited volumes of water, giving rise to ever greater temperature levels. Table 54 shows some estimated water requirements for a 1000MW plant; Table 55 presents some projections for total water needs for cooling purposes for just one region the northeastern United States (3). As the plant size increases, growing difficulties arise in dissipating its waste heat economically and in a manner least injurious to the local ecology. Additional problems are created by the water losses due to evaporation, reducing the multiple use of all available water along most major river systems.

Environmentally, the effects of raising water temperatures on the ecology of the receiving stream has attracted the greatest attention; however, it

Table 54. Estimated Cooling Water Requirements for 1000 MWe
Steam-Electric Plant Operating at Full Load (4)

	Type of Plant			
	Fossil		Nuclear	
	(1980)	(1990)	(1980)	(1990)
Plant heat rate,[a] Btu/kWh	9,500	9,000	10,500	10,000
Condenser flows—cfs, for various temperature rises across the condenser				
10°F	2,080	1,890	2,920	2,710
15°F	1,390	1,260	1,950	1,810
20°F	1,040	950	1,460	1,360
Consumptive Use, cfs for various types of cooling				
Once through	12	10	17	15
Cooling ponds[b]	16	14	22	20
Cooling towers[c]	28	26	40	35

[a]For fossil-fueled plants in operation in 1970, a heat rate of 10,000 Btu per kWh and a temperature rise of 13°F were assumed, except where reported heat rate data were available.

[b]Where appropriate, an additional allowance was made for natural evaporation from the pond surface.

[c]Evaporative towers; includes blowdown and drift.

also becomes increasingly evident that significant climatic effects result from adding vast amounts of heat to the atmosphere, as well as from injecting substantial amounts of moisture by evaporation from cooling ponds or in the plume of cooling towers (5). For comparison, Table 56 shows the heat input to the environment by common people-related activities (3); Table 57 shows average power densities for global and industrialized areas (6). Jaske et al. (3) have estimated that projections for energy consumption, such as those in Figure 3, lead to concomitant heat rejection rates from all forms of energy use that would range from 15-50% of the incident solar energy for the Boston-Washington urban conglomeration by the year 2000. Clearly any such large-scale addition to the atmospheric heat budget over such a large geographical area is bound to have severe climatic consequences. To this must be added effects of industrial pollution, giving rise to the counter-effective phenomena of solar heat filtering by airborne particulate clouds and the "greenhouse effect" associated with rising CO_2 levels (7). Thus for industrialized nations the problem of heat release to the atmosphere is closely linked to all other aspects of pollution control and maintenance of a liveable environment.

Table 55. Projected Cooling Water Requirements[a] for the
Northeastern United States by Decades (3)

	1970	1980	1990	2000
Ohio River Valley				
Total Expected Generation	18.3 GWe	36.5 GWe	71.0 GWe	99.0 GWe
Direct Cooling Flow	15,100 cfs	30,100 cfs	59,000 cfs	83,000 cfs
Consumptive Loss with Cooling Towers	550 cfs	1,046 cfs	2,130 cfs	2,960 cfs
Consumptive Loss with Direct Cooling	220 cfs	420 cfs	850 cfs	1,190 cfs
Northeastern States				
Total Expected Generation	33 GWe	56 GWe	117 GWe	190 GWe
Direct Cooling Flow	27,600 cfs	46,500 cfs	96,000 cfs	158,000 cfs
Consumptive Loss with Cooling Towers	990 cfs	1,680 cfs	3,150 cfs	5,700 cfs
Consumptive Loss with Direct Cooling	395 cfs	670 cfs	1,400 cfs	2,270 cfs
Upper Mississippi River Basin				
Total Expected Generation	16.7 GWe	33.1 GWe	67.2 GWe	96 GWe
Direct Cooling Flow - cfs[a]	13,800 cfs	27,600 cfs	56,000 cfs	80,000 cfs
Consumptive Loss with Cooling Towers[b]	500 cfs	1,000 cfs	2,000 cfs	3,000 cfs
Consumptive Loss with Direct Cooling	200 cfs	400 cfs	800 cfs	1,200 cfs

[a]Direct cooling figured at 18°F rise
[b]Consumptive use estimated at 30 cfs/GWe

Although the above considerations raise a number of questions of long-term importance affecting major parts of the globe, attention at present is focused on the more local and readily identifiable consequences of heat release to impounded or flowing bodies of water. Such effects can be classified in terms of impact on water quality, aquatic life and water utilization. Whether such effects are considered harmful or beneficial may be debatable in each case, but for the present we will assume that any major change in preexisting conditions in any of these categories would be regarded as objectionable.

Table 56. Heat Rejection Rates of Common Activities, Btu/hr (3)

One human being (at rest)	450
Electric lighting for a typical office	
Incandescent	2500
Fluorescent	900
Theoretical, 100% efficient	82
Air-conditioning for typical office	600
Heating to maintain 25°F ΔT for typical office	
Electrical resistance[a]	1000
Heat pump[a]	350
Furnace, oil or gas	1600
4000 lb automobile at 70 mph	
Internal combustion	750,000
Electric[a]	150,000

[a]Does not include generation or transmission losses.

Table 57. Representative Natural and Man-Made Power Densities (6)

Human Consumption	Present	Projected
Global average	0.033 W/m^2	1.35 W/m^2
F. R. Germany	1 W/m^2	5 W/m^2
Industrial area (Ruhr)	17 W/m^2	1000 W/m^2
Large nuclear park	–	20000 W/m^2
Heat balance on the surface of earth (average)		100 W/m^2
Latent heat density of rainfall on the continents		55 W/m^2
Sensible heat density for 1°C of rainfall water on the continents		0.1 W/m^2
Winds, waves, convections and currents (All globe)		0.7 W/m^2
Photosynthesis		0.075 W/m^2

ECOLOGICAL EFFECTS OF THERMAL DISCHARGES

Water Quality

Nearly every physical property of water is affected by temperature changes (Table 58). The viscosity decreases with temperature, and since the velocity of settling particles in a nonturbulent medium is described by Stokes' law:

$$v = \frac{d^2 g}{18\eta} (\rho_s - \rho_w) \tag{43}$$

where: v = settling velocity, cm/sec
η = viscosity of liquid, poises
d = particle diameter, cm
g = acceleration of gravity, 981 cm/sec^2
ρ_s ≡ density of settling particle
ρ_w = density of water, g/cc.

both η and ρ_w contribute to increased settling rates at elevated temperatures. This may have a significant effect on the location and amount of sediment and on sludge deposition in reservoirs, slow-moving rivers and estuaries.

Very slight differences in density are sufficient to cause stratification in quiescent bodies of water, but stratification stability also depends on depth and water movement. Stable stratification is common in lakes and

Table 58. Water Properties as a Function of Temperature

Temperature °C	Vapor Pressure torr	Viscosity centipoise	Density g/ml	Surface Tension dynes/cm	Oxygen Solubility mg/l	Oxygen Diffusivity ft^2/hr	Nitrogen Solubility mg/l
0	4.579	1.787	0.99984	75.6	14.6		23.1
5	6.543	1.519	0.99997	74.9	12.8		20.4
10	9.209	1.307	0.99970	74.2	11.3	61	18.1
15	12.788	1.139	0.99910	73.5	10.2	71	16.3
20	17.535	1.002	0.99820	72.8	9.2	81	14.9
25	23.756	0.890	0.99704	72.0	8.4	92	13.7
30	31.824	0.798	0.99565	71.2	7.6	106	12.7
35	42.175	0.719	0.99406	70.4	7.1		11.6
40	55.324	0.653	0.99224	69.6	6.6		10.8

reservoirs and may inhibit vertical mixing and oxygen transfer to lower waters; it may also lead to cool-water currents traveling at depth through a reservoir without mixing with overlying layers.

Evaporation increases as rising water temperatures raise the vapor pressure. The evaporation rate can be expressed as:

$$F = \frac{CW}{L} (e_s - e_a) \tag{44}$$

where: F = evaporation flux, lb/ft²-day
L = latent heat of vaporization = 1050 Btu/lb
W = wind speed, miles/hr
e_s = water vapor pressure of saturated air at the surface temperature, torr
e_a = water vapor pressure of overlying air, torr
C = empirical evaporation coefficient, ~11.4

The effect of changing water temperature on evaporation rates can be shown by holding all terms constant except e_s. For example, for a wind speed of 10 miles/hr, and a value of e_a = 7.5 torr (appropriate for an air temperature of 70°F with a relative humidity of 40%), and the above values for C and L, the following values are obtained [1]:

Water Temperature (°F)	e_s (torr)	F (lb-ft⁻² day⁻¹)
50	9.2	0.19
60	13.3	0.64
70	18.8	1.24
80	26.2	2.06
90	36.1	3.15

Similar relative changes in F with respect to water temperature will occur for other values of the independent variables.

Both the evaporation rate and the heat balance at a water surface depend on a large number of interdependent factors, as illustrated in Figure 45. Basically the net rate of heat transfer at an air-water interface is proportional to the water surface area, the difference between the temperature of water surface, T_s, and the equilibrium temperature, T_e:

$$\text{rate of heat dissipation} = -K\,A\,(T_s - T_e) \text{ (Btu/day)} \tag{45}$$

where the surface heat exchange coefficient, K, ranges typically from 100 to 200 Btu/ft²-day-°F, and A = water surface area in square feet. Figure 46 shows a chart of values of K as a function of wind speed and surface

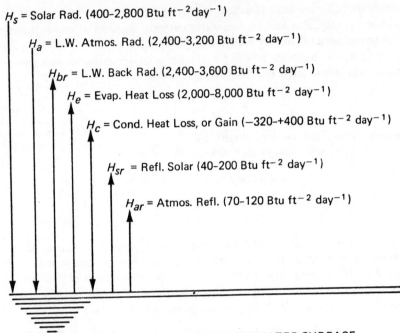

H_s = Solar Rad. (400–2,800 Btu ft^{-2}day^{-1})

H_a = L.W. Atmos. Rad. (2,400–3,200 Btu ft^{-2} day^{-1})

H_{br} = L.W. Back Rad. (2,400–3,600 Btu ft^{-2} day^{-1})

H_e = Evap. Heat Loss (2,000–8,000 Btu ft^{-2} day^{-1})

H_c = Cond. Heat Loss, or Gain (−320–+400 Btu ft^{-2} day^{-1})

H_{sr} = Refl. Solar (40–200 Btu ft^{-2} day^{-1})

H_{ar} = Atmos. Refl. (70–120 Btu ft^{-2} day^{-1})

NET RATE AT WHICH HEAT CROSSES WATER SURFACE

$$\Delta H = (H_s + H_a - H_{sr} - H_{ar}) - (H_{br} \pm H_c + H_e) \text{ Btu ft}^{-2} \text{ day}^{-1}$$

H_R Temp. Dependent Terms

Absorbed Radiation
Independent of Temp.

$H_{br} \sim (T_s + 460)^4$
$H_c \sim (T_s - T_a)$
$H_e \sim W(e_s - e_a)$

Figure 45. Mechanisms of heat transfer across a water surface (8).

temperature as prepared by Brady, Graves and Geyer (9). The equilibrium temperature T_e can be calculated approximately (9) from Equation 46:

$$T_e = T_d + \frac{H_s}{K} \tag{46}$$

where H_s is the mean daily solar radiation, and the T_d is the dewpoint temperature at which air would become saturated with water vapor when cooled at constant pressure. The equilibrium temperature can be measured directly by means of an electronic instrument or calculated from wet and dry bulb air temperatures. For the northeastern United States, the mean

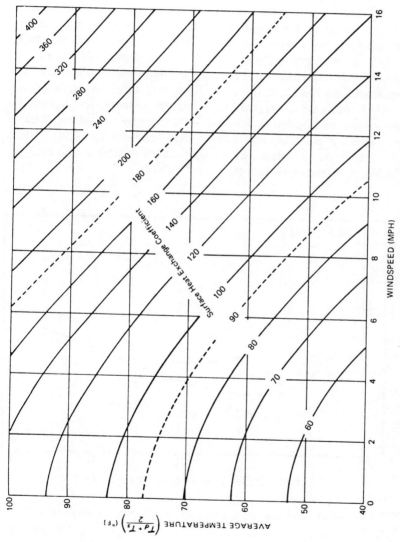

Figure 46. Design chart for surface heat exchange coefficient (8).

daily insolation in July is approximately 2000 Btu/ft² day, and T_d = 60°F. Thus if K = 150 Btu/ft²-day-°F, T_e = 73°F.

Of the parameters listed in Table 58, the solubility of oxygen is probably the most important since dissolved oxygen in water is necessary to sustain many forms of aquatic life. The lower solubility induced by higher temperatures, if combined with an organic load and an increased bacterial respiration rate, could lead to such low oxygen concentrations that fish could not survive. In addition the reaeration coefficient, which is important in determining the waste assimilative capacity of streams, increases with temperature. This change in stream waste assimilation has been studied by Krenkel, Cawley and Minch (10) and is shown in Figure 47, which shows the deoxygenation and oxygen recovery, for different temperatures, under selected flow and waste discharge conditions. Free flow conditions promote faster recovery than impoundment (1,11). Under severe conditions this effect may be counteracted to some extent by deliberate injection of air into the heated effluent.

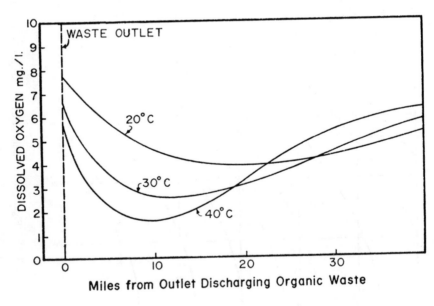

Figure 47. Relation between temperature and oxygen profile (1).

The addition of heat to a body of water may increase rates of chemical solubility and biochemical reactions, causing effects on aquatic organisms in regions of higher temperature (12). This may be significant in view of the wide range of plant chemicals, such as detergents, algicides, wood preservatives and corrosion inhibitors, that may be contained in the plant effluents, as shown in Tables 59 and 79 (see Chapter 7), which are representative of many power plants. Thus, the addition of heat to a water

Table 59. Chemical Wastes from Oconee Nuclear Station

	Pounds per Year (3 units)	Concentration in Water Released to Environment (ppm)	
		Average[a]	Maximum Possible[b]
Reactor coolant			
Boric acid	60,000 (startup)	2.4×10^{-5} [c]	8.8×10^{-4}
Lithium hydroxide	180	d	d
Hydrazine	300	e	e
Steam generator feedwater			
Hydrazine	1,800-13,000	d	d
Regeneration of deborating demineralizers			
Sodium hydroxide	4,100	f	f
Regeneration of water treatment demineralizers			
Sodium hydroxide	440,000	0.12 [g]	4.4 [g]
Sulfuric acid	150,500	0.068 [h]	2.5 [h]
Laundry and cleaning detergents			
Floor cleaning (liquid)	10,000	0.0046 [i]	0.17 [i]
Laundry (solid)	4,760	0.0022 [i]	0.081 [i]

[a] Total per year diluted by average tailrace flow of 1100 cfs (9.823×10^{14} cm^3/year).
[b] Total per year diluted by minimum tailrace flow of 30 cfs.
[c] 53 lb/year from evaporator overheads.
[d] Most will probably be removed by the demineralizers and evaporators.
[e] Normally, hydrazine is reacted chemically and is not discharged.
[f] Most of this material will be sent to the waste drumming facility as evaporator bottoms.
[g] Sodium released; Keowee River normal concentration is 1.2-2.8 ppm.
[h] Sulfate released; Keowee River normal concentration is 0.7-2.5 ppm.
[i] Processing of these wastes through the sanitary waste system may significantly reduce this value.

body can alter the water quality indirectly and heat may be regarded as a potential pollution agent. On the other hand, higher water temperatures might reduce the amount of chemicals required for the treatment of public water supplies. It has been estimated that the resulting savings would range from 30 to 50 cents per million gallons of water treated for each 10 degrees Fahrenheit rise in temperature. However, increases in summer water temperatures make drinking water less palatable and, by increasing the proportion of bluegreen algae in the algal population, may produce objectionable tastes and odors in the water supply.

Effects on Aquatic Life

Water provides the environment for many species of organisms and changes in its temperature, chemical content and flow rate may affect the number and kinds of such organisms in a given body of water. Although studies have been conducted on a variety of organisms, predictions as to the effects of temperature changes and maximum temperatures still cannot be made with certainty, especially if the changes are gradual, occur for only a short time, or affect limited areas or volumes of large bodies of water. Of particular importance is the ability of ecosystems to adapt to or to recover from such changes, and it is important to realize that many of the effects observed result from subtle changes in behavior or conditions affecting various links along an often very complex ecological chain.

Bacteria

It should be noted that a distinct difference exists between the ability of a microorganisms or a higher organism to endure a given temperature and its ability to grow well under the same conditions. Many bacteria, such as *E. coli*, the prime indicator of fecal pollution, obtain optimum growth conditions at increased temperatures. The rate at which biological oxidation takes place, usually expressed as biochemical oxygen demand (BOD), increases with temperature, reaches a maximum around $30^{\circ}C$ ($86^{\circ}F$) and then decreases.

Biodegradable organic material entering water exerts a biochemical oxygen demand that must be satisfied before assimilation of the material is completed. When the temperature of such a receiving water rises, the intensified action of microorganisms causes the BOD to be satisfied in a shorter distance from the discharge than would be accomplished at a lower temperature. It is apparent from the curves in Figure 47 that the deoxygenation effect caused by waste assimilation is exerted over a shorter stream distance at higher temperatures and that oxygen depletion occurs to a greater extent, since the sag point is lower at elevated temperature. Hence, it is possible that the discharge of an organic waste that previously had not caused excessive oxygen depletion could pose problems at an elevated temperature (1).

Since bacteria have short life cycles, many generations may occur within a relatively short time. If a gradual temperature increase occurs over several life cycles, each successive generation is subjected to only a small portion of this total temperature increase. Thus, bacteria can adapt to slow temperature changes more easily than higher forms such as fish.

The effect of temperature on bacteria cannot always be considered separately from other environmental factors. Some species are more abundant in winter, while others abound in the summer when different environmental conditions are encountered. Rising stream temperatures can be favorable for those bacteria that multiply in water by inducing the recurring cycles of life and death more rapidly. Higher temperatures in an organically polluted stream generally result in increased bacterial numbers, and low temperatures are not conducive to rapid growth. Temperatures of 1-8°C (33.8 to 46.4°F) may suppress growth and multiplication, but bacteria persist longer at these cool temperatures.

Increases in bacterial populations are not necessarily harmful. For example, those bacteria that play an active role in stream self-purification do perform a useful function. These include the bacteria that aerobically oxidize organic material, as well as those responsible for nitrification and anaerobic decomposition of bottom sediments. However, increases in pathogenic bacteria should always be avoided (1). A recent study on the survival of bacterial indicators of pollution showed that the lower the temperature, the longer the survival.

In summary, the temperature of natural waters, even during the summer, is usually below the optimum for pollution-associated bacteria. Increasing the water temperature increases the bacterial multiplication rate when the environment is favorable and the food supply is abundant. Increasing temperature within the growth range causes a more rapid die-off when the food supply is limited.

Algae and Other Aquatic Plants

Successive changes from continual temperature conditions may result in the elimination of desirable species along with the establishment of undesirable or nuisance organisms. Some species of algae and other aquatic plants are nuisances because they are unsightly or produce undesirable tastes and odors. In addition, the biological oxygen demand of the dead cells depletes dissolved oxygen resources. Blue-green algae are especially undesirable from the standpoint of taste and odor problems in municipal water supplies.

There appear to be particular temperature ranges that are tolerated by each algal species and by closely related species or groups of species.

For example, in an unpolluted stream, diatoms grow best at 18-20°C (64.4-68°F); green algae at 30-35°C (86-95°F); and blue-green algae at 35-40°C (95-104°F). If temperatures are increased from 10 to 38°C (50 to 100.4°F), the predominant species groups change correspondingly from diatoms to green algae to blue-green algae (Figure 48). Some of the more high-temperature-tolerant species belonging to algal groups other than the blue-green may persist with the predominant blue-greens at increased temperatures, and several less tolerant species of the blue-green algae may succumb with the diatoms and green algae as the temperature rises (1). These populations also vary seasonally as is shown rather dramatically in Figure 49, reflecting observations on the Hudson River (14).

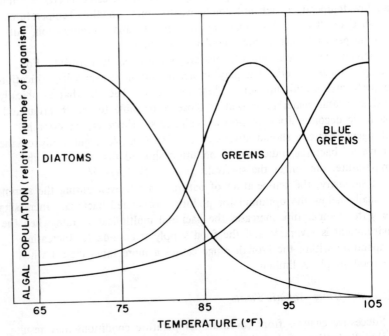

Figure 48. Algal population shifts with temperature (13).

Fish Life

Temperature changes normally play an important and highly regulatory role in the growth of aquatic plants and in the growth and physiology of fish and other cold-blooded aquatic animals. Reproductive cycles, digestive rates, respiration rates, and other processes occurring in the bodies of aquatic animals are often temperature-dependent. These effects are

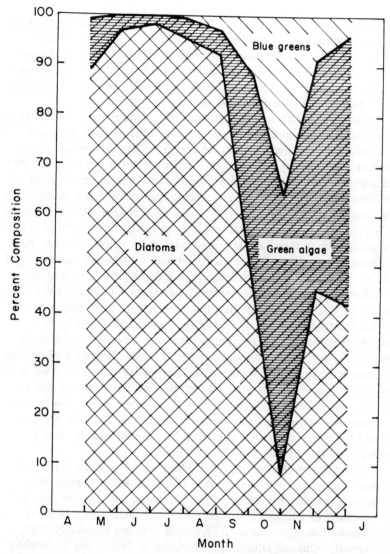

Figure 49. Relative proportions of diatoms, green, and blue-green algae in the standing crop at Indian Point, 1970 (14,15).

not consistent among species, however, so thermal constraints are among the most difficult to define and establish. It is known that temperatures higher than those normally experienced, particularly during summer months, can be detrimental in a variety of ways. The survival of individual organisms can be jeopardized, they may be more susceptible to disease or to the effects of toxic agents, their food supply or their ability to catch food may diminish, and the inability to reproduce or to compete successfully with other organisms may indirectly eliminate a species. There may also be unpredictable synergistic effects, related to dissolved heavy metals, for example. The elimination of one species in the food chain may change the ecological balance and cause significant changes in the composition of the plant and animal life that remains.

All aquatic species have an optimum temperature range. If the water temperature varies above or below this range, the chances of survival for a particular species decrease. Rapid change in temperature caused by thermal plant start-up and shut-down can be lethal to organisms in the affected area, although adult fish generally have the ability to avoid undesirable temperatures. For some aquatic species the effect produced by the warming of initially cold water could be beneficial. But the chances of survival of this same species would be progressively diminished if the temperature is further increased, especially if the change is rapid. Since different species favor different temperatures, a warming trend may lead to the population decline of one species and the growth of another.

Table 60 (1) illustrates the temperature ranges for survival for a few species of fish. It is evident that for many species these ranges can be shifted substantially if the fish have been transferred gradually to a warmer environment or have been acclimated. Since survival depends also on the availability of accustomed food, many fish will tend to seek or avoid warmer regions depending on the level of nutrients. Fish in the estuarine environment are more susceptible than those in fresh water. Acclimation to decreasing temperatures by estuarine fish is slower than acclimation to increasing temperatures; this fact implies that the shut-down of a power generating plant may be more detrimental than its normal discharge of heat (16,17).

The incubation of eggs and development of fry generally have more critical temperature requirements than either fingerling or adult fish. Table 61 (1) shows the minimum and maximum temperatures reported for the successful hatching of a few species of marine, estuarine and anadromous fish eggs. Successful hatching does not necessarily imply fry survival; Chinook salmon eggs incubated at 16.1°C (61°F) hatched successfully, but suffered severe mortality in the late fry stage (1).

Studies indicate that oxygen consumption of aquatic vertebrates increases with rises in temperature up to a limiting temperature beyond which the physical exchange of oxygen in the blood is no longer possible. This increased need for oxygen is coupled with the decreased ability of the water to hold oxygen at higher temperatures. Changes in temperature can cause certain gases dissolved in the water to change their selective toxicity toward fish. Low concentrations of carbon dioxide can be lethal

Table 60. Lethal Temperature Limits for Adult Marine, Estuarine, and Freshwater Fishes (1).

Species	Acclimation Temperature °C	Acclimation Temperature °F	Lower Lethal Temperature[a] °C	Lower Lethal Temperature[a] °F	Upper Lethal Temperature[a] °C	Upper Lethal Temperature[a] °F
Alewife	–	–	–	–	26.7-32.2	80.0-90.0
Bass, striped	–	–	6.0-7.5	42.8-45.5	25.0-27.0	77.0-80.0
California killifish	14.0-28.0	57.2-82.4	–	–	32.3-36.5	90.1-97.7
Common silverside	7.0-28.0	44.6-82.4	1.5-8.7	34.8-47.8	22.5-32.5	73.3-90.3
Flounder, winter	21.0-28.0	69.8-82.4	1.0-5.4	33.8-41.6	–	–
Flounder, winter	7.0-28.0	44.6-82.4	–	–	22.0-29.0	71.6-84.2
Herring	–	–	-1.0	30.2	19.5-21.2	67.1-70.1
Northern swellfish	14.0-28.0	57.2-82.4	8.4-13.0	47.1-55.4	–	–
Northern swellfish	10.0-28.0	50.0-82.4	–	–	28.2-33.0	82.9-90.4
Perch, white	4.4	40.0	–	–	27.8	82.0
Salmon (general)	–	–	0.0	32.0	26.7	80.0
Bass, largemouth	20.0	68.0	5.0	41.0	32.0	89.6
	30.0	86.0	11.0	51.8	34.0	93.2
Bluegill	15.0	59.0	3.0	37.4	31.0	87.8
	30.0	86.0	11.0	51.8	34.0	93.2
Catfish, channel	15.0	59.0	0.0	32.0	30.0	86.0
	25.0	77.0	6.0	42.8	34.0	93.2
Perch, yellow	5.0	41.0	–	–	21.0	69.8
(winter)	25.0	77.0	4.0	39.2	30.0	86.0
(summer)	25.0	77.0	9.0	48.2	32.0	89.6
Shad, gizzard	25.0	77.0	11.0	51.8	34.0	93.2
	35.0	95.0	20.0	68.0	37.0	98.6
Shiner, common	5.0	41.0	–	–	27.0	80.6
	25.0	77.0	4.0	39.2	31.0	87.8
	30.0	86.0	8.0	46.4	31.0	87.8
Trout, brook	3.0	37.4	–	–	23.0	73.4
	20.0	68.0	–	–	25.0	77.0

[a]Values are LD_{50} temperature tolerance limits, *i.e.*, water temperatures survived by 50% of the test fish after 1-4 days' acclimation.

above certain temperatures. An increase in the temperature of water saturated with nitrogen may result in the water becoming supersaturated. As supersaturation increases, the nitrogen dissolved in the water changes more easily to the gaseous state and forms bubbles. Such conditions can be lethal to salmonoid and other fish (4). Figure 50 illustrates some of the complex interactions that may cause fish mortality (13).

Table 61. Temperature Ranges for Successful Egg Hatching of Various Marine, Estuarine, and Anadromous Fishes (1)

Species	Lower Limit		Upper Limit	
	°C	°F	°C	°F
Bass, striped	12.8	55.0	23.9	75.0
California killifish	16.6	61.9	28.5	83.1
California grunion	14.8	58.6	26.8	80.1
Salmon, Chinook[a]	5.8	42.4	14.2	57.6
	9.4	48.9	14.4	57.9
	5.6	42.1	14.4	57.9
Salmon, sockeye	4.4-5.8	39.9-42.4	12.8-14.2	55.0-57.6
Sea lamprey[a]	15.0	59.0	25.0	77.0
	15.6	60.1	21.1	70.0

[a]Different ranges reflect reports by different observers.

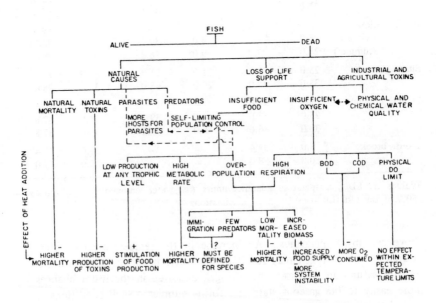

Figure 50. Fault analysis for fish mortality (13).

When properly controlled thermal discharges have resulted in an increase in the ability of certain commercially valuable aquatic species to multiply, while at the same time decreasing the time for the species to reach maturity. Experience has shown that at a number of plant locations the discharge of waste heat to a stream or reservoir has improved fishing in the vicinity of the discharge during the cooler months of the year. This increase in fish population in heated discharge regions due to higher availability of nutrients has led to the apparent paradox that most fish kills have occurred not during the release of heated effluents but whenever a plant shutdown occurs suddenly due to load dumping or equipment failure. This phenomenon, known as "cold shock," results from the sudden cooling of the water plume to whose temperature the fish have become acclimated, leaving them with a suddenly reduced body temperature, in a state of stupor, and suffering a loss of equilibrium (17, 18). Unless the temperature is raised again quickly, the fish may die in a matter of minutes or a few hours. The effect is most pronounced during the cold season when ambient water temperatures are low; it can be minimized either by lowering the temperature slowly to give the fish a chance to adjust to the new temperature or to seek other pastures, by diffusing the heated water well in the first place, or by combining the discharge mixing zones of several plants so that the shutting down of any one plant has a less drastic effect on the temperature of the receiving water.

Another major cause of fish kills is the entrainment of small fish in the intake water at velocities (> 0.5 cfs) too high for them to escape, leading to death either by impingement on the trash screens or by thermal shock as they pass through rapidly changing temperatures while they are carried through the condenser pipes (18). Thus, the use of water for cooling purposes at steam-electric plants may have effects on aquatic organisms other than those resulting from purely thermal discharges. The destructive effects of passing fish and their larvae or eggs through pumps and condensers may indicate the need for intake screens, preferably with traveling screens having little or no impingement velocity. The most visible environmental impact of some power plants has been due to fish kills resulting from improperly designed intake structures, leading to impingement and mutilation of passing fish on the trash screens (14). Most recent designs incorporate wide-intake, low-velocity systems with trash racks whose main function is to remove driftwood and other solids, with finer screens to remove algae and other entrained plants and animals. Before entering the condenser system the water has to be purified, softened and demineralized, so that the water leaving the plant will be significantly cleaner than the inlet stream. In addition, chemicals used intermittently for defouling the condensers could adversely affect fish and fish food organisms (19).

WASTE HEAT UTILIZATION

Considering the trouble and expense involved in rejecting the waste heat to the atmosphere, it would seem highly desirable to do this in as useful and constructive a fashion as possible, to avoid wasting the residual energy content. Considerable thought and ingenuity has been devoted to this question, but many proposed solutions appear to be of doubtful practicality, either because they contradict current trends for isolating nuclear power plants from industrial or populated areas or because they require sizable additional capital investment or involvement of the plant operator in various, quite unrelated enterprises. Diversion of heat into beneficial uses in no way changes the amount of heat ultimately rejected to the environment; it merely modifies the rate and density of energy release and may make the impact more acceptable in societal terms.

Ideally, beneficial uses for waste heat should be both economical and capable of utilizing most of the warm water discharge from an operating power plant, particularly during conditions of low flow and/or high natural water temperatures (normally during the summer). Unfortunately, most of the beneficial uses proposed so far do not meet either of these conditions. Many proposed beneficial uses of waste heat, such as space heating, frost protection for crops, extending the growing season, and maintaining ice-free channels for navigation, are suitable only for the winter months in the northern parts of the U.S. and probably cannot be employed during the critical summer months.

Of all proposed beneficial uses of waste heat, the most promising appears to be the employment of waste heat for residential and industrial space heating in the winter and absorption-type air conditioning in the summer. Such waste heat utilization for district heating has been demonstrated successfully in Sweden (21), West Germany, Finland (19), and the United States, although the summer use for air conditioning has not been implemented in practice. Residential heating use is practical only where pipe runs and heat losses can be minimized by moderately close spacing between plant and population center. The basic problem with space heating systems is their uneven load factor, necessitating auxiliary plant cooling devices for at least part of the year (23).

One approach to the problem is the design of dual systems that provide high quality steam taken from the steam cycle instead of from the cooling water. This system is used in Tapiola with apparently good results. But "waste" heat is no longer used since this steam would still be of value for production of electricity.

The price is especially high if one takes the steam from a power plant designed strictly for generation of electricity. Garton and Christianson (23) give the following example. If we postulate a typical nuclear plant at 100% capacity, the steam at the end of the cycle, after passage through the turbines, will be 92°F at a pressure of 0.74 psi (518 kg/m^2). This plant will have a heat rate of about 10,390 Btu/kWh of electricity produced. Using the same system but taking the steam at 161°F increases the backpressure to 4.42 psi (3108 kg/m^2) and increases the heat rate to 12,000 Btu/kWh. If we use a typical fuel cost of $0.20 per 10^6 Btu, we determine that fuel cost at a heat rate of 10,390 Btu/kWh is 2.08 mills/kWh. If we want 161°F steam, the heat rate is 12,000 Btu/kWh and the fuel cost is 2.40 mills/kWh. This is an increase in fuel cost of 0.32 mills/kWh, which will be added to bus-bar cost. If the plant generates 1000 MW for 7000 hours/year it produces 7 x 10^9 kWh/yr of electricity. At an additional cost of 0.32 mills/kWh, the increase in cost is (7 x 10^9 kWh) x (0.32 mills/kWh) = 2.24 x 10^9 mills or $2,240,000 per year. So, if an industry or housing complex takes all the available steam at 161°F from this plant, it will have to pay all distribution costs plus $2,240,000 to the power company to make the operation economical. Of course, the cost of steam will decrease proportionally with a decrease in amount used (23).

The purpose of this exercise is to show that the heat in water or steam that is above the normal condenser temperature of 90-120°F (32-49°C) is not waste and will not come without cost. Special designs for production of electricity and high-quality hot water or steam imply production of a specific product, and no longer merely waste heat utilization.

Similar considerations apply to many proposed uses of low-grade heat in the chemical industry. Most chemical processes require water or steam at higher temperatures than 90-120°F (32-49°C). On the other hand, low-temperature processes, such as might be used in the food processing industry, would be incapable of using more than a small fraction of the heat available and the demand would be seasonal.

Agriculture is a potential user of waste heat. Irrigation with heated water could promote winter seed germination and growth and extend the growing season. Hot houses could be used to grow tropical or subtropical crops in the more temperate regions of the country. However, a number of problems need to be solved before large-scale use of heated water for irrigation could become common practice. Also, the effects of any plant shut-down on such uses of warm water has to be explored.

Although warm water irrigation of crops on a year-round or nearly year-round basis appears to be somewhat more promising, a 1000 MWe nuclear plant would require 100-200 square miles of farm land to use this method for dissipation of its waste heat. Furthermore, irrigation is not usually necessary 24 hours a day, particularly during periods of rainfall; therefore some provision must be made to dissipate the waste heat contained in the cooling water when the land does not need this water. Also, the 1000 MWe nuclear plant would require 1000-3000 cfs of water for condenser cooling and, since sprinkler irrigation would probably be used, this withdrawal would represent a consumptive use of river water; only a few rivers in the U.S. could supply such volumes. Because of all these factors, waste heat dissipation by warm water irrigation can probably be used in only a few sections of the world (14,22,24).

An obvious advantage of raising the temperature of the receiving body of water is in the provision of ice-free shipping lanes and ports. Heat release may help to keep open significant portions of the St. Lawrence Seaway the year around, with judicious selection of reactor sites, resulting in considerable savings in transportation costs. However, the range of waterway affected may be limited to 10-20 miles of ice-free water, and the adverse effects on the ecology would be most pronounced during the summer months.

Another potential use of condenser discharge water is in aquaculture. Marine and freshwater organisms may be cultured and grown in channels and ponds fed with heated water. For example, it may be possible to grow commercially valuable oysters in areas where they cannot normally reproduce or survive due to low water temperatures. Studies are being made of the possibility of increasing lobster production in Maine with the use of waste heat. Consideration is being given in the Puget Sound region of Washington to the use of warm water to promote the spawning and growth of oysters, crabs, and mussels. Proposals have been made in Wisconsin to use waste heat to warm sport-fish hatchery waters and increase growth rates.

Elsewhere, studies are under way on the use of warm effluent water to farm fish, shrimp or high-protein algae. However, none of these studies are sufficiently advanced or on a large-enough scale to promise an appreciable utilization of the waste heat available and all require a continual, uninterruptible supply. One serious limitation remaining is the present necessity to site reactors away from heavily populated areas, exactly where waste heat could be used most effectively. Locations near population centers might open up a variety of as yet unexplored uses of waste heat.

DISCHARGE OF HEATED WATER

The ecological effects of the release of heated water from a plant depend on the mode of release, the temperature difference, the flow rate and volume of the receiving stream, and the degree of mixing. In the simplest case of "once-through" cooling, cool water of temperature T_1 is withdrawn from a stream or reservoir, pumped through the condenser system and released at a higher temperature, T_2, back into the stream or reservoir. Heat dissipation may occur either (a) by evaporation from the water surface if the effluent is added to the surface layer only, with little vertical mixing, or (b) by thorough turbulent mixing throughout an extended *mixing zone*, leading essentially to a dilution of the heat content and a consequent increase in the mean water temperature of the receiving stream, T_3, where (T_3-T_1) depends on the total volume of the mixing zone. The second approach is that traditionally used for industrial plant effluents and for ocean shore locations. For the very large quantities of heat rejected by the present generation of power plants the resulting temperature rise may always be excessive; in many states the size of mixing zones may be severely restricted or prescribed (14). The following recommendation has been made in Ohio and may be considered representative (14):

Where mixing zones are allowed, the following are recommended:

1. As a guideline, the maximum distance of the mixing zone in any direction should not exceed that obtained by multiplying the square root of the discharged number of million gallons per day times 200 feet (*i.e.*, 25 mgd $\rightarrow \sqrt{25}$ x 200 = 1,000 feet), and in no case exceed 5,280 feet.

2. Mixing zones for single or cumulative discharges should be limited to 25% of the cross-sectional stream area or volume of flow; when density differences between waste waters and receiving water produce stratification, the stratified layer shall not extend beyond 25% of the width of the stream.

3. At no place in the mixing zone should the 96-hour mean lethal time (TL$_m$) to aquatic life be exceeded.

4. Mixing areas shall be:
       free from substances attributable to municipal, industrial or other discharges that will settle to form putrescent or otherwise objectionable sludge deposits;
       free from floating debris, oil, scum and other floating materials of other discharges in amounts to be unsightly or deleterious;
       free from discharged materials that produce color, odor, or other conditions in such degree as to create a nuisance;
       free from substances and conditions or combinations thereof in concentrations that produce undesirable aquatic life.

More frequently, regulations limit primarily the effluent temperature, T_2, with some legal variations, as shown in Table 62 (14). A consequence of limiting $\Delta T = T_2 - T_1$, rather than $T_3 - T_1$, in many regulations has been the erection of some probably unnecessary cooling installations and a tendency to provide cooling ponds or canals on plant property, inside the fence, so that a fence-line temperature within the prescribed value of ΔT could be obtained. An example of this approach is the 25-mile long cooling canal built at Turkey Point, Florida, which resulted in a significant environmental impact of its own.

Table 62. Summary of Selected Approved State Temperature Standards (as of March, 1971) (14)

A. Shellfish and Tidal Waters

Alaska	Less than 68°F
Connecticut	85°F maximum and 4°F maximum change
Delaware	75°F maximum and 5°F maximum change
Georgia	93.2°F maximum and 10°F maximum change
Louisiana	96.8°F maximum and 5.4°F maximum change
Maine	4°F maximum change except 1.5°F maximum change during July, August and September
Maryland	90°F maximum, elevation of temperature not to exceed 20°F or 10°F depending on whether the natural water temperature is below or above 50°F, respectively, with a maximum of 60°F and 90°F, respectively
Mississippi	93°F maximum and 10°F maximum change
New Jersey	86°F maximum and 5°F maximum change (Delaware Bay and Estuary)
New York	86°F maximum and 5°F maximum change
North Carolina	Same as South Carolina
Pennsylvania	Same as New Jersey
Puerto Rico	93°F maximum and 4°F maximum change
South Carolina	4°F maximum change (fall, winter and spring) and 1.5°F maximum change in summer
Texas	Same as South Carolina
Virginia	Same as South Carolina
District of Columbia	90°F maximum and 5°F maximum change
Hawaii	1.5°F maximum change
Virgin Islands	Same as South Carolina with 90°F maximum
Massachusetts	No increase to exceed limits on most sensitive use
Oregon	No significant increase above natural background temperature, etc.
Rhode Island	No increase over recommended limits for the most sensitive use

Table 62, Continued

B. Fish and Wildlife Propagation

Alaska
: 10% change for fresh water (5% change for salt water), 60°F maximum, 0.5°F maximum hourly change

Arizona
: 93°F maximum; 5°F maximum change cold water fish - November-March, 55°F maximum. April-October, 70°F maximum. Not more than 2°F change

Arkansas
: 20°C (68°F) maximum in trout streams, 30°C (86°F) maximum in smallmouth bass streams, 35°C (95°F) maximum in other streams

Colorado - warm water fish
: 90°F maximum, cold water fish - 70°F maximum

Connecticut
: 85°F maximum and 4°F maximum change

Delaware
: 5°F maximum change

District of Columbia
: 90°F maximum and 5°F maximum change

Georgia
: 93.2°F maximum and 10°F maximum change

Guam
: 1.5°F change limit

Idaho
: No measurable temperature increase when stream temperature is 68°F or above, or more than 2°F increase when river temperature is 66°F or less

Indiana
: Warm water fish - 5°F maximum change. Cold water fish - not to exceed 65°F or 5°F maximum change

Kansas
: 90°F maximum and 5°F maximum change (streams and rivers) or 3°F maximum change (lakes and reservoirs)

Kentucky
: May-November, 93°F maximum. December-April, 73°F maximum

Louisiana
: 96.8°F maximum and 5.4°F maximum change

Maine
: 84°F maximum for warm water fish and 68°F maximum for trout and salmon waters, 5°F maximum change (rivers and streams) or 3°F maximum change (lakes).

Maryland
: Trout waters - not to exceed 72°F. All other waters - not to exceed 93°F; elevation of temperature not to exceed 20°F or 10°F depending whether the natural water temperature is below or above 50°F, respectively, with a maximum of 60°F and 93°F, respectively.

Massachusetts
: No increase except where temperature will not exceed the recommended limit on the most sensitive receiving water use and in no case exceed 83°F in warm water fish, and 68°F in cold water fish, or in any case raise the normal temperature more than 4°F.

Minnesota
: 86°F or 90°F maximum

Missouri
: 90°F maximum and 5°F maximum change

Montana
: Salmonid fish - increases - 32°F to 67°F - 2°F maximum; above 67°F - 0.5°F decreases; over 55°F - 2°F maximum/hr; 55°F to 32°F - 2°F maximum, provided that water temperature must be below 40°F in the winter season and above 44°F in the summer season. Non-salmonid fish - increases - 32°F to 85°F - 4°F maximum; above 85°F - 0.5°F maximum, decreases - same as Salmonid fish.

Table 62, Continued

Nebraska	Trout streams - 65°F maximum, 5°F change limit. Warm water streams - 90°F maximum, change limits same as PWS Missouri River - 85°F maximum, 4°F change limit.
Nevada	68-86°F maximum (summer), 57.2°F maximum (winter)
New Mexico	Warm water fish - 93°F maximum, 5°F change limit. Cold water fish - 70°F maximum, 2°F change limit
New York	Trout waters - no thermal discharge which will cause adverse effects on trout. Non-trout waters - 90°F maximum within mixing zone; 86°F maximum, and 5°F change limit outside mixing zone, plus a 2°F maximum change limit/hr, and/or 9°F maximum change limit for a 24 hr period (fresh waters)
North Dakota	90°F maximum, 5°F maximum rise.
Oklahoma	5°F change limit, provided the maximum manmade temperature does not exceed 70°F in trout streams, 75°F in smallmouth bass streams, or 93°F in warm water streams.
Pennsylvania	Trout waters 58°F maximum, 5°F change limit. Warm water fish - 5°F change limit above natural or 87°F maximum, whichever is less; 2°F change limit per hour.
Puerto Rico	93°F maximum, 4°F change limit
Rhode Island	68°F and 83°F maximum for cold and warm water fish, respectively; 4°F change limit
South Dakota	Cold water permanent fish - 68°F maximum, 4°F change limit. Warm water permanent fish - 85°F maximum, 4°F change limit. Warm water semi-permanent - 90°F maximum, 8°F change limit.
Texas	96°F maximum, 5°F change limit
Utah	Cold water fish - 68°F maximum, 2°F change limit. Warm water fish - 80°F maximum, 4°F change limit
Vermont	83°F maximum (warm water fish only)
Virginia	87°F or 90°F maximum, 5°F maximum change
Wisconsin	84°F maximum, 5°F change limit, 2°F change limit per hour. Where fishing is an additional use - 89°F maximum, 5°F change limit. In addition, authorization is required for proposed installations where thermal discharges may increase natural temperature by 3°F. Trout streams - no temperature change that will adversely affect trout.
West Virginia	Not to exceed 93°F May-November; not to exceed 73°F December-April
Wyoming	Streams where natural temperatures exceed 70°F, 4°F change limit over natural. Streams where natural temperatures are less than 70°F, 2°F change limit over natural. (Until study is made, maximum temperature limit is daily average plus allowable rise.)

Run-of-the-river cooling, which used to predominate in the United States, is rarely possible for the power stations in the ranges of 3000-4000 MWt now being ordered. For an average condenser rise of 15°F (8.3°C), a new fossil-fueled plant may require flow rates of 5200 cfs (147 m^3/sec), while a nuclear plant may require about 7000 cfs. To limit the temperature rise to 5°F (2.8°C) would require 14000 cfs and 21,000 cfs, respectively, flow rates that are far in excess of average flows for most of the rivers involved. Nonetheless, even for closed-loop cooling systems some heated water from blowdown and overflow must be released to a receiving stream, and tighter release standards may not always reflect the long-range optimum environmental solution.

Merriman (25), in a well-known study on the impact of heat release on the Connecticut River, showed that biological effects are of minor consequence if there is an adequate water flow available, a substantial mixing zone, and an undisturbed portion of the stream flow along the opposite bank of the river. As a rule these conditions can be met only if appreciable water flow pertains all the year round and only for wider rivers. For most of the larger power plants situated on fresh water streams, the need to assure the continual availability of adequate cooling water has produced a requirement for a pond or reservoir; having a storage pond encourages the use of a closed-loop cooling system, where water is recirculated between the plant condensers and the water body. When cooling ponds and lakes are combined with pumped storage or conventional hydroelectric developments, there are additional possibilities for multiple use of water resources.

For this reason environmental impact studies need to consider the dispersion of heated discharges in lakes and rivers. This subject has been reviewed concisely by Parker and Krenkel (26) and a National Academy of Engineering Committee (8). Although extensive hydraulic model studies have been described in the literature and several computational models have been developed, experience has shown considerable discrepancies between predicted and measured temperature profiles, and in most cases detailed measurements under actual conditions are essential to establish thermal characteristics, water quality and biological effects (27).

An important use for once-through cooling is at seashore locations where auxiliary cooling alternatives are most limited, primarily due to objectionable consequences of salt-water drift from cooling towers. Except in shallow embayments, tidal and wind-driven currents are generally available for rapid dispersion and dissipation of the added heat.

The primary advantages of once-through cooling are the conservative use of water, the ability to tailor the temperature distribution field in the receiving water to meet biological objectives, and heat dissipation to the atmosphere over a large area. Factors that must be carefully investigated in a judicious design of discharge structures include entrainment of aquatic organisms, discharge jet effects on benthic organisms, bottom erosion, and effects on navigation.

Thermal Discharge Analysis

In order to predict the mixing and dissipation of heated water discharged from a power plant, a theoretical analysis that takes into account the nature of the discharge structure and of the receiving water must be carried out. These parameters include the discharge type, single port or multiport, its shape and location, its direction to the stream, its position with respect to the water surface, volume flow rates, and the effluent excess temperature, as well as the receiver water characteristics, such as type, depth at outfall and currents.

The interaction mechanisms comprise one or several of the following: (a) jet entrainment, characterized by the *initial densimetric Froude number*, F_o, (b) cross flow, characterized by the velocity ratio of the initial discharge to the current velocity; (c) natural turbulence, often described by the eddy diffusivity, (d) buoyancy, and (e) recirculation, where partly diluted effluent may be re-entrained into the discharge plume. Depending on which mechanism is presumed to predominate, several mathematical models may be developed to describe the process (11,28,29). A concise summary of the mathematical formulation of thermal dispersion may be found in US NRC Regulatory Guide 4.4 (30), which also reviews some of the analytical models for solving the hydrodynamic and thermodynamic equations.

When heated water is discharged into a water body, the resulting temperature field can be divided into two distinct zones: (a) an initial, or near-field, region in which temperature changes are governed primarily by the geometry and hydrodynamics of the discharge and (b) a far-field region. The mechanisms affecting the temperature reduction in the near-field region are the dilution and entrainment due to the momentum of the discharge jet and the buoyancy effects due to the temperature difference between the discharge and the receiving water.

For the case of a heated jet, the Gaussian approximation of the transverse temperature profile along the jet takes the form

$$T_n = T \exp \left[-\frac{n^2}{2\lambda^2 \sigma_T^2} \right] \tag{47}$$

where T_n and T are the temperature in the transverse direction and the jet centerline temperature, respectively, n is the distance normal to the jet centerline, σ_T^2 is the temperature variance along n, and λ is an adjustable diffusion parameter. According to Taylor's theory, heat diffuses more rapidly than momentum in the transverse flow direction; therefore, the value of λ is always greater than unity. Under most circumstances, heated effluent is less dense than the receiving water in the immediate vicinity of the outfall. As a result of the density disparity, there is a buoyant force on the jet acting both vertically and horizontally. A submerged buoyant jet tends to rise to the surface; a buoyant jet at the surface forms a stable density layer. For either case entrainment is reduced, especially in the vertical direction, and the dilution rate decreases. In the far-field region the temperature distribution is governed by conditions in the receiving water body. The important properties of the receiving body of water are natural temperature stratifications, advection, diffusion and dispersion due to tidal currents, wind-driven currents and wave action, and heat dissipation from the water surface (31,32).

From an engineering viewpoint, the near-field region is the most important one since the temperature in this region can be controlled by proper design of the discharge structure. The basic design parameter controlling entrainment of the discharge time is the densimetric Froude number of the discharge, defined as

$$F_o = \frac{v_o}{\sqrt{\dfrac{gh\Delta\rho}{\rho}}} \tag{48}$$

where: v_o =   mean velocity at the exit section of the condenser water discharge structure

       g =   acceleration of gravity

       $\Delta\rho$ =   density difference due to temperature difference between condenser water at exit and receiving water

       ρ =   density of receiving water

       h =   vertical thickness of the jet as formed by the discharge structure.

For surface discharge systems for heated water, analytical and experimental investigations have been described by Stolzenbach and Harleman (33) and others (29,34), to predict the three-dimensional temperature distribution in the near-field region. Such calculations show the importance of the initial densimetric Froude number. For a higher Froude number, centerline temperatures tend to be reduced, due to the greater entrainment of the receiving water.

When sufficient water depths are available, submergence of the outlet is an effective means of increasing the path of a buoyant jet. This has the effect of increasing the total entrainment of the cooler ambient water.

By contrast with the surface discharge, lower water surface temperatures are produced; hence the rate of surface heat dissipation is relatively small. Single jets require a greater path length and hence greater submergence and water depths than multiport diffusers in order to achieve the same temperature reduction at the surface of the receiving water. Thus, if the available depth is limited, it may be necessary to use a multiport diffuser in order to prevent surface temperatures from exceeding a specified rise above the ambient temperature (Figure 51).

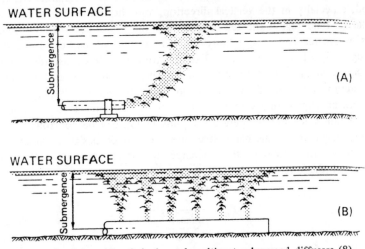

Figure 51. Schematic of single- and multiport submerged diffusers (8).

Multiport diffusers are large pipes, placed on the bottom of the receiving water body, from which heated condenser water is discharged through a number of openings. The net dilution can be large provided that the diluted water is removed from the discharge area by river, tidal, or wind-driven currents in the receiving water. If this is not the case, a build-up of temperature in the vicinity of the diffuser will occur as a result of reentrainment of the heated mixture.

For a single, round, buoyant jet discharging horizontally near the bottom of a stagnant body of receiving water, the basic parameters that determine the temperature rise at the water surface are the densimetric Froude number of the jet at the point of discharge and the relative submergence, y/D, where y = depth of receiving water above centerline of discharge port, D = diameter of jet at discharge, and $\ell = D$ = characteristic length in terms of the densimetric Froude number F_o.

Curves of maximum dilution S_m as a function of these parameters are shown in Figure 52 from both experimental and analytical sources. The

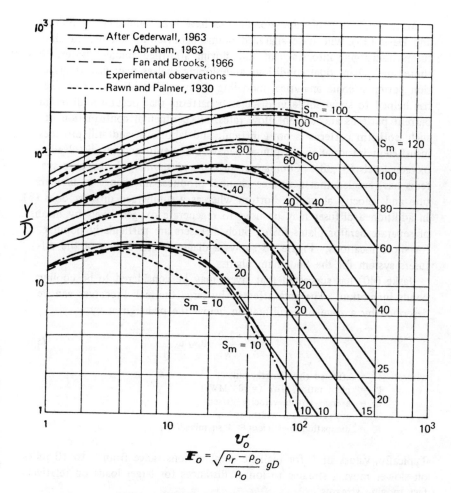

$$F_o = \sqrt{\dfrac{\rho_r - \rho_o}{\rho_o} gD}}$$

Figure 52. Theoretical solutions of dilution S_m for horizontal buoyant jet in stagnant receiving water of uniform density (8).

temperature rise at the surface of the receiving water, ΔT_c, at the center-line of the trajectory is related to the dilution and the initial temperature difference ΔT_o by

$$\frac{\Delta T_c}{\Delta T_o} = \frac{1}{S_m} \qquad (49)$$

The practical range of initial densimetric Froude numbers is between 20 and 50. The single jet is not a very efficient dilution device at relative submergences below $y/D = 15$.

More detailed graphs and nomograms have been published by Shirazi and Davis (35), who have analyzed plume behavior for submerged diffusers discharging into stagnant and moving stratified ambient water. For shore sites where tidal conditions and sloping shelves may distort dispersion patterns, some analytical modeling, such as that of Koh *et al.* (36), has helped to show shifts in dispersion patterns that occur for different current patterns and current flow rates. Experimental confirmation of such isotherm patterns is both difficult and laborious, and still provides insufficient information on plume behavior at depth and on dissolved oxygen levels (37,38). Comparing surface and submerged discharge systems has revealed that a surface channel produces no change in temperature and velocity along the bottom whereas a submerged diffuser does (8). In addition a diffuser may be used to regenerate oxygen levels by means of separate aerator plates (39). Such diffusers are particularly popular for shore locations; Figure 53 shows a diagrammatic sketch of the diffusers system for the Shoreham plant.

Jaske (40) has suggested that rivers may be considered to be cooling ponds in motion. The area affected by large thermal discharges can be described by a zone of equilibration below the plant given by:

$$L = K' \frac{5280}{\overline{w}} P \sqrt{\overline{u}} \qquad (50)$$

where:
L = affected river length, miles
P = plant rating, GWe (= 10^3 MWe)
$\overline{u}$ = average stream velocity, ft/sec
$\overline{w}$ = average stream width, ft
K' = dissipation coefficient ≈ 4 sq miles/GWe

Typically, values of L for realistic combinations range from 8 to 10 miles on slower moving streams to longer distances for larger loads on relatively fast moving streams.

In summary, where adequate water flow is available, once-through cooling may be possible, provided temperature rises at the outlet not exceeding 20°F are acceptable and a mixing zone of a few hundred feet in length to the 0.5°F isotherm can be tolerated. If these conditions are not attainable or are in conflict with applicable regulations, special cooling systems must be considered.

Two other proposals for dissipating waste heat may be mentioned in passing. One of them involves the blasting by a nuclear explosion of an underground cavity intersecting a natural aquifer that would serve as a source of fresh water. By pumping cooling water from the cavity and back again heat would be delivered to the underground waterflow for

Figure 53. Sketch for cooling water system for Shoreham plant shown submerged diffuser (15).

dissipation to subsurface rocks. Careful examination of this proposal reveals a number of undesirable features, primarily the weakening of the rock formations underlying the plant by the explosion and the release of fission products from the explosion into an active aquifer. In addition, it is doubtful if enough heat can be dissipated by this means or if adequate water flow can be maintained.

Another proposal utilizes a heat pump as a means of rejecting heat to the subsoil. In this case it is questionable if the thermal conductivity of the ground is high enough to permit the required flow of heat into it in order to maintain a workable temperature gradient. Considering the amounts of heat to be rejected, such a system would need to tap an exceedingly large ground volume to serve as an adequate heat sink.

COOLING SYSTEMS

As the previous discussion has shown, simple once-through cooling with dissipation of heat by evaporative surface cooling or dilution in a major body of water will be unacceptable for the majority of the large power plants on account of the high ΔT value reached at the effluent discharge point. In order to dissipate heat directly to the atmosphere before water is discharged to public waters, supplemental cooling by means of cooling towers or cooling ponds has to be provided, both in once-through systems and in closed-loop cooling. Supplemental cooling systems allow partial heat rejection to the atmosphere and may be used continually or seasonally as required.

Closed-loop systems theoretically entail no warm-water discharge to a receiving water body, such as a public stream, and are designed to dissipate into the atmosphere the entire amount of waste heat carried by the condenser cooling water. In practice, cooling towers and spray ponds require periodic flushing to prevent buildup of salts and plant chemicals, and this "blowdown" water will be released to the receiving stream and may amount to as much as 5% of the water circulating through the condenser. Both cooling towers and cooling ponds have inherent limitations in the extent to which the temperature of the warm water can be reduced, due to the decrease in evaporative cooling as the wet bulb temperature is approached. In addition, power requirements to provide the extra pumping power for circulation further reduce the effective efficiency value of the power plant. Water losses due to entrainment and drift may also severely limit the effectiveness of the cooling system.

Economic and environmental assessment of the various cooling systems is usually made in comparison to simple once-through cooling.*

Cooling Ponds

Cooling ponds are widely used as a source of circulating water for steam power plants when land is available at a reasonable price. Ponds may be constructed by simply pushing up an earth dike, and they may be operated for extended periods with just enough make-up water to balance water losses. They lend themselves to multiple use as water reservoirs to assure a supply of water for cooling purposes as well as for fire protection, recreational uses, fish farming, and sometimes for load flattening by means of pumped storage. Unfortunately, ponds and lakes have low heat transfer rates, require large surface areas, are susceptible to silting, and collect chemical impurities and dissolved solids, which must be removed by periodic blowdown. Area requirements are of the order of 1-2 acres (0.4-0.8 hectare) per megawatt of installed capacity, with recommended minimum depths of 5-12 feet (2-4 m). In cooler climates there is a possibility of fog on cold days at distances up to 600 ft (200 m) downwind from the pond.

The major factors affecting pond performance are surface air temperature, relative humidity, wind speed, wind fetch, solar radiation, aquatic growth and erosion. An "approach" of 3-4°F is the lowest practicable amount of cooling in ponds of reasonable size. Deep cooling ponds can provide water at colder temperatures than can draft towers in colder climates.

Typically the cost of building a cooling pond is approximately $1000 per acre of land and excavation, and $0.5/kW for piping and dams. Table 63 gives sample calculations of pond sizes and costs for a 1000 MW plant with an inlet temperature of 65°F and a discharge temperature of 80°F. Table 64 shows the seasonal variation in isotherm areas expected for the cooling lake of one particular power station (Oconee) in South Carolina. Similar data must be developed for extreme meteorological conditions.

Spray systems added to cooling ponds increase the heat loss per unit area and hence reduce the land required for a pond by a factor of about 20, to 50-100 acres/GWe. However, due to the enhanced evaporation, water losses require additional makeup water. These average around 1%, with total water losses in summer weather reaching 1.5 to 2% of the

---

*In order to evaluate different types of cooling devices, certain specialized definitions are required; these are listed in the Glossary (p. 666).

Table 63. Comparative Costs (1971) of Cooling Ponds

Type of Pond	Acreage Required	Cost
Fully mixed	11,700	$11,600,000
Flow through	3,360	$3,360,000
1.2 acres/MW	1,200	$1,200,000
2.0 acres/MW	2,000	$2,500,000

water sprayed (41). Spray ponds are usually unsuitable for a cooling range greater than 18°F (10°C) and are considered uneconomical for large plants. As the size of the pond increases, sprays on the downwind side become less effective since the air passing them is saturated already.

Average heat dissipated by a spray pond is 127 Btu/ft^2 per degree F temperature difference. Power required to operate a spray pond is 1.5% of the total power generated in the plant, about half that required for a mechanical draft cooling tower of comparable capacity. Figure 54 shows a typical spray pond layout (41).

Radioisotope build-up in cooling ponds should also be monitored. The concentration of radioisotopes as a function of time of operation in a cooling pond is calculated by the formula:

$$C(t) = \frac{Q}{V\Lambda} (1 - e^{-\Lambda t})$$

where: Q = activity emission rate for the individual isotopes, Ci/yr
         V = pond volume, cm^3
         Λ = $\lambda + r/V$, yr^{-1}
         r = rate of water loss from pond, cm^3/yr
         λ = radioactive decay constant for each radioisotope, yr^{-1}

The resulting concentration of radioisotopes in the cooling pond water can be used to compute radiation exposure through drinking water, and with the appropriate food chain concentration factors through the eating of fish taken from the water. For a 1000 MWe PWR operating in conjunction with a 2000-acre cooling pond having a net flow of one pond volume every two months, the exposure to the public that could result would depend on whether this small lake is used for drinking water, swimming and/or sport fishing. At the equilibrium concentrations of radioisotopes released in the pond, a person obtaining his average fish diet totally from this small pond might receive an annual exposure of 0.001 mrem; if, in addition, he obtained all of his drinking water from

Table 64. Seasonal Variations in 3°F ΔT Isotherm Areas for a Cooling Lake (Lake Keowee)

Month	Isotherm Temperature (°F)	Acres	Percent of Total Area
January	75	0	0
	70	160	1.0
	65	370	2.2
	60	650	3.9
	55	2540	15.2
	52	5100	30.5
	49	Ambient	
March	70	160	1.0
	65	430	2.6
	60	1240	7.4
	57	4020	24.2
	54	Ambient	
May	72	0	0
	71	Ambient	
July	86	0	0
	83	Ambient	
September	90	280	1.6
	85	3440	14.1
	83	4090	23.6
	80	Ambient	
November	80	160	1.0
	75	530	3.2
	70	1230	7.4
	66	5520	33.0
	63	Ambient	
February	70	0	0
	65	190	1.1
	60	430	2.6
	55	930	5.6
	50	4800	28.8
	47	Ambient	
April	70	430	2.5
	66	2170	12.8
	63	Ambient	
June	80	0	0
	79	Ambient	
August	86	1410	8.0
	83	Ambient	
October	90	0	0
	85	330	2.0
	80	2020	12.0
	76	4560	27.0
	73	Ambient	
December	70	160	1.0
	65	610	3.7
	60	2050	12.3
	57	5520	33.0
	54	Ambient	

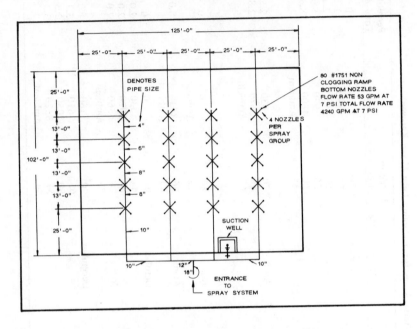

Figure 54. Typical spray pond layout (41).

the same source, his environmental exposure would be an additional 0.12 mrems per year (13).

Cooling Towers

Where tight water-quality standards or scarcity of water or suitable land preclude the use of once-through cooling or of large-area cooling lakes, cooling towers are generally employed for waste heat dissipation. They may be used to provide full cooling requirements only during certain periods of the year and may be combined with cooling ponds as back-up systems. Cooling towers have been in common use in Europe for many years, particularly in densely industrialized areas where land was at a premium and year-round water availability was not assured.

Cooling towers are classified as *dry* or *wet* according to whether the warm water is contained in pipes and air flow cooling occurs primarily by conduction across the pipe interface or whether the warm water is in direct contact with a flow of air and heat is dissipated principally by evaporation (1,42,43). Cooling towers can also be classified by the means of producing the air flow; this can be *induced mechanical draft*, using one or more powerful fans, or *natural draft*, using a tall, usually hyperbolically shaped "chimney," to provide a natural updraft. Figure 55 illustrates diagrammatically the main types of cooling towers in use.

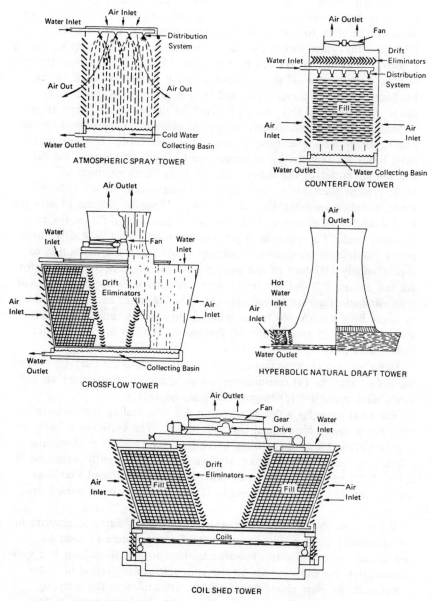

Figure 55. Principal types of cooling towers (8).

In wet towers the warm water is distributed at the top of a dispersal section, called the *fill* or *packing* that is generally made of cement, asbestos, treated fir or redwood, or plastic material such as polyvinylchloride. The fill may be of the splash type or film type. Splash-type fill breaks the water into droplets that subdivide as they descend, thereby exposing large surface areas to the air for the evaporative heat transfer. Film-type fill allows the water to descend as a thin continuous film so as to expose a large area to the air for the heat transfer. The cooled water is collected in a basin under the fill section from which it is pumped back to the condenser to pick up more heat and then is returned to the cooling tower. The limit of this cooling is fixed by the wet bulb temperature.

Small droplets of water may be carried upwards in the towers by the cooling air plume. This spray is generally called "drift" or "carryover." Drift losses are reduced by an arrangement of louvers between the fill and the air outlet of the tower, referred to as "drift eliminators."

At present there are no technological limitations as to the type of water, including seawater, that can be used. However, the use of seawater is of doubtful value owing to the effects of salt-laden drift on the surrounding land. For example, a wet cooling tower for a 1000-MWe nuclear power plant using seawater for make-up would cause the deposition of approximately 100 tons of salt per year even if the concentration factor did not exceed 2.0 and the tower were equipped with the most efficient drift eliminators presently available (8). On the other hand, experience in Great Britain with two seawater-cooled towers has resulted only in some browning of grass within 50 feet of the bottom of the 250-foot high towers (44).

Under certain circumstances, other factors limiting the use of wet cooling towers may be (a) consumptive use of water, (b) restrictions on blowdown discharges, and (c) fogging and icing hazards.

The total make-up water requirement for wet cooling towers includes evaporative losses, blowdown, and drift losses. The evaporative losses are approximately 1% of the cooling water flow for each 10°F of cooling effected in the tower. Drift or windage losses are currently estimated at 0.1 to 0.2% of the cooling water flow, but it is anticipated that they may be reduced in the near term to 0.05% by means of improved drift eliminators.

Solids from the dissolved chemicals in the source water accumulate in the circulating cooling water as a result of evaporation and must be periodically or continuously removed by blowdown. Blowdown is a nonconsumptive use and depends on the allowable concentration of water constituents to meet either water quality restrictions in the receiving water body or nonscaling requirements in the circulating water system.

Make-up water must be added to replenish the losses due to evaporation, drift, and blowdown. The added water may require chemical treatment to protect the fill from deterioration, to prevent the spray nozzles from clogging, or to protect the condenser tubes from corrosion. It is also necessary to assure appropriate sustained flows and acceptable water quality in the stream from which the cooling water supply is drawn and to which the blowdown is released. For a 1000 MWe nuclear unit, the consumptive loss of water at full load is approximately 32 cfs (906 l/s), and the total make-up water requirement is on the average approximately 40 cfs (1133 l/sec).

Whenever necessary, blowdown can be treated prior to discharging into a receiving water body. In certain cases, the cooling tower blowdown can be discharged into a closed reservoir or lagoon where solids are precipitated and then removed.

The use of dry cooling towers is now limited to relatively small generating units, but technological progress in this field is anticipated. Because of their need for additional pumping capacity they impose a penalty on thermal efficiency, and their cost is estimated to be three to five times larger than for wet cooling towers. On the other hand water losses are reduced by their use. Dry-type cooling towers use either air-cooled condensers or air-cooled heat exchangers; they are not in current use for nuclear power plants. In practice, for large power plants the choice lies between natural draft wet towers and mechanical draft wet towers.

It should be noted and stressed that selection of cooling water devices is intimately associated with condenser and turbine design and operation. Though one can make approximate statements on the effect of decreased or increased pressure, initial temperature, increased reheat temperature, and increased feed water temperature, the efficiency of the turbine and the resultant heat rate for any specific installation, for the most economic solution it is necessary to take into account the interrelationships of the following factors:

1. average annual duration of high temperature and humidity at the site
2. condenser heat loads and vacuum correction curves for the turbines specified
3. thermal performance of a number of cooling towers
4. rough calculation of the pump and pipe network to determine costs and hydraulic characteristics
5. performance of a number of condensers designed for the heat loads and cooling ranges under study
6. costs of each component and thereby the amortization charges, fuel costs, plant lifetime load factor, and incremental capability credit.

In choosing one type of cooling tower over another, both technical and economic criteria have to be applied; in addition, several environmental impact criteria have to be considered. The latter are usually expressed in relation to once-through cooling and take into account such factors as relative impact on aquatic biota in the absence of supplementary cooling devices, water losses, plume formation, and noise levels. The technical criteria involve power requirements for tower operation, pumping head, reliability, and thermal performance. Economic factors are capital costs, operating costs, land requirements and cleaning requirements.

Mechanical draft towers have lower capital costs but higher operating costs than natural draft towers. Induced-draft towers are favored over forced-draft towers, and crossflow towers over counterflow towers because of better thermal performance, lower static losses and less likely recirculation. Fan horsepower requirements can amount to as much as 1/2% of a power plant's generating capacity (45). These factors have been compared extensively in the literature (8,26,45-50) and only environmentally relevant factors will be covered here.

Mechanical draft towers must be located in an area at some distance from the plant proper. This is necessary not only for air supply considerations, but also because of problems associated with fogging and drift from the discharge air. Noise considerations also tend to dictate careful, distant siting for mechanical draft towers, since the noise level due to fan motors, air movement through the tower, and water flow can easily exceed 95 db at close distances.

In contrast hyperbolic, natural draft towers can be built adjacent to plant buildings, on centers 1.5 times their base diameter, so that piping costs to and from the towers are greatly reduced. Hyperbolic tower heights in excess of 500 ft (160 m) are becoming a common sight and their position must be carefully chosen so as not to interfere with the power plant exhaust plume. The natural draft tower has no fans, thus reducing maintenance expenses and annual operating costs. Disadvantages of natural draft towers include a decreased ability to design as precisely as for mechanical draft towers and the inability to control outlet temperatures as closely.

The performance of a natural draft tower is described in terms of a duty coefficient, C_D, which defines the overall capabilities of a tower under all operating conditions.

$$C_D = \frac{W_L}{90.59 \frac{\Delta Q}{\Delta T_w} \sqrt{\Delta T_a} + 0.3124 \Delta Q} \tag{51}$$

where: W_L = water load, in lb/hr

ΔQ = total heat of the air passing through the tower, Btu/lb dry air

ΔT_w = change in temperature of water passing through tower, °F

ΔT_a = difference between dry bulb and wet bulb air temperatures, °F.

The draft is due to the difference between the density of the air leaving the tower and that entering the tower, and to the aerodynamic lift of the wind passing over the top of the tower. Increases in loading, cooling range, and humidity all tend to improve cooling tower performance (8).

Hyperbolic towers are usually constructed of thin concrete shells that have good wind resistance and show little deterioration over the lifetime of a plant. Occasionally their rather stark appearance has been modified by a light paint treatment.

Mechanical draft towers may be constructed of wood or metal, or occasionally concrete. The structure must be designed to withstand a specific wind or earthquake stress, dead loads such as the weight of the tower and circulating water, and operational load and vibrations from the mechanical equipment. Wooden towers may present a fire hazard and procedures may be needed to keep them wetted down while on standby.

The cooling tower fill must be low in cost and easily installed. It should promote a high rate of heat transfer, offer low resistance to air flow, maintain uniform water and air distribution through the life of the tower, and must be highly resistant to deterioration. Wood is the most popular and economical material for the purpose and is easily installed. Many types treated with various wood preservatives are commonly used; as the preservatives leach out they form a major portion of the chemicals appearing in low concentrations in the blowdown water in the plant effluents. Plastics, stainless steel and asbestos board are also used but are more expensive; the latter may give rise to low but undesirable traces of asbestos fibers in the effluents.

IMPACT ASSESSMENT OF COOLING DEVICES

Climatic Effects

As mentioned before, the amount of heat rejected from our cities can be a significant fraction of the solar heat reaching the earth's surface for that region. One or more power station complexes may well increase further any local climatic effects. Dissipation of that heat in bodies of water, instead of directly to the atmosphere, may reduce this effect somewhat. The water diversion by evaporation from cooling ponds and towers is probably small enough to have only minor influence on local climate.

Localized ground fog does occur around many cooling towers when the cooling-air water-vapor saturation exceeds the capacity of the ambient air for water vapor. One way to reduce this fog formation is to increase the height of the discharge stack to enable the fog to diffuse before reaching the ground. The size and path of the fog plume from cooling towers can be predicted by applying the plume diffusion theory (Equation 12). The angle of spread of the plume usually varies from 18 to 24 degrees, and the length of the visible plume is

$$x_p = 5.7 \sqrt{\frac{Q}{102\pi u} \left(\frac{T_{ex} - T_{in}}{T' - T_{in}} - 1 \right)} \tag{52}$$

where: x_p = plume length, ft
u = wind speed, ft/sec
Q = air emission rate, m^3/hr
T = air or plume temperatures, $^\circ$C, at inlet and exit
T' = temperature at plume tip, $^\circ$C

Longer and lower plumes occur when air temperature is low, humidity high, and wind speed moderate to high. Such climatic conditions encourage the use of natural draft towers with their higher emission height.

Water Quality Effects

Water quality effects of cooling towers are due primarily to the chemicals used for corrosion and fouling control. Phosphates and chromates may be released in objectionable concentrations, as are zinc, bromates, organic reagents, as well as asbestos fibers where used in fill. Table 65 shows the toxicity and reconcentration factors of some elements commonly used in cooling tower technology. Obviously, as far as possible, chemicals used should be chosen for low toxicity in the environment in the concentrations released. If the blowdown material cannot be rejected directly to the stream, ion exchange resins may be used to remove the chromates on stream and this may even eliminate the need for blowdown release altogether. In passing through wet cooling towers, the temperature is reduced, the water is oxygenated, the BOD reduced, and the chlorine demand lowered. For a large, modern power station, blowdown can be as much as one million gallons per hour; this can constitute a sizable fraction of a receiving stream flow. Since mostly pure water is evaporated, it would be expected that the concentration of inorganics such as chloride,

Table 65. Toxicity and Concentration Factors of Elements used in Cooling Tower Operations (15)

Element	Concentration Factor[a]		Functions	Environmental Toxicity[b] (not injected)
	Plankton	Brown Algae		
As[c]		2,500		Carcinogenic; moderately toxic to plants, highly to mammals—especially as AsH_3
B		6.6	Essential for green algae, angiosperms	Moderately toxic to plants, slightly to mammals
Br[c]		2.8	Essential for marine organisms; amino acids	Br_2 is very toxic; Br^- is relatively harmless to organisms
Cl[c]	1	0.062	Essential for mammals and angiosperms	Cl^- is relatively harmless; Cl_2, ClO^-, ClO_3^- are highly toxic
Cr[c]	17,000	6,500	May serve some physiological function	Cr(III) is moderately toxic; Cr(VI) is highly toxic to organisms and is probably carcinogenic (by inhalation)
Cu[c]	17,000	920	Essential to all organisms	Very toxic to algae, fungi, and seed plants; highly so to invertebrates; moderately so to mammals
Hg[c]		250		A cumulative poison in mammals; very toxic to fungi and green plants; highly to mammals in some forms
N	19,000	7,500	Essential as structural atom	Relatively harmless; concentrations higher in plankton and fish
P[c]	15,000	10,000	Vital in many ways	
Pb[c]	41,000	70,000	None	Very toxic to most plants, moderately so to mammals; cumulative poison
S[c]	1.7	3.4	Essential to some plants	S_2 highly toxic to bacteria and fungi, relatively harmless to green algae, seed plants, and mammals; H_2S is highly toxic to mammals, SO_2 moderately to highly; SO_4^{2-} is relatively harmless
Si[c]				Scarcely toxic, but large amounts in mammalian lung harmful, used by foraminifera, porifera, etc.
Sn[c]	2,900	92	None	Very toxic to plants and green algae
Zn[c]			Essential to all organisms	Moderately toxic to plants; slightly toxic to mammals; uptake by plant roots not linked to metabolic process

[a] ppm in fresh organism divided by ppm in seawater

[b] Toxicity terms: very, 1-10 ppm; highly, 10-100 ppm; moderately, 10-1000 ppm; slightly, over 1000 ppm (as 24-hr median tolerance limit in moderate-sized organisms—e.g., fish).

[c] Accumulator species or genera known.

sodium, sulfate, and totally dissolved solids grows directly as a function of the number of recirculations, as does conductivity.

The main chemical changes in water passing through cooling towers are the loss of carbon dioxide, consumption of some of the oxidizable matter, and nitrification of ammonia. These processes are accelerated by the heated waters over their natural rate of occurrence in streams (26).

Cost Factors

Regardless of the method adopted, any cooling system for a power plant in the 500-2000 MWe range involves a sizable financial investment. To justify the higher costs entailed in more elaborate cooling systems some commensurate social or environmental benefit must be demonstrated. Environmental impact statements for new U.S. nuclear power stations are required to analyze formally the cost and environmental consequences of alternative cooling schemes. In the absence of any agreed cost equivalents of certain environmental effects, such as a reduction of species variety, it is difficult to arrive at a true cost-benefit balance and often such analyses serve little useful purpose.

Certain cost figures can be identified though a direct comparison of published cost figures is often difficult, since the systems referred to are rarely fully specified. For example, one comparison of cost estimates for cooling towers for the *same* plant done by a private company and a government agency in 1970 arrived at total costs as disparate as $6.5 million and $116.8 million (51)! Table 66 presents 1974 cost data believed to be representative of eastern U.S. prices. These costs usually include the condenser system and associated piping; both substantially higher and lower costs have appeared in the literature.

It is evident that these costs are similar to the cost of the nuclear steam supply system itself and will have to be reflected in final power prices to the consumer (51). Table 67 shows a breakdown of the annual costs for different cooling systems and the cost to the consumer for $\Delta T = 10°F$ (46). Among hidden operating costs, the power demand of 4-5% of installed capacity to operate the cooling system is the dominant factor that requires careful assessment. In drier climates evaporation losses and water consumption may be the overriding consideration in favoring cooling towers over ponds (38).

Cost-Benefit Assessment

In evaluating the overall desirability of different cooling devices one runs into several intangibles. Once it is accepted that the cheapest method, either in terms of investment costs or in terms of power costs, is not

Table 66. Cooling Device Costs (1974) for 1000 MWe Reactor Systems (52)

	Once-Through Cooling	Cooling Ponds	Mechanical Draft Towers	Natural Draft Towers
Heat dissipated	6.6×10^9 Btu/hr	6.6×10^9 Btu/hr	6.8×10^9 Btu/hr	6.9×10^9 Btu/hr
ΔT from condenser	20°F	20°F	31.6°F	29.5°F
ΔT after mixing	3°F	3°F	5.0°F	5.0°F
Water requirements				
Flow rate	600,000 gpm	600,000 gpm	374,000 gpm	373,000 gpm
Drift and evaporation			17,000 gpm	15,000 gpm
Blow down			61,320 gpm (with mixing)	4,000 gpm
Make-up			24,050 gpm	19,000 gpm
Land use				
Acreage required	30 acres	2000 acres		6 acres
Cost	$9,000	$60,000		
Chemical treatment costs	$50,000/yr		$100,000/yr	$100,000/yr
Water costs	$196,245/yr			
Cost of the system[a]	$5,750,000	$2,500,000	$22,000,000	$23,500,000
Operation and maintenance costs	base	base	$4,640,000	base
Number of cooling towers			3	1
Size of the cooling towers			55'x60'x180'	400'x500'
Capability loss ($1000)			105	69
Plume size			10 miles downwind max.	Touch down 3-15 miles 1000 above tower
Noise pollution		—	66 dB	negligible

[a]Cooling tower costs for a typical 1100 Megawatt plant:

Type of Tower	Nuclear	Fossil
Wood mechanical draft	$4 million	$2.5-3 million
Concrete mechanical draft	$6-8 million	$4.4-7.5 million
Natural draft	$8.75-13.5 million	$10-16 million
Wet/dry mechanical draft	$10-15 million	$16-20 million

Table 67. Comparative Costs of Alternative Cooling Systems ($\Delta T = 10°F$) (46)

Cooling System	Equipment Capital Cost $/kW	Operating Cost $/kW-yr	Maintenance Cost $/kW-yr	Total Cost $/kW-yr	Additional Cost to Consumer %
Once through run-of-river	5.88	0.99	0.99	2.51	0
Once through estuary	6.88	0.99	0.99	2.60	0.05
Cooling pond	7.50	1.24	1.24	3.16	0.37
Spray pond	8.10	1.98	0.99	3.70	0.67
ND wet tower; river makeup	12.50	1.98	1.98	6.16	2.08
ND wet tower; reservoir makeup	15.00	1.98	1.98	6.38	2.21
MD wet tower	9.40	2.34	2.34	6.23	2.12
ND dry cooling tower	22.0	2.00	2.00	7.38	2.78
MD dry cooling tower	15.0	2.34	2.34	7.43	2.82
Evaporative condenser	13.0	1.40	1.40	4.67	1.34
Air cooled condenser	17.0	1.00	0.50	4.43	1.10

necessarily the most acceptable one, there are several options open for evaluation (49). These are subject, however, to legal restraints through water quality standards, such as those listed in Table 62, that restrict plant effluent temperatures in terms of maximum ΔT values without specifying any permissible mixing zones. By concentrating simply on temperature rises, without taking into consideration other environmental factors, many power plant operators have been forced to build cooling towers where this either was not warranted in terms of the fish population to be protected or where other environmental impact arising from chemical releases, water losses, noise levels, plume effects, aesthetic impact of tall towers, and land use considerations might otherwise have been considered relatively more objectionable. In one case the cooling tower installation ostensibly costing $38 million was intended to save a fish catch of an estimated worth of $76 per year! In a large measure this concern with the preservation of fresh water fish reflects a situation unique to the United States, where the value of the subsidiary business related to the amateur angler exceeds the value of the commercial fisheries (Table 68). To a certain extent the magnitude of sportfishing-related activities may be adversely affected by impending fuel shortages, resulting probably in some reordering of priorities.

Table 68. U.S. Business Volume Associated with Thermal Pollution Effects (12)

Commercial Fisheries, 1963:	
Gross Receipts, United States Total	$277,144,000
Pacific Northwest	$102,691,000
Sport Fishing, 1965:	
No. of Persons who fished	28,348,000
Expenditures by Sport Fishermen	$2,925,000,000
Pleasure Boating, 1967:	
Value of Boats and Motors Sold	$552,000,000
Electric Light and Power Industry, 1967:	
Revenue from Ultimate Customers	$17,223,000,000
Required Investments for Thermal Pollution	
Abatement in Power-Generation (through 1973)	$1,300,000,000

By giving exclusive priority to an absolute reduction in effluent temperature, cost-benefit assessments may have been grossly distorted. For most large power plants, problems of ensuring continual availability of water would increasingly favor a closed-loop cooling system. However, it would appear (52) that substantial savings can be achieved in many cases with reasonable environmental effects by permitting slightly higher discharge temperatures, $\Delta T = 10 - 15°F$, with the use of diffusers and a well-defined mixing zone and, to the extent possible, a by-pass zone for fish permitting them to seek optimum temperatures and to acclimate themselves. In regions where water resource management is important to protect the rights of down-stream users, minimizing drift losses may be an overriding consideration and this should be reflected in all applicable regulations.

In summary, waste heat dissipation poses severe environmental problems that can be solved technically as long as a flexible approach is maintained and a fair evaluation is made of the relative costs and benefits of various proposed systems.

REFERENCES

1. *Industrial Waste Guide on Thermal Pollution* (Corvallis, Oregon: Federal Water Pollution Control Administration, 1968).
2. Summers, C. M. "The Conversion of Energy," *Scient. Amer.* 224 (3), 149 (1971).
3. Jaske, R. T., J. F. Fletcher and K. R. Wise. "A National Estimate of Public and Industrial Heat Rejection Requirements by Decades Through the Year 2000 A.D.," Paper No. 37A, 67th National

Meeting Am. Inst. of Chem. Engrs., Atlanta, 1970. Report BNWL-SA-3052, (Richland, Washington: Battelle Northwest, 1970).

4. *The 1970 National Power Survey.* (Washington, D.C.: Federal Power Commission, 1971).

5. Hanna, S. R. and S. D. Swisher. "Meteorological Effects of the Heat and Moisture Produced by Man," *Nucl. Safety* 12, 114 (1971).

6. Häfele, W. "Energy Systems," *Bull. IAEA* 16, 3 (1974).

7. Bolin, B. "The Carbon Cycle," *Scient. Amer.* 223(3), 125 (1970).

8. Committee on Power Plant Siting. *Engineering for Resolution of the Energy-Environment Dilemma* (Washington, D.C.: National Academy of Engineering, 1972).

9. Brady, D. K., W. L. Graves and J. C. Geyer. "Surface Heat Exchange at Power Plant Cooling Lakes," Publ. No. 69-901, (New York: Edison Electric Inst., 1969).

10. Krenkel, P. A., W. A. Cawley and V. A. Minch. *J. Water Poll. Control Fed.* 37, 1216 (1965).

11. Polk, E. M., B. A. Benedict and F. L. Parker. "Dispersion of Thermal Discharges in Bodies of Water," *Chem. Eng. Progress Symp. Series* 67 (119), 111 (1971).

12. Morgan, P. V. and H. C. Bramer. "Thermal Pollution as a Factor in Power Plant Site Selection," *Nucl. News* 12(9), 70 (1969), from *Proc. Am. Power Conf.* 31 (1969).

13. Wright, J. H., J. B. F. Champlin and O. H. Davis. "The Impact of Environmental Radiation and Discharge Heat from Nuclear Power Plants," in *Environmental Aspects of Nuclear Power Plants* (Vienna: International Atomic Energy Agency, 1971).

14. "Thermal Effects and U.S. Nuclear Power Stations," Report WASH-1169 (Washington, D. C.: U.S. Atomic Energy Commission, 1971).

15. U.S. Atomic Energy Commission. "Final Environmental Statement, Shoreham Nuclear Power Plant," (1972).

16. Coutant, C. C. "Cold Shock: Tolerable Temperatures and Guidance for Power Plants," *Trans ANS* 18, 47 (1974).

17. Coutant, C. C., H. M. Ducharme and J. R. Fisher. "Effects of Cold Shock on Vulnerability of Juvenile Catfish and Largemouth Bass to Predation," *J. Fisheries Res. Bd. Can.* 31, 351 (1974).

18. Coutant, C. C. "Biological Aspects of Thermal Pollution, II. Scientific Basis for Water Temperature Standards of Power Plants," *CRC Crit. Rev. Environ. Control.* 3, 1 (1972).

19. Dynatech R/D Co. "Total Community Considerations in the Utilization of Rejected Heat," Report 16130 DHS 11/70 Water Pollution Control Research Series (Washington, D.C.: U.S. Environmental Protection Agency, 1970).

20. Becker, C. B. and T. O. Thatcher. "Toxicity of Power Plant Chemicals to Aquatic Life," Report WASH-1249, U.S. Atomic Energy Commission (1973).

21. Josefsson, L. and J. Thunell. "Nuclear District Heating, a Study for the Town of Lund," in *Containment and Siting of Nuclear Power Plants* (Vienna: International Atomic Energy Agency, 1967).

22. Thomas, K. T., N. S. Sunder Rajan, and M. P. S. Ramani. "Prospects and Planning for Nuclear-Powered Agro-Industrial Complexes in India," *Proc. 4th Internat. Conf. Peaceful Uses of Atomic Energy* **6,** 131 (1971).

23. Garton, R. R. and A. G. Christianson. "Beneficial Uses of Waste Heat—An Evaluation," Report 16130FHJ 109/70 (Corvallis, Oregon: U.S. Environmental Protection Agency, 1970).

24. Delyannis, A. A. "Nuclear Energy Centres and Agro-Industrial Complexes," Technical Reports Series No. 140 (Vienna: International Atomic Energy Agency, 1972).

25. Merriman, D. "The Calefaction of a River," *Scient. Amer.* **222**(5), 42 (1970).

26. Parker, F. L. and P. A. Krenkel. *Physical and Engineering Aspects of Thermal Pollution* (Cleveland: CRC Press, 1970).

27. Koss, R., Ed. "Environmental Responses to Thermal Discharges from Marshall Steam Station, Lake Norman, North Carolina," Interim Report, EEI Project RP-49 (Baltimore, Maryland: Johns Hopkins University, 1971).

28. Brady, D. K., J. C. Geyer and J. R. Sculley. "Analysis of Heat Transfer at Cooling Lakes," *Chem. Eng. Progress Symp. Series* **67** (119), 120 (1971).

29. Mehta, B. M. and R. C. Ahlert. "Analysis of the Dispersion of Thermal Effluents," *Chem. Eng. Progress Symp. Series* **67** (119) 126 (1971).

30. "Reporting Procedure for Mathematical Models Selected to Predict Heated Effluent Dispersion in Natural Water Bodies," Regulatory Guide 4.4 (Washington, D.C.: U.S. Atomic Energy Commission, 1974).

31. Campbell, J. F. and J. A. Schetz. "Analysis of the Injection of a Heated Turbulent Jet into a Moving Mainstream," Report VPI-E-72-24 (Blacksburg, Va.: Virginia Polytechnic Institute, 1972).

32. Stefan, H., N. Hayakawa and F. R. Schiebe. "Surface Discharge of Heated Water," Report 16130FSU 12/71 (Washington, D.C.: U.S. Environmental Protection Agency, 1971).

33. Stolzenbach, K. D. and D. R. F. Harleman. "Analytical and Experimental Investigation of Surface Discharges of Heated Water," Technical Report No. 135, Parsons Laboratory for Water Resources and Hydrodynamics (Cambridge, Massachusetts: Massachusetts Institute of Technology, 1971).

34. Jaske, R. T., W. L. Templeton, J. R. Eliason and J. C. Sonnichsen. "Improved Methods for Planning of Thermal Discharges before Site Acquisition with a Specific Case Example on the Columbia River," in *Environmental Aspects of Nuclear Power Stations, Proceedings New York Symposium* (Vienna: International Atomic Energy Agency, 1971).

35. Shirazi, M. A. and L. R. Davis. "Workbook of Thermal Plume Prediction: I-Submerged Discharges," Report EPA-R2-72-005a (Washington, D.C.: U.S. Environmental Protection Agency, 1972).

36. Koh, R. C. Y., N. H. Brooks, E. J. List and E. J. Wolanski. "Hydraulic Modeling of Thermal Outfall Diffusers for the San Onofre

Nuclear Power Plant," Report No. KH-R-30 (Pasadena: California Institute of Technology, 1974).

37. Adams, J. R. "Thermal Power, Aquatic Life and Kilowatts on the Pacific Coast," *Nucl. News* **12**(9), 75 (1969) from *Proc. Am. Power Conf.* **31** (1969).

38. Scofield, F. C. and R. A. Fazzolare. "Nuclear Power Plant Heat Rejection in an Arid Climate," *Nucl. Technol.* **20**, 140 (1973).

39. Whipple, W., J. V. Hunter, F. W. Dittman, S. L. Yu, B. Davidson and G. Mattingly. "Oxygen Regeneration of Polluted Rivers: The Passaic River," Water Poll. Control Research Series, Report 16080 FYA 03/71 (Washington, D.C.: U.S. Environmental Protection Agency, 1971).

40. Jaske, R. T. "The Need for Advance Planning of Thermal Discharges," *Nucl. News* **12**(9), 65 (1969); from *Proc. Am. Power Conf.* **31** (1969).

41. "Spray Ponds: The Answer to Thermal Pollution Problems," (Burlington, Massachusetts: Spray Engineering Co., n.d.).

42. "Cooling Towers," *Power* **117**(3), 17 (1973).

43. Wordson, R. D. "Cooling Towers," *Scient. Amer.* **224**(5), 70 (1971).

44. "The Safety of Nuclear Power Reactors (Light-Water-Cooled) and Related Facilities," Report WASH-1250 (Washington, D.C.: U.S. Atomic Energy Commission, 1973).

45. Marley Co. *Cooling Tower Fundamentals and Application Principles* (Kansas City, Missouri: The Marley Co., 1967).

46. Dynatech R/D Co. "A Survey of Alternate Methods for Cooling Condenser Discharge Water-Operating Characteristics and Design Criteria," Water Pollution Control Research Series Report 16130 DHS 08/70 (Washington, D.C.: U.S. Environmental Protection Agency, 1970).

47. Jedlicka, C. L. "Nomographs for Thermal Pollution Control Systems," Report EPA-660-273-4 (Washington, D.C.: U.S. Environmental Protection Agency, 1973).

48. Kolflat, T. D. "Cooling Tower Practices," *Power Eng.* **78**(1), 32 (1974).

49. Mar. B. W., J. A. Crutchfield, E. B. Welch, M. C. Bell, N. M. Geitner, R. M. Bush, T. Meyer, R. Porter and A. H. Saad. "Analysis of Engineering Alternatives for Environmental Protection from Thermal Discharges," Report EPA-R2-73-161 (Washington, D.C.: U.S. Environmental Protection Agency, 1973).

50. Rossie, J. P., E. A. Cecil, and R. O. Young. "Cost Comparison of Dry-Type and Conventional Cooling Systems for Representative Nuclear Generating Plants," U.S. Atomic Energy Commission Report TID-26007 (1972).

51. *Nucl. Indust.* **17** (11-12), 67 (1970).

52. Eichholz, G. G., R. M. Heublein, J. S. Bland. Georgia Institute of Technology (unpublished) (1974).

ADDITIONAL REFERENCES

Ahlert, R. C. "Mathematical Description of Biological and Physical Processes in Heated Streams," *Chem. Eng. Progr. Ser. Water, 1971*, **68** (124), 191 (1972).

Beall, S. E. "Reducing the Environmental Impact of Population Growth by the Use of Waste Heat," Annual Meeting AAAS, Chicago (1970).

Brown, D. H., G. T. Bonk and W. A. Hartman. "Thermal Discharges and the Environment," General Electric Co. (1971).

Davis, W. S. "Conditions for Coexistence of Aquatic Communities with the Expanding Nuclear Power Industry," *Nucl. Safety* **10**, 292 (1969).

E. G. and G. Inc. "Potential Environmental Modifications Produced by Large Evaporative Cooling Towers," Report 16130 DNH 01/71 (Boulder, Colorado, 1971).

Eisenbud, M. and G. Gleason, Eds. *Electric Power and Thermal Discharges* (New York: Gordon and Breach, 1970).

Foster, R. F., R. T. Jaske and W. L. Templeton. "The Biological Cost of Discharging Heat to Rivers," *Proceedings 4th Internat. Conf. Peaceful Uses of Atomic Energy* **11**, 631 (1971).

Jirka, G. and D. R. F. Harleman. "The Mechanics of Submerged Multiport Diffusers for Buoyant Discharges in Shallow Water," Report No. 169, R73-22, Department of Civil Engineering (Cambridge, Massachusetts: Massachusetts Institute of Technology, 1973).

Lowry, W. P. "The Climate of Cities," *Scient. Amer.* **217**(2), 15 (1967).

Margen, P. H. "Thermal Energy Storage in Rock Chambers," *Proc. 1971 Geneva Conf. Peaceful Uses of Atomic Energy* **4**, 177 (1971).

McKelvey, K. K. and M. Brooke. *The Industrial Cooling Tower* (Amsterdam: Elsevier, 1958).

Middlebrooks, E. J., M. J. Gaspar, R. D. Gaspar, J. H. Reynolds and D. B. Porcella. "Effects of Temperature on the Toxicity to the Aquatic Biota of Waste Discharges—A Compilation of the Literature," Report PRW6-105-1 Utah Water Research Laboratory (Logan, Utah: Utah State University, 1973).

Problems in Disposal of Waste Heat for Steam Electric Plants (Washington, D.C.: Federal Power Commission, 1969).

Ryan, P. J. "Heat Dissipation by Spray Cooling," in *Themal Pollution Analysis, Proc. Blacksburg Symposium.* (Blacksburg, Virginia: Virginia Polytechnic Institute and State University, 1974).

Stobbs, J. J. and H. J. Pekarek. "The Influence of Thermal Pollution on Nuclear Power Plant Site Selection," in *Containment and Siting of Nuclear Power Plants* (Vienna: International Atomic Energy Agency, 1967).

Winiarski, L. D., B. A. Tichenor and K. V. Byram. "A Method for Predicting the Performance of Natural Draft Cooling Towers," Report 16130 GKF 12-70 (Corvallis, Oregon: U.S. Environmental Protection Agency, 1970).

CHAPTER 7

TREATMENT OF RADIOACTIVE EFFLUENTS

INTRODUCTION

In a perfect reactor system the only radioactive material leaving the plant would be the spent fuel assemblies. In a real plant, however, with many pumps, rotating systems, heat exchangers and valves, a small but finite leakage of contaminated steam and fluids into the plant environment has to be expected. Plant systems must be designed to collect and concentrate the radioactive contaminants to facilitate their safe disposal and to purify the air and water that can then be discharged to the surroundings.

For many years it was considered sufficient if the purification systems ensured that residual radioactive concentrations in the effluent gases and liquids were well below the maximum permissible concentrations (MPC) in air and water, to cause minimal exposure of the general population in the vicinity for the mixture of radioactive isotopes involved; in fact, the operating records of the various commercial reactors around the world showed that such low effluent concentrations could be obtained in normal operations (1-3). More recently, largely as a result of the efforts of Gofman and Tamplin and various environmental groups, the U.S. Atomic Energy Commission has adopted guidelines requiring the effluents to have concentrations of radioisotopes that are "as low as practicable." In theory, a level so-defined can be decreased indefinitely, given unlimited economic resources and further advances in technology, but in practice "as low as practicable" is interpreted to mean less than 1% of MPC for most radioisotopes, with a total annual release of iodine-131 not to exceed one curie per reactor at any given site (4).

In order to achieve very low release levels, two steps have to be taken: the first involves reduction of radioactivity in the primary source, the reactor coolant loop, by means of continuous and efficient purification.

This operation was described in Chapter 5. The second step involves a choice of treatment for each waste component among three possible strategies: (1) dilute and disperse, (2) delay and decay, and (3) concentrate and contain. The first mode has been the traditional method of disposing of low-concentration wastes. The second method has limited applicability and can be costly. The third involves the collection of most of the activity remaining in the waste liquids and the building air, and processing it to the highest practical degree to ensure that the waste water and off-gases finally released to the surroundings have as low a level of activity "as practicable." It must be recognized that the quantities of radioactivity finally released, except for tritium and noble gases, are insignificant compared with the activities handled routinely in the nuclear medicine department of any major hospital. Note the important distinction between *concentration* of radioactive isotopes in coolant and effluents (in $\mu Ci/ml$) and *total quantities* contained or released (in Ci/yr).

LIQUID WASTES

As was pointed out in Chapter 5, the pathways by which fission and activation products may escape from the primary coolant depend on the system design and are radically different for PWR's, BWR's and GCR's. Figures 56 and 57 identify the main routes of escape in a schematic diagram of a BWR and PWR. Figure 58 identifies pathways of all stray radioactive materials from their point of origin to their release to the environment for an HTGR (6). Such contaminated streams are referred to as radioactive wastes (*radwastes*). In essence, radioactive wastes escape from the reactor coolant through leaks in the steam system, the purification system, air ejectors, pump shafts and valve seals, and sampling locations. Other sources of radioactive wastes are the spent resin used in the coolant system purification, laundry wastes, laboratory drains and floor drains. In the latter cases the volume of radwastes is greatly increased by dilution in other sources of wastewater such as those listed in Table 69.

By reducing water volumes used for washing floors and equipment, considerable savings may be achieved in suspect waste tank storage and low-level treatment systems. Because of the nature of the steam supply systems, one would expect for the PWR that the bulk of the activity escaping would be in liquid form (except for the gaseous elements), whereas for the BWR and GCR types the activity would appear mainly in the airborne effluents. Most liquid wastes can be divided into two broad categories: "clean" or low conductivity wastes, and "dirty" or high conductivity wastes (7). These terms do not relate to the quantity

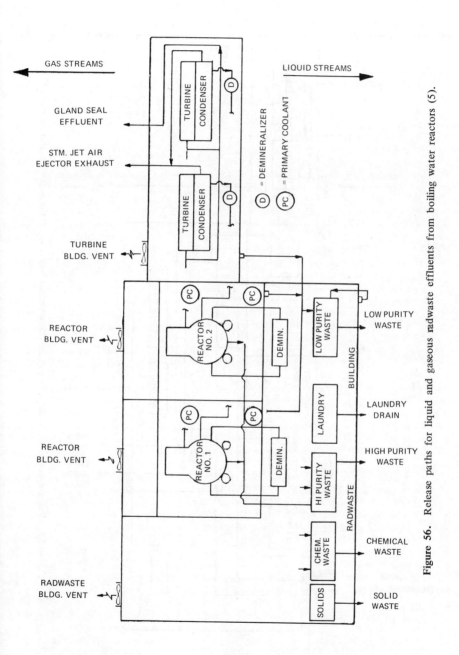

Figure 56. Release paths for liquid and gaseous radwaste effluents from boiling water reactors (5).

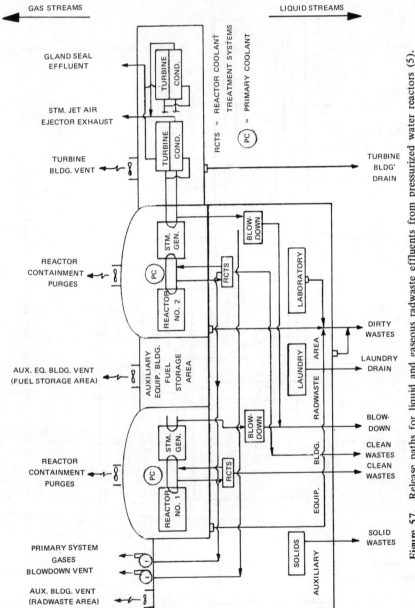

Figure 57. Release paths for liquid and gaseous radwaste effluents from pressurized water reactors (5).

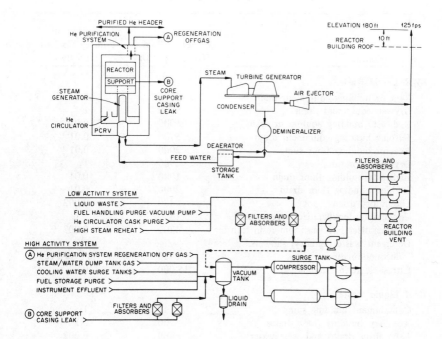

Figure 58. HTGR gaseous radioactive waste system (Fort St. Vrain) (6).

of radioactive material contained in the wastes; they are based on the conductivity of liquid, which is a measure of total dissolved solids and hence ionic or particulate concentration. This affects the different treatment methods that can be used to reduce or remove radioactivity.

BWR Liquids

Denton (1) classified liquids from a BWR into four source categories according to concentration of dissolved inert solids:

(1) controlled leakage from primary and secondary equipment that is generally of low conductivity but high radioactivity (clean wastes)

(2) floor drains that have high conductivity and low radioactivity (dirty wastes)

(3) chemical wastes that contain the greatest concentration of dissolved inert materials from the regeneration of condensate demineralizers and certain floor drains (dirty wastes)

Table 69. Sources of Liquid Waste (5)

Source	Flow Rate Gallons per Day	Fraction of Primary Coolant Activity
BWR Liquid Waste		
Reactor building equipment drain	2000	0.01
Drywell equipment drain	5800	1
Radwaste building equipment drain	1000	0.01
Turbine building equipment drain	5700	0.01
Reactor building floor sump	2000	0.01
Drywell floor sump	2900	1
Radwaste building floor drain	1000	0.01
Turbine building floor drain	2000	0.01
Laboratory drains	400	0.002
Condensate demineralizer regeneration	1800[a]	b
Demineralizer backwash and resin transfer	4200	0.002
Detergent waste (laundry, showers, decontamination)	450	10^{-4} μCi/cc
Ultrasonic resin cleaning	15,000	0.02
PWR Liquid Waste		
Containment building sump	40	1
Auxiliary building floor drains	200	0.1
Laboratory drains and wastewater	400	0.002
Sample drains[c]	35	1
Turbine building floor drains[d]	7200	Main steam activity
Miscellaneous sources	700	10^{-2}
Detergent waste (laundry, decontamination, showers)	450	

[a]Based on one demineralizer being regenerated every 5 days.
[b]Activity levels based on radionuclide inventory on resins using DF for condensate demineralizers and a constant regeneration efficiency.
[c]15 gpd for continuous purge recycle.
[d]Use 3200 gpd for once-through steam generator systems.

(4) miscellaneous wastes from equipment, washdown, decontamination processes, laboratories (dirty wastes).

Each of these sources is itemized in Table 69. Their volumes will vary from plant to plant depending on the design and type of equipment used but these values are representative of the nuclear industry. Apparent inconsistencies will be found in this and subsequent tables between data from various government and industry sources, and all such tables should be regarded as indicative only of magnitudes involved.

PWR Liquids

Denton (1) also classified PWR liquid wastes into four source categories:

(1) deaerated wastes (clean)
(2) aerated wastes (dirty)
(3) laundry and hot shower wastes (dirty)
(4) steam generator blowdown (dirty).

Since the PWR is a closed system, the primary coolant and leakages from the primary loop are considered deaerated. Steam generator blowdown results from opening a valve that permits steam generator pressure to remove sedimentation or scale that builds up inside. Fresh water is charged into the generator, thereby reducing total dissolved solids. This procedure is conducted either at frequent intervals or continuously to reduce carryover and improve heat transfer. PWR sources are also itemized in Table 69; much of the liquid listed can be recycled after decontamination.

HTGR Liquids

Liquid wastes in an HTGR are produced as a result of decontamination of fuel-handling machinery and control rod drives following the annual plant refueling. During refueling potentially contaminated water from showers and laboratory operations is also present. During normal operation there is very little liquid radioactive waste generated. Estimated quantities of liquids released are shown in Table 70 (8). This low-volume waste presents little problem in cleaning up and removing radioactivity prior to release to the environment.

GASEOUS WASTES

BWR Gaseous Wastes

A review of principal escape pathways for gaseous effluents from BWR's was given by Michels and Horton (9); they are:

(1) steam jet air ejector: the principal release point for gaseous radioactive effluents from a BWR during normal operation; 99% of the total gross activity from a BWR is estimated to be released here. On newer plants with delay holdup and charcoal bed filters this is no longer the case.
(2) mechanical vacuum pump
(3) turbine gland seal exhaust
(4) purging of the dry well and suppression chamber.

Table 70. Sources of Liquid Wastes and Off-Gases from HTGR's[a] (6,8)

Liquid Wastes

Source	Quantity	Activity Concentration
Decontamination during refueling	15,000 l/yr	5 μCi/ml
Showers during refueling	95,000 l/yr	10^{-4} μCi/ml
Other annual releases (showers)	190,000 l/yr	10^{-4} μCi/ml

Off-Gases

Source	Quantity	Frequency	Activity	Remarks
He purification system regeneration off-gas	1450 SCF (41 m^3)	once/6 months	3.49x10^3 Ci	–
He purification system unit regeneration	800 SCF (23 m^3)	once/month	630 Ci	activity is essentially H^3
Instrument effluent	0.014 SCFM (0.024 m^3/sec)	continuous	4x10^2 μCi/min	–
Fuel handling purge system vacuum pump	3390 SCF (96 m^3)	once/year	100 Ci	–
Core support floor vent	21 SCFM (36 m^3/sec)	continuous	0.58 Ci/min	primarily noble gas activity

[a]3,000 MWt HTGR.

All off-gases from a BWR are retained within the containment facility for treatment. Air ejector off-gas is channeled directly to the treatment system.

PWR Gaseous Wastes

Due to the difference in plant design, the principal PWR escape paths for gaseous wastes are slightly different from those of the BWR. Gaseous wastes result from:

(1) reactor plant leakages
(2) reactor coolant expansion due to heatup on start-up
(3) bleed water required to maintain the proper boron concentration if chemical shim is used
(4) degasification of primary coolant
(5) reactor plant drains.

HTGR Gaseous Wastes

The pathways for radioactive gas emission from an HTGR together with the quantities involved and projected activities are shown in Table 70 (6). Again further improvements should be obtained in plants now under construction.

Off-gas from the helium purification system results from scheduled regeneration of the system adsorbers. Instrument effluent results from analytical instrumentation provided to monitor activity in the primary coolant. Fuel handling off-gas results from evacuation of the helium atmosphere inside the fuel handling equipment.

Fast Breeder Reactors

For fast breeder reactors, both gas-cooled and liquid-metal-cooled, the radwaste system would be similar to that of a PWR since both types utilize a secondary steam loop for power generation. However, the source terms and leakage factors will be different to account for the slightly different coolant isotope inventory and for the fact that many fission products, such as iodine or plutonium, form less volatile compounds with liquid sodium than with water. On the other hand the radwaste system for an LMFBR must have adequate capacity to remove the large amount of 15-hour sodium-24 that may be liberated in case of a leak in the primary coolant and a consequent sodium fire. Although radiation damage effects would be expected to be more severe in a fast reactor than a thermal one, in actual fact no higher fuel failure rate could be tolerated in this case, and hence fission product leakage to the coolant would not be significantly higher. However, employment of stainless steel cladding instead of zircaloy leads to appreciably higher leakage of tritium from the fuel. A very detailed review of these factors and some unresolved issues can be found in ERDA-1535 (10).

SOURCE TERMS

In trying to estimate the quantities of radioactive material to be treated or released in plant effluents, obviously it is important to obtain a realistic estimate of the initial concentrations and forms present in the coolant and the proportion of each that is likely to escape into the building air, the system vents or the drainage system. Such estimates as published have varied widely, based as they are on various assumptions as to fuel leakage and to coolant purification. For example, Table 71 (5) shows some assumed escape rates from the fuel into the coolant assuming 0.25% failed fuel for PWR's and 0.16% for BWR's. Table 72 shows

Table 71. Principal Parameters Used in the Source Term Calculations
for the BWR and PWR 1100 MWe model reactors (5)

Reactor Power Level	3500 MWt
Plant Capacity Factor	80%
Fraction of Fuel Releasing Radioactivity to the Primary Coolant (Zircaloy-Clad Fuel)	0.25% (PWR)
Noble gases	100,000 μCi/sec for 3400 MWt after 30-min delay (BWR)
Iodine-131 (independent of power level)	5×10^{-3} μCi/ml (BWR)

Equilibrium Primary Coolant Radionuclide Concentration

Fission products released to the primary coolant

	Escape Rate Coefficients (sec^{-1})	
Element	BWR	PWR
Br	1.0×10^{-9}	1.3×10^{-8}
Rb, Cs	2.5×10^{-10}	1.3×10^{-8}
Sr, Ba	2.5×10^{-10}	1.0×10^{-11}
Tc	4.0×10^{-11}	1.0×10^{-9}
Mo	1.0×10^{-11}	2.0×10^{-9}
Others	1.6×10^{-12}	1.6×10^{-12}
Xe, Kr, gases	–	6.5×10^{-8}

Corrosion Products Released to the Primary Coolant

Element	Release Rate Coefficients (day^{-1})
Fe	8.3×10^{-2}
Cr	3.3×10^{0}
Ni, Co	1.3×10^{-2}
Mn	1.3×10^{-2}
Mo	4.3×10^{-5}
Si	1.3×10^{-2}
C, S, P	1.3×10^{-2}
V, Ti, W	1.3×10^{-2}
Na, Li, Sn, Zn	1.3×10^{-2}

consequent concentrations in the coolant for a BWR, except for tritium
and the noble gases. The total inventory of volatiles in the core and
coolant of a 1000 MWe plant is shown in Table 73 (11).

The coolant concentrations can be controlled to some extent by ion
exchange systems that remove dissolved ions, radioactive or otherwise,
and the laundry wastes can be similarly decontaminated. Table 74 shows
the decontamination factors (DF) reported for ion exchange systems and

Table 72. Concentrations of Radionuclides in BWR Primary Coolant (5)

Nuclide	Concentration μCi/cc	Nuclide	Concentration μCi/cc	Nuclide	Concentration μCi/cc
Fission Products					
Sr-89	0.0029	Ru-103	0.000022	I-133	0.029
Sr-90	0.00014	Ru-106	0.0000062	I-135	0.04
Sr-91	0.018	Rh-103m	0.000099	Cs-134	0.00028
Y-90	0.00012	Rh-105	0.000033	Cs-136	0.0001
Y-91	0.0024	Rh-106	0.00001	Cs-137	0.00019
Y-92	0.018	Te-127m	0.0000057	Ba-137m	0.00036
Y-93	0.0093	Te-127	0.0001	Ba-140	0.0052
Zr-95	0.000033	Te-129m	0.000026	La-140	0.00042
Zr-97	0.000093	Te-129	0.00066	Ce-141	0.000057
Nb-95	0.000027	Te-131m	0.00016	Ce-143	0.00017
Nb-97m	0.0001	Te-131	0.0012	Ce-144	0.000018
Nb-97	0.0002	Te-132	0.00098	Pr-143	0.000028
Mo-99	0.0083	I-130	0.0003	Pr-144	0.000062
Tc-99m	0.0084	I-131	0.005	Nd-147	0.000012
		I-132	0.044		
Corrosion and Activation Products					
Na-24	0.0015	Fe-59	0.00058	W-185	0.000013
P-32	0.000024	Co-58	0.003	W-187	0.0037
P-33	0.000087	Co-60	0.00034	U-237	0.00001
Cr-51	0.00045	Ni-63	0.000028	Np-239	0.00075
Mo-54	0.00004	Nb-92	0.000093		
Fe-55	0.0015	Sn-117m	0.000023		
		All others[a]	0.5		
		Total[a]	0.94		

[a]Except tritium and noble gases.

evaporators (11). With the decontamination factors shown it is evident that a fair amount of residual activity will still remain in the coolant. In addition any tritium, present as T_2O or HTO, will continue to build up in the coolant loop except when the coolant is drained off deliberately to control tritium levels. Noble gases, too, obviously will not be removed and are free to leak into the building air by any available route. Some iodine and noble gases are also released to the atmosphere through steam dump valves intended to relieve high pressures in the secondary system of PWR's from various abnormal operations (such as load rejection).

Table 73. Volatile Radionuclide Inventory in a 1000 MWe PWR Plant (11)[a]

Radionuclide	Half-Life	Reactor Core (MCi)	Fuel-Cladding Gap (MCi)	Primary Coolant (Ci)
Iodines				
I-131	8.05 d	74.9	0.76	465
I-132	2.3 h	114.0	0.14	186
I-133	21 h	171.0	0.64	766
I-134	52 m	206.0	0.12	117
I-135	6.7 h	158.0	0.34	420
Kryptons				
Kr-85	10.8 y	0.66	0.067	334
Kr-85m	4.4 h	33.5	0.95	439
Kr-87	76 m	64.4	0.076	261
Kr-88	2.8 h	93.0	0.149	775
Xenons				
Xe-133	5.3 d	164.0	4.17	52,290
Xe-133m	2.3 d	4.0	0.019	692
Xe-135	9.2 h	43.6	0.084	1,488
Xe-135m	15.6 m	46.4	0.016	42

[a]Parameters: 3040 MWt PWR, operating at full power for 500 days; 1% of the fuel rods assumed to release radioactivity to primary coolant.

The magnitude of such releases via the main steam relief valves can be on the order of many tens of thousands of pounds of steam in one minute, but the total activity released this way per year has been estimated as 12.7 millicurie per year, mostly due to Xe-133 (11). In general, load-following plants produce higher release levels than base-loaded plants. Table 75 shows some parameters used for calculations on model reactors (5) to estimate concentrations and leak rates.

If no special further treatment is applied, the remaining activity may be released to the environment after appropriate dilution to obtain concentrations well below MPC levels; this was standard practice with earlier reactors. With the increase in reactor size and a tightening of permitted effluent release levels, further treatment is invariably required to maintain not only the concentrations in the effluents at low enough levels, but also to reduce the total quantity of activity thus released. In Tables 76 and 77, which show the effluent levels reported to the U.S. Atomic

Table 74. Liquid Radwaste System Component Decontamination Factors (11)

Components	Decontamination Factors (DFs)				
	I	Cs,Rb	Y	Mo	Others
Demineralizers					
PWR					
Mixed Bed (Li_3BO_3 form, CVCS)	10	2	1	1	10
Mixed Bed (steam generator blowdown)	10	2	1	1	50
BWR					
Mixed Bed (H-OH form, clean waste)	10^2	10	1	1	10^2
PWR & BWR					
Mixed Bed in Evaporator Condensate	10	10	1	1	10
Evaporators (PWR & BWR)					
Waste	10^3	10^4	10^4	10^4	10^4
Laundry	10^2	10^2	10^2	10^2	10^2
Boric acid evaporators	10^2	10^3	10^3	10^3	10^3
Other					
Removal of Mo and Y by plating out, filtration, demineralization, etc.	–	–	10	10^2	–

Decontamination Factors for Demineralizers[a]	Cation[b]	Anion	Cs, Rb
Mixed Bed (H^+OH^-)			
Reactor Coolant	10	10	10
Condensate	10^3	10^3	10
Clean Waste	$10^2(10)$	$10^2(10)$	10(10)
Dirty Waste	$10^2(10)$	$10^2(10)$	1(10)
Powdex (any system)	10(10)	10(10)	1(10)

[a]For two demineralizers in series, the DF for second demineralizer is shown in parenthesis. For polishing demineralizer after an evaporator, use value for second demineralizer in series.

[b]Does not include Cs, Mo, Y, Rb, Tc.

Energy Commission for 1972 for liquid and airborne wastes discharged by commercial reactors (2), one can see the significant improvement obtainable in newer plants employing state-of-the-art equipment. Similar reductions have been reported for Canadian and European reactors.

There are several aspects of transcendent importance shown in Tables 76 and 77. First of all, out of the whole enormous fission product inventory in the reactor core, only a tiny fraction, a few curies per year,

Table 75. Release Parameters Used in Source Term Calculations for
Two-Reactor Plants (5)

	PWR	BWR
Tritium Released in Liquid Waste	3500 Ci/yr/reactor	20 Ci/yr/reactor
Weight of Water in Primary System	5.0×10^5 lb	1.6×10^6 lb
Weight of Water in Secondary System	4.1×10^5 lb	–
Number of Steam Generators	4	
Weight of Steam in Each Steam Generator	8.2×10^3 lb	
Weight of Liquid in Each Steam Generator	9.7×10^4 lb	
Total Steam Flow	1.5×10^7 lb/hr	1.5×10^7 lb/hr
Letdown Flow Rate	75 gpm	
Shim Bleed Flow Rate	1 gpm	
Primary-to-Secondary Leakage Rate	110 lb/day	
Reactor Building Leakage Rate	240 lb/day	500 lb/hr
Auxiliary Building Leakage Rate	160 lb/day	
Turbine Gland Seal Steam Leakage Rate	–	1.5×10^4 lb/hr
Turbine Building Steam Leakage Rate		
No special design features to reduce the leakage from steam line valves	1700 lb/hr	1700 lb/hr
Special design features to reduce leakage from steam line valves 2-1/2 in. and larger	340 lb/hr	340 lb/hr
Steam Generator Blowdown Rate	9000 lb/hr	
Primary System Volumes Degassed	2/yr	
Containment Building Volume	2×10^6 ft^3	
Frequency of Containment Purge	4/yr	
Partition Factors for Radioiodine		
Steam Generator, Internal Partition		0.01
Recirculating U-Tube	0.01	
Once-Through	1	
Steam Generator Blowdown Tank Vent	0.05	
Auxiliary Building Liquid Leakage	0.005	
Turbine Building Secondary Steam Leakage	1	
Containment Building Leakage	0.1	
Primary Coolant–Hot	0.1	
Primary Coolant–Cold	0.001	

Table 76. 1972 Summary of Liquid Effluents from U.S. Power Reactors (2)

Facility	Mixed Fission and Activation Products			Tritium		
	Curies Released	Average Concentration (μCi/ml)	Percent of MPC Limit	Curies Released	Average Concentration (μCi/ml)	Percent of MPC Limit
Boiling Water Reactors						
Oyster Creek	10.0	8.62×10^{-9}	32.3	61.6	5.31×10^{-8}	0.00177
Nine Mile Point	34.6	7.80×10^{-8}	3.18	27.8	6.28×10^{-8}	0.00209
Millstone 1	51.5	8.35×10^{-8}	7.04	20.9	3.39×10^{-8}	0.00113
Dresden 1	6.75	2.32×10^{-8}	23.2	43.3	1.48×10^{-7}	0.00493
Dresden 2,3	22.1	1.49×10^{-8}	14.9	25.9	1.81×10^{-8}	0.0006
LaCrosse	48.5	1.46×10^{-7}	6.39	120.0	3.61×10^{-7}	0.012
Monticello	2.90×10^{-6}	2.78×10^{-16}	2.05×10^{-8}	7.60×10^{-5}	7.31×10^{-15}	2.44×10^{-10}
Big Rock Point	1.09	1.06×10^{-8}	0.88	10.4	1.07×10^{-8}	0.0034
Humboldt Bay	1.40	7.78×10^{-9}	0.11	13.0	7.22×10^{-8}	0.00241
*Pilgrim	1.45	5.58×10^{-8}	7.98	4.18	1.61×10^{-7}	1.61
Quad Cities 1,2	2.41	1.73×10^{-9}	1.73	4.70	3.38×10^{-9}	0.000113
Pressurized Water Reactors						
*Maine Yankee	0.0169	1.75×10^{-10}	0.0919	9.22	9.57×10^{-8}	1.91
Palisades	6.81	8.50×10^{-9}	0.005	208.0	2.60×10^{-7}	0.430
Yankee	0.0206	1.28×10^{-10}	0.0249	803.0	4.97×10^{-6}	0.166
Indian Point 1	25.4	5.12×10^{-8}	2.88	574.0	1.16×10^{-6}	0.040
R.E. Ginna	0.375	5.18×10^{-10}	0.00234	119.0	1.64×10^{-7}	0.00547
Connecticut Yankee	4.78	6.20×10^{-9}	0.233	5890.0	7.64×10^{-6}	0.255
H.B. Robinson	0.826	3.82×10^{-8}	3.82	405.0	1.88×10^{-7}	0.000625
San Onofre	30.3	5.22×10^{-8}	1.49	3480.0	5.99×10^{-6}	0.2
Point Beach 1,2	1.53	3.11×10^{-9}	0.0052	563.0	1.13×10^{-6}	0.037
*Surry 1	0.0252	1.83×10^{-10}	0.00395	5.03	3.64×10^{-8}	0.091
Nonwater Reactors						
Peach Bottom 1	0.0209	1.57×10^{-9}	1.57	1.7	1.28×10^{-7}	0.0427
Fermi	0.222	9.53×10^{-9}	7.53	None		

*Operated less than one year.

Table 77. 1972 Summary of Airborne Effluents from U.S. Power Reactors (2)

Facility	Noble Gases			Halogens and Particulates (Half-life greater than 8 days)		
	Curies Released	Average Release Rate (µCi/sec)	Percent of MPC Limit	Curies Released	Average Release Rate (µCi/sec)	Percent of MPC Limit
Boiling Water Reactors						
Oyster Creek	8.66×10^5	2.74×10^4	10.4	6.48	2.05×10^{-1}	5.13
Nine Mile Point	5.17×10^5	1.64×10^4	2.01	0.969	3.08×10^{-2}	2.01
Millstone 1	7.26×10^5	2.33×10^4	2.91	1.32	4.21×10^{-2}	1.37
Dresden 1	8.77×10^5	2.97×10^4	5.30	2.75	8.73×10^{-2}	3.65
Dresden 2,3	4.29×10^5	1.32×10^4	1.51	5.89	1.86×10^{-1}	3.60
LaCrosse	3.06×10^4	9.72×10^2	5.13	<0.712	2.26×10^{-2}	43.4
Monticello	7.51×10^5	2.38×10^4	8.8	0.589	1.83×10^{-2}	1.6
Big Rock Point	2.58×10^5	8.19×10^3	0.819	0.148	4.92×10^{-3}	0.416
Humboldt Bay	4.30×10^5	1.37×10^4	27.3	4.79×10^{-1}	1.52×10^{-2}	8.41
*Pilgrim	1.81×10^4	1.15×10^3	0.343	0.0319	2.03×10^{-3}	5.71×10^{-5}
Quad Cities 1,2	1.32×10^5	4.19×10^3	3.89	0.747	2.37×10^{-2}	0.454
*Vermont Yankee	5.52×10^4	5.26×10^3	2.38	0.171	1.63×10^{-2}	1.63
Pressurized Water Reactors						
*Maine Yankee	2.13	2.71×10^{-1}	4.19	3.71×10^{-6}	3.53×10^{-7}	0.189
Palisades	5.05×10^2	1.60×10^1	0.0208	9.7×10^{-3}	3.08×10^{-4}	0.291
Yankee	1.83×10^1	5.81×10^{-1}	0.0263	7.77×10^{-4}	2.46×10^{-5}	0.0607
Indian Point 1	5.43×10^2	1.72×10^1	0.00337	0.928	2.95×10^{-2}	1.25
R.E. Ginna	1.18×10^4	3.74×10^2	0.678	3.50×10^{-2}	1.11×10^{-3}	3.80
Connecticut Yankee	6.45×10^2	2.05×10^1	0.252	1.81×10^{-2}	5.75×10^{-4}	8.71
H.B. Robinson	2.57×10^2	8.16	0.54	2.68×10^{-2}	8.51×10^{-4}	11.04 Halogen; 3.6 Particulates
San Onofre	1.91×10^4	6.06×10^2	1.09	4.74×10^{-4}	1.50×10^{-5}	0.00255
Point Beach 1,2	2.81×10^3	8.92×10^1	0.0875	2.97×10^{-2}	5.37×10^{-5}	0.000108
*Surry 1	1.26×10	7.97×10^{-4}	1.67×10^{-6}	1.75×10^{-4}	8.29×10^{-6}	0.09
Nonwater Reactors						
Peach Bottom 1	5.82×10^1	1.85	0.031	None measurable above background		
Fermi	1.84×10^2	5.84	19.5	1.02×10^{-3}	3.24×10^{-5}	1.84

was released in the liquid effluents. As shown in the case of the Monticello reactor, which was one of the first to have additional radwaste systems installed, even these low amounts can be further reduced to levels that are becoming standard for the newer reactors. Second, the largest amount of activity released by far consisted of noble gases, for which no simple removal system has been available even though MPC values were nowhere exceeded. This noble-gas release is particularly noticeable for boiling water reactors in which one expects to find most of the activity released in the gaseous wastes, whereas it is several orders of magnitude less in pressurized-water and gas-cooled reactors. In pressurized-water reactors tritium emerges as the major effluent in total amounts of activity, even though again concentrations were quite low.

Consequently, to reduce effluent levels to the lowest practicable levels every nuclear facility must include provision for removal of particulates, halogens, and chemical impurities from waste streams. Beyond that, increasing attention is being paid to the reduction of noble gases and tritium and these methods will be discussed separately.

BASIC RADWASTE SYSTEMS

Liquid Wastes

The technology for purifying liquids is well advanced and the removal of particulates and low-level dissolved ions is accomplished by fairly standard procedures. However, it must be recognized that removal below concentrations of the order of 0.1 ppm or less from waste waters is rarely practical or justified, since at that point the predominant impurities will be those dissolved ions and suspended materials present in the intake water, or the chemical substances added to the plant water for control of corrosion or fouling. In any case, except for any tritium content, the "waste" water released is apt to be considerably cleaner and freer of bacterial substances, albeit also warmer, than the intake water. As an example of the quantities involved, Table 78 shows the design estimates for the annual liquid discharges from the Watts Bar station.

Water Treatment Methods

Demineralizers

Ion-exchange methods, using both mixed resins and separate cation and anion exchangers, are used routinely to remove dissolved ions from waste waters. They have already been described on page 178; the main difference at this stage would lie in the nature of the liquids to be

Table 78. Design Estimates of Annual Liquid Waste Quantities from
Two 1200 MWe Nuclear Units (12)

	Volume (gal/yr)	
Source	Recycle	Processed and Discharged or Shipped Offsite
Equipment drains and leakoffs	140,000	
Laundry and shower		240,000
Hot lab equipment rinses		32,000
Equipment drains, leaks (nonrecyclable reactor grade and nonreactor grade)		39,000
Decontamination		16,000
Fuel cask decontamination		350,000
Reactor coolant discharged for tritium control		45,000[a]
Totals	140,000	722,000

[a]This is the estimated maximum quantity that must be removed annually to maintain a maximum tritium concentration of 2.5 $\mu Ci/ml$.

treated, which may contain appreciable quantities of detergents. On the other hand, the radioactivity level is much lower, thus resulting in less degradation of the ion exchange resin. The resin can be eluted and regenerated periodically and will have to be discarded as contaminated waste ultimately (13).

The efficiency of ion exchange treatment of waste streams depends on the type and composition of the waste liquid, type of exchanger, regeneration methods, radionuclides present, and operating procedures. Decontamination factors (DF's) as low as 2 and as high as 10^5 have been reported. Only low-level waste can be processed in an ion exchange system because bed exhaustion occurs rapidly for liquids with a high total dissolved solids content. Also, suspended solids will clog an ion exchanger, causing channeling and preventing the proper exchange of ions, thereby restricting the use of ion exchange treatment to low-level radioactive wastes with low total dissolved solids and low suspended solids. Generally, reactor effluents can meet these requirements, but most other effluents cannot without some pretreatment prior to ion exchange.

Mixed-bed exchangers are more efficient than cation exchangers alone due to the removal of radioactive anions. Higher decontamination factors are obtained when using separate beds of cation and anion exchange material, and elution is also much easier than with mixed-bed units. If use is intermittent as opposed to continuous, the advantage of a mixed-bed unit may be that it appears to have a lower ionic breakthrough rate following shutdown. Table 74 shows average decontamination factors assumed in source term calculations (5) and Table 75 some liquid flow rates. Note the wastewater contribution from demineralizer regeneration and backwash.

All experimental information obtained to date has been concerned with reduction of radionuclide concentration from above current allowable discharge limits to lower limits. For further reduction of activity to even lower levels, ion adsorption is markedly dependent on concentration of calcium ions and other ions that are strongly adsorbed (13).

Holdup Storage

Since a large part of the radioactive contaminants in the wastewater is relatively short-lived, many plants pump their contaminated wastes to holdup tanks where they are retained from a few hours to several weeks before being discharged to the environment. This permits decay of much of the activity, but may involve provision of rather extensive tank capacity whose cost has to be balanced against alternative removal procedures. The efficiency of holdup decay for liquid wastes is governed by the half-lives and concentrations of the various nuclides in the effluent stream and also by the waste liquid generation rate.

Much of the published information on waste activities indicates that the contribution to the total hazard is minimal for the shorter-lived fission and corrosion products, which are precisely the ones benefiting from waste storage. These include such nuclides as Sr-92 (2.71 hr), Y-93 (10 hr), Zr-97 (16.8 hr), Ru-103 (39.6 d), Te-132 (78 hr), I-133 (20.8 hr), Ba-139 (83.3 min) and La-142 (92.4 min). Relating holdup to anticipated flow rates, it is evident that retention for 24-48 hours may be worthwhile, but to obtain any significant reduction in the concentration (and activity) of the long-lived nuclides, extremely large volumes of tank capacity would be necessary to accommodate long-term storage of millions of gallons for periods of years. Relative to other available methods, increased holdup time is thus not an efficient and economical means of activity reduction for large volumes of already low-activity-level liquid waste containing low concentrations of long-lived nuclides.

Filtration

Filtration as a sole means of radioactive waste treatment is seldom utilized and thus little information is available concerning filter performance. Filtration is employed with other types of waste treatment as a necessary pre- or posttreatment step for the process liquid so that the potential of the principal treatment process (*e.g.,* ion exchange or chemical separation) can be fully realized. When prefiltration is used, the objective is removal of suspended solids to prevent interference of particulates in the following processes and clogging of process liquid flow channels, such as with ion exchangers. In posttreatment applications, filters perform such tasks as collection of sludges from chemical treatment and collection of resin "fines" escaping from ion exchangers.

Except for removal by simply straining or by sorption on biological life contained in the "Schmutzdecke" (top layer of material on a filter), sand filters have not been very effective in removing radioactive materials, unless they were previously incorporated in floc particles (13). Filter types that have been used include (1) natural filtration using sand or anthracite media, (2) vacuum and pressure precoat-type filtration, (3) filters utilizing woven cloths of various materials, such as nylon, and (4) "knife-edge" filters using a slotted screen element cleaned by a movable knife-edged device. Activated carbon used as a filter has proved fairly effective.

For processing large volumes of wastes, the easily handled cartridge-type filter may not be feasible because of limited capacity. Dresden 1 experience indicates that the filters initially installed in that liquid waste system would clog after processing between 10,000 and 50,000 gallons. The amount of labor involved to change the filter at this frequency was economically unacceptable and a shift was made to a precoat type of filter. While this also clogged in about the same amount of time, it could be backwashed and returned to service in a short time with less labor. Although cartridge type filters may not be attractive for high-volume applications such as at Dresden, there are some applications where this filter is preferable. For liquid streams where large volumes are not encountered normally, cartridge units may be more economical compared with precoat-type filters, due to the relatively small capital cost involved (14). Cartridge-type filters are changed when either high pressure drop or high radioactivity conditions occur. To avoid manual handling of radioactive filter elements and media, filters can be provided with a backwash cycle to permit the flow to be reversed or clean water to be introduced to break the accumulated cake or sludge away from the filter elements. Pressure drops caused by bridging of solids on the filter can be controlled by periodically backwashing the filter and bubbling steam into the filter tank to break up the solid bridges. All backwashing fluids must be accommodated in some effluent processing system at the facility (15).

Because laundry wastes contain such low levels of radioactivity, the associated filters can usually be handled manually. Cartridge-type filters are normally used for this application. The cartridges are changed when the pressure drop across them becomes high.

Centrifugation

Centrifuges separate suspended solids from liquids through the use of centrifugal force. Basically, a centrifuge is a cylinder or bowl that is rotated at high velocity about its axis. Usually, liquids containing suspended radioactive materials are fed continuously to the centrifuge. The separated solids are retained at the walls of the bowl while the clarified liquid overflows and passes to a surge tank. Periodically, the accumulated solids are removed by interrupting the flow of feed liquid and washing the solids into a hopper; such solids subsequently are packaged for solid wastes disposal.

Centrifuges are normally used in waste treatment systems to dewater sludges (that result from evaporation and filtration) as a means of preparing the solids for packaging and shipment.

Evaporation

Evaporation is used to concentrate radioactive contaminants in aqueous wastes by boiling off the water, leaving behind most of the dissolved and suspended solids. Evaporators for radioactive waste can vary from simple pots with steam heating pipes coiled inside to elaborate devices that have pumps to circulate the feed through outside heaters and compressors to provide more heat efficiency from the hot vapors. Satisfactory operation can be obtained from simple evaporators equipped with adequate auxiliaries to achieve the decontamination factor required. Depending on the amount of dissolved solids in the waste feed to an evaporator, a volume reduction of 10 to 50 times can usually be achieved in the radioactive concentrate. A detailed review of the various types of evaporators used in the nuclear industry has been published by Godbee (16). The efficiency of evaporation for radioactive waste treatment can vary widely, depending on the radioactive materials present. Overall decontamination factors of between 10^4 to 10^5 have been experienced, depending on the mass velocity of the vapor in the evaporator, and provided the radioactive contaminants are nonvolatile. If volatile radioactive materials such as tritium, iodine, or ruthenium are present, the overall DF may be substantially reduced (17).

Foaming can also greatly reduce DF; thus the concentration of soaps and detergents in the thick liquor should be kept as low as possible. If it is not possible to maintain low concentrations of these materials, there are many methods of preventing or depressing foam: (a) foam breakers, (b) chemical antifoam agents, or (c) acoustic vibrations (17). Experimental tests over the last forty years have proved to be fairly consistent for most types of evaporators. In general, as mass velocity of vapor in an evaporation chamber increases, liquid entrainment in the vapor increases and thus DF decreases. The DF for the boiler proper is generally on the order of 10^3 to 10^4. With the addition of cyclone separators and packing towers, the DF can be increased with increasing mass velocity up to a point where the efficiency of the boiler decreases rapidly, as can be seen in Figure 59.

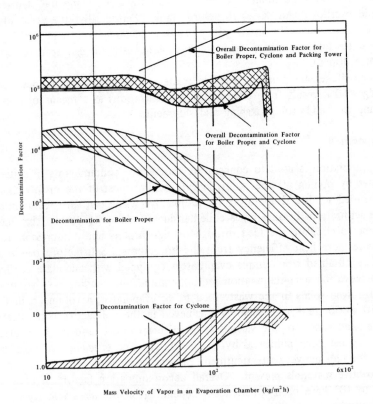

Figure 59. Decontamination factor for standard evaporators (17,18).

Evaporation is a preferred method of liquid waste treatment because streams of relatively high dissolved-solids content can be accommodated. Consequently, it is a suitable process for use together with subsequent ion exchange treatment. Care must be taken, however, to provide pretreatment of any feed streams, such as laundry wastes, that contain foaming agents.

Evaporation is a fairly expensive means of purification, but is competitive with ion exchangers for low-purity wastes. Figure 60 shows the relation between evaporation rate and process costs (14); the cost per unit volume decreases with rising evaporation rates. Figure 61 illustrates two types of evaporators used in processing liquid radwaste (16).

Water Recycle Process

The water recycle process is any combination of chemical treatment, filtration and ion exchange. It has achieved DF's of 10^3 to 10^4 for all major radionuclides. A block diagram of the system is shown in Figure 62. Only three nuclides for which measurements were taken showed a DF less than 10^3 (^{144}Ce, ^{125}Sb, and ^{95}Zr-Nb), and in all three cases the radioactivity in the processed water for these isotopes was reduced to the analytical limits of detection. The reported DF's in these instances were thus minimum values set by limits of detection (19). This system has considerable promise as a means for appreciable reduction of activity levels in liquids released to the environment (approaching a so-called zero release system). There are some restrictions on the operation, however; pH must be in the range of 7 to 8 prior to coagulation-clarification treatment and the process liquid must have a low salt content. Such a system appears to be advantageous for current reactor designs.

Gas Stripping

Gas stripping of liquid waste streams is a standard chemical process in use at several plants. Gas strippers are identical in purpose to deaerators, and similar apparatus will serve if properly rated. The operation consists of heating the feed under pressure, flashing off steam, and passing the flashed steam to a vent condenser that condenses most of the steam but allows the gas to escape (20). The gas concentration in the water leaving the gas stripper depends on the quantity of steam flashed and the degree to which the vent condenser reflux is denuded of gas.

The use of a side arm calandria or an external recirculated heater permits greater steaming without raising the feed to an uncomfortable temperature and pressure. It also allows steam production to be

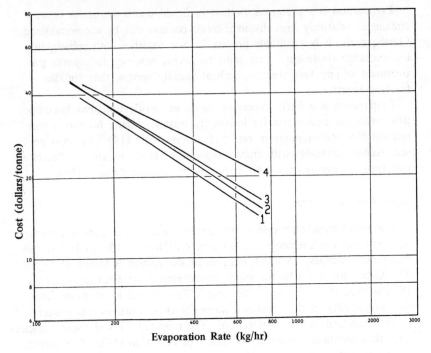

Curve

1   Natural and forced circulation
2   Multiple effect
3   Coil and wiped film
4   Vapor compression

1971 Evaporator Operating Costs

| | Cost (Dollars/Tonne) | |
| | Evaporation Rate | |
Type	200 kg/hr	500 kg/hr
Coil	35.52	20.40
Natural circulation	32.64	18.60
Forced circulation	32.40	18.60
Vapor compression	38.52	24.60
Wiped film	35.88	20.76
Multiple effect	35.40	19.80

Figure 60. Evaporator operating costs (14).

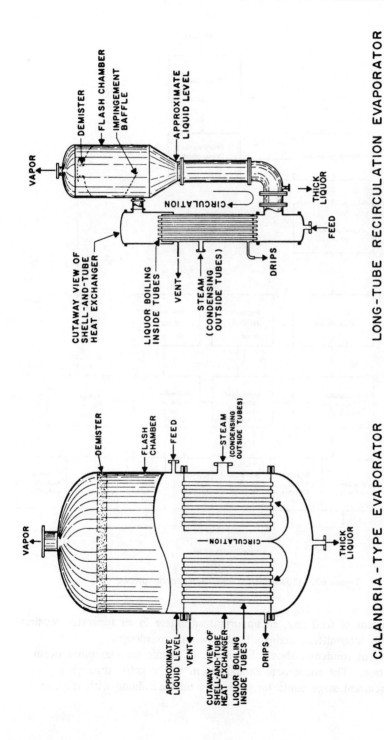

Figure 61. Typical evaporators used in processing liquid radwaste (16).

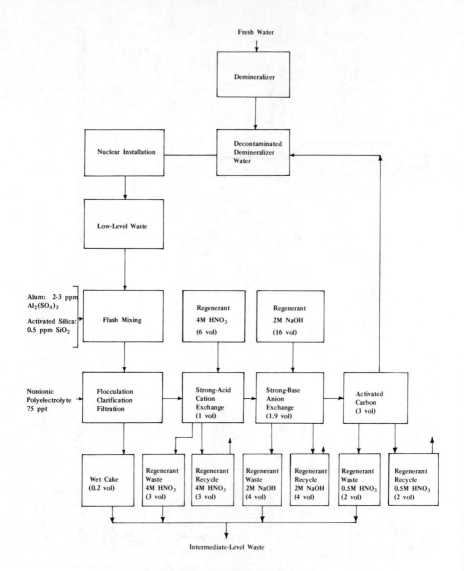

Figure 62. Flowsheet for the water recycle process (19).

independent of feed rate, an important advantage in an apparatus required to handle intermittent drains.   It also permits overdesign.

The vent condenser should be chosen to handle the maximum steam production.   The condensate should return to the kettle through a countercurrent stage contactor, such as a bubble column with trays or packing, through which the condensate descends and the steam rises. Sprays may serve the purpose. The kettle can have a large capacity and serve as a drain tank (20). A less complex design provides a spray header in a tank and a vent to the gas delay tank.

Electrodialysis

Electrodialysis combines electrolysis with dialytic diffusion in transporting ionic impurities through semipermeable membranes under the influence of an applied electric potential. Until the development of permselective ion exchange membranes, which permit either anions or cations (not both) to pass through them depending upon whether the ion exchange material is anionic or cationic, electrodialysis had few highly specific applications. The use of these membranes with continuous regeneration by means of an electric current at right angles to the solution flow has encouraged a restudy of this process to determine its applications to the treatment of low-level liquid radioactive wastes (21).

A simple electrodialytic cell consists of three compartments: a central, an anode, and a cathode compartment. As shown in Figure 63 these compartments are separated by the anion- and cation-selective semipermeable membranes. Electrodes of corrosion-resistant conducting material are inserted in the anode and cathode compartments (13).

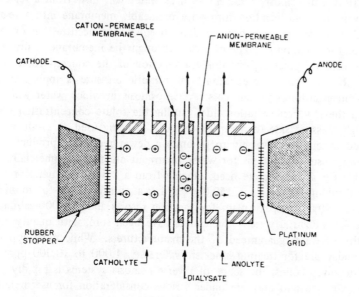

Figure 63. Cross-sectional view of tubular Lucite electrodialysis cell (13).

In spite of the simplicity of the three-compartment cell, a multicompartment cell is usually preferred in actual operation. The advantage of the latter is that less total electrode surface is required; this reduces polarization losses, undesirable electrode reactions, and capital costs. Furthermore, in such a cell fewer compartments are required per separation unit. The number of compartments is 2n+1, where n is the number of separation units. Thus five separation units are obtained from eleven compartments (2.2 compartments per separation unit), as compared to only one in a three-compartment cell.

To study the application of the permselective-membrane electrodialysis process to radioactive wastes, ^{137}Cs, ^{131}I, ^{89}Sr and ^{95}Zr-Nb were removed from distilled or tap water. Under optimum conditions removal of strontium and cesium was more than 99%, and removal of iodine was about 97%. Removal of zirconium-niobium, however, was poor. This was attributed to the failure of the process to remove colloidal materials. When mixed fission products were removed by electrodialysis, much of the residual activity was present in colloidal form, as shown by removal on passage through a sand filter or by coagulation with iron salts (13).

Reverse Osmosis

Osmotic phenomena have been known since the classical work of van't Hoff. In osmosis a solvent such as water will flow from a region of low solute concentration through a permeable membrane into a region of higher solute concentration, tending to equalize concentrations on both sides of the membrane. The diffusion through the membrane is driven by the osmotic pressure, essentially a function of the concentration gradient. If a reverse force exceeding the osmotic pressure is applied to the high concentration side, one gets reverse osmosis in which water will flow toward the low concentration side, causing the solute concentration on the high pressure side to rise. This process is being used to produce purified water, or conversely to generate a dewatered waste product.

Reverse-osmosis systems for water treatment are now commercially available for any moderate need, ranging from a 2.5-gal/day unit for a home drinking supply, to a 150,000-gal/day auxiliary source for municipal water, a 350,000-gal/day plant for vacation resorts and 800,000-gal/day plants for industrial water. Such units have been sold, with quality and quantity of water guaranteed by the manufacturers. While most applications today are for treating brackish water, *i.e.*, 1,000 to 10,000 ppm total dissolved solids, the scope of reverse osmosis systems is rapidly being extended and they are under serious consideration for use in radwaste systems.

Numerous polymers have semipermeable characteristics and exhibit good selectivity. Cellulose acetate (CA) has become the most widely studied membrane material and is the standard against which other developments are measured. It has good selectivity, dope formulations amenable to variation, good availability of raw materials, and relatively low cost.

Water molecules move through the membrane by an applied pressure that pushes the water from a bond with one acetyl group to the next. Only a moderate force is necessary because the bonds are transferred, not broken. Dissolved ions or molecules that do not hydrogen-bond cannot enter into attachments with bonding sites (acetyl groups) and are left to concentrate at the membrane surface (22). In order to obtain large membrane surfaces and systems capable of withstanding appreciable pressure, the polymer material is shaped in the form of spirals or hollow fibers.

The flux through a particular membrane is determined by its physical characteristics (thickness, chemical composition, porosity) and by the conditions of the system (temperature, differential pressure across the membrane, salt concentration of solutions touching the membrane, and velocity of the feed moving across the membrane). In practice, the properties of the membrane and the solutions are relatively constant, and water flux becomes a simple function of pressure. Quantitatively, it is described by:

$$F_{H_2O} = A(\Delta P - \Delta \pi) \tag{53}$$

where: A = a permeation coefficient for a unit area of a membrane. This term includes the physical variables of the membrane and is relatively constant.

ΔP = (P_F-P_P) = pressure exerted on the feed solution, P_F, less the pressure on product, P_P.

$\Delta \pi$ = $(\pi_F - \pi_P)$ = osmotic pressure of the feed solution, π_P, less the osmotic pressure of the product, π_P.

The equation can be simplified for high pressures (above 600 psi) by lumping all the constants into the coefficient A'. Flux for brackish water at constant temperature is approximately:

$$F_{H_2O} \approx (A') (P_{Feed}) \tag{54}$$

Flux falls off during the life of the membrane for several reasons. First, there is an initial decrease because capillary water is squeezed out as the operating pressure collapses the porous layer of membrane. CA has some "memory" and if the pressure is released before the membrane is permanently set, it will swell and regain some of the capillary water.

Following this initial compaction, there is a long-term compaction that slowly reduces the water flux. This results from densification of the thin, air-dried, membrane layer and probably corresponds to narrowing of pores through which water must pass. As the channels narrow, flow decreases.

Another cause of flux decline is the hydrolysis of acetyl groups that continually occurs during the life of the membrane. This reaction results in a loss of hydrogen bonding sites, which further reduces the water transport. The reaction is also a source of salt leakage because there are fewer water bridges blocking the passage of foreign materials through the pores.

Membranes age and flux decreases steadily. Fouling of the membranes causes flux degradation, and regular cleaning is essential for consistent operation. Typically such maintenance may involve cleaning twice a week for hollow fiber units and once every two weeks for spiral systems. Scaling by insoluble salts such as calcium carbonate and calcium sulfate can occur frequently. Some of the precautions taken to prevent scaling are: (1) pH adjustment by adding acid to the feed, which breaks down the bicarbonate ion and prevents formation of carbonates in the boundary layers, (2) keeping the product recovery (product-to-feed ratio) low to assure that a saturated calcium sulfate solution is not formed in the boundary layer, and (3) adding antiscale substances such as sodium hexa-metaphosphate to prevent deposits. Deposits can be dissolved by chelating agents but after each cleaning the membrane always shows some permanent damage.

Summary

From the foregoing descriptions it is evident that no single method of purification is adequate to provide the high decontamination factor required before waste waters can be discharged. In practice, therefore, several combinations of methods are employed, as shown in Figure 64, which schematically presents the liquid radwaste system of the Shoreham plant (23). One needs to remember, too, that the chemical, nonradioactive wastes may be a somewhat more severe environmental problem than the low-level residual radioactivity. This may be particularly important in closed-loop operation where most of the released contaminants come from condenser and cooling-tower blowdown. Some of these materials need removal, but most of them are simply diluted in the receiving stream in concentrations that meet local water quality standards.

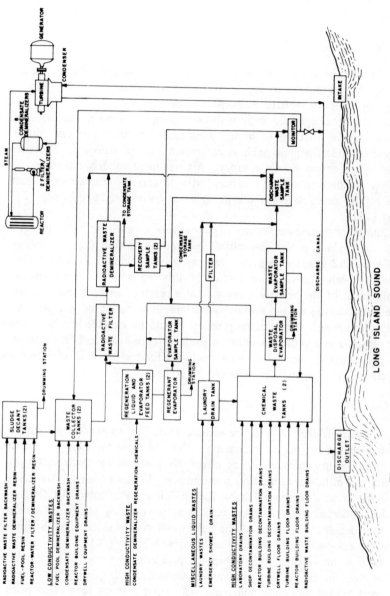

Figure 64. Flowsheet of liquid radwaste system (Shoreham plant) (23).

Figure 65 shows the quantities and pathways of these liquids for the Fort St. Vrain HTGR Station (6) and Table 79 the type and quantities discharged.

Airborne Wastes

Most of the fission products, being soluble or in the form of oxide precipitates, will tend to appear in the liquid wastes; only the noble gases or the more volatile elements will be airborne. The airborne pollutants can be categorized in four distinct groups: aerosols and particulates of radioactive materials or having them sorbed to their surfaces; noble gases; volatile materials such as ruthenium and the halogens; and tritiated hydrogen gas or water vapor. Air treatment processes tend to follow standard industrial practice, but with certain refinements to obtain the highest practical removal efficiencies at correspondingly higher cost (24).

Since the immediate impact of any release of airborne radioactivity on the surroundings is less controllable or predictable than that for liquid effluents and a wider perimeter can easily be affected, particular attention has been paid to the reduction of airborne activity. Furthermore, while most of the activity leaking from a reactor can be channeled into the liquid wastes during routine operations, it is during or just after a major accident that any activity release would be expected to be primarily airborne and provision must be made to cope with and retain most of that activity. Several different methods for treatment of effluents are in current use and will be reviewed here (25).

Filtration

Gaseous effluents in practice include some fine suspended particulate matter that may also be radioactive. Most of these particles are too fine or insufficiently dielectric to be effectively removed by electrostatic precipitators, although these have been installed in some plants. The most effective means of removal is by High-Efficiency Particulate Air (HEPA) filters, followed in some cases, such as at reprocessing plants, by scrubbers. A very detailed discussion of the design and performance of such filter systems has been provided by Burchsted and Fuller (26).

Even in nonnuclear applications the problems of achieving a reliable, truly efficient air cleaning system are substantial. The best filters used in conventional air-conditioning and ventilation systems have efficiencies of no more than 80 to 85% for 0.3μ particles; that is, at least 1500 to 2000 of every 10,000 entering particles of this size would penetrate the

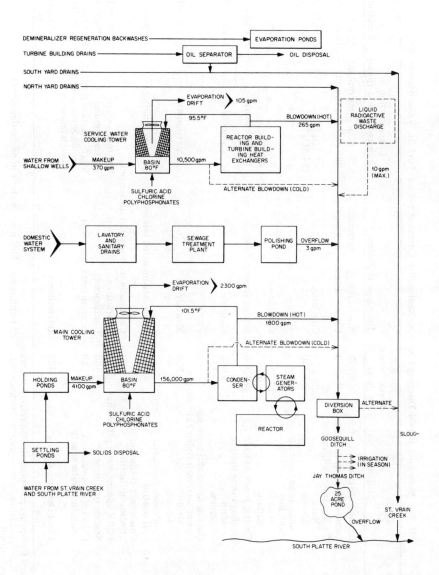

Figure 65. Flowsheet for non-radioactive liquid waste system—
Fort St. Vrain Station (6).

Table 79.   Estimated Liquid Discharges of Non-Radioactive Chemical Wastes—Fort St. Vrain Station (6)

Chemical	Use	Annual Consumption lb/yr	Discharge Procedure	Concentration in Discharge Liquid (ppm)	Receiving Stream
H_2SO_4 (93%)	Demineralizer regeneration	20,930	Batch (100 gpm maximum)	5780 SO_4^{2-} (maximum)	Evaporation ponds
NaOH	Demineralizer regeneration	12,500	Batch (100 gpm maximum)	1880 Na^+ (maximum)	Evaporation ponds
H_2SO_4 (93%)	Cooling tower pH control		Continuous	347 SO_4^{2-}	S. Platte River[d]
	Main	3,358,000		114 SO_4^{2-}	
	Service	153,300			
Nalco 347[a]	Cooling tower corrosion inhibitor		Continuous		S. Platte River[d]
	Main	109,500		12.4	
	Service	5,293		4.3	
Chlorine	Cooling tower biocide	69,400	Several times per day	1 (maximum)	S. Platte River[d]
Nalco 321[b]	Cooling tower biocide		Monthly	50 (maximum)	S. Platte River[d]
	Main	5,666			
	Service	432			
Nalco 71-D5[c]	Cooling tower anti-foraming agent used with Nalco 321		Monthly	5 (maximum)	S. Platte River[d]
	Main	567			
	Service	43			
Ammonia	Condensate feedwater pH control	4,088	Batch (accumulated leakage)	0.8 (maximum)	Evaporation ponds
Hydrazine	Condensate feedwater hydrogen control	1,300	Chemically used up; not discharged		
NH_4OH	PCRV cooling water pH control	3	Intermittent	100 (maximum)	S. Platte River[d]
NaOCl	Sewage treatment	80	Continuous	1.5 (maximum)	S. Platte River[d]
Na_2SO_3	Chlorine removal from domestic water supply	10	Intermittent (as needed)	SO_4^{2-} reaction product, probably removed by demineralizer	St. Vrain Creek & S. Platte River
Detergent	Floor cleaning, etc.	420	Intermittent		St. Vrain Creek & S. Platte River

aNalco Chemical Company, phosphated, ethoxylated glycerine.
b20% 1-alkyl (C_6-C_{18}) amino-3-aminopropane monoacetate; 30% isopropanol; 50% inert (water + nonionic detergent).
cFatty acid—polyglycol.
dAn alternate is St. Vrain Creek.

filter. Most of the filters used have substantially lower efficiencies, some as low as 5% for these small particles. By contrast, the HEPA or "absolute" filter required in high-efficiency air cleaning systems has a minimum efficiency of 99.97% for 0.3μ particles (*i.e.*, a maximum penetration of three particles in 10,000). Similarly, the activated-charcoal adsorbers associated with some reactor and radiochemical plant systems also require efficiencies much higher than those needed for nonnuclear applications. It is obvious that design and installation practices suitable for lower-efficiency air conditioning and ventilation system air cleaning would be highly questionable for high-efficiency systems, and they have, in fact, proved grossly inadequate.

The principal costs in operation of an HEPA filter system are for the replacement filters and labor, and the principal factor affecting these costs is the frequency of changing HEPA filters. Replacement filter and labor cost may contribute as much as 70% of the total cost of such a system (including capital costs) over a 20-year period. Measures such as the use of high-efficiency building supply filters, the use of prefilters, operating the filter system below its rated capacity, and operation of filters to a high pressure drop before replacement all tend to decrease filter change frequency and thereby reduce costs (26).

By definition, an HEPA filter is a throwaway, extended-medium, dry-type filter having (1) a minimum particle removal efficiency of no less than 99.97% for 0.3μ particles, (2) a maximum resistance when clean of 1.0 inch of water (25.4 kg/m^2) when operated at rated air flow capacity, and (3) a rigid casing extending the full depth of the medium.

HEPA filters are intended primarily for submicron particle removal and are not coarse dust collectors. They have low dust capacity and may plug rapidly when exposed to high concentrations of dust, smoke, or lint. The HEPA filter is also the most important element in the exhaust system from the standpoint of particle removal, and its failure will result in failure of the system. Prefilters, installed either locally at the entrances to the duct or in the central exhaust filter housing, extend the life of the HEPA filters and provide a measure of protection against damage. The actual increase in filter life, of course, depends on the quality of the prefilter selected and the nature and concentration of dust.

The measured efficiency of most HEPA filters passing through the Quality Assurance Stations at this time is close to 99.99% (a penetration of one particle in 10,000; DF = 10^4). The resistance and air flow capacity of the filters are determined at the same time as the efficiency, and all three values are stamped on the filter casing. The five sizes of HEPA filters listed in Table 80 are standard for contaminated exhaust service. The 1000-cfm size is recommended for bank systems. Although some

Table 80. Standard HEPA Filter Sizes (26)

Capacity at Clean-Filter Resistance of 1.0 in. H_2O (scfm)	Filter Face Dimensions (in.)	Filter Depth less Gaskets (in.)
25	8 x 8	3 1/16
50	8 x 8	5 7/8
125	12 x 12	5 7/8
500	24 x 24	5 7/8
1000	24 x 24	11 1/2

manufacturers rate their filters differently, the capacities shown are recommended for design purposes (26).

In the construction of wood-cased and steel-cased open-face HEPA filters, the core (filter pack) is made by pleating a continuous web of fiber glass or cellulose-asbestos paper (the medium) back and forth over corrugated separators, which add strength to the core and form air passages between the folds. The core is sealed into a full-depth wood or steel casing (frame), usually with an elastomeric adhesive. The filter papers consist of very fine (submicron) fibers in a matrix of larger (1 to 4μ) fibers. An organic binder (5% by weight maximum) is added to hold the fibers during the papermaking process. Fire-resistant papers are made from glass fibers. Combustible papers required for certain special applications are made from cellulose and asbestos fibers. A minimum thickness of 0.015 in. and a minimum basis weight of 48 lb per 3000 ft^2 are sometimes specified to ensure adequate abrasion resistance. The filter paper is extremely weak and fragile, and the filters must be handled with care to avoid damage. Tensile strength is sharply reduced (about 50%) at temperatures above 400°F (200°C), probably due to burnoff of the binder. Improved performance is obtained if entrained moisture is removed by a moisture separator preceding the filters. Such moisture separators are necessary in the air cleaning systems of many reactors to protect the HEPA filters and absorbers. Large drops are effectively removed by wave-plate separators whose efficiency is practically 100% for drops over 400μ but it falls below 60% for droplets in the range of 10-50μ. For the removal of small and intermediate droplets (1-100μ), knitted-fabric and nonwoven fiber mat separators are the most acceptable devices.

Relative costs of various open-face HEPA filter constructions are given in Table 81 (26). Unit costs of 1000 cfm filters were about $60-80 (1970 prices). The standard filter units are assembled into modules having a typical cross section of 24 x 24 in. and a depth of 11 1/2 in.

Table 81. Relative Costs of Various HEPA Filter Constructions (26)

Casing Material	Separator Material				
	Untreated Asbestos	Aluminum Foil	Treated Asbestos[a]	Plastic	Stainless Steel
Wood	1.00	1.02	1.08	1.71	7.73
Carbon steel	1.13	1.16	1.22	1.89	
Stainless steel	1.96	1.99	2.04	2.69	8.69

[a]Qualified for moisture and corrosion resistance in accordance with procedures established by the U.S. Atomic Energy Commission.

To prolong the life of the HEPA filters, a bank of roughing or prefilters normally precedes them. Such filters are classified by efficiency and construction as panel filters of the viscous impingement type or as extended medium-dry type units. In radwaste system applications, an important function of filters is to collect radioactive particulates formed when a gaseous parent nuclide decays to a radioactive daughter which is nonvolatile. For maximum effectiveness, HEPA filters should therefore be placed where the particulate concentration is the highest. For off-gas systems in water-cooled reactors, this location is just prior to release of the gas through the stack, thus allowing a maximum of transport time for gaseous radionuclides to decay to a particulate daughter product before filtration.

Even though data on the fraction of total gas stream activity in particulate form are not available, installation of filters appears advisable because radionuclides in the decay chains from the noble gases are both solid and of biological importance, *e.g.*, Cs and Sr. Similarly, the filters can retain radioactive iodine (normally a gas) that has been adsorbed on small solid particles.

Activated-Carbon Adsorbers

Protection against the release of radioiodine is normally accomplished by installing a bank of charcoal adsorbers immediately after the bank of HEPA filters. These adsorbers, containing activated charcoal, are also made in modular units that have basically the same size and air flow capacity as the HEPA modules.

Activated-carbon adsorbers, often referred to as gas filters, are the most satisfactory devices available at present for trapping fission product gases from nuclear reactors and radiochemical operations. These adsorbers are tightly packed beds of adsorbent carbon granules through which air and gases are passed before being released to the atmosphere. The carbon is activated by controlled heating under a steam atmosphere, which drives off organic matter and generates large internal surfaces or "sites" on which adsorption can take place. The effective area of activated carbon varies from 700 to 1800 m^2/g. The carbon is often impregnated with chemicals to increase its affinity for certain gases.

The gas of primary concern in the design of adsorber systems for nuclear applications is elemental radioiodine, which is the most abundant (from the standpoint of radioactivity) sorbable gas released in a reactor accident or in operations with nuclear fuel. Under some conditions, organic radioiodine compounds, principally methyl iodide, may also be formed. Although the quantity is small relative to the amount of elemental radioiodine present, it represents a possible health hazard, and it is usually necessary to design the system to remove it. Carbons impregnated with potassium iodide (KI), triethylenediamine (TEDA), or other compounds are necessary for trapping organic radioiodine compounds in humid air. Future reference to impregnated carbons will mean those impregnated for trapping of methyl iodide and other organic iodine compounds.

The important properties of an adsorber are trapping (adsorption) efficiency, holding capacity or "activity," retentivity or the ability to prevent desorption of the sorbed iodine, ignition temperature, air flow capacity, and resistance to air flow.

The efficiency of a carbon adsorber is a function of (1) the degree of activation, ash and moisture content, and impurities in the charcoal, (2) the type and quantity of impregnant (for impregnated carbons) (3) granule size (the separation factor varies inversely with granule size), (4) the contact time between the gas and the carbon (*i.e.*, the residence time in the carbon bed, which is a function of bed depth and velocity) and (5) the temperature and humidity of the gas stream. Special high-purity low-ash grades of carbon are required for the high levels of performance required of nuclear plant exhaust applications. The efficiencies of nuclear-grade charcoals, based on 1-in. (2.5 cm) bed depth for elemental iodine efficiency and 2-in. (5 cm) bed depth for methyl iodide efficiency, can be summarized as follows:

(1) Nonimpregnated charcoals: Efficiency for elemental iodine is satisfactory (over 99%) even after extended operation in high-temperature (260-280°F, 127-140°C) environment containing steam and water droplets. Efficiency for methyl iodide is satisfactory at relative humidities less than 70%, but nil at high (over 80%) humidity.

(2) Impregnated charcoals: Efficiency for elemental iodine is satisfactory (99%) under all temperature and humidity conditions, up to

270°F (135°C) and 100% relative humidity. Flooding of the carbon due to condensation or impingement of free water may reduce efficiency to as low as 20%.

The efficiency of both impregnated and nonimpregnated carbons, for both elemental iodine and methyl iodide, is reduced somewhat when the iodine loading in the gas stream is less than about 0.01 mg/m^3 (26).

Holding capacity is a function of the amount of charcoal in the system, the number of remaining active sites on which adsorption can take place, and, for impregnated charcoals, the nature, quantity, and condition of the impregnant. The mechanism for capture of elemental iodine is physical condensation (adsorption) of iodine molecules on the active surfaces of the charcoal. The mechanism for capture of organic radioiodine is apparently a combination of chemical reaction and isotopic exchange, in which the stable iodine of the impregnant substitutes for the radioactive iodine of the methyl iodide; the nonradioactive methyl iodide goes on through the adsorber.

Holding capacity for both iodine and methyl iodide decreases with time because of physical or chemical "poisoning" of the carbon by impurities, particularly by hydrocarbons and water, that occupy active sites on the surface or react with the impregnant. The holding capacity of impregnated carbon may also decrease with time due to loss of the impregnant by volatilization. Tests with KI and TEDA-impregnated coke-base carbons indicated a loss of 50% of the initial holding capacity for methyl iodide during 18 months exposure to flowing air, a 50% loss in three years when the beds were exposed to static air (as would be the case in a standby system), and a loss of 50% in five years when the beds were sealed in closed containers. Experience with coconut-shell carbons used in the United States shows a slower but still significant loss of holding capacity (26).

Holding capacity can be severely limited when the carbon is wet, as would be the case in some types of reactors if adequate moisture separators were not provided. Because adsorbers may have to be operated for several days or weeks following a reactor accident, consideration must be given to retention of the trapped gases. The retentivity for a sorbed chemical is usually about 35 to 45% of the holding capacity. As long as the integrated quantity of iodine or methyl iodide charged to the bed is less than the retentivity limit, desorption is not a significant problem at temperatures below 250 to 300°F. Some iodine loss will take place at higher temperatures, particularly from impregnated carbons, which will also desorb the impregnant and further reduce the capacity for CH_3I.

To determine the amount of charcoal required for a standby gas treatment system, the quantity of radioiodine that may exist in the containment following an accident must be estimated. It is generally assumed that one-half of the potential iodine inventory in the fuel can escape to the containment space and that one-half on this amount will plate out on the walls and floor. On this basis, one-quarter of the inventory could theoretically get to the adsorbers. Using these assumptions, the amount of charcoal required can be estimated from the following formula:

$$C = 0.22Q \tag{55}$$

where: C = pounds of charcoal required in the adsorber system
         Q = potential iodine inventory that could be released in an emergency (grams).

Using this formula, a 1000 MWe power reactor, having a potential iodine inventory of 15,000 g, would require a minimum of 1470 kg of carbon to provide standby protection against both elemental iodine and methyl iodide. Half of this amount is required if it can be shown that only elemental iodine would be present.

Adsorbers are available in a variety of bed and cartridge designs, of which the bed types are of primary interest for large systems. A bed-type adsorber consists of a layer of carbon granules tightly compressed between perforated metal screens that are enclosed with a metal casing. Because thin beds are subject to channeling of the charcoal, a minimum bed thickness of 1 in. (+1/8, - 0) and a maximum velocity of 60 fpm through the bed (*i.e.,* perpendicular to the screen face) are recommended.

When it is necessary to consider only elemental iodine, a single 1-in. (2.5 cm) bed is sufficient. When methyl iodide must also be considered, bed thicknesses of at least 2 inches or two 1-inch beds in series are required to keep space requirements of the system to a minimum (a residence time of the gases in the carbon of 0.2 sec is necessary to develop the required methyl iodide efficiency); up to 8 inches (20 cm) have been used. The use of two 1-inch bed units in series is recommended over a single 2-inch bed unit, both for higher efficiency in operation and as protection against failures. The cost of two 1-inch bed units is not appreciably more than that of a single 2-inch bed unit. Installation costs, on the other hand, are substantially higher because two mounting frames and considerably more housing space are required. Since the upstream unit alone may usually need replacement, lower replacement and maintenance costs may result in lower total costs over a given period of time (20 years).

The carbon granules may be arranged in multiple-tray units or in pleated bed systems. The latter seem to be more popular on account of their higher weight, lower bulk and easier installation. Charcoal adsorbers have also been found to be effective in the removal of noble gases from waste gas streams. It has been found that the transport of such gases through a charcoal medium is a function of the concentration of the gas, flow rate of the sweep (carrier) gas, length of the adsorber, amount of the adsorber and the dynamic adsorption coefficient (27).

An advantage of charcoal adsorbers for off-gas treatment is that the adsorption process provides an appreciable holdup time for the radioactive noble gases, thus facilitating decay to particulate daughters that should be retained in the charcoal element acting as a filter. If there is enough filter material of proper configuration, then the efficiency of noble gas removal could be made as high as desired (except for long-lived ^{85}Kr) (14).

A theoretical expression that is useful in designing an adsorber system is (5):

$$M = \frac{Ft_r}{0.53 \ K} \tag{56}$$

where:
- M = weight of charcoal, tons
- F = volume flow rate of the sweep gas, cfm
- t_r = holdup time, hours
- K = dynamic adsorption coefficient (depends on humidity)

Typical values for K are 50 cm^3/g at room temperature for krypton in most carrier gases and 1000 cm^3/g for xenon; water and CO_2 affect the adsorption of the noble gases adversely (27).

For maximum adsorption effectiveness, humidity control must be accomplished in the gas stream to the adsorber unit. Although the trapping efficiency for noble gases increases as the temperature of the adsorber is reduced, a reduced-temperature system would have to provide adequate safety provisions in the event of loss of refrigeration. This would include means for stopping the flow of gas and ample gas storage capacity as the adsorber warms to room temperature.

Delay-Decay Storage

Iodine, krypton and xenon constitute the major radiation hazards in effluent gases from commercial LWR's and only the iodine is chemically reactive; consequently, physical methods must be used to delay the release to the atmosphere of these gases.

The amount of radioactive gaseous materials released to the environment can be significantly reduced by storing the gases for a sufficiently long period to allow the short-lived radionuclides to decay to very low levels. This is accomplished at BWR and some PWR plants by retaining the gases for a minimum of 30 min in large holdup pipes or by adsorbing the radioactive gases on large charcoal beds for approximately 16 hr for radioactive krypton and 9 days for radioactive xenon; at PWR plants the gases from the primary coolant are retained in storage tanks for 30 to 60 days before release. In the case of BWR's the gases retained for the shorter periods of time are released to a tall stack to obtain a greater dilution and dispersion of the gases (Figure 66). In the case of PWR's, the gases retained for the longer periods are released from roof vents.

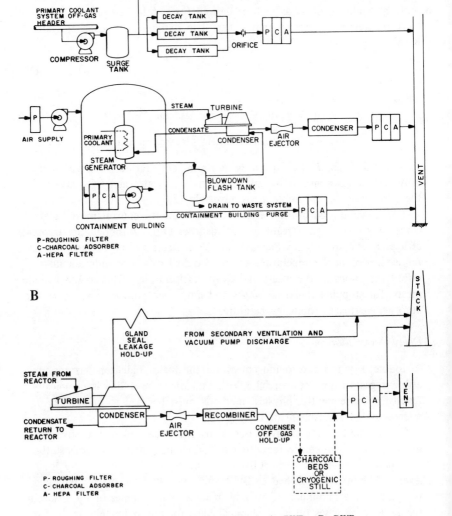

Figure 66. Gaseous waste systems. A: PWR; B: BWR.

Recombination

The intense gamma radiation in the core of an LWR decomposes a small portion of the coolant water to produce hydrogen and oxygen gases continually as the reactor operates. In a PWR, the primary loop is maintained under a pressure of hydrogen and nitrogen and the radiolytic hydrogen and oxygen gases continuously recombine to form water. In a BWR, the radiolytic gases and the radioactive gases from leaks in the fuel cladding travel with the gaseous effluent from the turbine, making it desirable to provide a device in the gaseous waste treatment system to reduce the explosion hazard by recombining the gaseous hydrogen and oxygen to form water.

While recombination of oxygen and hydrogen does not reduce the quantity of radioactive gases, it does reduce total gas volume by about 90%, leaving a more concentrated, smaller volume of gas to handle. As generated, the off-gas from a BWR is an explosive mixture of oxygen and hydrogen; the percentage of hydrogen in the total gas volume must be reduced to less than 4% in order to avoid possible explosion. This can be accomplished by dilution with another gas, but such added gas must subsequently be removed in order to obtain maximum advantage of a reduced total volume after recombination. Since hydrogen makes up about 60% of the total volume of the off-gas, the volume of diluent gas is large, and a large recombination chamber is necessary. If steam is used as a diluent, it can be condensed out after the recombination takes place, thus removing the diluent as required. A concurrent benefit of recombination is the conversion of gaseous tritium, as HT, to HTO transferring it to the liquid effluent. This makes its release safer and easier to control, and also greatly reduces the diluent air that might otherwise be required in the stack gases (28).

Typically the BWR mainstream may contain 30 ppm H_2 and 15 ppm O_2 at a temperature of 540°F (282°C) and a pressure around 1015 psia (29). It may also contain about 400 μCi/sec as radioactive nitrogen and oxygen. A typical recombiner consists of a replaceable cartridge or bed containing a catalyst in a steel tank. The catalyst is generally an array of metal strips or a ceramic material (*e.g.,* alumina pellets) that has been precoated with finely divided particles of platinum or palladium. Provisions are made for heating and cooling the bed to maintain its temperature between approximately 250 and 900°F (120 and 480°C) as dictated by the specific flowsheet. At lower temperatures the recombination is less effective and liquid water may condense in the bed and inhibit diffusion of gas to the catalyst. At higher temperatures the catalyst has a shorter life. Typically, the feed gas to a recombiner contains a small amount of hydrogen (controlled to less than 4 vol %), along with oxygen, inert diluents (steam, nitrogen) and small quantities of radioactive materials. The hydrogen-to-diluent ratio is monitored and controlled to minimize the possibility of uncontrolled combustion of hydrogen and oxygen upstream of the recombiner. The catalyst beds in recombiners gradually lose their effectiveness if they become chemically poisoned by materials such as iodine. To avoid this, multiple recombiners in parallel are used and provisions are made for periodic replacement of the cartridge or bed of catalyst (15,20).

The most significant factors to be considered when selecting one of these catalysts are temperature and wetting potential. For high-temperature applications, or where the catalyst could be flooded on start-up or by misoperation, all-metal supported catalysts are recommended (29). These materials are rugged and permit operation at temperatures as high as 1200°F (650°C) without encountering mechanical problems.

At temperatures below 250°F (120°C) the limitation in gross surface area for unit volume limits the reaction rate, and initiation of oxygen-hydrogen reactions is unreliable. For low-temperature applications a ceramic catalyst provides excellent activity and may be used at temperatures as low as 50°F (10°C). If sustained operation at high temperature with little or no cyclical temperature variations or temperature shock conditions can be assured, ceramic catalyst may be used, but it is not recommended for applications where temperatures exceed 1100°F (590°C).

Physical conditions of the inlet gas stream such as temperature, pressure and hydrogen concentration have a decided effect on the quantity of catalyst required to obtain a given conversion level. Since the outlet concentration is essentially inversely proportional to inlet concentration, one can see that by throttling steam diluent the catalyst inventory can be reduced. The recombination of $H_2 + O_2$ is an exothermic reaction. The heat release in terms of temperature rise is about 125°F (69°C) per percent H_2 concentration in the inlet gas. The higher the inlet gas temperature the less catalyst is required (29).

While recombination is straightforward in concept, the additional processes (such as dilution) required in the system designed for application to BWR off-gas treatment substantially increase the complexity of the operation. Because of the proprietary nature of the design information on catalytic recombination systems, no information concerning details of processes is readily available in the literature. As a result of this lack of pertinent literature, no data have been published that substantiate the claims of approximately 90% volume reduction by catalytic recombination that are implied in several Safety Analysis Reports (14). It is of course, important to design the system to ensure that the condensate from the recombiner system will pass through the liquid radwaste system.

SPECIFIC WASTE COMPONENTS

Iodine

Although iodine is chemically reactive and theoretically readily removable, even a reduction by a DF of 10^4 may leave measurable activity behind if the initial source activity is high enough. The U.S. Atomic Energy Commission has stated that experience has shown that the limiting dose in achieving "as low as practicable" releases of radioactivity in power reactor effluents is the potential thyroid dose due to radioiodine taken into the body of a child by ingestion (4). This assumes that the radioiodine is released in elemental form in the gaseous effluents of the reactor and is dispersed by atmospheric forces in the air before entering the food chain. Any reduction in levels of released iodine, therefore, will help to reduce any potential population exposure or conversely permit a reduction in the plant's exclusion zone.

The environmental effects and population dose calculated for routine releases and accident conditions depend clearly on the assumptions made concerning the leakage of iodine from the fuel, the chemical form in which it occurs in the coolant and the atmosphere in the various buildings, the fraction absorbed inside the building on paint and metal surfaces, and the fraction assumed to escape from the building after decay-holdup or under accident conditions.

It has been found that some iodine appears to be in organic form for reasons not clearly understood for light water reactors, though more evident in the case of reprocessing plants. For CO_2-cooled reactors the organic iodine concentration may be much higher than for LWR's, and filter systems should be designed for the efficient removal of methyl iodide amounting to 10% of the total iodine in the reactor core (30).

Iodine may be found as any of several chemical compounds such as molecular (I_2), hypoiodous acid (HIO), metal iodides, hydriodic acid (HI), iodic acid (HIO_3), methyl iodide (CH_3I) and several other organic iodide compounds (ethyl-, isopropyl-, n-propyl-, tert-butyl-, sec-butyl-, n-butyl-). The physical properties of the iodine compounds vary over a wide range, and the compounds may interact with other materials. The incomplete understanding of specific mechanisms for the formation of iodine compounds and the complex chemical behavior of the compounds in the LWR stations combine to yield a considerable degree of uncertainty in quantifying the capability of the radwaste equipment to provide treatment of gaseous effluents and in predicting and characterizing the "source term" in gaseous effluents from the LWR power stations.

Many compounds of iodine are not stable and conversion from one chemical compound or form to another may occur depending on the environment. Ozone and nitrous oxides in the atmosphere can also modify the organic forms of iodine.

Elemental iodine in gaseous form deposits on surfaces of materials much more readily than does methyl iodide. Other compounds of iodine exhibit deposition characteristics between those of elemental and methyl iodide. It is this range in tendency to deposit on surfaces exhibited by the various chemical compounds of iodine and the instability of these compounds that make the radiological impact difficult to evaluate quantitatively. Most of the organic iodine is in the form of methyl iodide or related compounds and is more readily assimilated into the body in this form than as an inorganic compound. This difference in chemical characteristics is important also in evaluating the removal effectiveness for iodine of various filters and charcoal adsorber systems.

For purposes of predicting the potential radiological consequences of a loss-of-coolant accident for an LWR, the U.S. Atomic Energy Commission (31) has recommended, as a conservative assumption, that 25% of the equilibrium radioactive iodine inventory produced from maximum full-power operation of the reactor core shall be presumed to have leaked from the primary reactor containment. They suggest that 91% of this activity should be assumed to be in the form of elemental iodine, 5% particulate iodine, otherwise unspecified, and 4% in the form of organic iodides. Final release to the surrounding environment is suggested to occur without mixing in the containment building atmosphere and by way of an elevated plume from a stack, where a tall stack is provided. These assumptions are probably unduly pessimistic and merely symbolize an assumed worst release condition. Dispersion conditions are then computed by procedures indicated in Chapters 4 and 8.

For comparison purposes an MPC limit of 1×10^{-10} $\mu Ci/cm^3$ is commonly used in specifying annual effluent limits for iodines and particulates. The iodine isotopes listed in Table 82 are of interest in this connection. Heavier iodine isotopes are very short-lived.

Except for iodine-129, the fission product iodines will reach equilibrium concentrations in the fuel proportional to the product of fission yield x half-life in normal power operation, so that I-131 is effectively the predominant species and is used exclusively in most dose calculations. Table 83 shows assumed radioiodine source terms for a two-unit, 2000 MWe BWR power plant (11). Because of its low specific activity, I-129 has been considered until recently to be of no particular importance;

Table 82. Main-Fission Product Iodine Isotopes

Name	Half-Life	Fission Yield	Main Radiations Emitted (MeV)	
I-129	1.59 x 10^7 yr	0.9%	β 0.15	γ
I-131	8.04 day	2.77	β 0.6	γ 0.36
I-132	2.28 hr	4.13	β 0.8	γ 0.67
I-133	20.8 hr	6.77	β 1.27	γ 0.53
I-134	52.6 min	7.19	β 2.42	γ 0.85
I-135	6.58 hr	6.7	β 1.0	γ 1.28

however, in view of its long half-life some new attention has been directed to its cumulative effects on population exposure over the next 50 years or longer (32). Even on that scale it would seem that the I-129 dose is small compared with the potential radiation dose from other effluents.

A rough estimate (33) for the relative thyroid dose can be derived from the equation

$$D_{129} = 50 \ R \ x \ D_{131} \tag{57}$$

where: D = thyroid doses from I-129 or I-131
R = release rate ratio of I-129 to I-131 in pCi/sec.

In any case, any effective method of iodine removal will also control iodine-129 build-up in the surroundings (31).

As mentioned in Chapter 5, ambient air concentrations of iodine released in reactor accidents can be reduced by adsorption on paint and metal surfaces (34); in addition most of the iodine would also be

Table 83. Radioiodine Source Term: Two BWR 1000 MWe Units (Ci/yr) (11)

	Elemental		Organic	
	I-131	I-133	I-131	I-133
Gland seal	0.0058	0.033	0.0058	0.033
Reactor building	0.017	0.099	17.200	99.000
Radwaste building	0.0014	0.004	0.120	0.680
Turbine building	1.000	5.700	1.000	5.700
Totals	1.024	5.836	28.326	105.413

entrained in steam and water vapor and condense on the floor (35). However, any airborne iodine or iodine compounds would have to be recovered by the normal radwaste treatment system. When fuel melt-down occurs in an oxidizing atmosphere the iodine is primarily I_2; in a reducing atmosphere it is primarily HI (or solid compounds, such as CsI). CH_3I is not formed during meltdown but by subsequent reactions within the containment shell. The various methods employed for the control of iodine in the nuclear industry have been reviewed in an IAEA report of that title (30) and will be discussed here briefly. Most of the methods for iodine removal promote the reaction of gaseous iodine with silver or other metals to produce a nonvolatile compound, or entrain the iodine in steam or liquids to convert it into a soluble compound that can subsequently be more easily concentrated by ion exchange methods. In filter systems, HEPA units should be followed by an iodine absorber, since iodine may desorb from filter particles after being trapped.

Liquid Scrubbers

Liquid scrubber systems for removing radioiodine from gaseous streams are widely used. To ensure efficient removal, special chemicals, such as NaOH or LiOH are added to the water to convert elemental iodine to nonvolatile alkali iodides. Though the reaction is different, sodium thio-sulfate has the same effect and is often preferred over other reagents. There is evidence that systems containing hydroxides or thiosulfates do not remove methyl iodide efficiently (30).

Alkaline scrubbers produce an additional liquid effluent that must either be allowed to decay or else treated before disposal. Mercuric salts are often added to dissolver solutions in reprocessing plants to in-hibit volatilization of the iodine. Typically, caustic scrubbers pass the off-gas over packing in a tower that is continuously wetted with an aqueous caustic solution. A mercuric iodate precipitate can result when the off-gas is scrubbed with a mixture of 8-14 molar nitric acid and 0.2-0.4 molar $Hg(NO_3)_2$ (11). This technique is effective for the removal of large quantities of elemental iodine and trace quantities of other isotopes. Due to its limited efficiency, in a fuel reprocessing plant a scrubbing system should be placed just before a silver zeolite bed to remove most of the iodine and help reduce the zeolite bed loading to extend its lifetime.

High removal efficiencies (DF = 10^4-10^5) may be obtained by passing the gases through a 30% solution of tributyl phosphate in kerosene; the iodine may be recovered from the solvent by scrubbing with 1M NaOH (25). A variation is the Iodex Process, illustrated in Figure 67, which

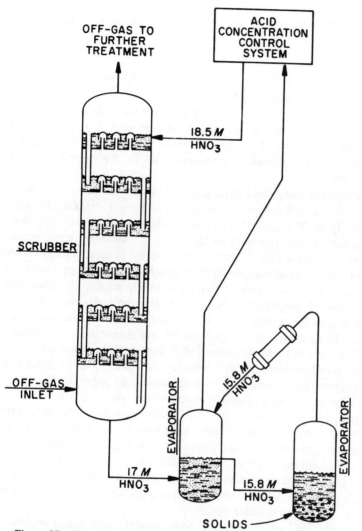

Figure 67. Diagram of the Iodex process for removal of iodine from gaseous effluents (15).

consists of a bubble-cap scrubber in which concentrated nitric acid is contacted with the gases (15). The iodine is separated in an evaporator as a solid (I_2O_5) and the acid is recycled.

Silver Reactors

Elemental iodine reacts very strongly with silver surfaces to produce silver iodide at moderately-high gas temperatures. Appreciable removal of radioiodine has been observed in the heterogeneous exchange reaction

$$2AgI + {}^*I_2 \rightleftharpoons 2Ag^*I + I_2 \qquad (DF \sim 10^3)$$

with rates at 250-300°C sufficiently great for this method to be practicable at concentrations as low as 10^{-7} g iodine/ft^3 (25). The silver reactor can take the form of heated silver nitrate beds or a tower of unglazed porcelain Berl saddles that have been soaked in a solution of silver nitrate. At 200°C, removals were of the order of 99.9% in large scale trials; the maximum temperature is set by the melting point of AgI at 212°C.

Silver-plated metals and silver-plated rockwool are being used as a means for reducing iodine concentrations in hot air streams with efficiencies up to 99.9% (24). Decontamination factors of 10^3-10^5 have been achieved with synthetic zeolites, used as molecular sieves, provided the gases are dried first; this may be done by passing through a column of "finer pore" molecular sieves that absorb water vapor in preference to iodine. The silver form of zeolites is widely employed; silver zeolite beds, $AgNO_3$-impregnated in an alumina-silica molecular sieve, have an excellent removal capability for both elemental and organic forms of iodine. This material is not combustible and decontamination factors of 10^4 for methyl and elemental iodide have been reported on gas streams at velocities of 20-30 ft/min for 2-in. thick beds of silver zeolite. One disadvantage of zeolite is its susceptibility to poisoning by contaminants such as halides and sulfur compounds in the off-gas stream. An advantage of zeolite beds is the high loading capacity of zeolite. Pence and Maeck (36) report that zeolite beds have 20 times the absorbent capacity of ordinary activated charcoal. This suggests the use of small filter beds with an excellent removal efficiency. Silver zeolites show good operating characteristics in high temperature, high humidity, and strongly oxidizing gases. This makes them particularly useful in off-gas streams of fuel reprocessing plants.

Retentions of over 99.99% for both elemental iodine and organic iodides have been demonstrated, and silver zeolite beds are being installed as an additional "polishing" step to reduce traces of iodine left after other treatment processes (11,15).

Charcoal Beds

Activated charcoal is the most widely used material for the removal of iodine, in both organic and elemental forms, and as a means of decay storage. The surface areas of activated charcoals run up to 1800 m^2/g.

It is a very effective material for removing molecular iodine, even at high humidities. The design of such systems has been described already.

The removal efficiency of this method depends to a great extent on the number of charcoal modules in the flow stream and the type of flow mesh used. At room temperature, the desorption of iodine from charcoal is about 0.001%, but as the temperature of the charcoal increases, the iodine mobility increases and greater desorption results.

Organic compounds of iodine, such as methyl iodide (CH_3I), will be trapped less efficiently by activated charcoal. In the case of high-capacity purification plants it is essential to ensure that iodine and its compounds are trapped in thin layers. For this purpose the charcoals used are impregnated with substances that react chemically with compounds of iodine or enter into isotopic exchange with gaseous compounds of iodine. At the same time, adsorption takes place. An example is represented by charcoal impregnated with potassium iodide (KI), where isotopic exchange occurs between KI, which contains a stable isotope of iodine, and CH_3I, which contains one of its radioactive isotopes. Another example is charcoal impregnated with triethylenediamine. Triethylenediamine undergoes a chemical reaction with alkyl halides to form quarternary ammonium salts. Both types of charcoal are commercially available.

Low volatility of the impregnant and the final reaction product, good chemical and radiolytic stability, and the price are factors in the selection of an impregnant.

Important factors affecting the efficiency of trapping are the relative humidity and temperature of the off-gas. At low relative humidities both unimpregnated and impregnated charcoals may be used. At high relative humidities impregnated charcoals are, as a rule, more effective. Some impregnants give good results for the trapping of methyl iodide at high humidity. Liquid must not be allowed to reach the charcoal. Activated charcoal will age and can be seriously poisoned by impurities such as organic vapors and NO_2 so that the useful life of the charcoal may be limited to a few months.

Flammability and relatively low desorption temperatures limit the use of impregnated activated charcoal for the removal of fission-product iodine. Spontaneous ignition of activated charcoal is a distinct disadvantage. Typically ignition temperatures range from 250°C to 500°C (37). Normally, spontaneous ignition should not be a problem if the charcoal beds are operated properly since the flow of gases over the beds keeps them cool. However, various devices have been developed to provide automatic cooling sprays in the event of charcoal ignition.

Cost of Radioiodine Removal

It is evident that most reactor facilities will require several stages of air cleaning equipment for iodine recovery, both to absorb routine releases and to provide spare capacity to recover iodine releases inside the containment building in case of a fuel meltdown accident. Even more stringent iodine removal systems will be needed for reprocessing plants, especially if the long-lived iodine-129 is taken into account.

Such removal facilities do not come cheap. This is illustrated in Table 84, which shows the capital cost and annual costs for successive additions in control equipment for both elemental and organic iodine (11). A value of 10 is assumed for the DF of a charcoal bed absorber, and an incremental DF of 10 for a deep-bed absorber. These may be unduly pessimistic values but may be justified as representing the average performance over long runs. The right-hand columns indicate the corresponding remaining iodine release rates to the environment. A reduction by a factor of about 10^3 is obtained over the currently used off-gas systems, consisting of a 30-minute delay pipe and a 100 m stack, but at a cost of close to eight million dollars for a 2000 MWe station. However, significant iodine removal already may be obtained at lesser cost.

Noble Gases

The noble gases—xenon, krypton and to a lesser degree argon—pose a special problem in radwaste treatment planning. Apart from argon-41, which results from activation of air in the vicinity of the reactor core, both xenon and krypton isotopes are produced with a high fission yield and constitute the major gaseous fission activity in the core inventory and the primary coolants (Figure 42). Owing to their chemical inertness they diffuse fairly readily through cracks and pores in cladding and piping, they are not removed by the normal decontamination procedures, and until recently they were treated only to the extent of a delay-decay system before being released up the stack, diluted with an adequate amount of makeup air. In accident analyses, most design base accidents are assumed to result in a 100% loss of noble gases, which, therefore, account for the greater part of the external radiation exposure to the surrounding population, from a passing cloud or plume of radioactivity immediately following such an accident. Since the uptake of noble gases into the human body is fairly insignificant, only the immediate lung dose from the inhalation of contaminated air is usually considered.

Experience with the release of gaseous effluents from operating power reactors, shown for 1972 in Table 77 (2), has demonstrated that the

Table 84. Annual Costs (1973) for Radioiodine Removal from BWR Gaseous Effluents (Two Units, 1000 MWe Each) (11)

Control Option Added	Estimated Capital Cost (Cumulative)	Estimated Annual Cost (Cumulative)	Radioiodine Release (Ci/yr)		
			I-131	I-133	Total
A. Elemental Iodine					
Present operation[a]	$ 0	$ 0	(30.0)	(176.)	(206.)
None[b]	0	0	1.02	5.84	6.86
Clean steam: valves > 2.5 in. diameter	1,800,000	300,000	0.224	1.28	1.50
Turbine building charcoal adsorber	4,300,000	750,000	0.044	0.250	0.295
Upgrade to deep bed charcoal adsorber: turbine building	5,100,000	1,200,000	0.026	0.147	0.173
Reactor building charcoal adsorber	7,100,000	1,600,000	0.017	0.058	0.075
Radwaste building charcoal adsorber	7,350,000	1,640,000	0.010	0.055	0.065
Turbine gland seal clean steam	7,950,000	1,840,000	0.004	0.022	0.026
B. Organic Iodine					
Present operation[a]	$ 0	$ 0	(178.)	(955.)	(1133.)
None[b]	0	0	28.3	105.	135
Reactor building charcoal adsorber	2,000,000	400,000	2.85	16.3	19.2
Upgrade to deep bed charcoal adsorber: reactor building	2,500,000	800,000	1.30	7.40	8.70
Radwaste building charcoal adsorber	2,750,000	840,000	1.19	6.79	7.98
Clean steam: valves > 2.5 in. diameter	4,550,000	1,140,000	0.390	2.23	2.62
Turbine building charcoal adsorber	7,050,000	1,590,000	0.210	1.21	1.42
Turbine gland seal clean steam	7,650,000	1,790,000	0.204	1.17	1.37

[a]Illustrates projected effects of two 1,000 MWe BWR's operating with presently used off-gas system (i.e., 30-minute delay and 100 m stack for air ejector noble gases).

[b]Reflects source term for sources other than air ejector as "augmented" BWR noble gas treatment systems (charcoal adsorption, selective absorption, or cryogenic distillation) will effectively remove air ejector radioiodines.

quantity of noble gas activity released annually easily exceeds the sum of all other activities taken together. It is also seen that the quantity of noble gas activity from BWR's is several orders of magnitude higher than that from other reactor types and rather closer to the MPC values than would be acceptable as being " as low as practicable." However, since most of the xenon isotopes and all krypton isotopes except krypton-85 are short-lived, the radiation hazard from these emissions is lower than would be the case for comparable activities of other radioisotopes (28,38-40).

All these considerations make it desirable, though not imperative, to devise systems to reduce the quantity of radioactive noble gases released to the environment and to find means of storage and disposal of this material. There is a cost-benefit problem associated with this question, in view of the common attitude that readily diluted noble gas emission poses little radiation hazard to the general population and hence constitutes largely an ethical dilemma associated with the prospect of a more than tenfold increase in the global krypton-85 inventory over the next 25 years if unrestricted release is permitted to continue (33,41). This projected increase in krypton-85 concentrations in the atmosphere is shown in Figure 68; Table 85 shows some estimated annual doses to the

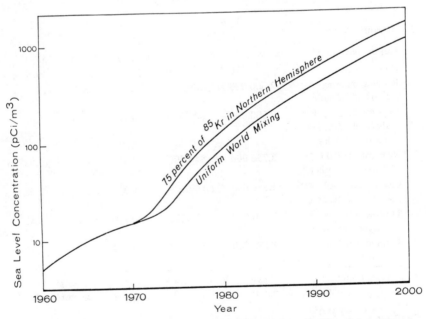

Figure 68.   Rise in estimated Kr-85 concentration in the northern hemisphere from nuclear power generation (33).

Table 85. Estimated Annual Doses to the U.S. Population from World-Wide Distribution of Krypton-85 (33)

Year	Dose			
	Whole-Body		Skin	Lung
	(mrem/person)	(man-rem)	(mrem/person)	(mrem/person)
1960	0.0001	20	0.005	0.0002
1970	0.0004	80	0.02	0.006
1980	0.003	700	0.1	0.005
1990	0.01	4,000	0.6	0.02
2000	0.04	12,000	1.6	0.06

general population in the northern hemisphere (33). It is evident that the individual dose is insignificant and the importance of the cumulative population dose is marginal. In these calculations a concentration of 3×10^5 pCi/m^3 of air was considered to give 7 mrem/yr to the whole body, 310 mrem/yr to the epidermis and 12 mrem/yr to the lungs. A detailed review of the literature and an analysis of the radiation hazards from Kr-85 has been published by Kirk (42), who estimated that 0.2 kCi of Kr-85 is produced for each megawatt of power output.

Many discussions of this topic have assumed that krypton-85 is uniformly distributed in the environment. However, practically speaking, concentrations are greater in the vicinity of fuel reprocessing plants. Based on this assumption, the annual dose to persons living near reprocessing plants may be higher by a factor of 10 or more (43).

Although individual exposures seem relatively small, one must remember that krypton-85 affects the entire world population. It is customary to calculate the man-rem dose and use the linear hypothesis of radiobiological damage to estimate radiation-induced cancer deaths on an annual basis. On that basis, it has been projected that an additional average skin dose of 10 mrem/yr and a whole body dose of 0.1 mrem/yr may result in about 2,000 additional cancer deaths per year on a world-wide basis. This estimation seems to justify the need for investigating possible removal and storage techniques, even though such levels are nowhere being approached.

Table 86 lists the half-lives and source terms of the noble gases for a model boiling water reactor (11). Taking the product of half-life and source level as a measure of environmental significance, S, one sees that only Xe-133 and Kr-85 are of predominant interest.

Table 86. Noble Gas Source Terms for Two-Unit Model BWR Plant (Ci/yr) (11)

Radionuclide	$T_{1/2}$ Half-Life	Air Ejector Effluent	Turbine Gland Seal Effluent	Turbine Building	Mechanical Vacuum Pump	Q Totals	$S = Q \cdot T_{1/2}$
Kr-83m	1.86h	1.5(5)[a]	3.5(2)	2.9(1)	—	1.5(5)	2.79(5)
Kr-85m	4.4h	2.8(5)	6.1(2)	4.9(1)	—	2.8(5)	1.23(6)
Kr-85	10.76y	7.6(2)	2.4(0)	4.2(-1)	—	7.6(2)	7.16(7)
Kr-87	76m	7.6(5)	1.9(3)	1.7(2)	—	7.6(5)	9.6 (5)
Kr-88	2.80h	9.1(5)	2.0(3)	1.7(2)	—	9.1(5)	2.55(6)
Kr-89	3.2m	9.1(3)	3.8(3)	1.0(3)	—	1.4(4)	7.47(2)
Kr-90	33.0s	—	3.1(3)	2.0(2)	—	3.3(3)	3.03(1)
Xe-131m	11.8d	7.6(2)	2.0(0)	1.2(0)	—	7.6(2)	2.15(5)
Xe-133m	2.26d	1.5(4)	3.0(1)	2.3(0)	—	1.5(4)	8.14(5)
Xe-133	5.27d	4.1(5)	8.6(2)	6.6(1)	4.4(3)	4.1(5)	5.19(7)
Xe-135m	15.6m	3.5(5)	2.4(3)	2.1(2)	—	3.5(5)	9.1(4)
Xe-135	9.2h	1.1(6)	2.2(3)	1.8(2)	6.6(2)	1.1(6)	1.01(7)
Xe-137	3.9m	3.4(4)	1.1(4)	1.3(3)	—	4.6(4)	3.0 (3)
Xe-138	17.0m	1.1(6)	8.6(3)	5.9(2)	—	1.1(6)	3.12(5)
Xe-139	43.0s	—	4.7(3)	9.9(2)	—	5.7(3)	6.81(1)
Totals		5.1(6)	4.2(4)	5.0(3)	5.1(3)	5.1(6)	

a1.5(5) = 1.5 x 10⁵ or 150,000

Methods for the treatment and recovery of noble gases have been reviewed by Slansky (44), Dunster and Warner (43), Bendixsen and Buckham (45), Till (46) and Leonard, Baer and Eckart (14). These methods can be classified as delayed storage, cryogenic processes, use of permselective membranes, selective absorption in fluorocarbons, and entrainment processes. Many of them are still at the experimental stage and their economic feasibility has yet to be demonstrated. Complete removal of noble gases from the secondary containment atmosphere is usually regarded as impractical because of their low concentrations and the large volume of air involved (24).

Holdup-Decay

The only currently practiced method of minimizing the discharge of noble gases to the environment routinely or after an accident is by holdup for radioactive decay followed by controlled discharge through a high stack. The utility of this method depends, of course, on the particular isotopes present. To illustrate the magnitude of activity reduction attainable in boiling water reactors, consider the mixture of fission product noble gases as shown in the Shoreham PSAR (Table 87). If a delay line is considered in which the average time delay for gas transport prior to release is 30 minutes, an overall decontamination factor of 48 is computed for this particular gas mixture. Using the 30 minute delay time as a reference point, DF's on the order of 10^3 could be attained for all krypton nuclides with a time delay of just under 3 days. A similar DF could be attained for all xenon nuclides with about 35-40 days of time delay, assuming the gas mixture as previously stated. The xenon activity is largely determined by the ^{133}Xe decay after 3 days of delay. By then the only krypton isotope of significant activity is long-lived ^{85}Kr. Thus the efficiency of increased time delay as a means for reducing gaseous activity release is limited by the xenon decay constant.

Reservoirs intended to achieve gas holdup times of several days and more must be designed in such a manner to assure that an appreciable fraction of the gas is not released prematurely as a result of mixing of gases generated at different times. Thus the design should utilize either very long, narrow passages to minimize mixing effects, or alternately a series of separate chambers that are filled, left undisturbed for a period of time, and then discharged sequentially. In either case, the volumetric storage capacity (and cost) of such a system should be roughly proportional to the delay time selected (14). Current design trends favor a 60-day holdup before release of gases to the atmosphere.

Table 87. Effect of Holdup-Decay on Release Rates of Noble Gases (23)

Isotope	Half-Life	Emission Rate (μCi/sec)						
		Decay 0	Decay 30 min	Decay 10 hr	Decay 24 hr	Decay 3 day	Decay 9 day	Decay 40 day
Xe-143	1 sec	502,000						
Kr-94	1 sec	1,500,000						
Kr-93	2 sec	2,440,000						
Xe-141	3 sec	2,700,000						
Kr-92	3 sec	4,050,000						
Kr-91	10 sec	3,030,000						
Xe-140	16 sec	2,450,000						
Kr-90	33 sec	2,280,000						
Xe-139	41 sec	1,920,000						
Kr-89	3.2 min	8,881,000	1,320					
Xe-137	3.8 min	1,020,000	5,290					
Xe-135m	15 min	158,000	41,600					
Xe-138	17 min	450,000	132,000					
Kr-87	1.3 hr	102,000	77,800	493				
Kr-83m	1.86 hr	15,100	12,500	397	2			
Kr-88	2.8 hr	96,600	85,400	8,120	254			
Kr-85m	4.4 hr	31,200	28,500	5,480	480			
Xe-135	9.2 hr	89,600	89,500	43,900	15,100	387		
Xe-133m	2.3 day	930	924	820	688	376	62	
Xe-133	5.27 day	24,900	24,800	23,600	21,900	17,000	7,860	130
Xe-131m	12 day	74	74	73	70	63	44	7
Kr-85	10.4 yr	37	37	37	38	38	37	37
Approximate Total		24,000,000	500,000	83,000	39,000	18,000	8,000	174

aBased on diffusion mixture. These rates are indicative of substantial fuel failures.

Ambient-temperature charcoal adsorption is currently being used to delay noble gas releases from reactors. One advantage of charcoal adsorption is simplicity (1). However, very large charcoal beds are required, and there is a fire and explosion hazard. The quantity of charcoal required for a typical nuclear facility is in the tens-of-tons range; thus process equipment would be large and bulky, although operation would be relatively simple (45). By lowering the temperature of the charcoal bed, the adsorption capacity is increased and the bed volume is reduced by a factor of 2 to 5. A fire and explosion hazard still exists. Such a system was used at the Idaho Chemical Processing Plant at one time but proved to be unreliable and was prohibitively expensive due to refrigeration costs.

Low-Temperature Charcoal Adsorption

Since the adsorption of xenon and krypton increases as the temperature of an adsorber is reduced, fission-gas adsorption systems are often designed for operation at or near the temperature of liquid nitrogen. The principal benefit obtained is the small size of the adsorber, which is a factor of considerable importance in mobile installations. On the other hand, low-temperature adsorbers have certain disadvantages. Their design is influenced by provisions for safety in the event of loss of refrigeration. Means must be available for stopping the flow of radioactive gas, and the systems must be designed to withstand the pressure buildup that will occur as the adsorber warms to room temperature. Precautions must be taken to avoid accumulation of explosive mixtures of oxidizing gases in the adsorber, and water and condensable materials must be removed from the inlet gas stream to avoid plugging the refrigerated parts of the system.

In a study of the removal of fission-product noble gases from helium by low-temperature adsorption on charcoal in support of the development of the high-temperature gas-cooled reactor of the OECD Dragon Project, experiments were conducted on the breakthrough of krypton, xenon, and other single gaseous impurities in helium carrier-gas streams on both laboratory and semiplant scales. Gas-pore diffusion was found to govern the rate of breakthrough in the laboratory, whereas external fluid-phase diffusion appeared to predominate in the semiplant-scale equipment during the early stages of breakthrough. Rate determinations confirmed that gas-pore diffusion control is also applicable in the absence of helium carrier gas (47). Despite their drawback in necessitating a readily available supply of liquid nitrogen, such systems are under consideration at several power plants to reduce noble-gas releases (30).

An alternative approach, which may be more effective and convenient, is the use of low-temperature molecular sieves. Such systems have good selective adsorption characteristics and slow desorption rates on warming and may provide a convenient means of storage and disposal of krypton wastes (44,48-51).

Cryogenic Distillation

In cryogenic separation the noble gases and part of the air or other carrier gas are first liquified. Then the noble gases are separated from the bulk gases and concentrated by fractional distillation. As in all low-temperature operations, water and other condensable gases that would form solids must be essentially removed prior to the treatment of the noble gases. Solids in the system cause physical difficulties, and the presence of liquid ozone, which is formed from the radiolysis of oxygen, creates an explosion hazard. Such explosions can be avoided if O_2 and O_3 are excluded from the low-temperature region (47).

Figure 69 shows the flow sheet for a rare-gas recovery facility at a reprocessing plant (44,47). There is some question whether such a system can be justified for the separation of Kr-85 from desorbed gases from a charcoal-bed system to facilitate long-term storage at nuclear power plants, though it is being strongly advocated at some of the bigger plants.

The major advantage of the cryogenic distillation process for adsorption of noble gases is its present high technological level. Liquid air plants have been in existence for decades, and thus considerable knowledge has been assembled on materials of construction, valves, compressors, distillation column design, modes of operation, and reliability. Specific cryogenic processes recovering natural krypton from air have also been operated for some time. Because of this, the projected capital and operating costs for a fission-product noble-gas removal system are well defined. In addition, the cryogenic distillation process is the only process to have operated on a significant scale at a fuel reprocessing facility for the recovery of production quantities of krypton-85. A number of U.S. companies are offering similar units for use at nuclear facilities and have projected krypton-85 recovery efficiencies typically in excess of 99%. Adequate control of krypton-85 leakage rates to achieve high recovery efficiencies has not been demonstrated yet, but appears achievable (45).

Fluorocarbon Absorption

Selective absorption processes for the separation of krypton and xenon from radioactive gaseous effluents have been investigated at several

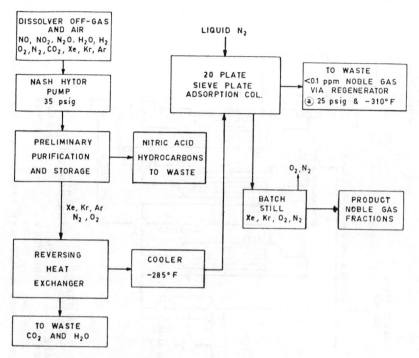

Figure 69. Flowsheet of ICPP cryogenic distillation process for recovering xenon and krypton from dissolver off-gas (44).

installations. These processes exploit differences in solubility between various gaseous components in organic solvents. Specifically, krypton and xenon are markedly more soluble in fluorocarbon refrigerants, such as Freon 11 or 12, than are argon, oxygen or nitrogen. The solvent can be operated at temperatures of -30 to -90°F and at pressures up to 400 psig, leading to much lower refrigeration costs than the cryogenic distillation process, low solvent cost and good stability. Figure 70 shows a flow sheet of the freon process. The contaminated air stream is contacted with solvent in a multistage absorber for removal of the noble gases. The gas-laden solvent is then routed to one or more stripping units, where the dissolved gas is ultimately desorbed completely. The concentrated noble gas is then compressed and bottled for storage (44,47). Explosion hazards with ozone or xenon tetrafluoride are estimated to be very low. Process system size, mode of operation and capital cost are comparable to the cryogenic distillation process (45,52).

Other solvent processes that have been explored on a pilot scale are a kerosene-CO_2 system, in which noble gases are carried by CO_2 and

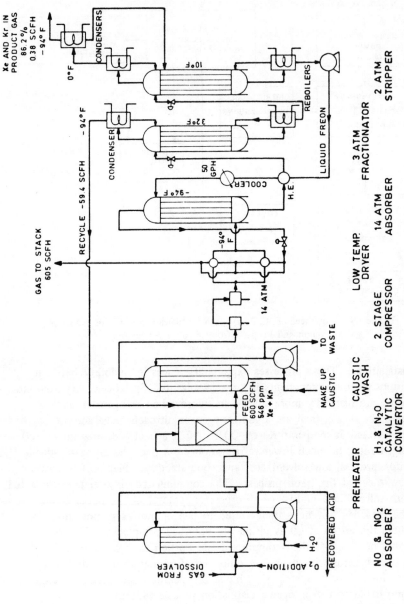

Figure 70. Freon process for noble gas absorption (44). (Gas flow rates shown as SCFH = standard cubic feet per hour.)

at the end the CO_2 is removed by a caustic scrubber, a carbon tetra-chloride process, and several kerosene-based solvents, all of which have proved feasible (44).

Permselective Membranes

The transport process for permselective membranes is quite different from the transport of gases through a porous membrane. The process involves four successive steps: contact of the gas with the high-pressure surface of the membrane, dissolution of the gas in the membrane, diffusion of the gas through the membrane, and evaporation of the gas from the low-pressure surface of the membrane. The separation is proportional to the ratio of the products of the solubilities in the membrane and the diffusivities through it. The effective solubility of krypton is about 5-10 times that of nitrogen and approximately twice that of oxygen (47).

Using these differences at high pressures (150 psi) with a silicone rubber membrane, Kr and Xe can be separated from bulk off-gases. Large membrane surface areas and high pressures are required, resulting in high capital and power costs. The membranes are sensitive to ozone, iodine and NO_x damage and can be affected by thermal and pressure shock. The system appears economical for low decontamination factors only (45).

Rare-Gas Clathrates

A clathrate is an inclusion compound whereby the molecules of one component, the noble gas, are contained (imprisoned) in the crystalline structure of a second component without the formation of a classical chemical bond. Various rare-gas clathrates have been formed over the years in various crystal lattices, primarily as a source of radiotracer gases. For gas collection, hydroquinone $C_6H_4(OH)_2$ crystallized from organic solvents was found to be the best crystalline compound, providing a large available space for rare gases within the crystal. Release of the enclosed gas can be accomplished only by breaking the hydrogen bonds of the crystals, by heating, dissolving or crushing them (53).

Clathrates do not seem to be a practical means of removing noble gases in the ppm range of concentration because of the high pressures required, the slow rate of crystallization and poor removal from very dilute streams. However, they may be useful for storing krypton as solids, after recovery and concentration by other processes (44).

Other Processes

Other recovery processes, mostly still at the laboratory stage, that have been proposed include electrostatic diffusion, formation of kryptonates, and fluorine reactions (45,54). Table 88 compares some of the more promising methods. All of the methods are capable of producing decontamination factors of greater than 100. The costs of increasingly more sophisticated methods of rare-gas removal in conjunction with holdup beds become appreciable and have been presented in Reference 11 for PWR's and BWR's.

The cryogenic distillation of krypton removal is the best-developed method at this time but that is only because it has been used commercially for many years. It is very likely that fluorocarbon absorption will prove to be more economical and give better decontamination. Full investigation of all methods should be expanded to pilot plant and commercial operation for a more valid judgment of performance.

During all of the removal processes reviewed there is no selective removal of krypton-85. It is assumed (55) that krypton gas collected from a typical fuel reprocessing plant is about 7.76% krypton-85. Therefore, if krypton is separated, there is no way to further isolate the radioactive krypton-85 and the entire quantity collected must be stored.

Containment and Storage

All of the removal processes assume storage on-site of the noble gases to ultimate decay. No adequate processes exist at this time to fully immobilize the gases to permit safe underground disposal.

Methods proposed for containment are storage in low pressure tanks, storage in high-pressure cylinders, encapsulation, and injection underground. Low-pressure storage presents the advantage of low leakage probability (43). However, this method involves large tanks and a large permanent storage facility. High-pressure cylinders have already been used to ship radioactive gases and give proven reliability. This method greatly reduces the total storage volume. Ozone removal is necessary to prevent corrosion, and approximately 80 containers per year of radioactive noble gas would be produced by the turn of the century (56).

Encapsulation has the advantage of providing double containment. This idea has been tested (55) where the gases were trapped in molecular sieves and then mixed with epoxy resin, glass, metals, and plastic. Although the total storage volume is increased, an added measure of safety is provided. Injection of radioactive gases into permeable basalt strata or empty gas wells has been tested with favorable results. This method of disposal is inexpensive; however, public reaction probably prohibits

Table 88. Comparison of Processes for the Removal of Krypton-85 from Dissolver Off-gas from a Fuel-Reprocessing Plant (44)

Process	Kr Recovery (%)	Development Status (1971)	Advantages	Disadvantages
Room-temperature charcoal beds or molecular sieves	99	Bench scale completed; scale-up feasible	Simple operation; accepts dilute feed gas	Large-volume adsorber beds; charcoal can ignite; strong oxidizing gases must be removed prior to adsorption
Low-temperature charcoal beds or silica gel	99	Development completed; plant operated	Small-volume beds; uses dilute feed gas	Charcoal can ignite; oxidizing gases, CO_2 and H_2O must be removed; large consumption of liquid nitrogen: adsorbers must withstand high pressure; high operating cost
Cryogenic distillation	98	Developed and operated on a significant scale	Low capital cost and low operating cost	Explosion hazard in forming and concentrating ozone
Liquid extraction	99	Bench scale completed; demonstration needed for large scale	Using Freon-12: low refrigeration costs; low solvent costs; no explosion hazard; might eliminate pre-treatment	The absorber column operates at 200 psig; the volume of extractant is large if operated at 15 psig
Clathrate precipitation Kryptonates	Unknown	Laboratory studies only; no engineering	^{85}Kr is collected as a solid; storable	Needs concentrated feed gas; crystallization step slow
Perm-selective membranes	99	Bench-scale work; need engineering tests		Membranes sensitive to chemicals; high power costs
Thermal diffusion		Little pertinent data		
Electrostatic diffusion	L	Limited; technical feasibility not proven		Poor economics for disposal of dilute ^{85}Kr waste

its use. Final disposition of the krypton could be at an underground engineered storage facility located in a remote area (56).

In terms of potential environmental impact, a question still remains whether more of a hazard may be associated with the accumulation of a large inventory of radioactive krypton at any one location in the form of high-pressure gas bottles, that may develop leaks, than with the steady release of highly diluted krypton-85 to the environment from the top of a tall stack. In this connection studies have been conducted by Underhill to predict the consequences of the rupture of a pressurized noble-gas adsorption bed (57).

Tritium

Tritium is a major component of the off-streams of nuclear reactors, and even more so of reprocessing plants. As was explained in Chapter 5, it arises mainly from ternary fission and from the boron used as a reactivity control in PWR's. Either directly or through the use of recombiners, most of the tritium will appear in the liquid effluents, probably as HTO. Table 76 shows the quantities of tritium released from U.S. power reactors in 1972; as one would expect, the release from PWR's greatly exceeds that from other reactor types. Tritium will also form in high concentration in heavy-water reactors, creating a serious contamination hazard around any leak in the coolant system. In HTGR's most of the tritium in effluents arises from neutron capture in helium-3, which is present in concentrations of 137 ppm in natural helium.

There are two ways of being exposed to tritium: ingestion in food and drinking water, and absorption through the skin. Since the tritium readily exchanges with other hydrogen, it largely follows the water cycle, both inside the body and in the environment; no reconcentration would be expected with respect to ordinary hydrogen and most of the tritium will ultimately end up in the oceans (58). In the body, three effective biological half-lives have been identified—8.6 days, 30 days and 450 days—corresponding to the turnover of body water, organic compounds and organically bound hydrogen, respectively, with the water component accounting for about 84% of the total. Owing to the low energy of the beta particles emitted and the rapid dilution and elimination of the tritium, its maximum permissible concentration in drinking water is the highest for any radioisotope, 0.03 $\mu Ci/cm^3$; for air concentrations, where the skin absorption is a limiting factor, the MPC is 2 x 10^{-6} $\mu Ci/cm^3$. As Table 76 shows, even with the large amounts of tritium discharged, none of the reactors listed had any trouble staying well below these concentrations.

As the number of reactors grows the total amount of tritium generated will become substantial and by the year 2000 may equal the natural equilibrium inventory from cosmic ray production. Current tritium levels in the atmosphere due to nuclear weapons tests are well above that level and it is to be hoped that they will never be permitted to rise that high again (Figure 71). Since tritium will be needed to fuel fusion reactors,

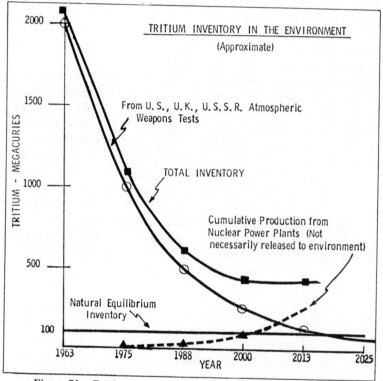

Figure 71. Total amounts of tritium from all global sources (15).

there is some incentive to recover it; however, recovery of the low concentrations of coolant tritium is uneconomical at present and there is little inducement to reduce the low effluent concentration. In 1970 the average annual radiation dose from all sources of tritium was 0.04 mrem per person (33); yet the philosophy of maintaining effluent concentrations as low as practicable would make it desirable to contain the tritium instead of releasing it freely.

Reduction of tritium levels can proceed by one or more of the following steps: (1) reduction at source, (2) containment, and (3) recovery and reconcentration.

Reduction at Source

Reduction at source in light-water reactors may be accomplished by replacing ordinary LiOH by ^{7}LiOH in coolant-demineralizer systems, by using other reactivity controls than boron compounds, and by reductions in coolant volume. For light-water reactors, annual tritium production rates are about 19 Ci/MWe for uranium fuel, and 36 Ci/MWe for plutonium fuel (33), depending somewhat on the load factor assumed and the core life. For gas-cooled reactors, reduction in tritium production by isotopic separation of He-3 from the helium coolant has been proposed but appears economically unattractive.

Heavy-water reactors, which account for about 5% of the world's production of nuclear power, contain typically about 430 kg of D_2O per MWe, with a tritium concentration of 11 Ci/kg, of which 2.5% is assumed to be lost to the environment annually (33). Three reactions play a role in the tritium chemistry in such a system: (1) the DTO formed by neutron irradiation can dissociate to form tritium radicals that further react with deuterium radicals to form DT; (2) D_2O can react with DT to form DTO and D_2; and (3) the probability of deuterium-radical formation in the radiolysis of DTO is about twice that of tritium-radical formation, so the hydrogen molecules formed will be depleted in tritium. The first process would lead to tritium discharge as a gas, and the last two processes would favor tritium retention (58).

Recycle

Containment of tritium to reduce discharge levels is accomplished by a recycling of tritiated water from all expected sources by segregating drains containing HTO from those that do not. At the TVA Watts Bar plant any liquid showing a tritium concentration with higher than one-tenth of the concentration in the primary coolant is recycled and not released. This effectively recycles more than 90% of the tritium collected as liquid. Plant operating procedures are designed to assure that during all periods tritium levels in the condenser vacuum pump effluent, steam generator blowdown distillate, and any other effluents are within acceptable limits. Tritiated water is recycled back into the primary system until the primary coolant tritium concentration reaches a maximum desirable level from an operating personnel dose standpoint. The exact concentration that may be considered the maximum safe level will be determined largely by doses that could be received by plant personnel during refueling operations. Tentatively, this concentration has been set at 2.5 μCi/ml for analysis purposes, and, based on the assumptions used for estimating routine releases, this level would be reached about eight years after startup. Tritiated water will be periodically extracted from the primary system to maintain the maximum safe level (12). Tritiated water bled from the primary system must be shipped offsite as necessary, as low-specific-activity liquid waste, for retention at an approved disposal site.

It is impossible to prevent totally the release of tritium from a nuclear power plant by reasonable means. Vaporization from the refueling canal and spent fuel pit carries off significant amounts of tritium. In addition, some secondary liquids and vapors must be released from the plant, and at times these will contain small amounts of tritium that have leaked into the secondary system from the primary system. A major source of this type of tritiated discharge will be steam generator blowdown during periods of primary-to-secondary leakage in the steam generator. Other expected sources of tritium release include purging of the containment and fuel storage areas, condenser vacuum pump effluent during periods of steam generator leaks, and any leakage that is at a tritium concentration too low to be recycled. Secondary system blowdown, even though it contains tritium due to a leak, must also be discharged. With a liquid radwaste system modified for tritium recycle (12), the volume of water in the primary system is approximately 280,000 gallons at any time, except during refueling. This includes water in the reactor coolant system, the primary makeup water storage tank, the CVCS holdup tank, the CVCS monitor tank, and the tritiated waste holdup tank. The tanks are assumed to be partly full at normal operating levels. This means that that particular reactor at equilibrium will contain a tritium inventory in the cooling loop of $2.5 \times 10^{-3} \times 280,000 \times 3.785 = 2.65 \times 10^3$ Ci. At this point one has to decide whether a sudden release of a sizable fraction of that amount of tritium during an accident may be a lesser hazard compared with the steady, though dilute, discharge of much of the tritium in the liquid effluents.

Tritium Recovery

Unless the tritium can be recovered at source, it will appear in such highly diluted form in plant effluents that separation is bound to be inefficient and expensive. The difficulty lies in finding a process of isotopic separation from ordinary hydrogen that is sufficiently simple, capable of treating large volumes of water or steam, and reasonably cheap to operate. Various processes have been developed for tritium enrichment or separation for weapons purposes or for the production of tritium on a laboratory scale (58), but few of these processes lend themselves to waste treatment. For instance, electrolysis is well established as a method of separating the hydrogen isotopes, but to recover the tiny proportion of tritium-containing molecules from vast quantities of wastewater would necessitate the consumption of a prohibitive amount of electricity. Most of the methods have been reviewed by Lin (59) and Burger (60). Several methods being investigated for tritium reduction in HTGR plants include (1) chromatographic adsorption-separation of He^3 from He^4, (2) separation of He^3 from He^4 by diffusion through a semipermeable membrane, and (3) removal of H^3 from primary coolant by means of isotope exchange with a fixed charge of hydrogen (61). The success of these tests has not been reported.

Among the small-scale methods for tritium enrichment reported by Jacobs (58) and Lin (59) are fractional distillation, thermal diffusion, electrolysis and gas chromatography. The principles of operation, their relative advantages, and estimated cost are presented in Table 89. Table 90 is Lin's estimate of the relative costs involved (59). He concluded that only the dual-temperature water-H_2S exchange process and the cryogenic distillation process have economic potential.

Rogers and Michalek (62) have described a tritium removal system that may be capable of being used on nuclear plant effluent. This process catalyzes tritium to form tritiated water, which is then adsorbed on a desiccant. This method is reported to reduce tritium concentrations to less than 1 ppb from an initial concentration of 1000 ppm, or a decontamination factor of 10^6. The capacity of this system was not mentioned, but it appears suitable for off-gas such as that from the voloxidation process. Burger (60) also considers solvent extraction methods worth investigating, as well as a process based on selective excitation of molecular vibrations, but neither seems to have reached a sufficient stage of development for practical application.

Thus the choice for reduction in tritium discharges lies between long-term holdup by recycling or storage and a relatively expensive isotope separation process of limited efficiency, or a combination of both. Leonard and Kueck (63) have attempted to define optimum treatment conditions and to determine cost factors in balancing storage and treatment options. Assuming a liquid waste volume of 3000 gallons/MWe year and a tritium concentration of 2.2 $\mu Ci/ml$, they showed that the cost of storage alone and the different treatment processes varied in a similar fashion and represented a significant addition to power costs. As Figure 72 shows, a trade-off occurs between enrichment costs and tankage costs as the product volume varies and one can find an optimum combination. Reduction in tritium activity by less than 50% does not seem worth the cost, and the alternative of 50-100 years' storage sounds equally unattractive. Also, under the worst conditions the tritium in 10^5 gallons

Table 89. Isotope Separation Techniques[a] for the Concentration of Tritium from Tritiated Water (59)

Separation Technique	Principles of Operation	Estimated Cost ($/gal)	Advantages	Disadvantages
Dual Temperature Exchange	Enrichment of tritiated water with respect to ^3H based on variation in the extent of ^3H exchange between water and H_2S gas at different temperatures.	0.10-0.20	Relatively high separation factor[b] and feed throughput.[c]	Corrosive property and toxicity of H_2S. Difficulty in process control.
Fractional Distillation	Separation based on the difference in the boiling point and the volatility between isotopes either of liquid hydrogen or of water.			
(a) Cryogenic distillation	Conversion of tritiated water to gaseous isotopes of hydrogen followed by distillation at cryogenic temperature ($\sim$-250°C)	0.20-0.40	High separation factor.	Requirement for prior conversion of water to gaseous hydrogen. Possible problems in handling a large quantity of liquid hydrogen.
(b) High temp. vacuum distillation	Direct distillation of tritiated water near boiling point under vacuum.	0.30-0.60	Simplicity of operation.	High thermal energy consumption. Relatively low separation factor.
Gas-Solid Chromatography	Separation utilizing characteristics of solid materials to preferentially adsorb certain isotopes of hydrogen.		Simplicity of operation.	Requirement for prior conversion of water to gaseous hydrogen. Relatively low throughput of feed.
(a) Solid: palladium	Uses Pd-black which adsorbs lighter isotopes more strongly.	> 25	High separation factor.	Excessively high inventory costs.
(b) Solid: zeolites	Uses synthetic zeolites that preferentially adsorb heavier isotopes.	> 1	High capacity and low inventory costs.	Relatively low separation factor.
Electrolysis	Separation by electrolysis of tritiated water that produces gaseous hydrogen isotopes at the cathode, having an isotopic composition different from that of water (^3H concentrates in water).	> 1	Very high separation	High energy costs. Low feed throughput.

Table 89, Continued

Separation Technique	Principles of Operation	Estimated Cost ($/gal)	Advantages	Disadvantages
Thermal Diffusion	Separation by confining a gaseous isotopic mixture of hydrogen in the narrow annular space between the hot and cold cylindrical walls under the influence of the horizontal temperature gradient and the natural convection current.	> 3	Simplicity in design and operation of equipment.	Very low separation factor. High thermal energy consumption. Very low feed throughput. Not suitable for large-scale operation.
Solvent Extraction	Selective extraction of tritiated water by an organic liquid ion exchanger, followed by evaporation of the organic phase to recover ^{3}H-rich water.	> 1	Relatively simple to operate.	Separation factor and rate of mass transfer may be low, and inventory costs may be high depending on the type of liquid ion exchanger.

aAll isotope separation techniques separate tritiated water into tritium-enriched and -depleted portions. The depleted portion is released to the environment, while the enriched portion is disposed of by some other methods. When the enriched portion is a gas, it would probably have to be converted to water before disposal.

bSee text for the definition of separation factor.

cA high throughput of feed generally implies requirement of a small plant size.

Table 90.   Comparative Cost for Various Isotope Separation Processes (59)

Process	Capital Investment $(10^6$ $)$	Total Annual Cost [d] $(10^6$ $)$	Cost ($/gal Water)	
			As Calculated	Probable Range
Dual-temperature water-H_2S exchange	80	26.0	0.16	0.10-0.20
Cryogenic H_2 distillation[a] (H_2 from water gas reaction)	79	29.6	0.27	0.20-0.40
Vacuum fractional distillation of water	103	36.9	0.38	0.30-0.60
Electrolysis[b]	–	–	> 1	–
Chromatography[b]				
(a) w/Pd-black	–	–	> 25	–
(b) w/zeolites	–	–	> 1	–
Thermal diffusion[b]	–	–	> 3	–
Solvent extraction[c]	–	–	> 1	–

[a]Cost of cryogenic distillation of H_2 is estimated at $0.12/gal water. The remaining costs are for conversion of water to H_2 gas, and are dependent upon the process selected.

[b]Economic feasibility of these processes is negative at present.

[c]Insufficient data are available to perform a good economic evaluation of this process, but it appears that costs will be greater than $1/gal.

[d]Includes 28% annual fixed charges on capital.

of reactor coolant would occupy a volume of only a few milliliters if extracted as T_2O.

Since the natural deuterium level in water is about 10^4 times that of the tritium, the water inventory may contain 1000-2000 lb of heavy water, with a DTO impurity. It may be economically attractive to ship any DTO concentrate to a heavy water recovery plant for reprocessing. An alternative suggestion has been to store it in an ice field (60). Controlled storage may also involve conversion to a form with better radiation and thermal stability than water, such as certain metal hydrides.

In the meantime, closed-loop cooling and dilution of the liquid effluents appear to be the only readily available means of dealing with the tritium problem. Further research in this area is clearly essential.

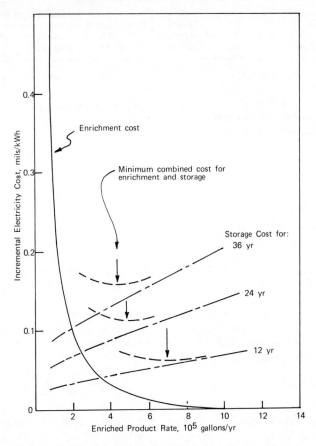

Figure 72. Trade-off between enrichment and storage cost elements for tritium (63).

Ruthenium

Of the various fission products, ruthenium is one of the most trouble-some and one deserving, perhaps, rather more attention than it is receiving at present. Its major fission product, Ru-103, has a fission yield of 3.1% and a half-life of 39.6 days. Ruthenium-106 and its short-lived daughter, Rh-106, have half-lives of 369 days making them one of the more persistent fission products. Ruthenium is volatile, readily forms complexes, and on contact with air forms nitro- and nitrosocompounds. For this reason it has been found unusually elusive, escaping readily from many treatment and conversion systems and proving disconcertingly mobile at disposal sites. It is taken up preferentially by seaweed and many fish and will absorb only partially at exchange sites on clay particles (25,64).

For this reason, whenever the loss of volatile ruthenium seems possible, such as in incinerators, calcination and conversion systems, or reprocessing plants, additional treatment stages may have to be included for ruthenium recovery. These may consist of silver zeolite filter beds, charcoal bed absorption, or compounding with metallic iron, magnesium or zinc. Both solvent-extraction or ion exchange can be employed with suitable backup and monitoring (13).

The removal of ruthenium from waste solutions is dependent upon its ionic state. Whether a specific treatment precipitates part or all of the ruthenium activity is determined by the ratio of anionic and cationic ruthenium. The ferrocyanide-carrier precipitation technique applicable to wastes containing cesium and ruthenium is not hindered by the ionic species of the ruthenium.

At a pH of 2.5, it is possible to carry 80-95% of the ruthenium on an iron hydroxide precipitate with less than 1% contamination by strontium, cesium, or rare-earth activity initially present in the waste. This method has been in use on a pilot-plant scale for some time for the removal of iron and ruthenium at reprocessing plants. Since it separates the bulk of the iron and ruthenium from the waste stream, the subsequent concentration of strontium and rare-earth radionuclides is simplified.

Scrubbing of the waste with iron powder in the pH range 0.5 to 4.2 results in a decontamination factor of 10^2 for ruthenium. Reduction of the ruthenium to its metallic state by the iron is given as the explanation for this effective removal. In such a reaction an apparent equilibrium arises between the rate of metal dissolution and the surface available for ruthenium deposition (13).

COST-BENEFIT CONSIDERATIONS

It is evident that technically it is feasible to reduce the levels of radioactivity in liquid and gaseous effluents considerably below those customary in the past when effluent concentrations had to be maintained merely below maximum permissible concentrations with little regard to the total quantities involved. With a mandate to keep these concentrations as low as practicable, questions arise: how low is practicable and is it sensible to reduce levels ever lower as further technical advances are made? One also needs to weigh the comparative risks or hazards from the steady release of low concentrations of some of the radwaste components against the possibility of the accidental release, briefly, of large quantities that have accumulated during routine operations (32,65).

One way of arriving at a reasonable answer to these questions is to compute the expected release levels for each of the radioisotopes of interest and the consequent radiation dose received by an individual at the plant fence line as well as the integrated population dose from that component. Iodine, krypton, and tritium doses are of particular concern in this connection, with cesium, strontium, and ruthenium following closely.

Certain radwaste procedures, such as use of HEPA filters, holdup pipes for off-gases, and demineralizer systems and holdup tanks for liquid wastes may be considered as standard, minimum treatment systems. Beyond these, various plants have been designed somewhat differently, bearing in mind any need for and availability of a tall stack or the requirement for a large exclusion area. Reduction of effluent levels clearly reduces the population dose at the fence line or permits a corresponding reduction of the required exclusion area. In view of the cost of additional equipment, a point can be reached when it is difficult to justify further expenditure once the estimated population exposure has dropped substantially below the natural radiation background. Figure 73 shows the additional improvement obtainable by the installation of subsequent treatment stages for gaseous waste systems (65) with any consequent increase in power cost for the case selected. Such an analysis has to be carried out for each individual plant and should be reflected in the environmental impact statement to demonstrate that the system selected represents a reasoned choice and a deliberate, *bona fide* attempt to maintain effluent levels and population exposures as low as practicable. Figure 73b shows one method of evaluating the cost effectiveness of the various processes in terms of residual estimated dose levels.

Graphic examples of the relationship between the cost of various treatment options and their effectiveness in reducing population dose can be found in the U.S. AEC Report WASH-1258 (5). Without going into details on the system options represented by the points on the graphs, Figure 74 shows a case where an enormous reduction in total body dose was achieved by two additional treatment stages for a PWR station at very little extra cost, whereas in Figure 75 a choice was offered between a small improvement in iodine dose at high cost for roof level discharge or a spectacular reduction in thyroid dose accomplished with a tall stack at significant extra cost. Some of these decisions involve more than a simple cost comparison but may relate also to projected land use, plant appearance, and long-range development prospects for the surrounding countryside. Another way of determining cost-effectiveness is to plot the ratio $\Delta(\text{cost})/\Delta(\text{dose})$ against the dose reduction as shown in Figure 76 for the case of gaseous waste delay systems (65).

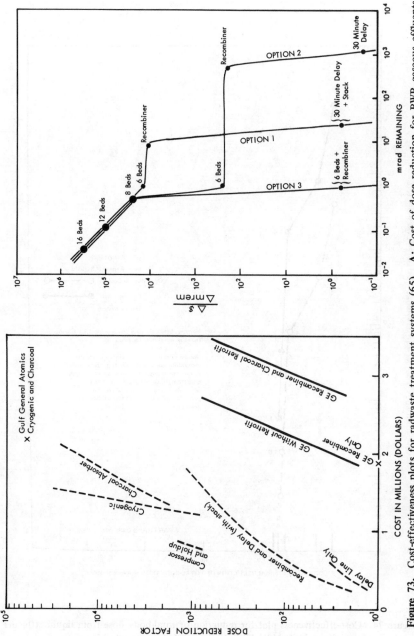

Figure 73. Cost-effectiveness plots for radwaste treatment systems (65). **A:** Cost of dose reduction for BWR gaseous effluents; **B:** Cost-effectiveness of incremental gas treatment components to achieve annual dose reduction to individuals offsite.

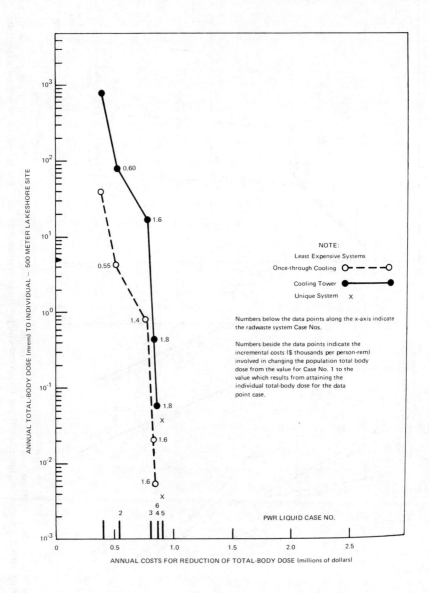

Figure 74.   Cost-effectiveness plot for reduction of total-body dose from liquid effluents from 2400 MWe PWR station at Lakeshore site (5).

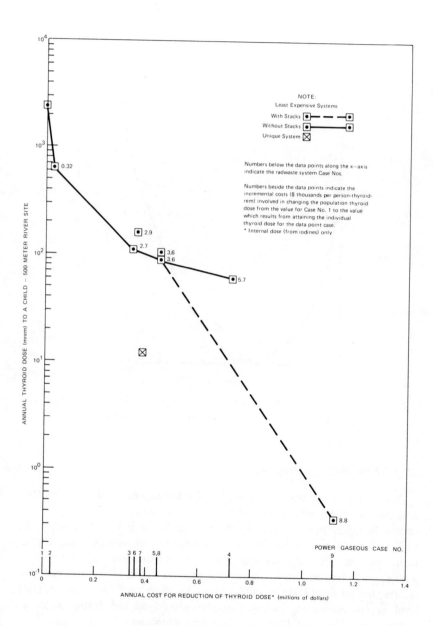

Figure 75. Cost-effectiveness for reduction of population thyroid dose from iodine in gaseous effluents (5).

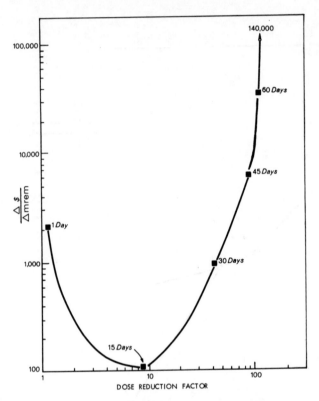

Figure 76. Cost-effectiveness of PWR gaseous waste treatment (recombiner and holdup tanks) for various retention times **(65).**

ON-SITE WASTE HANDLING

The various treatment processes for the decontamination of effluents result in the accumulation of radioactive waste materials that require careful handling and storage. Since most of the activity will continue to be contained within the fuel and be shipped out as spent fuel, the plant operator will be concerned primarily with low-level radioactive wastes, consisting of laundry wastes, contaminated glassware and containers, lab coats and gloves, and various handling tools, and intermediate-level wastes, comprising mainly the spent resin from the demineralizer systems, air filters and filter bed materials, evaporator bottoms, and centrifuged sludges, as well as any bottled radioactive gases.

Liquid wastes include those from various facility drains, laundry, laboratory decontamination, ion exchange regeneration backwash and condensate from waste concentration. These waste streams are filtered, processed through ion exchange columns, or neutralized as required and concentrated. The liquid portion is then either returned to one of the reactor water systems or discharged to surface waters as appropriate, and the evaporator concentrates are processed through the solid radwaste system for burial.

Solid wastes include the evaporator bottoms mentioned above, filters, sludges, spent demineralizer resins and miscellaneous dry wastes such as air filters, paper, rags, clothing, tools and small items of equipment.

Dry wastes are collected and compressed, if feasible, for storage in 55-gallon drums and ultimate burial. The wet solids are mixed with cement or other immobilizing materials, drummed and stored for burial.

Table 91 lists estimates for the annual quantities of solid radioactive wastes discharged by a 1000 MWe light-water reactor (15,71). At an

Table 91. Low-Level Solid Wastes from LWR Power Plant[a]

Volume/year	2000-4000 cu ft/yr; 56-110 m^3/yr
Weight	60-100 ton/yr
Number of drums	270-540
Storage (burial) area used	1800-3500 sq ft/yr; 165-330 m^2/yr

[a]These figures are assumed to be halved for an LMFBR (71).

assumed handling and burial charge of about $10 per cubic foot, the volume and value in 1980 (with two years lag time after startup, 108,000 MWe installed in 1978, and a BWR/PWR ratio of 1 to 2) is estimated to be about 250,000 cubic feet and $2.5 million. Table 92 shows the characteristics and quantities of radioisotopes found in solid waste; they are strongly dominated by cesium and cobalt isotopes (5).

Low-level liquid wastes are monitored and may be stored for limited periods to permit decay before being released. Figure 77 shows a diagram of a liquid waste storage tank. Various precautions have been taken in the design of these installations to permit the early detection of tank leakage and to handle any leakage problems that arise. For example, the tanks rest in a steel "saucer" which serves to collect any minor leakage and is equipped with built-in monitoring instruments. The same applies to the concrete vault in which the tank and its saucer are housed. Below

Table 92.   Calculated Radioactivity Content of On-Site Solid Waste (5)

		PWR Activity	BWR Activity[a] (Ci/yr/reactor)	
Nuclide	Half-Life	(Ci/yr/reactor)	With HEPA Filter	With Charcoal Absorber System
Sr-89	52.7d	2.0	29	49
Sr-90	27.7d	2.7	77	90
Zr-95	65d	0.64	0.7	1.1
Ru-103	39.5d	0.10	1.7	2.1
Ru-106	368d	1.7	–	–
Te-127m	109d	11	0.42	1.6
Te-129m	34.1d	1.3	–	–
Cs-134	2.0y	2.8×10^3	110	130
Cs-137	30.0y	2.7×10^3	100	120
Ce-141	32.5d	–	0.13	0.25
Ce-144	284d	4.2	4.1	5.1
Cr-51	27.8d	0.35	0.30	0.60
Mn-54	303d	22	9.6	13
Fe-55	2.6y	200	690	800
Fe-59	45.6d	1.8	3.6	6.6
Co-58	71.4d	120	79	130
Co-60	5.3y	280	170	200
Total		6.1×10^3	1.3×10^3	1.6×10^3

[a]After 180-day decay

the floor of the vault is a four-foot bed of gravel that is also monitored for radioactivity and is connected by standpipes to the ground surface to facilitate recovery of any material that should find its way out of the vault.

Low-level waste material is usually compacted and drummed for disposal.   Incineration is rarely practiced because it would involve collection of particulates and off-gases again.   Burial of low-level wastes in slit trenches has been the practice at some sites;   however, because of problems of monitoring and surveillance, that type of disposal is gradually superseded by shipment to recognized disposal sites.

Solidification

Intermediate level wastes are composed of evaporator bottoms, cartridge filters, spent demineralizer resins and condensate sludges.   They form the largest quantity of radioactivity that has to be handled as

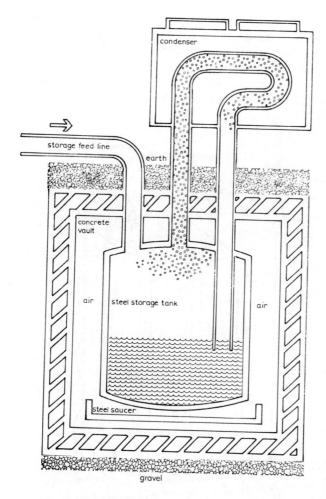

Figure 77. Cross section of liquid waste storage tank.

waste material and may require remote handling and conversion. For safe shipping and disposal this material is almost invariably solidified, and the waste solidification operation may necessitate a sizable on-site facility, usually housed in a separate building.

The basic purpose of solidification is the immobilization of the radionuclides, the reduction in volume to be shipped, and conversion to a form and container size that can be stored and loaded remotely and can be transported in compliance with radiation standards. Three basic operations are involved in most solidification processes: dewatering, conversion to nonleachable form, and container loading.

Intermediate-level wastes must be dewatered and solidified to the greatest possible extent before shipping out. Figure 78 shows a drumming device used to process solid waste from water systems such as spent resins and evaporator bottoms. After being automatically loaded

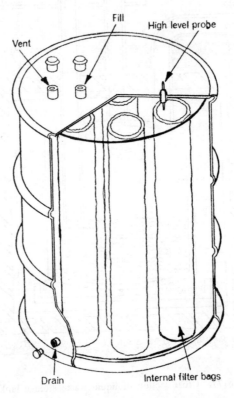

Figure 78. Solid waste shipping drum (66).

into drums containing an internal filter, the waste is filtered by gravity drain. The drums are used to recycle solid-containing liquid wastes until the filter contains a cakelike mass of solid waste. Then the top of the drum is sealed, and the package is ready for storage or to be shipped to a waste-processing site. Thus, the drum serves as a collection tank and as a shipping container, and handling and shipping operations are minimized (66).

Most plants are installing much more elaborate systems than that (67, 68). Usually the wastes are dewatered in a settling tank, by evaporation or by centrifuging (67). They are then mixed with a setting compound, such as cement, urea formaldehyde or asphalt. Urea formaldehyde requires a catalyst, typically sodium disulfide, for setting. The degree of prior drying required depends on the process selected and is obviously less in the case of cementation. In many cases vermiculite is added as a filler to the cement mixture to improve fixation. There are various commercial solidification systems for low and intermediate wastes capable of continuous operation with a bitumen or plastic matrix.

The still-fluid cement or resin mixtures are then transferred to a drum or concrete liner that serves as the storage or shipping container. The leachability of such solidified waste products in asphalt and cement has been compared and projected for storage periods up to 300 years. The leachability of asphalt mixtures is about two orders of magnitude lower than cement mixtures (68). Table 93 indicates solidification practices at various research establishments and the proportion of waste to cement being used; specific activities are of the order of 10^{-5}-10^{-2} curies per liter (67). Both the packaging plant and the drum storage must be capable of being operated by remote control, and plant layout must include shielding walls around all equipment that may require frequent maintenance.

An alternative system coming on the market uses a fluid bed calciner to produce a low-volume anhydrous waste powder that is then solidified by means of a thermal setting resin. The proponents of this system claim considerable savings resulting from the volume reduction obtained.

Using averaged annual utility data for liquid disposal costs, expenses for the packaging, transportation, burial (without cask rental costs) have been calculated for the three systems and compared. For the cement/vermiculite process it was found to cost $335,818; for the urea-formaldehyde process, $160,706; and for the fluid-bed process, $58,202 (1974 costs).

Waste Storage

At most plants waste shipments will take place at monthly or quarterly intervals and provision must be made for safe and secure storage of the solidified wastes in the meantime. Such storage may occur in walled-off portions of the radwaste treatment building or in separate block houses (66). Of course, regular monitoring for leaks or contamination is essential. At some sites, storage is done in asphalt- or cement-lined trenches for indefinite periods up to 10-20 years. Problems may

Table 93. Cementation Practices (1970) at Various Establishments (67)

Establishment	Nature of Waste	Composition of Mixture	Remarks
Grenoble (France)	Concentrate from evaporation, 400 g/liter	250 liters sludge; 300 kg cement (CPA); 40 kg vermiculite	Conventional concrete mixer used
Saclay (France)	Calcium carbonate and nickel ferrocyanide sludge	83 kg sludge; 55 kg CPAL cement; 10 liters water	
Karlsruhe (F.R. Germany)	Evaporator concentrate	(a) 100 to 110 liters fed to 200-liter drum containing 200 kg cement	Original practice
		(b) 100 liters concentrate; 150 kg cement	Later practice
Brookhaven (U.S.A.)	Evaporator concentrate (20% solids)	Liquor solidified by the addition of a mixture of exfoliated vermiculite (2.7 m^3) and Portland cement (0.68 m^3)	Solidification in reinforced concrete vaults (1.5 x 1.5 x 1.9 m)
Los Alamos (U.S.A.)	Neutralized concentrated waste	(a) 75 liters of concentrate added to 128 kg cement + 4 kg of exfoliated vermiculite	Batch process
		(b) Pug mill feed 20 to 35 liters/min of concentrate and 60 to 65 kg/min of cement	Continuous process
Rez (Czechoslovakia)	(a) Sludge from the chemical treatment of radioactive liquid waste (containing kaolinite, copper sulfate, ferric sulfate, potassium ferrocyanide, phosphoric acid, etc. Solids content 20 to 25%	35 liters of sludge; 110 kg of cement	-----
	(b) Evaporator concentrates neutralized to pH 6 to 8 (200 g/liter)	10 kg of sludge; 5 kg of evaporator concentrate; 22 kg of cement	Preferred mix
USSR (general recommended practice)	Evaporator concentrates (Max 150 g/liter)	Less than 130 g of salt of the sodium nitrate-type per kg of cement	Limit fixed by desirable strength of final product

arise then from occasional flooding of the trenches, attraction to rodents and other burrowing animals, and from other causes that may lead to bursting of the drums, though this is not serious if the waste has been properly solidified.

In those cases where low-level wastes are buried on site, burial grounds should be chosen on the basis of extensive prior geological, hydrological and geochemical investigations. Containment of the radioactivity can best be achieved if contact between the radioactive materials and percolating ground water can be prevented. The water table should therefore be well below the bottom of the trenches or wells; usually leaching can be prevented if the geological structure is impermeable. However, if leaching should occur, the ion exchange capacity of the geological materials should be adequate to restrict the migration of radionuclides through the ground. Selective sorption of many radionuclides by constituents of the soil, such as clay and shales, takes place and can provide a further obstacle to the migration of radioactivity (69).

Storage in the ground, at shallow depth, has a number of limitations. In particular, it should not be used for significant quantities of long-lived radioisotopes such as strontium-90 or cesium-137, unless satisfactory containment can be assured for the period required for decay, and is unsuitable for transuranium elements except as a temporary measure pending adoption of a permanent solution. There is also the factor that access to burial grounds may need to be limited for long periods. Central records and administrative procedures should be maintained to assure the long-term protection of the public. For these reasons few power plant operators are willing to engage in sizable disposal operations.

In summary, safe, well-monitored storage for spent resin and other low- and intermediate-level wastes must be provided at the plant site. However, it is rarely anticipated that any plant operator will have the space or the desire to be responsible for the permanent disposal of any but the lowest-level waste materials.

MONITORING AND CONTROL

The final requirement for effective radwaste treatment and disposal is an efficient system for monitoring and controlling all radioactive effluents. Monitoring implies measuring total radiation levels and concentrations of specific radioisotopes, either continually or at predetermined times. Controlling means the provision of warning signals whenever radioactivities exceed specified levels and the existence of methods and procedures that can be instituted immediately to reduce effluent levels below the alarm value. Such measures may include retention in tanks for decay, recycle of solutions for additional decontamination or evaporation, and cutting off all stack discharges of off-gases until the cause of the excess activity has been found and dealt with.

Apart from monitoring personnel exposures, radiation levels to be controlled include all effluents discharged to the environment, waste storage tanks, spent resin storage, suspect waste drains, and filter media. Table 94 lists the most important monitoring functions and procedures.

Table 94. Methods of Monitoring and Control (70)

Reactor water	Measure purity; clarify and deionize if needed; recycle
Regenerants	Neutralize and either (a) concentrate, drum bottoms and recycle distillate, or (b) release to canal
Spent solid resin	Package as solid and store
Condenser off-gas	Measure gross γ after two minutes. Retain one-half to one hour, filter and release to elevated point
Release to atmosphere	Monitor radiogases; sample and analyze for halogens and particulate matter
Housekeeping wastes	Measure and either (a) evaporate and recycle, or (b) filter and release to canal
Filter media	Dewater; package; store

To avoid plant shutdown due to monitor malfunction, certain atmospheric monitors and coolant activity monitors have to be designed as redundant systems. In addition, care must be taken to provide instruments capable of responding to both low and high levels of radiation and to have stack monitors with appropriate time constants to detect sudden surges in activity, without excessive fluctuations during normal operation. Thus, proper instrument specification and preventive maintenance form an important aspect of the reliability and serviceability of these safety systems.

Tritium levels must be measured periodically by batch sampling to obtain proper records and indications of coolant functions. Similarly iodine levels on filters may have to be measured at set intervals.

Maintenance of adequate records of total effluent activities, as well as iodine, noble gas, and tritium levels is required by law in most countries. Such records are also valuable in correlating positively or otherwise with any unsuspected activity found in the near or distant environment, as indicated at many recent public hearings and compensation lawsuits, and to provide the starting point for any procedure for monitoring fence-line exposure levels for routine operations.

REFERENCES

1. Denton, H. R. "Statement on the Sources of Radioactive Material in Effluents from Light-Water Cooled Nuclear Power Reactors and State of Technology of Waste Treatment Equipment to Minimize Releases," (Washington, D.C.: U.S. Atomic Energy Commission, 1972).

2. "Report on Releases of Radioactivity in Effluents and Solid Waste from Nuclear Power Plants for 1972," (Washington, D.C.: U.S. Atomic Energy Commission, 1973).

3. Wright, J. H., J. B. F. Champlin, and O. H. Davis. "The Impact of Environmental Radiation and Discharge Heat from Nuclear Power Plants," in *Environmental Aspects of Nuclear Power Stations*, Proc. New York Symp. 1970 (Vienna: International Atomic Energy Agency, 1971), pp. 549-559.

4. "Interim Licensing Policy on As Low As Practicable for Gaseous Radioiodine Releases from Light-Water-Cooled Nuclear Power Reactors," Regulatory Guide 1.42 (Rev. 1) (Washington, D.C.: U.S. Atomic Energy Commission, 1974).

5. "Numerical Guides for Design Objectives and Limiting Conditions for Operation to Meet the Criterion 'As Low As Practicable' for Radioactive Material in Light-Water-Cooled Nuclear Power Reactor Effluents," Report WASH-1258 (Washington, D.C.: U.S. Atomic Energy Commission, 1973).

6. U.S. Atomic Energy Commission. "Final Environmental Statement: Fort St. Vrain Nuclear Generating Station" (1972).

7. Clelland, D. W. and A. D. W. Corbet. "The Treatment of Inter-mediate-Level Radioactive Wastes by Evaporation: The Design and Performance of Evaporation Systems," in *Practices in the Treatment of Low- and Intermediate-Level Radioactive Wastes* (Vienna: International Atomic Energy Agency, 1966).

8. Gitterman, M. and A. J. Goodjohn. "Cryogenic Adsorption Systems for Radioactive Gas Cleanup," *ANS Trans.* 14, Suppl. 2 (1971); Report Gulf General Atomic Corp, San Diego, California (1971).

9. Michels, L. R. and N. R. Horton. "Improved BWR Off-Gas Systems," *Proceedings 12th AEC Air Cleaning Conference*, Oak Ridge, CONF-72023 (1972).

10. "Final Environmental Statement: Liquid Metal Fast Breeder Reactor Program," Report ERDA-1535, 10 vols. (Washington, D.C.: U.S. Energy Research and Development Administration, 1975).

11. "Environmental Analysis of the Uranium Fuel Cycle, Pt. II—Nuclear Power Reactors," (Washington, D.C.: U.S. Environmental Protection Agency, 1973).
12. "Environmental Statement, Watts Bar Nuclear Plant, Units 1 and 2," TVA Report OHES-EIS-72-9. (Chattanooga: Tennessee Valley Authority, 1972).
13. Straub, C. P. "Low-Level Radioactive Wastes: Treatment, Handling, Disposal," (Washington, D.C.: U.S. Atomic Energy Commission, 1964).
14. Leonard, J. H., T. S. Baer, and L. E. Eckart. "Techniques for Reducing Routine Release of Radionuclides from Nuclear Power Plants," Contract Report CPE-R-70-0015 (Cincinnati, Ohio: University of Cincinnati, 1971).
15. "The Safety of Nuclear Power Reactors (Light-Water-Cooled) and Related Facilities," Report WASH-1250 (Washington, D.C.: U.S. Atomic Energy Commission, 1973).
16. Godbee, H. W. and A. H. Kibbey. "Application of Evaporation for the Treatment of Liquids in the Nuclear Industry," Nuclear Safety 16, 458 (1975).
17. "Design and Operation of Evaporators for Radioactive Wastes," Technical Report Series 87 (Vienna: International Atomic Energy Agency, 1968).
18. Leonard, J. and T. S. Baer. "Costs Associated with Effluent Radiation from Nuclear Power Plants," Proceedings Environmental Progress in Science and Education, CONF-720511 (1972).
19. Yee, W. C., F. DeLora and W. E. Shockley. "Low-Level Radioactive Waste Treatment: The Water Recycle Process," Report ORNL-4472 (1970).
20. Chave, C. T. "Waste Disposal System for Closed-Cycle Water Reactors," Nucl. Technol. 15, 36 (1972).
21. Rauzen, F. V., N. F. Kuleshov, I. M. Smolkin, D. I. Trofimov and A. A. Dyakov. "Results Obtained in a Pilot Plant for the Treatment of Low-Level Radioactive Waste by Electrodialysis," Proc. 4th Internat. Conf. Peaceful Uses of Atomic Energy 11, 387 (1971).
22. Kaup, E. C. "Design Factors in Reverse Osmosis," Chem. Eng. 80(8), 46 (1973).
23. U.S. Atomic Energy Commission. "Final Environmental Statement: Shoreham Nuclear Power Station" (1972).
24. Keilholtz, G. W., C. E. Guthrie and G. C. Battle. "Air Cleansing as an Engineered Safety Feature in Light-Water-Cooled Power Reactors," Report ORNL-NSIC-25, Oak Ridge National Laboratory (1968).
25. Amphlett, C. B. Treatment and Disposal of Radioactive Wastes (Oxford: Pergamon Press, 1961).
26. Burchsted, C. A. and A. B. Fuller. "Design, Contruction and Testing of High-Efficiency Air Filtration Systems for Nuclear Application," Report ORNL-NSIC-65, Oak Ridge National Laboratory (1970).
27. Browning, W. E. "Removal of Fission Product Gases from Reactor Off-Gas Streams by Adsorption," U.S. Atomic Energy Report CF-59-6-47, n.d.

28. Hull, A. P. "Comparing Effluent Releases from Nuclear and Fossil-Fueled Power Plants," *Nucl. News* **17**(5), 51 (1974).

29. Larlee, W. D. "Recombiners for Nuclear Power Generation," (Darien, Connecticut: Air Correction Division, Universal Oil Products Co., 1973).

30. "Control of Iodine in the Nuclear Industry," Technical Reports Series No. 148 (Vienna: International Atomic Energy Agency, 1973).

31. "Assumptions Used for Evaluating the Potential Radiological Consequences of a Loss of Coolant Accident," Regulatory Guides 1.3 and 1.4 (Rev. 2) (Washington, D.C.: U.S. Atomic Energy Commission, 1974).

32. Rowe, W. D., Ed. "Environmental Radiation Dose Commitment: An Application to the Nuclear Power Industry," Report EPA-520/4-73-002 (Washington, D.C.: U.S. Environmental Protection Agency, 1974).

33. Klement, A. F., C. R. Miller, R. P. Minx and B. Shlein. "Estimates of Ionizing Radiation Doses in the U.S., 1960-2000," Report ORP/CSD 72-1 (Rockville, Maryland: U.S. Environmental Protection Agency, 1972).

34. Rosenberg, H. S., J. M. Genco, D. A. Berry, G. E. Cremeans and D. L. Morrison. "Research and Development on Coatings for Retaining Fission Product Iodine," *Reactor Fuel Processing Technol.* **12**, 115 (1969).

35. Keller, J. H., F. A. Duce, D. T. Pence and W. J. Maeck. "Iodine Chemistry in Steam Air Atmospheres," Proceedings Midyear Topical Symposium on Health Physics Aspects of Nuclear Facility Siting, (Idaho Falls: Health Physics Society, 1970).

36. Pence, D. T., F. A. Duce and W. J. Maeck. "Application of Metal Zeolites to Nuclear Fuel Reprocessing Plant Off-Gas Treatment," *Trans. ANS.* **15**, 96 (1972).

37. Milham, R. C. "High-Temperature Properties of Activated Carbon," in *Treatment of Airborne Radioactive Wastes* (Vienna: International Atomic Energy Agency, 1968).

38. "Results of Measurements of Iodine-131 in Air, Vegetation and Milk at Three Operating Reactor Sites," Directorate of Regulatory Operations, U.S. Atomic Energy Commission (1973).

39. Kolde, H. E., W. L. Brinck, G. L. Gels and B. Kahn. "Radioactive Noble Gases in Effluents from Nuclear Power Stations," Noble Gases Symposium, Las Vegas, Nevada (1973).

40. Tadmor, J. "Deposition of [85]Kr and Tritium Released from a Nuclear Fuel Reprocessing Plant," *Health Phys.* **24**, 37 (1973).

41. Bernhardt, D. E., A. A. Moghissi and J. A. Cochran. "Atmospheric Concentrations of Fission Product Noble Gases," Noble Gases Symposium, Las Vegas, Nevada (1973).

42. Kirk, W. P. "Krypton-85: A Review of the Literature and an Analysis of Radiation Hazards," (Washington, D.C.: Environmental Protection Agency, 1972).

43. Dunster, H. J. and B. F. Warner. "The Disposal of Noble Gas Fission Products from the Reprocessing of Nuclear Fuel," UKAEA Report AHSB(RP) R-101, Harwell, Berkshire (1970).

44. Slansky, C. M. "Separation Processes for Noble Gas Fission Products from the Off-Gas of Fuel-Reprocessing Plants," *Atomic Energy Rev.* **9**, 423 (1971).

45. Bendixsen, C. L. and J. A. Buckham. "General Survey of Techniques for Separation and Containment of Noble Gases from Nuclear Facilities," Noble Gases Symposium, Las Vegas, Nevada (1973).

46. Till, J. E. "Krypton-85 Releases from Nuclear Installations: Effect and Control," 1974 ANS Student Conference, Knoxville, Tennessee (unpublished).

47. Keilholtz, G. W. "Krypton-Xenon Removal Systems," *Nucl. Safety* **12**, 591 (1971).

48. Eichholz, G. G. and J. S. Simon. "Feasibility of Krypton Storage on Molecular Sieves," Georgia Institute of Technology, Atlanta (1973) (unpublished).

49. Kinsey, D. V. and C. Harper. "Helium Coolant Chemistry and Basic Design of Molecular Sieve Absorption Plant," in *Nuclear Engineering*, Pt. XXI, R. H. Moen, Ed. *AICHE Chem. Eng. Progress Symp. Series* **66** (104), 50 (1970).

50. McLain, M. E. "Release of Fission Product Xenon from High Surface Area Uranium Targets," Ph.D. Thesis, Georgia Institute of Technology, Atlanta (1972).

51. "Use of Local Minerals in the Treatment of Radioactive Waste," Techn. Reports Series No. 136 (Vienna: International Atomic Energy Agency, 1973).

52. Merriman, J. R., J. H. Pashley, and S. H. Smiley. "Engineering Development of an Absorption Process for the Concentration and Collection of Krypton and Xenon," U.S. Atomic Energy Commission Report K-1725 (Oak Ridge, Tennessee: 1967).

53. Shimojima, H., Y. Nakayama, K. Matsumoto and H. Hyodo. "The Capture of the Radioactive Noble Gases by the Clathrate Method," *Proc. 3rd Conf. Peaceful Uses of Atomic Energy*, Geneva, **14**, 314 (1964).

54. Stein, L. "Removal of Xenon and Radon from Contaminated Atmospheres with Dioxygenyl Hexafluoroantimonate $O_2S_6F_6$," *Nature* **243**, 30 (1973).

55. Clark, W. E. and R. E. Blanco. "Encapsulation of Noble Gas Fission Product Gases in Solid Media Prior to Transportation and Storage," Report ORNL-4473, Oak Ridge National Laboratory (1970).

56. Cohen, J. J. and K. R. Peterson. "Consideration in Siting Long-Term Radioactive Noble Gas Storage Facilities," Noble Gases Symposium, Las Vegas, Nevada (1973).

57. Underhill, D. W. "Effect of Rupture in a Pressurized Noble-Gas Absorption Bed," *Nucl. Safety* **13**, 478 (1972).

58. Jacobs, D. G. "Sources of Tritium and Its Behavior upon Release to the Environment," TID-24635 (Washington, D.C.: U.S. Atomic Energy Commission, 1968).

59. Lin, K. H. "Tritium Enrichment by Isotope Separation Technique," Report ORNL-TM-3976, Oak Ridge National Laboratory (1972).

60. Burger, L. L. "The Separation and Control of Tritium–State of the Art Study," (Richland, Washington: Battelle Northwest, 1972).
61. Gitterman, M. "Helium Purification for High-Temperature Gas-Cooled Reactors," Report GA-9185, Gulf General Atomic (1969).
62. Rogers, W. M. and R. Michalek. "Tritium Removal Systems," (East Newark, New Jersey: Engelhardt Industries).
63. Leonard, J. H. and J. J. Kueck. "Cost Potential for Reducing Tritium Activity in Aqueous Effluents," Paper, ANS National Meeting, Washington, D.C. (1972).
64. Lomenick, T. F. "Movement of Ruthenium in the Bed of White Oak Lake," Health Phys. 9, 835 (1963).
65. Martin, J. E., F. L. Galpin and T. W. Fowler. "Technology Assessment of Risk Reduction Effectiveness of Waste Treatment Systems for Light-Water Reactors," Radn. Data Reports 14, 1 (1973).
66. Graf, P. and W. Rottenberg. "Plans for Long-Term Storage of Low- and Medium-Level Radioactive Wastes in Switzerland," in Disposal of Radioactive Wastes into the Ground, Proceedings Vienna Symposium (Vienna: International Atomic Energy Agency, 1967).
67. Burns, R. H. "Solidification of Low and Intermediate Level Wastes," Atomic Energy Rev. 9, 547 (1971).
68. Kibbey, A. H. and H. W. Godbee. "A Critical Review of Solid Radioactive Waste Practices at Nuclear Power Plants," Report ORNL-4924, Oak Ridge National Laboratory (1974).
69. European Nuclear Energy Agency. Radioactive Waste Management Practices in Western Europe (Paris: OECD, 1971).
70. Kent, C. E., S. Levy and J. M. Smith. "Effluent Control for Boiling Water Reactors," in Environmental Aspects of Nuclear Power Stations (Vienna: International Atomic Energy Agency, 1971).
71. West, P. G. "Waste Management for Nuclear Power," Bull. IAEA 16, 78 (1974).

ADDITIONAL REFERENCES

Air Sampling Instruments for Evaluation of Atmospheric Contaminants, 4th ed. (Cincinnati: American Conf. of Governmental Industrial Hygienists, 1972).

Bigeleisen, J. "Isotope Separation Practice," in Isotope Effects in Chemical Processes, R. Gould, Ed., Advances in Chemistry Series No. 89, (American Chemical Society, 1969), pp. 1-24.

Budnitz, R. J. "Krypton-85: A Review of Instrumentation for Environmental Monitoring," Report LBL-1779 (Berkeley, University of California, Lawrence-Berkeley Laboratory, 1973).

Burns, R. W. "The Enrichment and Determination of Tritium in Water," Thesis, Ohio State University, Columbus (1965) (unpublished).

Charalambus, S. and K. Goebel. "Enrichment of Tritium by Diffusion Through Palladium," Z. Naturforsch. 20a, 1085 (1965).

Cochran, J. A., W. R. Griffin, E. Troianello. "Observation of Airborne Tritium Waste Discharge from a Nuclear Fuel Reprocessing Plant," Report EPA/ORP 73-1 (Washington, D.C.: U.S. Environmental Protection Agency, 1973).

Coleman, J. R. and R. Liberace. "Nuclear Power Production and Estimated Krypton-85 Levels," *Radiol. Health Data Rep.* 7, 615 (1966).

Daly, J. C., S. Goodyear, C. J. Paperiello and J. M. Matuszek. "Iodine-129 Levels in Milk and Water Near a Nuclear Fuel Reprocessing Plant," *Health Phys.* 26, 333 (1974).

Davis, J. S. and J. R. Martin. "A Cryogenic Approach to Fuel Reprocessing Gaseous Radwaste Treatment," *Trans. ANS.* 16, 176 (1973).

Delmas, R., P. Courvoisier and J. Ravoire. "Isotopic Exchange Between Hydrogen and Liquid Ammonia Catalyzed by Alkali Amides," in *Isotope Effects in Chemical Processes*, R. Gould, Ed. Advances in Chemistry Series No. 89, (American Chemical Society, 1969).

Diethorn, W. S. and W. L. Stockho. "The Dose to Man of Atmospheric [85]Kr," *Health Phys.* 23, 653 (1972).

Ekman, L., A. Eriksson, L. Fredriksson and U. Greitz. "Studies on the Relationship Between Iodine-131 Deposited on Pasture and Its Concentration in Milk," *Health Phys.* 13, 701 (1967).

Elwood, J. W. "Ecological Aspects of Tritium Behavior in the Environment," *Nuclear Safety* 12, 326 (1971).

Evans, E. A. *Tritium and Its Compounds*. Princeton: Van Nostrand, 1966).

Fisher, B. B., A. E. Norris and D. G. Rose. "Adsorption of Radiokrypton on Activated Charcoal in the Presence of Hydrogen," Noble Gases Symposium, Las Vegas, Nevada (1973).

Gabay, J. J., C. J. Paperiello, S. Goodyear, J. C. Daly and J. M. Matuszek. "A Method for Determining Iodine-129 in Milk and Water," *Health Phys.* 26, 89 (1974).

Haller, W. A. and R. W. Perkins. "Organic Iodine-131 Compounds Released from a Nuclear Fuel Chemical Reprocessing Plant," *Health Phys.* 13, 733 (1967).

Hilliard, R. K., A. K. Postma, J. D. McCormack and L. F. Coleman. "Removal of Iodine and Particles by Sprays in the Containment Systems Experiment," *Nucl. Technol.* 10, 499 (1971).

Hogg, R. H. "New Radwaste Retention System," *Nucl. Eng. Internat.* 17, 98 (1972).

Hornyik, K. "Tritium Generation in the Coolant-Moderator of Pressurized Water Reactors," *Nucl. Sci. Eng.* 49, 247 (1972).

Hull, A. P. "Reactor Effluents: As Low as Practicable or as Low as Reasonable?" *Nucl. News* 15 (11), 53 (1972).

Jauho, P and L. Mattila. "Distribution of Tritium in a Pressurized Water Power Reactor Plant," *Nucl. Technol.* 16, 472 (1972).

Karches, G. J., H. E. Kolde, W. L. Brinck, R. E. Shearin and C. R. Phillips. "Field Determination of Dose from ^{133}Xe in the Plume of a PWR," (Cincinnato, Ohio: U.S. Environmental Protection Agency, 1971).

Kirchmann, R., J. van den Hoek and A. Lafontaine. "Transfer and Incorporation of Tritium in the Constituents of Grass and Milk under Natural Conditions," *Health Phys.* 21, 61 (1971).

Knowles, F. E. and E. J. Baratta. "The Determination of Accuracy and Precision Limits for Tritium in Water," *Radiol. Health Data Reports* **8**, 405 (1971).

Koranda, T. J., J. R. Martin and L. R. Anspaugh. "The Significance of Tritium Releases to the Environment," *IEEE Trans.* **NS-19**, 27 (1971).

La Belle, D. W. "Post-Accident Hydrogen Generation and Control," *Nucl. Technol.* **10**, 454 (1971).

Langer, S., C. C. Adams, E. E. Anderson, J. N. Graves and T. Yamaguchi. "The Retention of Iodine by Pyrolytic Carbon-Coated Nuclear Fuel Particles," in *Radionuclides in the Environment*, E. C. Freiling, Ed., Advances in Chemistry Series No. 93 (Washington, D.C.: American Chemical Society, 1970).

Lewis, L. C. "Evaluation of Absorbents for Purification of Noble Gases in Dissolver Off-Gas," U.S. Atomic Energy Commission Report IN-1402, Idaho Nuclear Corp. (1970).

Lin, K. H. "Use of Ion Exchange for the Treatment of Liquids in Nuclear Power Plants," Report ORNL-4792 (1973).

"LWR and HTGR Radioactive Waste System Comparison," Gulf General Atomic Corp., San Diego (undated).

Magno, P. J., T. C. Reavey and J. C. Apidianakis. "Iodine-129 in the Environment Around a Nuclear Fuel Reprocessing Plant," (Washington, D.C.: U.S. Environmental Protection Agency, 1972).

"Management of Radioactive Wastes at Nuclear Power Plants," Safety Series No. 28 (Vienna: International Atomic Energy Agency, 1968).

McEwan, A. C. "The Critical Organ for Tritium Gas," *Health Phys.* **23**, 742 (1972).

Milioto, S. J., A. Sherman, R. L. Ritzman and J. A. Gieseke. "Analytical Studies of Elemental Iodine Removal by Sprays in the Donald C. Cook Nuclear Plant," *Nucl. Technol.* **16**, 497 (1972).

Moghissi, A. A. and M. W. Carter, Eds. "Tritium," CONF-710809 (Phoenix, Arizona: Messenger Graphics, 1973).

Moghissi, A. A., M. W. Carter and E. W. Bretthauer. "Further Studies on the Long-Term Evaluation of the Biological Half-Life of Tritium," *Health Phys.* **23**, 805 (1972), also *Health Phys.* **21**, 57 (1971).

Nakhutin, I. E., I. M. Smirnova, V. M. Makarov, G. A. Loshakov, G. A. Lauskhina and V. I. Yaroshinskii. "Removal of Radioactive Iodine from Gases," *Proc. 4th Geneva Conf. Peaceful Uses Atomic Energy* **11**, 399 (1971).

Oscarson, E. E. "Effects of Control Technology on the Projected Krypton-85 Environmental Inventory," Noble Gases Symposium, Las Vegas, Nevada (1973).

Parsly, L. F. "Pilot Plant Studies of Methyl Iodide Cleanup by Sprays," *Nucl. Appl. Technol.* **8**, 13 (1970).

Parsly, L. F. "Spray Program at the Nuclear Safety Pilot Plant," *Nucl. Technol.* **10**, 472 (1971).

Perrault, G., P. Thièblemont, Ch. Pasquier and G. Marble. "Cinétique. du Passage du Radioiode Soluble a Travers les Epitheliums Respiratoires après Inhalation," *Health Phys.* **13**, 707 (1967).

Peterson, H. T., J. E. Martin, C. L. Weaver and E. D. Harward. "Environmental Tritium Contamination from Increasing Utilization of

Nuclear Energy Sources," in *Proc. Vienna Symposium 1969, Environmental Contamination by Radioactive Materials* (Vienna: International Atomic Energy Agency, 1969), pp. 35-59.

Pinson, E. A. and W. H. Langham. "Physiology and Toxicology of Tritium in Man," *J. Appl. Physiol.* 10, 108 (1957).

Ratney, R. S. and D. W. Underhill. "The Effect of High Pressure and Low Temperature on the Absorption of Xenon and Krypton from Helium and Argon Streams," *Proc. 12th AEC Air Cleaning Conf.* (1973).

Ray, J. W. "Tritium in Power Reactors," *Reactor Fuel Process. Technol.* 12, 19 (1968).

Ray, W. H. "Krypton-85 Hazards in Perspective," *Proc. 12th AEC Air Cleaning Conf.* 856 (1973).

Rhinehammer, T. B. and P. H. Lamberger, Eds. "Tritium Control Technology," AEC Report WASH-1269 (Miamisburg, Ohio: Mound Lab., Monsanto Chemical Corp., 1973).

Ross, R. D., Ed. *Air Pollution and Industry.* (New York: Van Nostrand-Reinhold Co., 1972).

Schwibach, J. "Radiation Protection Guidelines for the Approval for the Release of Radioactive Materials," *Atomwirtsch.* 17, 196 (1972).

Shiojiri, M., Y. Hasegawa and K. Konishi. "Reaction of Iodine Vapor with Silver Films in the β AgI Phase Region," *J. Appl. Phys.* 44, 2996 (1973).

Smith, A. R., E. L. Field and R. L. O'Mara. "Cryogenic Absorption Systems for Noble Gas Removal," Noble Gases Symposium, Las Vegas, Nevada (1973).

Smith, J. M. and J. E. Kjemtrup. "Development in Nuclear Plant Effluent Management," Proc. Am. Power Conf. (1972).

"Status and Future Technical and Economic Potential of Light Water Reactors," Report WASH-1082 (Washington, D.C.: U.S. Atomic Energy Commission, 1968).

Stewart, J. E. and A. F. Desai. "Cryogenics: A Solution to Radioactive Pollution," *Electrical World* 176 (10), 39 (1971).

"Storage Tanks for Liquid Radioactive Wastes," Techn. Reports Series No. 135 (Vienna: International Atomic Energy Agency, 1973).

Tadmor, J. "Consideration of Stable Iodine in the Environment in the Evaluation of Maximum Permissible Concentrations for Iodine-129," *Radiol. Health Data Reports* 12, 611 (1971).

Tadmor, J. "Deposition of [85]Kr and Tritium Released from a Nuclear Fuel Reprocessing Plant," *Health Phys.* 24, 37 (1973).

Tadmor, J. and K. E. Cowser. "Underground Disposal of Krypton-85 from Nuclear Fuel Reprocessing Plants," *Nucl. Eng. Design* 6, 243 (1967).

Underhill, D. W. "A Mechanistic Analysis of Fission Gas Holdup Beds," *Nucl. Applic.* 6, 544 (1969).

Unruh, C. M. "Radiation Protection Practices for Tritium," Report BNWL-3390 (Richland, Washington: Battelle Northwest, 1970).

Watson, J. S. "An Evaluation of Methods for Recovering Tritium from the Blankets or Coolant Systems of Fusion Reactors," U.S. Atomic Energy Commission Report ORNL-TM-3794 (1972) (unpublished).

Weaver, C. L., E. D. Harward and H. T. Peterson. "Tritium in the Environment from Nuclear Power Plants," *Public Health Repts.* **84**, 363 (1969).

Whipple, G. H. "Possible Effects of Noble Gas Effluents from Power Reactors and Fuel Reprocessing Plants," Noble Gases Symposium, Las Vegas, Nevada (1973).

Wylie, K. F. "Tritium," *Health Phys.* **24**, 683 (1973).

Zittel, H. E. "Radiolytic Hydrogen Generation after Loss-of-Coolant Accidents in Water-Cooled Power Reactors," *Nucl. Safety* **13**, 459 (1972).

CHAPTER 8

ENVIRONMENTAL DISPERSION OF RADIOACTIVE EFFLUENTS

INTRODUCTION

As Tables 76 and 77 showed, a small but finite amount of radioactivity will still be released even with effective radwaste treatment systems designed to maintain effluent levels as low as practicable. In addition, one must postulate that under severe accident conditions there may be a transient but momentarily significant increment in the released activity, particularly in the airborne component. Therefore in order to evaluate the radiological impact of a nuclear power plant, one needs to evaluate the probable exposure of an individual residing near the plant and the total population dose for a region up to 50-100 miles surrounding the plant.

Such calculations presuppose a detailed knowledge of the transport phenomena affecting the movement and dispersion of the released material, any reconcentration process that may reverse the dilution mechanisms, and all pathways and exposure conditions that would contribute to the total dose received by humans from this source. As our experience grows in assessing the relative significance of all the fission products and activation products contained in the reactor core, as they relate to the quantities and concentrations expected to be found in the surrounding countryside during routine operations and following an accident, we can focus on a relatively small number of radioisotopes and follow them along their migration paths.

The annual dose to an individual residing near the fence line is the sum of four components:

1. the dose due to liquid effluents released in routine operations and effective through immersion (swimming), ingestion as drinking water, and ingestion via uptake in freshwater or marine organisms consumed as food

2. the dose due to airborne effluents released in routine operations and effective through inhalation or by ingestion as food via their deposition on soil or foliage and subsequent incorporation in the food chain

3. the dose due to airborne effluents released in a major accident and effective through inhalation of airborne matter or immersion in a passing radioactive cloud extending to ground level; some radioactive contamination of soil or food may occur but may be guarded against by proper subsequent monitoring

4. natural radiation background from soil and cosmic ray sources and, in some cases, gamma radiation from nitrogen-15 in the plant system.

Liquid effluents released under accidental conditions can be minimized by holdup and storage or release to receiving bodies under conditions in which concentrations can be monitored and use for drinking water can be controlled, much as for accidental releases of chemical poisons and pollutants. However, inadvertent releases may occur.

Except for some differences in composition, the movement in the atmosphere of routine effluents and of accidental releases will follow similar pathways and distribution patterns, unless a violent rupture of the containment is postulated in the latter case. Separate calculations have to be applied, of course, to nuclear ships, where all wastes are either retained or highly diluted (14); this subject is beyond the scope of this book.

Any calculational model must follow a scheme of the following type:

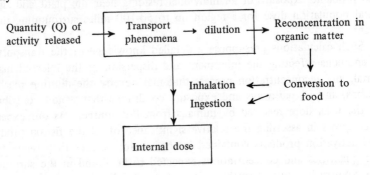

In practice the sequence of computations involves:

1. description of release source terms in terms of isotope composition and release rate

2. identification of dispersion mechanism effective for each component

3. calculation of concentration contours in all directions of interest

4. correction for reconcentration in food chains that are judged of importance
5. calculation of inhalation and ingestion rates
6. conversion to body dose commitment
7. comparison of isodose lines with exclusion zone boundaries and population patterns for accident releases
8. computation of dose distributions for routine releases and estimates of population dose
9. evaluation of radiological environmental impact by comparing population doses resulting from various radwaste treatment options, stack height, seasonal crop assumptions and exclusion area dimensions
10 projection of population doses over the lifetime of the plant.

For airborne releases the population dose can be expressed in the following mathematical form:

$$D_{(r,\theta)} = \sum_i Q'_{(t)} \cdot \frac{\chi}{Q}(r) \cdot C_i \cdot I_i \cdot f_{(\theta)} \cdot \rho_{(r,\theta)} \cdot K_{(D,i)} \tag{58}$$

where: $D_{(r,\theta)}$ = population dose for compass segment θ at distance r

$Q'_{(t)}$ = quantity of activity released by plant, during time interval t

$\dfrac{\chi}{Q}(r)$ = average gaseous dispersion factor per unit activity at the distance r in sector θ, calculated for average turbulence condition

C_i = reconcentration factors for isotope i through various pathways in the food chain

I_i = normal intake rate by inhalation and ingestion of element i for various population groups

$f_{(\theta)}$ = frequency factor for wind vector to lie in direction θ

$\rho_{(r,\theta)}$ = population density in the segment (r,θ)

$K_{(D,i)}$ = dose conversion factor for whole body dose or thyroid dose for a given body burden for isotope i

The quantity $Q'_{(t)}$ can be obtained either from published data on effluent release rates for comparable plants or from the primary coolant source term by allowing for reasonable leakage and decontamination factors:

$$Q'_{(t)} = \sum_{ij} A_i(t) \cdot G_{ij} \cdot L_{ij} \cdot \left(\frac{k}{DF}\right)_{im} \cdot e^{-\lambda_i t_i} \tag{59}$$

where: $A_i(t)$ = steady state or instantaneous activity of radioisotope i in the primary coolant

G_{ij} = partition factor for isotope i in system j

L_{ij} = leakage rate (liquid or gaseous) from system j into effluent stream

$\left(\dfrac{k}{DF}\right)_{im}$ = residual activity factor of isotope i after undergoing treatment in system m with decontamination factor DF and concentration/dilution in ratio k

λ_i = decay constant of isotope i

t_i = holdup or storage time before release of isotope i

To provide some additional safety margin, it is sometimes assumed, particularly in computing the consequences of severe accidental releases, that radwaste treatment systems and filters function only at half design-capacity on such occasions.

In view of the many variables involved, and taking into account all possible pathways, calculating the radiation intensities at specific distances from the plant can be very complicated and usually involves using a computer. Separate calculations have to be done for liquid and airborne effluents. Only rarely is there any need to consider direct external radiation exposure from any component of the power plant or any waste storage on the plant site to any member of the general public not occupationally exposed. The sole exception is potential exposure of the public during transportation of spent fuel and wastes; this aspect will be considered separately in Chapter 10.

Any transport calculation has to identify first its source term: the quantity of radioactive material released, separately as a liquid and as gaseous effluents in routine operations and under accident conditions. The source term must be specified as to isotopic composition, chemical form and rate of emission for each of the discharge options (continuous or batchwise) and radwaste treatment options (with or without delay storage, with or without cryogenic treatment, etc.) under consideration. The release source term can be determined either empirically on the basis of effluent observations such as those listed in Tables 76 and 77, allowing for technological improvements and reasonable margins of error, or by a step-by-step calculation based on the coolant inventory as modified by appropriate leakage terms and successive decontamination factors appropriate to all the radwaste facilities, with some downrating to allow for diminished efficiency after long service and for accidental malfunctions.

Since the iodine release and exposure levels in many cases represent the limiting dose levels to the surrounding population, a separate calculation must be done for all iodine pathways, including in particular the grass-cow-milk-man food chain. Other elements of particular importance are krypton, xenon, tritium and ruthenium in the gaseous effluents, and Zr-95, H-3, Cs-134,137, Sr-89,90 and Tc-99m in the liquid effluents, with Mn-54 and Co-60 where applicable. Some computation codes allow for all fission products in the coolant source term (Table 47) but a more restricted program is usually adequate and still quite sizable (5,6). Table 95 lists the principal pathways for radiation exposure; usually it is sufficient to restrict computations to the listed conditions.

Table 95. Principal Pathways for Radiation Exposure from Reactor Effluents (6)

Radionuclide	Discharge Mode	Principal Exposure Pathways	Critical Organ
Radioiodine	Airborne	Ground deposition - external irradiation	Whole body
		Air inhalation	Thyroid gland
		Grass-cow-milk	Thyroid gland
		Leafy vegetables	Thyroid gland
	Water	Drinking water	Thyroid gland
		Fish consumption	Thyroid gland
		Shellfish	Thyroid gland
Tritium	Airborne	Air inhalation and transpiration	Whole body
		Submersion	Skin
	Water	Drinking water	Whole body
		Food consumption	Whole body
Noble gases	Airborne	External irradiation	Whole body and skin
Cesium	Airborne	Ground deposition - external irradiation	Whole body
		Grass-cow-milk	Whole body
		Grass-meat	Whole body
		Inhalation	Whole body
	Water	Sediments - external irradiation	Whole body
		Drinking water	Whole body
		Fish consumption	Whole body
Transition metals (Fe,Co,Ni,Zn,Mn) Mn)	Water	Drinking water	G.I. tract
		Shellfish consumption	G.I. tract
		Fish consumption	G.I. tract
Direct radiation	–	External irradiation	Whole body

Exposure pathways to be included in the model can be represented pictorially or schematically (Figure 79). Such diagrams also have to be constructed to describe the pathways to and exposure of selected animal species. Two kinds of animal species will be referred to: specified species and endangered species. Specified species are either animal

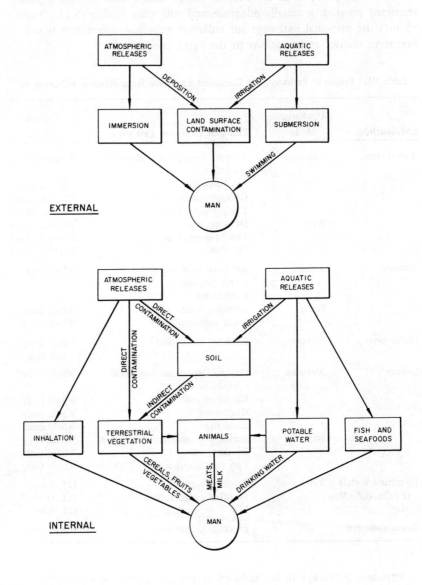

Figure 79. Pathways for radiation exposure of man.

species that can serve as indicators of any significant disturbance of the ecosystems surrounding the plant or species that are of commercial importance. Depending on the location of the plant, specified species may include ducks, deer, trout, muskrat, cows, shrimp, oysters, or reindeer, or any species that contributes or participates in the food chain or serves as an important link in existing environmental patterns. By selecting certain specified species for observation, surveillance and dose calculations, the problem of predicting and monitoring general environmental impact is reduced to manageable proportions.

Endangered species, as the name indicates, relates to rare plant and animal species, possibly existing at the plant site or the surrounding area either as permanent residents or during migratory seasons. Any species identified as endangered by law or regulation must be protected, and such need for protection must be incorporated in plans for the plant site and its associated structures (roads, water storage, power lines, cooling towers).

Another way of monitoring environmental impact is by observing species diversity. This usually involves a listing of all species (birds, mammals, fish, insects) seen at or near the site over any preceding period. Any reduction in species diversity is then considered an undesirable environmental impact. This criterion is hard to quantify and the presence or absence of a given migratory species, for instance, could be ascribable to so many unrelated causes as to make any correlation with the operation of the plant dubious. Usually, such lists are too large and diverse to permit tracing of specific exposure conditions and food chains or to make possible any sensible prediction of potential radioisotope uptake and body burdens.

In this chapter we are primarily concerned with the effects of radioactive effluents and other chemical toxicants released by the nuclear facilities on these various species and on man.

TRANSPORT MODELS

Essentially the calculation of population exposure involves three independent conditions: (1) the movement of radioisotopes released in liquid form, (2) the movement of routine releases in airborne effluents, and (3) the accidental release of higher-level short-term releases of gaseous material. It is therefore usually convenient to carry out three independent computations.

Liquid Effluents

Liquid effluents contain low concentrations of fission products that have not been fully removed by ion exchange or evaporation, as well as most of the tritium found in the secondary coolant loop (primary loop in BWR's). In many cases they are diluted by mixing with cooling tower blowdown water or in the general coolant flow where once-through cooling is used. As a result, they will be contained in the thermal plume and undergo dilution depending on the depth and velocity of the release jet (Figure 51).

In its simplest form, diffusion of the effluent takes the form of an exponential law

$$C_{(r,t)} = \frac{Q}{k} (1 - e^{-kt}) \tag{60}$$

where the quantities were defined in Equation 2.

In practice the effluent paths are more involved, and the concentration $C_{(r,t)}$ at a distance r from the point of release must include partition coefficients and reconcentration factors. By analogy with Equation 58 the dose from immersion in or consumption of the water or ingestion of any marine organism will take the form

$$D_{(r,t)} = \sum_i (C_{(r,t)} \cdot G_i \cdot I_i \cdot f \cdot K_{D_i}) \tag{61}$$

where the coefficients have the same meaning as before. The partition factors G_i now relate to the removal of activity by sedimentation or by anionic adsorption. Depending on whether effluent release in water is continuous or batch-wise, an equilibrium concentration in water and resident marine life may be reached far downstream, permitting the calculation of an average annual exposure. In reality, though, seasonal variations in water levels and in freshwater and marine life populations require separate monthly estimates if any accuracy is desired. Similarly, calculations of the external radiation dose due to submersion, *i.e.,* while swimming, have frequently overestimated the dose from this cause by assuming unrealistic stay times in water and a higher population participation than even Chamber of Commerce claims for seaside resorts would substantiate.

In general, dose calculations from liquid effluents are concerned mainly with two aspects: drinking water concentrations at downstream locations and reconcentration of radioisotopes through the food chain involving plankton, algae, shellfish, crustacea or fish.

Dose to Man from the Consumption of Water

This computation requires information on stream flow rate and on public and industrial water systems up to 100 miles downstream. It is assumed that the suspended portion of the activity will be mixed uniformly with the stream flow within a 5-10 mile reach from the release point. Dilution is calculated using annual flow data appropriate for all river stretches, both for average and minimum flow conditions.

Radioactive decay and build-up of daughter activity may be considered based on estimates of transport time to major downstream points. It may be assumed that each individual consumes 2.2 liters of water per day from all sources, including drinking water, food and bottled drinks. Unless definitive data are available on silt load and sedimentation losses, the partition factor G initially may be assumed to equal unity.

Internal doses, D_{ij}, for the jth organ from the ith radionuclide are calculated using the relation

$$D_{ij} = (DCF)_{ij} \times I_i \qquad (62)$$

where: $(DCF)_{ij}$ = the dose commitment factor for an average adult assuming that the dose can be accumulated over a 50-year interval, (mrem/μCi)

I_i = the activity of the ith radionuclide taken into the body annually via ingestion (μCi).

The dose commitment factors are derived from data given in References 7-9 and are defined by the equation

$$(DCF)_{ij} = \frac{C' \cdot f_{wij}\, \epsilon_{ij}\, [1-\exp(-\lambda_{ij}T)]}{m_j\, \lambda_{ij}} \quad (\text{mrem/Ci}) \qquad (63)$$

where: C' = $51.2 \times 10^3 \left(\dfrac{g\ rad}{MeV}\right)\left(\dfrac{disintegrations}{\mu Ci\text{-}day}\right)\left(\dfrac{mrem}{rem}\right)$

f_{wij} = fraction of the ith radionuclide taken into the body by ingestion that is retained in the jth organ

ϵ_{ij} = effective energy absorbed in the jth organ per disintegration of the ith radionuclide including daughter products, (MeV-rem/dis-rad)

λ_{ij} = the effective decay constant of the ith radionuclide in the jth organ, (day^{-1})

T = integration time (18,250 days)

m_j = mass of the jth organ (g).

In the absence of a detailed knowledge regarding solubility characteristics of the radionuclides, the dose for the gastrointestinal tract tends to be overestimated using the assumption that none of the radionuclides is removed from the G.I. tract by absorption. Estimates of the doses to bone, thyroid, and total body are based on fractional uptakes given by the International Commission on Radiological Protection (8,10). The highest estimated dose from any single radionuclide is the thyroid dose due to iodine-131 ingestion, as a rule.

Dose to Man from the Consumption of Fish

The role of fish and marine products in human diet will vary from place to place and country to country. For example, for the Tennessee River estimated annual fish harvests are 15.2 lb/acre for sport fish and 13.7 lb/acre for edible commercial fish (7). For Watts Bar, it was assumed that these rates would increase with population expansion, so that the dose calculations were based on harvests of 23 lb/acre for sport fish and 20 lb/acre for commercial fish in the year 2000. The Tennessee River was segmented into 17 reaches in order to facilitate the calculations of fish harvests and radioactivity concentrations. The radioactivity levels in the fish from each reach were estimated by the product of an average activity concentration in the reach and a concentration factor for each radionuclide (Table 96) (11,12,15).

It was assumed that the maximum annual consumption of fish by any individual is 45 lb. The population dose was calculated using the assumption that all of the edible fish harvested are consumed by humans. Radioactive decay is not usually considered between the time the fish is removed from the water and the time of consumption. Allowance will have to be made if some of the activity is concentrated in inedible portions of the fish; otherwise the entire mass of the fish is assumed to be eaten. Dose commitments are calculated with Equations 62 and 63 which were discussed for water ingestion in the previous section.

Dose to Man from Water Sports

Dose from water immersion (swimming) will depend on the amount of time spent in the water and the radioisotope concentration in that water. Assuming complete submersion, Soldat (14) suggests a dose factor $K_{(D)}$ for swimming, in mrem/hr per pCi/m^3:

$$K_{(D)_{sw}} = 2.13 \times 10^{-6} \, (\overline{E}_j + \frac{\overline{E}_\beta}{2}) \tag{64}$$

Table 96.   Bioaccumulation Factors for Aquatic Organisms (13) (pCi/kg per pCi/l)

Element	Freshwater (14)				Saltwater (11)			
	Fish	Crustacea	Mollusks	Algae	Fish	Crustacea	Mollusks	Algae
H	0.9	0.9	0.9	0.9	1	1	1	1
Na	100	200	200	500	1	1	1	1
P	100000	20000	20000	500000	10000	10000	10000	100000
Cr	20	2000	2000	4000	100	10000	1000	1000
Mn	400	90000	90000	10000	3000	4000	50000	10000
Fe	100	3200	3200	1000	1000	10000	20000	6000
Co	50	200	200	200	100	100	300	100
Ni	100	100	100	50	500	50	100	100
Rb	2000	1000	1000	1000	30	1	10	10
Sr	30	100	100	500	1		1	20
Y	25	1000	1000	5000	30	100		300
Zr	300	6.7	6.7	1000	30	100	100	1000
Nb	30000	100	100	800	100	200	100	100
Mo	10	10	10	1000	10	100	200	100
Tc	15	5	5	40	10	100	100	1000
Ru	10	300	300	2000	3	100	100	1000
Rh	10	300	300	200	10	100	100	100
Te	400	75	75	100	10	10	100	1000
I	15	5	5	40	20	100	100	10
Cs	2000	100	100	500	30	50	10	10000
Ba	4	200	200	500	3	3	3	10
La	25	1000	1000	5000	30	100	100	100
Ce	1	1000	1000	4000	30	100	100	300
Pr	25	1000	1000	5000	100	1000	1000	300
Nd	25	1000	1000	5000	100	1000	1000	1000
W	1200	10	10	1200	10	10	100	1000
U	2	60	60	0.5	10	10	10	100
Np	10	400	400	300	10	10	10	67
								6

where $\bar{E}_j$ and $\bar{E}_\beta$ are the "average" gamma and beta energies per disintegration, in MeV, for the radionuclide. A more convenient equation for the dose rate to the skin is (7):

$$R_i = C' \cdot C_{wi} \, (\bar{E}_\beta/2 + E_\gamma)_i \quad \text{(mrem/day)} \quad (65)$$

For the dose rate to the total body,

$$R_i = C' \, C_{wi} \, E_{\gamma i} \quad \text{(mrem/day)} \quad (66)$$

where: $C' = 51.2 \times 10^3$, as before

C_{wi} = water concentration for the ith radionuclide (μCi/g)

$E_{\gamma i}$ or $(\bar{E}_\beta/2 + E_\gamma)_i$ = average effective energy emitted by the ith radionuclide per disintegration (MeV-rem/dis-rad).

Dose rates for activities such as boating are assumed to be half the swimming doses. The U.S. Atomic Energy Commission has recommended (13) a more elaborate equation for calculation of external dose to the skin and total-body from swimming (water immersion) or boating (water surface), given as

$$R_{pr} = 1119 \, \frac{U_p M_p}{f \, K_p} \, \sum_{i=1}^{136} Q_i R_i \, D_{ipr} \, \exp\,(-\lambda_i t_p) \quad (67)$$

where K_p is 1 for swimming and 2 for boating; other definitions are found in Table 97, which summarizes the various dose relations for liquid pathways in the form recommended by the U.S. Atomic Energy Commission (15).

Doses to Organisms Other than Man

Radiation doses from liquid effluents to organisms other than man may constitute a significant portion of the total environmental impact. Free swimming organisms, such as fish, receive radiation doses from immersion in the cooling water as well as from internally deposited radionuclides. Crustacea or mollusks living on the bottom receive radiation doses from the mud and sediments as well as from internally deposited nuclides. For marine and fresh water organisms bioaccumulation factors listed in Table 96 may be used. Calculations for aquatic organisms (13) have shown that for bottom organisms the long-lived radionuclides in sediments represent the most important source of exposure.

It is usually appropriate to specify indicator species as representative of environmentally important ecological impact conditions. For wading

Table 97. Equations for Radiation Doses from Liquid Effluent Pathways (15)

1. Potable Water

$$R_{pr} = 1.1 \times 10^3 \, \frac{M_p U_p}{F} \sum_{i=1}^{n} Q_i \, D_{ipr} \, \exp\left(-\lambda_i t_p\right)$$

2. Aquatic Foods

$$R_{pr} = 1.1 \times 10^3 \, \frac{M_p U_p}{F} \sum_{i=1}^{n} Q_i \, B_{ip} \, D_{ipr} \, \exp\left(-\lambda_i t_p\right)$$

3. Shoreline Deposits

$$R_{pr} = 1.1 \times 10^3 \, \frac{M_p U_p W}{F} \sum_{i=1}^{n} Q_i \, t_i \, D_{ipr} \, \exp\left(-\lambda_i t_p\right)\left[1-\exp\left(-\lambda_i t_s\right)\right]$$

4. Swimming and Boating

$$R_{pr} = 1.1 \times 10^3 \, \frac{U_p M_p}{F K_p} \sum_{i=1}^{n} Q_i \, D_{ipr} \, \exp\left(-\lambda_i t_p\right)$$

where

R_{pr}	=	total dose rate to organ r from all of the nuclides i in pathway p (mrem/yr)
Q_i	=	release rate of nuclide i (Ci/yr)
F	=	flow rate of the liquid effluent (ft^3/sec)
t_p	=	transit time required for nuclides to reach the point of exposure. For internal dose, t_p is the total time elapsed between release of the nuclides and ingestion of food or water (hr)
U_p	=	usage, the exposure rate or intake rate associated with pathway p (hr/yr or kg/yr) (as appropriate)
M_p	=	mixing ratio (reciprocal of dilution factor) at the point of exposure (or the point of withdrawal of drinking water or point of harvest of aquatic food)
λ_i	=	radioactive decay constant of nuclide i (hr^{-1})
B_{ip}	=	bioaccumulation factor for nuclide i in pathway p, pCi/kg per pCi/liter
W	=	shoreline width factor, dimensionless
t_i	=	radioactive half-life of nuclide i (days)
t_s	=	period of time sediment is exposed to the contaminated water, nominally taken to be the operating lifetime of the facility (hr)
K_p	=	geometry factor, dimensionless, for swimming and boating.
n	=	number of radionuclides in dose factor library
D_{ipr}	=	dose factor: a number specific to a given radionuclide, pathway, and organ used to calculate radiation doses from the intake of a radionuclide (mrem/pCi intake) or from exposure to a given concentration in water (mrem/hr per pCi/l).

or water birds or mammals immersion may form a major exposure condition; for other birds and terrestrial animals doses received will be similar to those received by persons residing in adjacent locations, with appropriate adjustments for drinking water intake or skin dose. For example, a muskrat living in a burrow constructed in the bank downstream of a reactor site would be irradiated from the silt while in his den ($\sim$ 1/3 of the time). He would also receive small external doses from swimming ($\sim$ 1/3 of the time) and air submersion ($\sim$ 1/3 of the time). For estimating the annual internal total body dose it is assumed that a duck or muskrat has a mass of 1,000 g, an effective radius of 10 cm, and consumes 333 g of green algae per day. Long-lived radionuclides such as Sr-90 can deliver significant portions of the total dose commitment long after the time of ingestion. For these animals five years may be a reasonable time for the integration interval T.

In the absence of data specifically applicable to ducks or muskrats, ICRP data can be used for the fractional uptake in the total body and for the biological half-life of parent radionuclides. The use of human data for the biological half-lives is conservative because, in general, warm-blooded vertebrates that are smaller than man exhibit more rapid elimination rates. Equation 62 then assumes the form

$$D_i = 51.2 \times 10^3 \ I_i \ f_{wi} \ \epsilon_i \ [1\text{-}\exp(-\lambda_i T)]/\lambda_i m \quad \text{mrad} \tag{68}$$

where: $I_i \ = \ 333\left(\dfrac{g}{d}\right) x \ C_{wi} \ F_{pi} \times 365\left(\dfrac{d}{y}\right) \ (Ci/y)$

$C_{wi} =$ water concentration of isotope i ($\mu Ci/g$)

$F_{pi} =$ concentration factor for aquatic plants for isotope i (dimensionless)

$T \ = \ 1,825$ days

$m \ = \ 1,000$ g.

External doses may be estimated using the conservative assumption that the duck and muskrat are exposed by full immersion in the water one-third of the time.

Benthic organisms such as mussels, worms, and fish eggs may receive higher external doses if significant radioactivity is associated with bottom sediments. Accurate prediction of the accumulation of activity in sediment requires a detailed knowledge of a number of physicochemical factors including mineralogy, particle size, exchangeable calcium in the sediment, channel geometry, waterflow patterns, and the chemical form of the radiocompounds. Many of these factors must be obtained from extensive field experiments. In the absence of detailed knowledge, the doses may be calculated using the following assumptions:

1. Two-tenths of the activity in the liquid effluent is deposited uniformly in a sediment bed having dimensions of 10 cm x 100 m x 10 km
2. The radioactivity concentration in the sediment is calculated assuming a build-up over the plant life of 35 years at a constant rate of deposition.
3. Beta doses are based on a 4π geometry, and gamma doses assume a 2π geometry.

The doses calculated using these assumptions are probably overestimated (7).

Gaseous Effluents

Radiological consequences of the release of gaseous or airborne radwaste materials may be due to external exposure to a passing radioactive cloud, to inhalation of airborne containments, or to deposition on soil or foliage with subsequent movement through the food chain.

In all cases, for release above ground the Pasquill expression for ground level cloud concentration at a great distance x from the plant forms the starting point (Equation 11).

$$\frac{\chi(x)}{Q'} = \frac{\psi}{Q} = \frac{1}{(2\pi)^{3/2}} \frac{n}{\bar{u}\,\sigma_2 x} \exp\left[-\frac{h^2}{2\sigma_z^2} - \lambda t\right] \tag{69}$$

where: $\chi(x)$ = ground level airborne concentration, Ci/m^3
ψ = time-integrated ground level concentration, $Ci\text{-}sec/m^3$
Q' = source release rate, Ci/sec
Q = time-integrated release, Ci
x = distance from stack, m
h = effective stack height, m
$\bar{u}$ = average wind speed in $(22\ \tfrac{1}{2}^{\circ})$ sector
n = number of sectors
σ_z = standard deviation of vertical distribution of assumed Gaussian cloud, m
λ_i = decay constant of radioisotope i, sec^{-1}
t = transit time from stack to distance x, sec

The actual numerical computation is performed for each radionuclide and stability class by dividing the radial dimension into discrete mesh intervals with the assumption that the concentration within each mesh interval is uniform. The source in each mesh is corrected for nuclide decay and daughter build-up using the appropriate arrival time at each interval. Values of $\sigma_z(x)$ can be obtained from Figure 21, or can be evaluated in each interval using the expression

$$\log_e \sigma_z = \log_e c_1 + c_2 \log_e x + c_3 (\log_e x)^2 \qquad (70)$$

where c_1, c_2 and c_3 are parameters associated with each stability classification. Table 98 lists a set of these parameters obtained by curve fitting under representative dispersion conditions (10).

Table 98. Dispersion Coefficients (10)

Stability Classification	c_1	c_2	c_3
A	1100	-2.62	0.367
B	9.24	-0.752	-0.162
C	0.038	1.24	-0.0248
D	0.036	1.24	-0.0366
E	0.0082	1.56	-0.0603
F	0.00162	1.82	-0.0754

It should be noted that the Pasquill formulation should not be expected to provide precise results for locations many miles from the source. For such long distances and correspondingly long diffusion times, wind velocity, wind direction, and stability classifications could easily change. However, since there is no other simple model that adequately treats such situations, the Pasquill model is commonly used to compute the plume concentration at all distances. As partial justification, it may be observed that at large distances from the source the precise magnitude of the dose is not of critical importance since all doses are far below the illness and fatality thresholds. At these distances, only the man-rem calculation is of some significance with regard to health effects, and the population dose is insensitive to cloud concentration since a more dilute cloud normally exposes more people to a lower individual dose. Thus the product of cloud concentration and exposed population remains relatively constant.

Further refinements can be made to allow for finite cloud dimensions and differential deposition rates for aerosols and volatile components.

Equation 69 provides the ground level air concentration at a distance x from the release point. This concentration is then used to calculate the radiation dose from inhalation and transpiration (tritium) and the deposited activity on the ground surface that contributes to external whole body exposures and to food intake pathways. The inhalation and transpiration doses are computed directly from the ground level air concentration by the following expression:

$$R = (\frac{\chi}{Q'}) \cdot Q \cdot K_{D_{ipr}} \tag{71}$$

where R is the dose rate in mrem/year, (χ/Q') is the atmospheric dispersion factor as computed above (sec/m^3), Q is the annual release rate (curies/year), and $K_{D_{ipr}}$ is an appropriate dose conversion factor (mrem/year per Ci/sec/m^3) for the radionuclide and exposure mode of interest. Values of $K_{D_{ipr}}$ have been listed in various computer programs, such as HERMES (14); some were presented in Table 20. More detailed equations for the various pathways for airborne beta and gamma ray exposures are listed in Table 99 in the forms recommended by the U.S. Atomic Energy Commission (15). Such calculations should be performed for all radionuclides of significance; since the thyroid dose due to radioiodine is usually considered as the limiting dose, that case is discussed in more detail in a subsequent section.

The activity deposited on the ground surface can be computed from the ground level air concentration as follows:

$$\omega = \frac{\chi}{Q'} Q \nu_d \tag{72}$$

where ω is the deposition rate (Ci/m^2 sec), ν_d is the deposition velocity (m/sec), and the other quantities are as defined above. The deposition velocity is an empirical quantity defined as:

$$\nu_d = \frac{A'}{\chi t} \tag{73}$$

where A' is the accumulated deposit (Ci/m^2) and χt is the integrated air concentration over the period of measurement. The airborne concentration is depleted uniformly by the deposited activity using the continuity principle. Values for ν_d have been listed by Slade (16); typical values for iodine are given on page 389.

The accumulated deposit is given by:

$$A' = \frac{\omega}{\lambda_e} (1 - e^{-\lambda_e t}) \tag{74}$$

Here ω is the deposition rate as given above, A' is the deposited activity (Ci/m^2), λ_e is an effective removal rate constant (day^{-1}) and t is the

Table 99. Dose-Rate Equations for Gamma and Beta Exposures for
Gaseous Effluents (15)

Gamma dose rates, mrad/yr

| | Wind Speeds | Stability Types | Energy Groups Geometry Fns | Nuclides & Abundances |

$$R = \frac{287}{r\Delta\theta} \cdot \left[\sum_m 1/\bar{u}_m \cdot \sum_j f_{ijm} \cdot \sum_k \mu_{ak} E_k (\bar{I}_1 + k\bar{I}_2) \cdot \sum_n Q'_{mn} A_{kn} \right]$$

Beta air dose rates, mrad/yr

$$R = 460 \sum_{jm} f_{ijm}/\sigma_{zjr} \, \bar{u}_m \, r \cdot \sum_n Q'_{mn} \, E_n \exp\left(-\frac{h_e^2}{2\,\sigma_{zjr}^2} \right)$$

where

R = total dose rate in air at distance r in sector i, mrad/yr

f_{ijm} = fraction of year with wind speed u_m and stability type j in sector i

$\bar{u}_m$ = average wind speed group m, m/sec

σ_{zjr} = vertical plume dimension for stability j at distance r

r = horizontal distance from release point to receptor, meters

Q'_{mn} = activity of radionuclide n remaining after decay in transit to distance r at speed u_m during stability type j

E_n = average beta energy emitted by nuclide n, MeV

h_e = effective stack height with respect to receptor location, meters

$\Delta\theta$ = sector width over which atmospheric conditions are averaged, radians

μ_{ak} = energy absorption coefficient of air for gamma-ray energy E_k, m^{-1}

E_k = average energy of photons emitted in energy group k, MeV

A_{kn} = fraction of photons in energy group E_k from nuclide n, per disintegration

$\bar{I}_1 + k\bar{I}_2$ = results of numerical integration accounting for distribution of radioactivity according to meteorological conditions (wind speed group m and stability type j) as a function of distance r; tabulated by Slade (16).

time interval. For deposition onto foliage that leads to an ingestion dose, the effective removal rate is defined as:

$$\lambda_e = \lambda_i + \frac{\ln 2}{12} \tag{75}$$

where λ_i is the radiological decay constant (days^{-1}) and the remaining term accounts for physical removal by wash-off, wind, and plant growth of the radionuclide from plant surfaces, which is assumed to have a half-time of 12 days. Computation of the external whole body dose from deposited radionuclides is performed assuming a uniform semi-infinite plane source by:

$$R = A' K'_{D_{ipr}} \qquad (76)$$

where R is the annual dose rate (mrem/year), A' is the accumulated activity deposit (Ci/m^2) and K'_D is the dose conversion factor for a semi-infinite plane source (mrem/year per Ci/m^2).

Computer Models

The solution of the equations for dispersion of liquid and airborne effluents tends to get fairly complicated if all exposure paths, all compass directions, and all population distributions up to 50 miles from the plant are taken into account. For this reason several computer models have been developed that are capable of handling the source terms, various radwaste treatment systems and the particular meteorological conditions of a given site. Most of the models have been adapted to deal with airborne effluents only and the consequent radiation dose to the population, since this tends to be a more general situation involving a larger number of variables. Prediction of population dose due to ingestion or immersion in low-level radioisotope concentrations in water is much more straightforward and rarely justifies the use of elaborate computer codes.

An example of such a computer code is the HERMES code described by Soldat (14), which utilizes data inputs and subcodes specifying concentrations of radionuclides in air and deposited on soil (ARTRAN), concentrations in water (WTRAN), and demographic, dietary and recreational information (LPM). Figure 80 illustrates the overall flow of the dose calculation scheme, showing the air and water pathways to be considered. This includes various food pathways that are described in analytical form in Figure 81.

A similar though less elaborate program is the WEERIE code developed by Clarke in Great Britain (5), which yields cloud concentration and ground deposition values; it can then be followed by a dose conversion code called SAURON. This program has been correlated with deposition data obtained following the 1957 Windscale accident.

Such programs can be used in conjunction with an accident release code such as CORRAL, which is described by Postma, Owzarski and

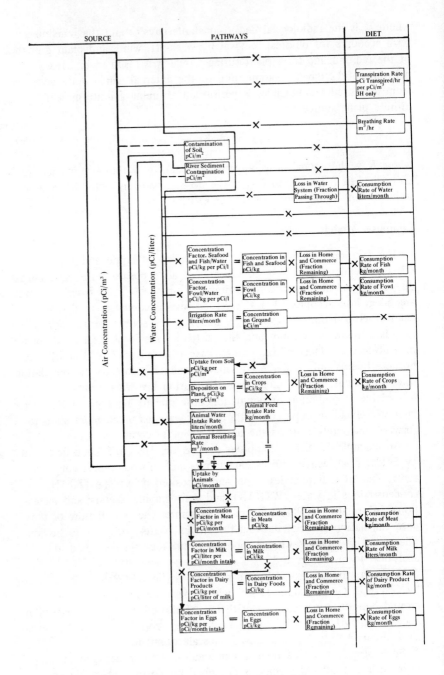

Figure 80. Dose calculation model (14).

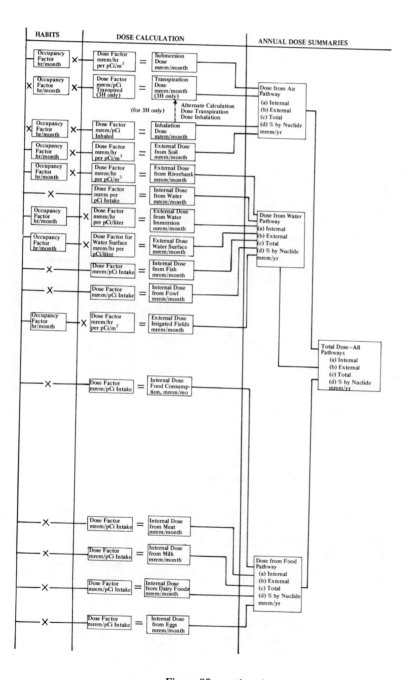

Figure 80, continued.

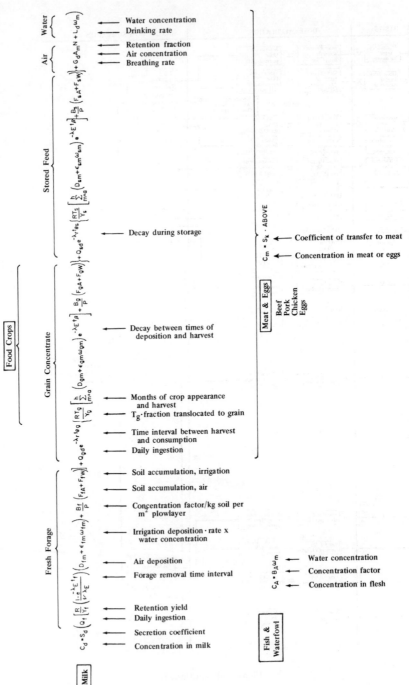

Figure 81. Analysis of food pathways (14).

Lessor (17) as part of the so-called "Rasmussen Report." This code describes the factors leading to deposition and retention of airborne fission products within the containment building for both PWR's and BWR's and identifies the form of all airborne components, some of which might escape to the surroundings.

Another code to calculate radiation doses from airborne effluents is the AIREM computer code (18). This code provides for a Gaussian or bell-shaped concentration profile in the vertical direction and a uniform concentration distribution in the horizontal crosswind direction. The vertical diffusion is limited to a finite mixing height and the technique of image sources is used to account for reflection from both the mixing layer and ground surface. The basic diffusion equation used in AIREM is a standard sector-average equation modified to include radionuclide decay by time flight, i.e., Equation 69.

This was further developed in the AIRDOS code (19), a Fortran IV computer code written to estimate population and individual doses resulting from the continuous simultaneous atmospheric release of as many as 36 radionuclides from a nuclear facility. Five pathways to man are considered: (1) inhalation of air containing radionuclide gases or particulates, (2) immersion in contaminated air, (3) exposure to surfaces contaminated by radioactive fallout, (4) ingestion of food produced on contaminated ground surfaces, and (5) immersion in contaminated water, as by swimming. Dose and dose commitments are estimated for each pathway and for the following 11 reference organs: whole body, G.I. tract, bone, thyroid, lungs, muscle, kidneys, liver, spleen, testes and ovaries.

The internal radiation dose to man following a continuous intake through inhalation or ingestion can be estimated using INREM (20), a Fortran IV computer code that estimates the cumulative dose to various human body organs.

Finally, detailed computer codes (SYSIN) used to calculate the radioactivity in gaseous effluents, based on Binford's STEFFEG programs, are presented in volume 2 of Report WASH-1258 (13), which also contains the necessary nuclear data to calculate the activity in the liquid effluents. A further development of that code called GALE, adapted for BWR's and PWR's, is described in full detail in Reference 15. The GALE code is a computerized mathematical model for calculating the release of radioactive material in gaseous and liquid reactor effluents. The calculations are based on a series of generically applicable parameters that describe the appearance rates and formation of fission and activation products in the primary coolant, the release and transport mechanisms resulting in their appearance in liquid and gaseous wastes, and the effectiveness of design considerations incorporated to reduce the quantity of radioactive materials that may be released to the environment.

Many other computer codes have been developed, some of them proprietary, to calculate diffusion conditions and population doses; among these are Baetsle's description of computational methods for the prediction of underground movement of radionuclides (21). Most of the codes deal with routine releases and may not necessarily be adaptable to include accident conditions.

Indirect Exposure Paths

So far we have discussed the dispersion of radioactive effluents in air and water and the subsequent exposure of people or other biota to average or momentary local concentrations of the various radionuclides. To complete the picture one needs to include also indirect exposure paths resulting from selective adsorption or deposition of some of the effluent components on foliage or river bottoms and from movement and reconcentration of certain elements through the food chain. This field is often referred to as radioecology and has been the subject of intensive investigations (2,22-27). Much of this work was sparked by the need to establish the effects of world-wide population exposure to fallout from atmospheric nuclear tests, and it has been concerned primarily with the environmental behavior of the longer-lived fission products, such as cesium-137, strontium-90 and cerium-144, and activation products such as manganese-54 and tungsten-185 (28-30). In the case of effluents from reactor and reprocessing plants the same nuclides are of significance but some of the shorter-lived and more volatile components, such as iodine-131, ruthenium-106 and barium-lanthanum-140, also need to be considered.

Such a discussion should include three subjects:

(1)   the food chain: the ingestion of radioactive contaminants by man or animals in higher than ambient concentrations owing to some preferential concentration process in the chain leading from air or water through one or more intermediate transformation links. Among the better known examples are:

> Sr or I deposition from airborne contamination → uptake in crop foliage → grazing by cow → preferential appearance in cow's milk → man

> Cs deposition on lichen → grazing by caribou or reindeer → consumption by man.

(2)   the internal dose to organisms, such as marine organisms, due to preferential uptake of certain elements and accumulation in bone or

tissue. (In many cases this second process also contributes to the first.)

(3) the selective movement of certain radionuclides through the soil or subsurface aquifers. Again this may be a contributory process to the other processes (31-34). It is of major significance in predicting possible mobility and subsequent population exposure from waste storage areas or contaminated water supplies. Since most of these processes tend to be slow in character, they are not usually considered a major contributor to calculations of population exposure from nuclear reactors, but they may be significant in waste disposal assessment.

The Food Chains

Considering the complex ecological relationships between plants, animals and man, there are clearly an infinite number of pathways leading from trace concentrations of certain elements in air and water to their appearance in the diet of man and animals (35). The relative importance of some of these relations will vary with geography and dietary habits. The above-mentioned appearance of cesium-137 in caribou and reindeer affects solely the population of Eskimos and Laplanders, but is exceedingly important in that context (22). A good summary of the main pathways will be found in Eisenbud (36) and the UNSCEAR and BEIR Reports (37,38). Most of the data reported come from observations on atmospheric fallout from atom bomb tests, but again it must be remembered that there is also a substantial uptake of natural radionuclides, especially potassium-40 and radium-226 and its daughters. Table 100 shows some representative intake figures for those nuclides. Doses resulting from uptake from power plant effluents are, therefore, additional to these.

In practice, depending on local dietary habits and the composition of plant effluents, only a few food chains will be of critical importance in a given locality or population group. The ICRP has suggested the use of the critical pathway approach to designate such food chains (39). In essence this procedure suggests that the complex pathways, by which discharged radioactivity may result in public radiation exposure, can in practice be reduced to one or two pathways that are overwhelmingly more important than all others. For any proposed release, this approach still requires a preliminary study to identify and evaluate these critical pathways, the critical radionuclides, critical materials and critical population group. An integral part of such a study is a habits survey that provides data on consumption rates of certain foodstuffs or the handling time of certain materials. Unless this is done with some judgment a

Table 100. Radium-226, Potassium-40 and Lead-210 Content of
Composite Diet Samples at Cities in the United States (25)

Item Analyzed	No. of Cities	Average per Day (Range)	Average per kg (Range)	Average per g Ca (Range)
^{226}Ra	25	1.6 pCi (nd-5.0)	0.44 pCi (nd-1.4)	0.83 pCi (nd-2.4)
^{226}Ra	5	3.02 pCi (2.17-4.35)	0.76 pCi (0.58-1.04)	1.85 pCi (1.35-2.40)
^{40}K	9	3400 pCi (2600-3800)	980 pCi (800-1100)	
^{210}Pb	9	2.4 pCi (nd-5.3)	0.7 pCi (nd-1.4)	
Ca	25	1.87 g (1.27-2.11)	0.52 g (0.44-0.59)	
Ca	5	1.64 g (1.54-1.82)	0.42 g (0.40-0.43)	
Diet	25	3.60 kg (2.89-4.10)		

nd = not detectable.

great deal of meaningless and redundant data can be produced at appreciable computer cost (40). Table 101 gives some examples of critical pathways in the aquatic environment for Great Britain (39).

The example of laverbread is an interesting illustration of such an identification of a critical population group with unusual dietary habits. Laverbread includes a particular kind of seaweed that concentrates ruthenium-106. The population group involved is of the order of 26,000, situated as shown in Figure 82. Surveys have shown a mean consumption of laverbread of 14.7 g/day for adults and 10.4 g/day for children with a roughly Gaussian distribution. The small fraction consuming significantly higher quantities, about 65-388 g/day, consisting of about 100 persons in this population would be considered as the critical group for this purpose. Based on the average consumption rate of this group, 160 g/day, and the ICRP permissible daily intake for ^{106}Ru (8), the derived working limit becomes 138 pCi/g of laverbread and relates to a dose limit for the lower large intestine of 1.5 rem/yr (39). Where the irradiation of an organ is contributed to by more than one radionuclide, limits must be prorated; for instance, in the laverbread case cerium-144 and other lesser

Table 101. Critical Pathways and Derived Working Limits for Some Typical UK Situations (39)

Site and Pathway	Critical Material	Exposed Population Group	Daily Consumption or Exposure (hours/yr)	Exposed Organ	Major Nuclides Contributing to Dose	Derived Working Level	ICRP Dose Limit mrem/yr
Windscale (Internal)	*Porphyra* (seaweed)	Laverbread consumers. Crit. Gp. 100	160 g laverbread (≈80 g seaweed)/day	GI tract (LLI)	^{106}Ru (critical)	300 pCi/g	1500
(External)	Estuarine silt	Fishermen, 10 persons	350 hours/yr	Total body	^{95}Zr, ^{95}Nb, ^{106}Ru	1.4 mrem/hr	500
Bradwell (Internal)	Oyster	Fishermen, 50 persons	75 g/day	Total body	^{65}Zn (critical)	2900 pCi/g	500
Trawsfynydd	Trout and perch	Lake anglers, ≈100 persons	100 g/day	Total body	^{137}Cs ^{134}Cs	440 pCi/g 200 pCi/g	500
Hinkley Point (Internal)	Fish and shrimp	Fishermen and families, ≯100 persons	90 g/day	Total body	^{137}Cs ^{134}Cs	490 pCi/g 220 pCi/g	500
(External)	Mud/silt	Fishermen, ≯10 persons	880 hours/yr	Total body	^{137}Cs ^{134}Cs	570μ rem/hr	500

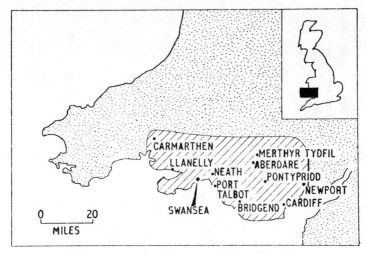

Figure 82. Area of sale of laverbread in South Wales with principal towns (39).

beta emitters may contribute to the irradiation of the lower intestine. Over a calendar quarter the limitation on discharge, in curies, was

$$\frac{^{106}\text{Ru discharged}}{15000} + \frac{^{144}\text{Ce discharged}}{90000} + \frac{\text{Total beta}}{300,000} \not> 1.$$

Cow's milk has long been recognized as a critical pathway for the ingestion of iodine and strontium radionuclides for atmospheric releases. Recent studies for ^{137}Cs, ^{54}Mn and ^{144}Ce-Pr have shown that ingestion of contaminated milk is the more restrictive pathway for ^{144}Ce-Pr, whereas direct radiation from radioactive material deposited on ground and plant surfaces was found to be the more limiting exposure route for ^{137}Cs and ^{54}Mn (41). The milk pathway depends on a variety of other factors such as the rate of foliar deposition, the composition of feed, and seasonal variations. For instance, Porter et al. (42) showed that cesium-137 levels in Tampa, Florida, milk were consistently higher than elsewhere and traced this back to one particular feed component, pangola hay. Table 102 shows the variations in ^{137}Cs concentrations in their dairy feed samples. The transfer of ^{137}Cs from feed to milk was estimated as 11-12% in that case, but varied considerably with overall feed conditions.

In general, for numerical evaluation, estimates must be made for each of the stages involved: foliar deposition, soil deposition, uptake in plants, grazing rates, conversion in meat or milk, and utilization in human diet.

Direct contamination of foliage leads to a much higher radionuclide content than uptake through roots when the fallout rate is high. Three

Table 102. Cesium Concentration in Tampa Dairy Feed Components (42)

Feed	^{137}Cs Concentration (Dried) (pCi/kg)	
	Average	Range
Brewers' grain	100	–
Cotton seed hulls	640	570-670
Ground snap corn	480	410-560
Mineral mix	100	–
Citrus pulp	950	870-1,090
Beet pulp	100	–
Peanut hulls and stems	210	a
Pangola hay	6,300	3,700-9,600
Spanish moss	18,700	18,000-19,000
Water cress	960	a
Salt	100	a
Hominy feed	800	a
Alfalfa	1,400	340-3,300
Wheat bran	2,000	a
Crimped oats	840	a
Cotton seed meal	1,400	a

[a]Value based on one feed component.

mechanisms of direct contamination have been recognized: (1) foliar contamination, which is retention and absorption through the leaves; (2) floral contamination, which is entrapment and absorption in inflorescends; and (3) plant-base absorption, which is entry into the basal tissues of shoots of superficial roots by material initially lodged on them or washed down by rain from the foliage. Material is deposited on plants by dust or other particulate matter, precipitation or sprays. Retention depends upon such factors as intensity and amount of precipitation, wind speed, particle size and density, wettability of leaves, leaf type and age, and thickness and continuity of the cuticle. To the extent that radionuclides are water-soluble, they may be absorbed through the leaves or basal tissues following much the same relationships described previously for root absorption. Once absorbed, processes of translocation influence distribution within the plant (25).

Much work has been done on radiocontamination of grazing lands because of the importance of pasture as a route of exposure of man. Both direct contamination and absorption from the soil are important. Perhaps the primary factor in soil uptake is the distribution of the nuclide in soil. Most pasture grasses obtain their nutrients from the top few inches of soil and in humid regions have shallow root systems; thus, the passage of time may remove contaminants from the rooting zone or otherwise reduce their accessibility. It appears that the rate of reduction of uptake of ^{90}Sr from soil is about 13-14% annually. The relative contribution of the soil ^{90}Sr to milk was reported in 1965 to vary from 20-50% in different years, being greatest in times of low fallout and making only a minor contribution in years of relatively high fallout.

In temperate regions, mineral soils that bind ^{137}Cs predominate. Studies have shown that only about 0.01% of ^{137}Cs artificially applied to an average bluegrass pasture was transferred from soil to grass. Experience with organic soils in Florida indicated a generally enhanced uptake from the soil in that region (37).

Most of the work on transfer of nuclides to milk has been confined to isotopes of iodine, strontium, and cesium. A frequently used parameter is the transfer coefficient, which is the percentage of the daily intake transferred to each liter of milk under steady-state conditions. Milk concentrations are also expressed in terms of concentrations in cut herbage or amount of nuclide per unit area of herbage. Use of area as a basis stems from the observation that the area from which a grazing cow obtains its daily intake is relatively more constant than the quantity of herbage consumed. In the case of strontium, the results are often expressed as observed ratios (OR), which denote the comparative behavior of strontium and calcium.

Average transfer coefficients for ^{131}I determined for cows under laboratory conditions range from 0.5 to 1.0% of daily intake per liter of milk. Values from field trials have ranged from 0.12 to 2.4, the large variance arising primarily from differences in physical properties of fallout and in physiology of the animals. Calculations have indicated that continuous grazing of cows on pasture carrying 1 μCi of ^{131}I/m^2 would lead to a milk concentration of about 0.2 μCi/liter. Field trials and also the experience of the Windscale accident have indicated that an average ratio of 10:1 would be more appropriate (37).

Numerous studies have been done with ^{90}Sr. An average value of 0.08 has been determined under laboratory conditions for the transfer coefficient of ^{90}Sr from diet to cow's milk. Under field conditions, values from 0.05 to 0.22 have been reported. However, the transfer coefficient is dependent upon dietary calcium and it is more meaningful to express results in terms of Sr/Ca ratios in diets and milk (OR milk/diet = ^{90}Sr/Ca in milk $\div$ ^{90}Sr/Ca in diet as measured at steady state). OR milk/diet values (overall ratios of deposition) usually fall in the range of 0.08 to 0.16.

A representative average value for the transfer coefficient of ^{137}Cs to cow's milk is 1.2, but field values have been reported as low as 0.25.

Low values are thought to be due to the binding of ^{137}Cs on clay particles associated with hay or by adsorption in the rumen contents. Data on transfer coefficients for other elements are scarce but some ranges can be indicated: 1 to 4 - Na, Zn, K; 0.1 to 1 - Ca, Fe, Co; 0.01 to 0.1 - Te, Ba, W, Po, Ra, U. The small amounts of very poorly absorbed elements (*e.g.,* ^{144}Ce, ^{239}Pu) found in milk are thought to occur from fecal contamination of the udders.

Limited studies have been done on the transfer of ^{131}I, ^{90}Sr, and ^{137}Cs to milk products (25). From 0.4 to 2.7% of ^{131}I in the original milk has been found per gram of skim milk, cream, butter, and cheese. In assessment of potential exposure from ^{131}I in such products, the time delay in consumption permitting radioactive decay must be taken into account. Relative concentrations of ^{90}Sr in butter, cottage cheese, and cheddar cheese following *in vivo* contamination of milk have been reported as 0.07, 6.8 and 0.34 respectively. Since the distribution of strontium follows that of calcium in milk products, there is nothing to be gained by substituting cheese for milk as a source of calcium in the human diet. For ^{137}Cs relative concentrations in butter vary from 0.03 to 0.11, in fresh cheese from 1.3 to 6.2, and cheddar cheese from 0.6 to 1.4.

Considerable attention has been given to the prediction of milk levels following pasture contamination. The objectives are to help in planning appropriate emergency measures following an uncontrolled release or to aid in the design of installations that release radioactivity under controlled conditions. For accurate estimates, there must be a field study at the time and place of contamination. Reliable models have been described for predicting the total intake of ^{131}I and ^{137}Cs by an average individual from knowledge of the milk level at a known time after a contaminating event (37,43).

Calculation of Thyroid Dose to the General Population

Most of these studies in the past have not distinguished adequately between inorganic and organic forms. Especially for iodine, it is important to indicate the rate processes involving elemental iodine or organic iodine, usually represented as methyl iodide. Except for air inhalation, the pathways from airborne discharges depend on the transfer coefficient between air and vegetation on the ground. This transfer coefficient is termed the *deposition velocity.* For iodine in the elemental form, the deposition velocity generally ranges between 0.002 and about 0.1 m/sec, depending upon the fraction absorbed on airborne particulates. Generally, deposition velocities in the range of 0.005 to 0.015 m/sec are considered typical for reactor effluents. The corresponding deposition velocity for the organic form, methyl iodide, has not been well characterized but has been shown to be several orders of magnitude smaller than for the elemental form. This results in negligible deposition of methyl iodide onto vegetation and the ground surface. Because of this, the principal exposure pathway for methyl iodide releases is likely to be direct inhalation rather than milk ingestion. For the elemental form, milk ingestion is likely to be the controlling pathway for iodine-131 if there are grazing animals (dairy cattle or goats) in the vicinity of the plant. This results from the ability of the cow (or goat) to concentrate the radioiodine by virtue of the large area of grass grazed daily (20-80 m^2/day). The exact ratio of the thyroid dose from milk ingestion to air inhalation depends upon parameters (primarily the mass of the thyroid gland, the ventilation or breathing rate, and the average milk consumption) associated with the age of the individual involved. The critical receptor is usually taken to be a young child because of a smaller thyroid mass and a greater daily milk consumption than for other age groups. These parameters are presented in Table 103 for selected age groups (6).

Table 103. Radioiodine Dose Conversion Factors per Unit Activity Intakes (6)

Age Group (years)	Age of "Typical" Individual (years)	Dose Intake Conversion Factor	
		^{131}I	^{133}I
Inhalation		(mrem/pCi inhaled)	
< 1	0.5	$1.5x10^{-2}$	$4.8x10^{-3}$
1-9	4	$1.2x10^{-2}$	$4.8x10^{-3}$
10-19	14	$2.4x10^{-3}$	$6.0x10^{-4}$
> 20	20	$1.2x10^{-3}$	$3.0x10^{-4}$
Ingestion		(mrem/pCi ingested)	
< 1	0.5	$2.0x10^{-2}$	$6.3x10^{-3}$
1-9	4	$1.6x10^{-2}$	$4.5x10^{-3}$
10-19	14	$3.1x10^{-3}$	$8.0x10^{-4}$
> 20	20	$1.6x10^{-3}$	$3.9x10^{-4}$

To allow for variations in milk consumption, thyroid mass and radiation sensitivities with age, thyroid dose computations have been performed separately for the age groups in Table 103 (6). Three intake modes were considered: air inhalation, vegetation consumption, and milk ingestion. The external whole-body dose from deposited radioiodine was found to be negligible compared to the inhalation and ingestion pathways. In analyzing the radiation exposure from organic iodides, a deposition factor of 1/1000 that of elemental iodine was used. Although that chosen value is somewhat higher than the best available estimate, it was felt that the use of the higher value is justified in view of the uncertainty in the chemical form and the possibility of a change in chemical form due to radiolytic or photodissociation of the methyl iodide in the environment or its attraction to airborne aerosols. Both of these processes could drastically affect the chemical form and, hence, the deposition velocity.

The relationship between the dose (equivalent) rate delivered to the critical organ by a radionuclide and the ambient air concentration of that radionuclide can be expressed as:

$$R = \chi K_D' P \tag{77}$$

where: R = dose (equivalent) rate in millirem per year
χ = air concentration of the radionuclide (pCi/m^3)
P = pathway transfer factor ($pCi/year$ per pCi/m^3) relating the intake rate to the ambient air concentration
K_D' = dose (equivalent) rate delivered per unit intake rate ($mrem/year$ per $pCi/year$).

The air concentration, χ, is determined by the dispersion models discussed in the preceding section. The pathway transfer factor, P, which relates the intake rate by an individual to the ambient air concentration, for iodine is presented in Table 104. For the inhalation pathway the transfer coefficient is the breathing rate in $m^3/year$.

Table 104. Pathway Transfer Coefficients, P (6)

Pathway:	Milk		Vegetables		Inhalation
Age Grouping	I-131	I-133	I-131	I-133	Breathing Rate (m^3/yr)
Average 6-month-old	8.41×10^4	7.66×10^2	0	0	1.15×10^3
Average 4-year-old	1.18×10^5	1.07×10^3	4.09×10^3	3.95	3.53×10^3
Average 14-year-old	1.11×10^5	1.01×10^3	6.55×10^3	6.33	6.44×10^3
Average adult	3.88×10^4	3.52×10^2	9.83×10^3	9.49	7.30×10^3

A slightly different approach, allowing for crop growing conditions, is outlined in Table 105, leading to slightly lower population doses. For quick conversion calculations, Table 106 presents some thyroid dose relations (44).

Table 105. Parameters Used for Calculation of Thyroid Doses to the General Population from Inhalation and Ingestion (13)

Parameter	Air	Milk	Leafy Vegetables
Growing season	–	6 months	4 months
Time between crop appearance and harvest	–	1 month[a] (pasture grass)	3 months
Concentration ratio for I-131	–	$560 \dfrac{\text{pCi/liter}^{b}}{\text{pCi/m}^{3}}$	$1100 \dfrac{\text{pCi/kg}^{b}}{\text{pCi/m}^{3}}$
Concentration ratio for I-133	–	$91 \dfrac{\text{pCi/liter}}{\text{pCi/m}^{3}}$	$170 \dfrac{\text{pCi/kg}}{\text{pCi/m}^{3}}$
Decay time between harvest and consumption[b]	–	3 days	7 days
Decay correction – I-131	–	0.772	0.547
– I-133	–	0.0916	0.0038
Loss in food preparation intake rates	–	–	50%
(0-9.9 yr) 4-yr-old	2560 m^3/yr	91 ℓ/6 mo	5.3 kg/4 mo
(10-19.9 yr) 14-yr-old	4930 m^3/yr	91 ℓ/6 mo	9.0 kg/4 mo
(≥ 20 yr) Adult	7300 m^3/yr	46 ℓ/6 mo	12 kg/4 mo
Dose (mrem/yr per pCi/m^3 Air) from I-131			
4-yr-old	12	370[c]	20[c]
14-yr-old	8.2	130[c]	12[c]
Adult	11	50[c]	12[c]
Population average	11	99[c]	
Dose (mrem/yr per pCi/m^3 Air) from I-133			
4-yr-old	3.2	15[c]	0.84
14-yr-old	2.1	5.3[c]	0.49
Adult	2.7	2.0[c]	0.50
Population average	2.7	0.49[d]	0.0022[d]

[a]The cow regrazes the same spot every 30 days
[b]See Reference 14.
[c]Without correction for decay between harvest and consumption.
[d]Calculated from individual factors by correcting for decay and for age distribution of population.

Table 106. Radioiodine Thyroid Dose Relations (15,44)

Appropriate Individual	Iodine Isotope	Source of Dose	Annual Thyroid Dose (mrem/yr)[a]
Infant[b]	I-131	Cow milk	$1.1 \times 10^8 \, (\chi/Q)_j \, Q_{131}R$
	I-133	Cow milk	$2.1 \times 10^6 \, (\chi/Q)_j \, Q_{133}R$
Infant[b]	I-131	Goat milk	$5.8 \times 10^8 \, (\chi/Q)_j \, Q_{131}R$
	I-133	Goat milk	$1.1 \times 10^7 \, (\chi/Q)_j \, Q_{133}R$
Adult	I-131	Leafy vegetables	$2.1 \times 10^6 \, (\chi/Q)_j \, Q_{131}R$
	I-133	Leafy vegetables	$8.3 \times 10^4 \, (\chi/Q)_j \, Q_{133}R$
Infant	I-131	Inhalation	$4.8 \times 10^5 \, (\chi/Q)_j \, Q_{131}R$
	I-133	Inhalation	$1.2 \times 10^5 \, (\chi/Q)_j \, Q_{133}R$
Adult	I-131	Inhalation	$4.0 \times 10^5 \, (\chi/Q)_j \, Q_{131}R$
	I-133	Inhalation	$9.8 \times 10^4 \, (\chi/Q)_j \, Q_{133}R$

[a]In these relations, $(\chi/Q)_j$ is given in seconds per cubic meter, Q_j is an iodine source term in curies per year, and R is a cloud depletion term and is dimensionless

[b]Dose relations provide doses appropriate to full year grazing.

The risk of health effects (principally thyroid cancer) resulting from radiation exposure of the thyroid appears to be age-dependent. For this reason, the population-integrated dose cannot be directly multiplied by a single dose-to-risk conversion factor. This can be accomplished, however, if the age-dependent risk per unit dose values are weighted by the fraction of the total population dose that is delivered to that age group. The resultant dose-to-risk conversion factors for an average individual are relatively independent of the radionuclide (I-131 or I-133) or the chemical form of the radioiodine (elemental or organic) (6). As an emergency measure it may be possible to remove the iodine from any affected milk by an ion exchange procedure (45); however, there are few conditions where this would be considered a worthwhile operation on the scale required.

A related problem is the radiation dose from iodine-129, a long-lived fission product (half-life 1.59×10^7 yr) of low specific activity. For reactors, I-129 would need to be considered only for severe accident releases. Since it is almost stable, its main effect would be of a long-term nature when an environmental equilibrium, involving all pathways, would exist between I-129 and total iodine. Tadmor (46) calculated the effects of such I-129 occurrences and concluded that the maximum exposure levels for the nonoccupational population should be of the order of

0.84 pCi/m^3 of air. As far as routine releases go, iodine-129 would be of concern primarily around reprocessing plants, where it would represent the residual iodine activity liberated from cooled spent fuel elements (47).

Aquatic Pathways

In view of the importance of seafood in the human diet, a careful assessment must be made of the contribution of aquatic pathways to the population dose. Many trace elements are upgraded in the food chains by algal components and bacterial action and their appearance in food stuffs cannot always be anticipated. In addition marine animals may be subjected to varying doses themselves, depending on the critical organs for each nuclide (3). For instance, a large proportion of radioactive liquids has been found reconcentrated in seabed silt and sediment, resulting in significantly higher dose rates to benthic and demersal organisms, such as bottom-feeding flatfish than the natural background (48). Polikarpov (27) has presented a detailed review of the accumulation of radioactive substances by marine organisms and the effect of radiation on marine organisms. The latter effect is minor at the concentrations normally encountered in the effluents from nuclear plants and most available data have been derived from laboratory or weapons tests (26).

The concentration of natural trace elements in certain organisms has been observed in many contexts; in recent years the reconcentration of mercury in fish has attracted particular attention. Among the more spectacular cases is the high concentration of iodine in red algae, iron in mollusks and diatoms, and manganese in fish and mollusks (27). The most convenient way of assessing the probable reconcentration of radioactive effluents in aquatic organisms is by means of the "bioaccumulation factor," C_f, expressed in pCi/kg of organism for a given concentration in water. Table 96 listed these factors for freshwater and saltwater organisms, based on the work of Freke (11) and Thompson *et al.* (14). These tabulations are based on dissolved activities; they do not take into account possible differences in uptake for organically bound radionuclides. Additional corrections may have to be made for the recycle of radionuclides, such as strontium-90, in the course of detritus formation after the death of the initial concentrator organism (27).

The population dose can be calculated, using one of the computer models mentioned (14) if adequate information is available on seafood consumption rates. In many cases a further distinction has to be made if only the meat is consumed, or the whole fish as fish meal, or the bone meal is recycled for chicken feed.

Alternatively, if the limiting hazard is likely to be internal dose to man due to ingestion of aquatic organisms (fish, crabs) a derived working limit, DWL, with respect to the critical seafood for the relevant radionuclides can be established based on the dietary habit of the critical group or individuals in the exposed population.

For a 70 kg standard man consuming 1 g protein per day per kg body weight, assuming that the entire dietary protein requirement is derived from the particular seafood, the DWL for the seafood is given by (11):

$$DWL_{sf} = \frac{MPC_w \times 2200 \, P_c}{7 \times 10^4} \; \mu Ci/g \tag{78}$$

where: MPC_w = maximum permissible concentration in water for 168 hr week occupational exposure (8)

P_c = protein content of given seafood

2200 = daily intake of water for standard man (cc).

With respect to seawater this becomes

$$DWL_{sw} = \frac{MPC_w \times 2200 \, P_c}{7 \times 10^4 \, C_f} \; \mu Ci/cm^3 \tag{79}$$

Values of P_c range from 6% for mollusks, through 16% (crustacea) and 17.5% (fish), to 50% w/w for seaweed (11).

A general review of the observed impact from the release of radionuclides at various installations and the different methods of exposure evaluation has been presented by the National Academy of Sciences (2).

Subsurface Movement of Radionuclides

Reference was made in Chapter 3 to the migration of trace element contaminants through the soil and through subsurface aquifers. In the present context we are mainly concerned with the movement of certain species insofar as they may add to the population dose arising from the operation of a nuclear facility. One general observation can be made: such movement tends to occur at a very slow rate, so its contribution to any accident exposure would be expected to be small and certainly delayed. Therefore, contamination of subsurface waters is of significance mainly in connection with small but steady liquid releases and with leakages from storage sites, as well as indirect fallout effects due to ground seepage of dissolved contaminants that may have been deposited on the soil surface (23,24,49,70).

Actual observations around nuclear power plants have shown that this effect is a negligible contributor to the total population dose. Although one can visualize various unlikely accident scenarios, it does not seem profitable to develop elaborate calculations to evaluate such situations. Soil surface contamination has been observed at various locations (51-53) and has almost invariably been due to weapons test fallout. However, there appears to be little, if any, evidence that this activity is sufficiently mobile to move into the water table. In practice, it has been difficult to measure any deposited activity from nuclear plant effluents against the background activity provided by weapons tests.

ENVIRONMENTAL SURVEILLANCE

As the previous sections have indicated, there are plenty of calculational models to predict the spread and distribution of low-level radioactive effluents in the environment around a nuclear power plant. However, inevitably simplifying assumptions have to be made regarding atmospheric dispersion, ground distributions and release rates and, to provide an adequate safety margin, many conservative assumptions tend to overestimate some of the factors involved. In the end, there are no real alternatives to actual measurement of radioactivity levels at all locations where population exposure may be envisaged. Because of the necessarily low dose-levels and concentrations, such measurements are difficult and it is important to discuss their effectiveness (54).

An environmental surveillance program must accomplish a variety of objectives: It must

1. provide check measurements to verify that radwaste treatment systems meet design objectives
2. measure actual radiation exposure to the general population at the fence line and beyond
3. warn against any sudden or unexpected rise in radiation levels in the vicinity of the plant
4. document for legal and regulatory purposes actual radiation exposure levels and radionuclide concentrations in air and water
5. monitor radionuclide occurrences in foods, soils and crops
6. provide records on preoperational and postoperational environmental conditions, including data on wild life, vegetation, background radiation and fish populations
7. ensure compliance with state and federal regulations on water quality, air pollution, waste management and land use
8. contain enough flexibility to provide monitoring services in emergency situations.

Only a well-designed, systematically operated program can meet these various objectives. Some existing programs, though meeting legal minimum requirements, may not always accomplish their nominal purposes.

Essentially, a surveillance program consists of the following types of operations:

1. radiation monitoring within the plant area and in its vicinity
2. monitoring of liquid and gaseous effluents (not strictly part of the surveillance program but an essential input and reference measurement)
3. ecological observations in any areas directly affected by the plant operations
4. measurement of beta and gamma emitters in cooling ponds and receiving streams
5. measurement and identification of low-level contaminants in air samples taken at selected locations
6. determination of ^{131}I, ^{137}Cs, ^{90}Sr-Y and other identified nuclides in milk, crops or local food products as appropriate.

Some of these measurements, such as effluent monitoring and detection of ambient airborne radioactivity, can be made continuously. Most of the others require careful sampling and concentration of the activity in the sample to obtain detectable levels. To take complete measurements of an area up to 50 miles around the plant site clearly becomes an enormous task and may be of questionable utility; it would also entail an excessive expense in equipment and manpower. For this reason most viable surveillance programs have attempted to select sampling points and sampling frequencies that might provide the most relevant information.

Variations in ambient radiation background must be allowed for, such as those due to atmospheric nuclear weapons tests, which unfortunately are still carried out. One way to discriminate between plant-related events and unrelated events is to divide any surveillance plan into annular zones centered on the plant, so that any increase in activity common to outer and inner ring zones can be clearly ascribed to external events. This is illustrated in Figure 83 where monitoring stations are located in inner and outer ring zones.

To indicate some of the samples and measurements normally considered in the vicinity of a nuclear facility, Table 107 shows some representative sampling and analysis categories employed (56); local considerations will determine the exact nature and extent of a given program. Such local factors would include the following:

1. type of installation and potential hazard associated with it
2. nuclides to be released, their activity, their physical and chemical form, and the method and route of release
3. the existing or expected presence of these nuclides from other sources
4. the behavior of the released nuclides in the environment

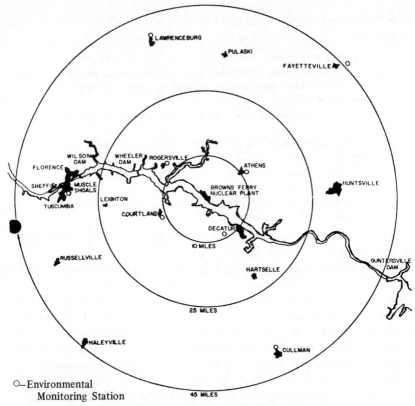

O—Environmental
   Monitoring Station   45 MILES

Note: The following samples are collected from each station:

                    Air Particulates   Rainwater
                    Radioiodine   Soil
                    Heavy Particle Fallout   Vegetation

Figure 83. Typical atmospheric and terrestrial monitoring network (55).

5. natural environmental features that affect the behavior of released
   radionuclides (climate, topography, pedology, geology, hydrology,
   hydrography, and vegetative cover)

6. man-made features of the environment that affect the behavior of
   released radionuclides (reservoirs, regulated streams or rivers, and
   harbor installations)

7. utilization of the environment for agriculture, fisheries, water and
   food supplies, industry and recreation

8. population distribution and habits.

Table 107. Some Samples and Measurements Conducted Around a Nuclear Facility (56)

Medium or Vector	Collection Frequency	Analyses	Location
Air	Daily	Gross beta	Several within a 1-kilometer radius, using wind data for guidance in siting
	Bulked monthly	Gamma spectra and gross alpha	
Water			
Surface streams	Weekly composite	Gamma spectra	Above and below installation, prior to other major stream use
	Bulked monthly	^{90}Sr, ^{3}H, U, Pu	
Precipitation	Monthly composite	Gamma spectra ^{90}Sr, ^{3}H	Several within a 1-kilometer radius, using wind data for guidance in siting
Potable	Monthly composite	Gamma spectra ^{90}Sr, ^{3}H, U, Pu	Supplies within 5-10 kilometers of installation
Bottom sediments	Six-monthly	Gamma spectra ^{137}Cs, ^{90}Sr, U, Pu	Above and below installation, prior to other major stream use
Aquatic biota	Six-monthly	Specific nuclides	Above and below installation, prior to major stream use
Vegetation	Six-monthly	Specific nuclides	Milk or food production areas within 10-20 kilometers of installation
Milk and other foodstuffs	Monthly composite	Gamma spectra ^{90}Sr	Farm areas within 10-20 kilometers of installation
Direct (area) radiation	Cumulative monthly	Integrating dosimeters	Several within 1-kilometer radius, using wind data for guidance in siting

By gathering information on these factors, an attempt should be made to identify: (a) the principal routes of exposure, (b) the population at greatest risk, and (c) the potentially most hazardous nuclides discharged.

In many cases a distinction is made between *plant monitoring*, which is the continuous determination of personnel exposure, ambient radiation levels within all structures of the facility, and radionuclide concentrations at the release points of liquid and airborne effluents, and *environmental surveillance*, which comprises continuous and batch-wise measurements of all environmental factors outside the facility. These two types of operations should be closely integrated and ideally would be performed by the same group. Detailed discussions on the organization and planning will be found in the transactions of various symposia (3,24,57-61). However, any program is only as useful as the quality and experience of the personnel involved will permit.

Types of Surveys

Surveillance programs can be classified by timing or by extent in several ways. The first classification distinguishes between preoperational surveys, initial operation and routine surveys.

Preoperational Surveys

These surveys include the collection and analysis of samples of the flora and fauna, as well as of air and water samples, during a period preceding start-up of the plant and ideally at least 10 months before the start of any construction or site-clearing. The preoperational investigations should establish baseline data on natural radiation background and its seasonal and diurnal fluctuations, meteorological data (wind, turbulence, precipitation), water flow and temperature, and critical population groups. This phase of the work should also include training of staff and testing of all equipment.

Much of the information collected may be needed for regulatory and legal reasons and it is important for it to be as complete as possible.

Initial Operation

Surveys during start-up and during the early months of routine operation of the plant are of particular importance in pinpointing any unexpected occurrences or any transient conditions. By carrying out high-density, fairly frequent measurements the effective impact of the plant can be evaluated and compared with predicted levels. On the basis of the results obtained, a sustained surveillance program can be devised that will provide continuous records of the most significant environmental factors, especially regarding radiological doses and effluent dispersion.

Routine (Postoperational) Surveys

On the basis of the early experience a surveillance program can be designed to collect significant data with optimum effectiveness, while retaining enough flexibility to respond to emergencies, or unexpected or abnormal occurrences. In many cases specifications call for a phased reduction in sampling schedules after start-up if continued low activity levels prevail.

The other kind of classification distinguishes between general, special and local surveys.

General Surveys

These surveys include area or national surveys, like the National Water Quality Network or the Milk Network, which may involve sources of pollution or environmental impact other than nuclear facilities.

Special Surveys

These are studies of selected topics or of critical pathways, such as certain fallout studies, studies on dietary habits of selected groups (Eskimos, Japanese fishermen, mine workers, or small infants), that deserve more detailed attention in specific circumstances.

Local Surveys

These surveillance programs identify local sources of pollution or radiation sources, conditions that would introduce special features into any predictional computation, or conditions that would affect the planning of the surveillance program. Such factors may include the existence of large coal-burning installations, mining operations or quarries causing unusually high radon evolution, seashore locations, and the existence or proposed location of major conurbations within the orbit of the plant.

Design of Surveillance Program

It is necessary for the objectives of an environmental surveillance program to be clearly established in order to have a satisfactory design. In a given situation one of the objectives will probably be of overriding importance and will control the elaboration required in the design of the surveillance program. An excellent review can be found in Reference 56.

Most of the objectives can be best realized if the environmental surveillance program includes the evaluation of environmental dose. The exposure of the local population to radiation evaluated in terms of estimated dose provides a sound method for demonstrating the plant's impact on the health of such populations. This method (a) is of assistance in meeting most regulatory requirements, (b) assists with the evaluation of a failure of the facility's containment systems, (c) provides a basis for assuring the public of their protection from harmful radiation, (d) becomes a necessary part of expert testimony in answering liability actions, and

(e) permits making optimum use of site and facility resources by establishing the relative range in which the environmental situation falls (61). Protection of people is the overriding design objective and the program must be capable of anticipating any hazardous situation that may arise from any sudden release of radioactivity and incorporate appropriate emergency measures, including means of notification of any threatened population groups and plans for an orderly evacuation. These plans should be clearly formulated even though they may never be needed, and they must be updated at periodic intervals.

Two types of measurements must be provided for: *continuous*, at fixed locations where radiation levels or other parameters are monitored continuously and are recorded there or remotely; and *periodic*, at selected locations where appropriate environmental samples are collected at set times or for set periods for subsequent measurement and evaluation at some central laboratory or counting facility. The latter type of sample may include measurements done by means of mobile equipment or a mobile laboratory, both of which are finding increasing use.

The choice of monitor and sample locations must be based on detailed knowledge of the area, its topographical features and the location of populated areas. Model pathways developed by theoretical dispersion calculations should be used to pinpoint the most significant locations. Availability of or access to reliable power may be important in selecting survey positions. The type, number and frequency of measurement must be specified as well as the sample size and method of analysis. The evaluation and assessment of the results should be done as soon as possible and provision must be made for program revisions. This is where an early start on the preoperational phase is important, since the first year's results often turn out to be useless while the operator gains experience in using the equipment, and the sample sites may turn out to be injudicious for year-round use; *e.g.,* they may get flooded out in spring or may be inaccessible in mid-winter. The evaluation of results will also show if the data satisfy program objectives or whether the objectives may have to be modified.

Finally the program must provide for assessment of the results in terms of their accuracy and consistency to permit valid comparison of preoperational and postoperational measurements. In general, this assessment will require correlation with effluent monitoring records and nationwide programs on such subjects as atmospheric fallout and losses of industrial or medical radioactive materials. For North America such data on fallout airborne radioactivity, and radionuclides in milk and drinking water used to be reported regularly in *Radiation Data and Reports.* The program design also has to determine the relative responsibilities of the plant operator, the surveillance group, and any public health authorities for record keeping, public information and ecological data.

Methods of Measurement

The method of measurement chosen for radiological survey work to a large extent determines the sampling procedure. It must be borne in mind that the levels of radioactivity encountered off-site are expected to be exceedingly low and only high-sensitivity detection systems or integrating dosimeters would be of any use here. Furthermore, to an increasing extent only specific quantitative results are acceptable in place of determinations that merely established that a given radiation level was less than some specified sensitivity limit or guideline maximum. Also, gross beta-gamma count rates are considered insufficiently specific and are gradually being replaced by measurements on identified radionuclides. The cost of survey measurements of this kind ranges from $150,000 per year for a single installation to $750,000 for a central laboratory to serve a number of nuclear plants regionally.

To test the sensitivity of the equipment employed it is usually best to test it first in or near the plant in relatively high levels of activity and gradually move it out to peripheral locations to confirm the rate of decrease with distance (inverse square or exponential, depending on energy). Regular, planned calibration checks are essential to maintain reliability and credibility of all measurements. For a general discussion of available methods the reader is referred to References 62-66.

A minimum detectable level can be defined for any detector as three times the standard deviation associated with the background measurement. For example, Brinck et al. (67) have described measurements with a commercial TLD system based on CaF_2:Mn dosimeters. The internal background of these dosimeters due to the radioactivity of the dosimeter materials was found to be 1.98 ± 0.09 μR/hr; the natural background exposure rate in a certain locality was measured to be 2.0 μR/hr. Such a dosimeter placed in the field for a 30-day monitoring period would have accumulated 1.43 ± 0.06 mR from internal background alone. Therefore, the minimum detectable exposure for a one-month monitoring period would be three times 0.06 mR or 0.18 mR for a single dosimeter.

For three dosimeters, the minimum detectable exposure would be less. Thus, a typical natural background radiation exposure level of 6 mR/month can be readily measured with TLD. However, the uncertainty in the measurement of an increase above natural background is greater. The minimum detectable increase above natural background radiation exposure contributed by a nuclear power station that can be measured is typically 0.9 μR/hr.

Measurement of Ambient (Airborne) Activity

Both for routine operations and in the assessment of the consequences of an accidental release it is important to have accurate survey results. For this purpose suitable sites have to be found, based on wind rose data, topology and population data (68). Power should be available for any active system and the equipment should have adequate protective housing. The predominant downwind direction is not necessarily the direction of highest concentrations, and preferably all 16 sectors of the wind rose should be covered, at least during early stages of operations and at 5-mile radial intervals. Sample positions should be about 10 ft (3 m) above the ground, at least 10 stack heights away from the plant, and be located to permit easy access for maintenance and sample collection but secure enough to prevent vandalism or interference.

Three types of sample may have to be analyzed: total air, particulates, and precipitation (rain or snow). At short distance direct shine, possibly from nitrogen-16, has been observed and must be allowed for, in addition to the usual background corrections for radon daughters and weapons fallout. As passive monitors, thermoluminescent detectors (TLD) are preferable to film badges because of their better sensitivity, less fading and improved consistency. Minimum sensitivity is of the order of 0.1 mR.

Pressurized ionization chambers (62) are finding increasing use, particularly as fence-line monitors of plume activity. Spherical high-pressure ionization chambers with sensitivities of the order of 1 μR/hr are available commercially. Systems are available for the use of compensating ionization chambers to subtract external gamma-ray background from the gaseous radioactivity reading. NaI scintillation detectors are used widely to monitor stack gases and fence-line activities; however, they are temperature sensitive and will show marked diurnal variations.

Particulate and precipitation samples can be collected at regular intervals and analyzed at a central laboratory for beta activity and specific gamma-emitting isotopes. This implies air sampling at a regulated flow through a particulate filter. In series with such a particulate filter a charcoal filter may be used to collect iodine. In some cases (55) each monitor location will have a collection tray and storage container to collect rainwater and a horizontal platform covered with gummed acetate to catch and hold heavy particle fallout. The gummed acetate sheet is ashed and counted for gross beta activity. Charcoal filters are analyzed for radioiodine in a multichannel gamma spectrometer often in coincidence with a beta detector. The various radioassay procedures have been described in detail in References 65 and 66. Air sampling instruments have been reviewed in References 54, 58 and 65.

Air filters are collected and analyzed for gross beta activity and specific gamma emitters. At least 3 days decay of radon daughters should be provided for after sample collection.

Rainwater can be analyzed for gross beta, specific gamma and ^{90}Sr activity. For the gross analysis, 0.5 ℓ of the sample is boiled to dryness and counted. To monitor loss of volatile components, the steam must be condensed and counted. The strontium isotopes are separated by ion exchange and counted in a low-background system. Usually the counting results of each method are averaged over the month or the year to yield an annual exposure value at the sample locations. This procedure has been questioned as obscuring data of real validity in terms of dose level fluctuations in specific directions at particular periods of time.

Atmospheric tritium can be analyzed by the method of Sax and Gabay (61) who employed a silica gel (200 g) collector to collect tritiated water vapor. Budnitz (69), who reviewed tritium monitoring techniques, sets practical limits of tritium detectability at 1 pCi/cc for ion chambers and gas proportional counters and advocates the use of silica gel collectors or bubblers for further concentration, coupled with liquid scintillation counting or the use of combustion techniques followed by HTO counting.

General gas counting is often done by means of Curran-type internal gas proportional counters, such as that described by Schell (70). Such counters are of the single-wire guard ring type and capable of counting tritium as hydrogen in a 3:1 H_2:CH_4 gas mixture, argon-37 in 9:1 argon-CH_4 mixtures, and carbon-14 as methane in either mixture. Discrimination is obtained by appropriate voltage biases.

Noble gas monitoring poses particular difficulties and in most cases seems to be limited to stack monitoring and the incidental inclusion in the TLD reading. Around reprocessing plants cryogenic distillation techniques have been employed. Collection of noble-gas daughters on the particulate filters is difficult and of low sensitivity, since most of them are either stable or long-lived cesium and rubidium isotopes, though monitoring of short-lived Rb-88 or Cs-138 may be possible. Analysis of the noble gases (krypton-85, xenon-133 and xenon-135) should also include argon-41 and possibly argon-37. For the HTGR, releases of argon-37 could approach 2 MCi/yr if holdup times are less than 120 days and can result in ^{37}Ar being dose limiting (71). The source of ^{37}Ar in the gas effluents is suspected to be the reaction ^{40}Ca (n, α) ^{37}Ar, which may take place in calcium impurities in the graphite moderator.

Collection of the noble gases most frequently is done by adsorption on activated charcoal, which will only partially adsorb N_2, O_2 and Ar. Successive desorption and adsorption will purify the krypton and remove traces of CO_2, O_2 and N_2. Using the train illustrated in Figure 84, with a 500 cm^3 counter volume, a detection limit of 4 pCi/m^3 has been reported for 100-liter air samples (72-74).

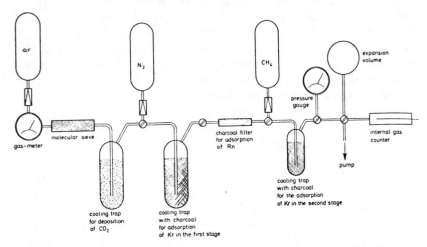

Figure 84. Apparatus of adsorption and measurement of krypton-85 (72).

Somewhat better sensitivities have been reported for gas chromatographic separation methods for the various gas fractions on a series of molecular sieve columns followed by gamma-ray spectrometric analysis of the gamma emitters and internal gas counting of the beta emitting gases, ^3H as hydrogen gas and methane, and ^{14}C as CO_2 or CH_4. Concentrations down to 10^{-8} μCi/ml have been counted from samples collected in 16-liter stainless steel vessels (71,75).

For specific determination of krypton-85, the gases can be adsorbed on a molecular sieve column. The krypton is separated from other gas components by passing the air sample, spiked with ^{83m}Kr, through a charcoal trap at liquid nitrogen temperature and reduced pressure. The noble gases are fractionated cryogenically and the krypton collected in a liquid scintillation void for counting. The ^{83m}Kr tracer permits estimation of the krypton recovery of the procedure (76).

Airborne iodine is usually determined by collection on charcoal filters and Ge(Li) or NaI detectors (77). The limit of detectability is of the order of 10^{-3} pCi/m^3 for ^{131}I; because of the low specific activity of ^{129}I (1.7 x 10^{-4} Ci/g) it is only rarely detectable in air samples and can be determined only indirectly in terrestrial or water samples (47).

To meet the more demanding requirements of recent NRC regulations, Goldstein et al. (78) have proposed a more elaborate procedure for low-level radioiodine determinations, which involves the use of 2-3 cartridges of triethyl diamine (TEDA)-impregnated charcoal through which air is passed for 23 hours at flow rates around 100 and 200 ft/min. After a day's delay, the cartridges are counted for many hours in a well-shielded Ge(Li) detector system. With longer collection and counting times, concentrations of ^{131}I down to 5 x 10^{-15} μCi/ml without enormously long collecting and counting times would be difficult, especially since its major gamma-peak is easily obscured by background X-rays. Charcoal cartridges impregnated with KI appear to have improved adsorption characteristics and are now employed to an increasing extent. However, they are not suitable for analytical procedures employing a distillation step.

Measurement of Water Samples

The main interest lies in the measurement of stream samples, domestic water supplies obtained from streams and wells, and in estuarine and shallow water oceanic samples. The location of sample sites is usually dictated by geography with sample sites at all significant downstream locations and at wells within a 10-30 mile radius of the plant. There does not seem to exist a good continuous water monitor capable of measuring the low concentrations expected. Consequently, periodic samples are collected using automatic samplers if available. Fewer, larger samples are preferred to provide the necessary analytical accuracy. Assessment of samples has to take into account stream flow conditions, variations in background, especially in tidal waters, and disturbance from river traffic. The sample location must be secure, out of the range of vandals or grazing animals, and protected from flooding or ice conditions. Silt and sediments that may sorb released radionuclides should be collected and screened for counting (79,80); container walls may also absorb trace elements.

Gross counting of water samples is rarely useful and concentration of sample material must be undertaken. To remove sedimentary material and colloids, a sizable fraction of the sample should be filtered through membrane filters and then analyzed. Freshwater samples should be allowed to settle; then 100-150 ℓ of water should be passed through ion exchange columns. This may take 10-12 hours. The extracted activity can then be counted in a gamma ray spectrometer, either still on the ion exchange resin or after selective elution. Specific radionuclides of interest include Na-24, Cr-51, Mn-54, Fe-59, Co-58, Co-60, Ru-106, Sr-90, I-131, I-133, Cs-134, Cs-137 and Sb-125 (51-53).

Tritium has to be analyzed separately, usually by liquid scintillation counting of an aliquot that had been distilled for purification. Electrolytic enrichment can be used (81) but is rarely justified.

Evaporation methods have also been used widely to concentrate water samples, but that method is tedious and apt to lose some of the more volatile nuclides.

Table 108 indicates the minimum detectable sensitivities routinely accessible in 1972 for environmental assays (82). It should be noted that for iodine, especially, there is a trend to require detectability at lower concentrations, though this is more marked for airborne sample analyses than for water (65,78,83). A more detailed listing of detection methods and capabilities will be found in Reference 63.

Table 108. Minimum Detectable Sensitivities, 1972 (82)

Air (pCi/m^3)	
Gross beta	0.003
Fallout $(pCi/m^2/day)$	
Ruthenium-106	270
Zirconium-niobium-95	54
Strontium-89	10.8
Strontium-90	2.7
Hydrogen-3 (pCi/liter)	100
Water and Milk (pCi/liter)	
Gross Alpha	1
Gross Beta	1
Cesium-137	10
Cesium-134	10
Cobalt-60	15
Ruthenium-106	50
Zinc-65	20
Antimony-125	20
Zirconium-niobium-95	10
Barium-140	15
Iodine-131 (isotopic scan)	10
Iodine-131 (ion exchange)	0.3
Plutonium-238	0.02
Plutonium-239, -240	0.02
Strontium-89	5
Strontium-90	1
Iodine-129	0.3
Hydrogen-3	100

The need for monitoring *ground* water depends primarily on local geological conditions and the mode of radioactive waste discharge utilized by a nuclear facility. It has been established, based on geological and hydrological studies for licensing of low-level waste disposal sites, that radioactivity moves at a very slow rate through geological formations. In those cases where the ground water may be contaminated by the liquid discharge from a nuclear facility, it should be sampled and analyzed for specific radionuclides to evaluate the potential health hazard. Sampling frequency usually need not be greater than at quarterly intervals and frequently annual samples are sufficient where public information is the principal objective for this phase of the environmental surveillance program (59).

The sampling and analysis of sediment frequently reveals the fate of a substantial fraction of the radionuclides in the liquid waste discharge from a facility. These radionuclides accumulate in sediment deposits and, during periods of flooding, can be reentrained, causing relatively high levels in downstream waters. The sampling frequency will vary with the characteristics of the facility's receiving waters, the quantity of waste released within a given time, and the chemical and physical characteristics of the waste. Usually an annual, semiannual, or seasonal collection interval will be sufficient. Two types of commercially available dredges for collecting sediment are the Ekman dredge and the Peterson dredge. The Ekman dredge is used for removing sediment from soft muddy bottoms, and the Petersen dredge is used for removing sediment from sandy and gravel-type stream beds.

Aquatic biota present in the receiving waters of the facility's liquid waste discharge should be sampled extensively when commercial fishing or harvesting of shellfish occur. The seasonal changes in the habits of biota should be considered. Because shellfish have the ability to selectively concentrate a number of radionuclides normally found in the reactor's liquid effluent, they are very sensitive indicators of the concentration of these radionuclides in water and should be sampled and analyzed to determine if they exceed permissible concentration guidelines. In some cases they are the critical vector for transporting the radioactivity from the liquid effluent via the food chain to man. Again, specific indicator species should be selected to provide consistent meaningful indications of environmental effects from plant operations (84).

Special sampling criteria may have to be applied to waters frequented by nuclear-powered ships. The primary objective in these waters may be to minimize reconcentration of wastes and to detect malfunction of the radwaste removal systems (4). Table 109 indicates the quantities of liquid radioactive wastes discharged into U.S. ports. Analysis of effluents has

Table 109. Summary of Liquid Radioactive Waste Discharged into Harbors by U.S. Navy Nuclear Powered Ships (2)

Shipyard or Naval Harbor[a]	Liquid Radioactive Waste Discharges[b]				
	1961-1965	1966		1967	
	(Ci)	1,000 gal	Ci	1,000 gal	Ci
Portsmouth, N.H.	0.48	155	0.01	265	0.01
Quincy, Mass.	0.15	0	< 0.005	–	< 0.005
Groton-New London, Conn. (Electric Boat Div., state pier and submarine base)	7.18	1,274	0.03	606	0.01
Camden, N.J.	0.01	0	< 0.005	–	–
Newport News, Va.	2.31	1,581	0.06	1,533	0.04
Norfolk, Va.	1.17	1,051	0.03	1,784	0.03
Charleston, S.C.	1.19	369	0.04	320	0.01
Pascagoula, Miss.	0	0	< 0.005	–	< 0.005
San Diego, Calif.	0.99	0	< 0.005	–	< 0.005
Vallejo, Calif.	2.33	270	0.19	–	< 0.005
Bremerton, Wash.	0.03	0	< 0.005	–	< 0.005
Pearl Harbor, Hawaii	3.20	654	0.03	683	0.01
Apra Harbor, Guam	0.01	0	–	–	< 0.005

[a]Other U.S. harbors have had less than 20,000 gal and less than 0.1 Ci discharged into them per year.

[b]Radioactivity measurements have been calibrated to ^{60}Co standard; volumes are prior to dilution.

indicated that the corrosion products ^{56}Mn and ^{60}Co accounted for an appreciable fraction of the total activity. Similarly, during its first year of operation, the *Savannah* discharged about one curie of mixed radioactive waste, most of it at sea (2).

Measurement of Soil Samples

Soil may accumulate radionuclides deposited in precipitation and from air and transfer these radionuclides to food and vegetation. Long-lived radionuclides such as ^{90}Sr and ^{137}Cs accumulate in soil over many years. Background activity from the naturally occurring nuclides ^{40}K, ^{232}Th and ^{226}Ra normally will be encountered. Procedures for analyzing soil involve either gamma-ray spectrometry of the dried sample or radiochemical analysis for ^{89}Sr and ^{90}Sr. Detailed separative procedures have been described in References 61, 66 and 85.

Soil samples should be collected once or twice a year in the sectors with the highest probability of deposition, preferably in the vicinity of food crops. Since the radionuclides of interest are also constituents of fallout from weapons tests, samples from the plant area should be compared with control samples collected beyond the influence of the reactor. The value of soil sample analysis is somewhat controversial, since it is clearly very sensitive to the nature of the sample, the depth of collection and the type of soil and vegetation. Since there is rarely any distinction possible between fixed and mobile radioactive compounds, the significance of the results may be questionable and some experts advocate the analysis of soil samples mainly for record purposes if sample size and depth are well defined.

To standardize soil sampling the U.S. Atomic Energy Commission has recommended (86) two soil sampling procedures, the HASL Method (65) that can be used for most soil types, and the Nevada Method, a more specialized procedure for sampling dry sandy soil.

In the HASL procedure a level site is selected that can be resampled at later times as necessary. The site should be nearly level with moderate to good permeability. Occasionally a depth profile is desired and an area free of rocks and stones should be selected. For the regular sample procedure, experience has shown that a total sample area of 460 to 930 cm^2 (0.5 to 1 ft^2) will provide a reasonably good estimate of total deposit if the area consists of a composite of ten or more individual plugs or cores. A tool for taking samples may be anything that takes a core or plug of equal area throughout its entire depth. A good pair of sampling tools is an 8.9-cm diameter (3-1/2 in.) topsoil cutter that takes a 5-cm-deep (2-in.-deep) sample and a 8.3-cm-diameter (3-1/4-in.) barrel auger that cuts an 8.9-cm-diameter (3-1/2-in.) sample.

The topsoil cutter is used to remove the sod to a depth of 5 cm, and the auger takes the remaining sample to the depth desired. The soil from ten cores sampled to a depth of 30 cm (12 in.) is composited to make a single sample weighing from 18 to 36 kg (40 to 80 lb). If desired, the 0 to 5-cm sod samples may be kept separate, making a sample of higher concentration. The amount of contaminant found in the upper 5 cm and that found in the remaining subsurface are added to give the total for the 620 cm^2 (0.67 ft^2) of surface represented by the ten cores.

Powdery, dry, loose, single-grain soils cannot be sampled in a simple, satisfactory way by the core method. It is best to take samples when soil contains enough moisture to be cohesive, even if this necessitates that the area to be sampled be wetted by sprinkling.

The Nevada Procedure involves either a ring method of sampling or a trench method to obtain 5-cm surface samples or depth profiles. In the ring method, a ring (12.7 cm i.d. x 2.5 cm deep) is pressed into the soil. Soil inside the ring is removed with a disposable plastic spoon to a depth of 2.5 cm and is bagged. Soil is next removed from outside the ring to the 2.5-cm depth, the ring is pressed into the soil another 2.5 cm, and another sample is taken with a second spoon. In this manner, profile samples can be taken to a desired depth; at lower depths, where radionuclide concentrations may be low, it may be desirable to increase increments to multiples of 2.5 cm.

A sample consists of soil taken from a minimum depth of 5 cm. A minimum of five separate samples should be taken along a straight line transect and composited for analysis. Since it may be desirable to resample the site at a later time, the coordinate distances of the transect should be measured to fixed landmarks to identify the relative position of the transect.

In the trench method, a rectangular trench of appropriate size is dug 15 to 25 cm deeper than the desired sampling depth. Samples are taken from a trench wall with a three-sided tray, 10 cm wide by 10 cm long by 2.5 cm deep.

The procedure for taking a sample is as follows: Push tray in from the side of the trench with the top edges of the tray flush with the surface of the ground. After the tray is pushed into position, press down with a trowel or a thin, flat piece of metal approximately 15-cm. wide above the open-sided front end of the tray. With the metal or trowel in place, scrape off the soil outside the tray down to the depth of the tray. Remove and bag the separated soil and repeat the process until blocks of soil have been removed to the desired depth. A sample consists of soil taken from a minimum depth of 5 cm. A minimum of five samples should be taken along a straight line transect, each from a separate trench, and composited for analysis. The trench method is often used for taking depth profile samples to obtain information on the distribution of a contaminant with depth.

If care is taken, either method works well in fine-textured soils; however, stony soils present difficulties. Larger scoops and hence larger soil volume will minimize perturbations caused by stones, which interfere with the progress of the edges of the scoop as it passes through the soil. A representative depth, however, is more important than a representative width (86).

Results should be expressed in nanocuries per gram of dry soil (total soil); the field bulk density should be recorded, as well as the area and depth sampled, to provide information necessary to calculate and express contaminant activity per unit area.

Analysis of Milk Samples

Since milk has been shown to be a major vector in the food chain leading to limiting doses from radioiodine to the thyroid, the occurrence and analysis of radiostrontium and radioiodine in milk has received much attention. It has been shown that their occurrence is related to the type of pasture fertilization employed (88), the particular kind of feed grain (29,42,43,89), and soil characteristics (22,89).

To monitor ^{131}I, ^{137}Cs and ^{90}Sr levels in milk, regular samples are analyzed to warn of any unusual appearance of activity. In the western hemisphere this is handled by the milk sampling networks shown in Figure 85, whose results were reported regularly in *Radiation Data and Reports*. In addition, as a check on possible environmental contamination, each power plant is required to monitor milk samples whenever commercial milk production takes place in the vicinity of the plant. Typically, weekly composites of daily samples are collected from such dairies as may be located at strategic distances and directions from the site. Weekly average concentrations of ^{131}I and monthly averages of ^{89}Sr, ^{90}Sr and ^{137}Cs have been reported (53,63).

Milk should be collected from dairy cows fed on fodder and pasturage grown within a 10-mile radius of the plant. If possible, one sample should be collected from cattle fed on vegetation grown in the downwind area of maximum predicted concentration. An additional sample should be collected from a local dairy representative of a milkshed for the area. Excessive dilution of samples with milk from unaffected areas should be avoided. At least a 1-gallon sample should be collected in polyethylene bottles and preserved with about 12 ml of 37% formaldehyde solution for later analysis. Alternatively, an ion exchange column may be used to separate radionuclides from the milk (87). The usual procedures involve separation of the iodine from raw milk by ion exchange. The concentrate is then counted with a scintillation detector or in a Ge(Li) gamma-ray spectrometer (45,66,87,90-93). Alternatively, iodine-131 is removed from milk by concentrating the iodine on an anion exchange resin column and subsequently removing it from the resin by batch extraction using NaOCl. After reduction to I_2 by hydroxylamine hydrochloride, the iodine is extracted into CCl_4, reduced with bisulfite, and back-extracted into water. The iodine is precipitated as palladous iodide. Chemical yield based on the added carrier iodine is determined gravimetrically. The I-131 concentration is determined by counting the palladous iodide precipitate in a low-background beta counter. The yield for the procedure is 70 to 85% (94).

Figure 85. Milk sampling networks in the western hemisphere.

The direct ion exchange method for separating iodine from raw milk requires that the iodine be in a readily exchangeable anionic form. Radioiodine tracer experiments on cows have shown that 0-10% of the iodine in milk may be protein-bound, rendering this fraction essentially unavailable for exchange with the ion exchange resin. However, this uncertainty of up to 10% in the I-131 determination will not significantly affect data interpretation and therefore is considered to be acceptable. Nevertheless, for the sake of accuracy, it is recommended that a fixed correction factor of 1.05 be applied to the counting data to compensate for this effect.

Also, there may be instances where milk samples may curdle to varying degrees in the interim between collection and analysis. Milk in this condition is unsuitable for analysis by the direct ion exchange method. A fresh sample should be obtained should this occur. If clabbering continues to be a problem, the ion exchange separation should be performed at the sample collection point and the ion exchange resin column shipped to the laboratory for processing. If this cannot be done, the milk sample should be frozen prior to shipment to the analytical laboratory.

The above procedure has been shown to be adequate to measure I-131 in milk at the 0.25 picocurie per liter concentration level. This sensitivity is based on using a 4-liter milk sample and beta counting for 1000 minutes or longer in a low-background counter with a nominal background count rate of 0.5-1 count per minute (cpm). The analytical sensitivity can be further improved by using more than 4 liters of milk and counting for longer than 1000 minutes.

A recount to check decay times should be performed 7-10 days after the first count. However, because it is not practical to do this for net counting rates of less than about 0.3 cpm, a recount should be made only when the initial net counting rate is greater than 0.3 cpm. If, after the second count, it is determined that the activity is not decaying with a half-life of 8 days, a third count should be made 4-5 days after the second count.

Various ion exchange processes have been tested for the decontamination of milk contaminated by radioactive fallout components in the forage (45,91) but such processes are not necessarily designed for quantitative extraction of the radionuclides, emphasizing instead undiminished palatability of the milk.

For field assays for the rapid determination of radionuclides in cases of suspected contamination, methods are preferred that can be performed with simple equipment or in a mobile trailer. Porter et al. (93) have described a method where EDTA (disodium ethylene diamine tetraacetate) is used to make up a prepared complexing solution. In the field 300 ml of this complexing solution are added to 1 liter of milk and mixed well; this complexes calcium ions and prevents loading of cation-exchange resin by calcium. The milk samples are then passed consecutively through 40 ml of anion exchange resin and 85 ml of cation exchange resin. The radioiodine will be retained by anion exchange, Sr, Ba and Ce by cation exchange. The resin columns are washed, sealed and shipped to the laboratory for analysis. I-131, Ba-140, Cs-137 and K-40 will be determined by gamma ray spectroscopy of the resin columns. Radiostrontium must be eluted from the resin, separated from barium, and then counted in a low-background beta-counter. Growth of the yttrium-90 daughter permits distinction between ^{89}Sr and ^{90}Sr. The chemical procedures are described in more detail in References 85, 92 and 93.

Evaluation of radioactivity data for milk samples can be done by relating them to the Radiation Protection Guide (RPG), defined as the radiation dose that should not be exceeded without careful consideration of the reason for doing so, and to the Protection Action Guide (PAG), defined as the projected absorbed dose to individuals in the general population that warrants protective action following a contaminating event. Table 110 lists the RPG and PAG values for the three radionuclides of interest in milk surveys; the ranges refer to various transient rates of daily intake that require increasingly stricter action responses (50).

Around a reprocessing plant, though less so around a reactor, release of long-lived iodine-129 may be of concern (46,47). Because of the low specific activity of I-129, a neutron activation method has been found satisfactory. The sample is solubilized by fusion with NaOH. The iodine is separated by a series of solvent extraction steps with CCl_4 and toluene. It is then irradiated in an ammonium sulfite solution in a reactor. After irradiation the sample is further purified and counted three times by gamma ray spectrometry after 1 hour to determine I-127 via I-128, after 6-12 hours for I-129 via I-130 and after 5 days to measure I-131 directly (47).

A conventional solvent extraction procedure followed by liquid scintillation counting of the low-energy beta emissions from I-129 has been described by Gabay et al. (83), who reported an analytical sensitivity of 0.3 pCi/l for a 4-liter sample for I-129.

Emergency Surveys

In the case of accidents leading to a significant release of radioactive material due to an accident involving the reactor system, the fuel handling facilities, or failure of all or parts of the radwaste treatment system, the surveillance program must be capable of rapidly obtaining data on the dispersion of radioactive material and on possible or prospective population exposures. In general, such an event will be immediately apparent to the plant operator who has to initiate appropriate action to protect plant personnel and prevent undue exposure of the surrounding population. Since the meteorological and hydrological conditions and the nature of the release will, in general, be known, it should be easier to pinpoint locations to obtain meaningful measurements than for routine surveillance,

Table 110. Radiation Protection Guides—Normal Peacetime Operation, and Protection Action Guides—Acute Contaminating Event (50)

Radionuclide	Critical Organ	RPG for Individual in the General Population (rad/year)	RPG (rad/year)	Guidance for Suitable Samples of Exposed Population Group[a]			
				Corresponding Continuous Daily Intake (pCi/day)	Range I (pCi/day)[b]	Range II (pCi/day)[b]	Range III (pCi/day)[b]
Strontium-89	Bone Bone marrow	1.5[c] 0.5[c]	0.5 0.17	2,000[d]	0-200	200-2,000	2,000-20,000
Strontium-90	Bone Bone marrow	1.5[c] 0.5[c]	0.5 0.17	200[d]	0-20	20-200	200-2,000
Iodine-131	Thyroid	1.5	0.5	100	0-10	10-100	100-1,000
Cesium-137[e]	Whole body	0.5	0.17	3,600	0-360	360-3,600	3,600-36,000

[a]Suitable samples which represent the limiting conditions for this guidance are: strontium-89, strontium-90—general population; iodine-131—children 1 year of age; cesium-137—infants.

[b]Based on an average intake of 1 liter of milk per day.

[c]A dose of 1.5 rad/year to the bone is estimated to result in a dose of 0.5 rad/year to the bone marrow.

[d]For strontium-89 and strontium-90, the Council's study indicated that there is currently no operational requirement for an intake value as high as one corresponding to the RPG. Therefore, these intake values correspond to doses to the critical organ not greater than one-third the respective RPG.

[e]The guides expressed here were not given in the FRC reports, but were calculated using appropriate FRC recommendations.

Continued

Table 110, Continued

| | | | Category I (Pasture-Cow-Milk) | |
| | | | Guidance for Suitable Sample, Children 1 Year of Age | |
Radionuclide	Critical Organ	PAG for Individuals in General Population (rads)	PAG (rads)	Maximum Concentration in Milk for Single Nuclide that Would Result in PAG (pCi/liter)
Strontium-89	Bone marrow	10 in first year; total dose not to exceed 15[a,b]	3 in first year; total dose not to exceed 5[a,b]	1,110,000[c]
Strontium-90	Bone marrow			51,000[c]
Cesium-137	Whole body			720,000[c]
Iodine-131	Thyroid	30	10	70,000[d]

[a]The sum of the projected doses of these three radionuclides to the bone marrow should be compared to the numerical value of the respective guide.

[b]Total dose from strontium-89 and cesium-137 is the same as dose in first year; total dose from strontium-90 is 5 times strontium-90 dose in first year for children approximately 1 year of age.

[c]These values represent concentrations that would result in doses to the bone marrow or whole body equal to the PAG, if only the single radionuclide were present.

[d]This concentration would result in the PAG dose based on intake before and after the date of maximum concentration in milk from an acute contaminating event. A maximum of 84,000 pCi/liter would result in a PAG dose if that portion of intake prior to the maximum concentration in milk is not considered. Children, 1 year of age, are assumed to be the critical segment of the population.

and the expected radiation levels should be more readily detectable. Furthermore, emergency plans should involve appropriate state and municipal authorities to divert traffic, control water use, and, if necessary, evacuate any threatened population (95). To assure public cooperation and to prevent panic, provision of clear and immediate public information on the extent of and possible hazard associated with the released radioactivity is an essential ingredient in any emergency plan.

Radiological surveillance in emergencies is essentially of two types: immediate and long-range. Immediate surveys must establish the extent, direction and characteristics of any radioactive cloud, determine its associated ground level dose and compute its projected pathway. Soil deposition and water contamination should also be determined in a qualitative fashion as quickly as possible. Long-range measurements would deal with contamination of pastures and its probable effect on milk and livestock, contamination of fresh- or saltwater and its effect on fish and marine organisms, and the radioactivity in streams and water supplies that may require some additional purification.

For practical planning purposes it is convenient to classify accidental releases as *incidents* and *accidents* according to the range of dose given to members of the public (96). This classification is shown in Table 111.

Table 111. Range of Doses to Members of the Public Associated with Accidental Release of Radionuclides (96)

Incidents		Accidents	
Type of Accidental Release			
Uncontrolled release below licensed limits for normal operation	Accidental release somewhat above licensed operational limits	Earlier range of maximum credible accidents of the safety analysis	Hypothetical accidents with release above reference level of the safety analysis
Range of Doses to Members of the Public			
<10-30 mrem whole body dose;	30-500 mrem whole body dose;	0.5-25 rem whole body dose;	>25 rem whole body dose;
< ca. 100 mrem thyroid dose	100-1500 mrem thyroid dose	1.5-25 rem thyroid dose	> 25 rem thyroid dose

The first 24 hours after an accident will usually be concerned with obtaining the results needed for immediate executive decisions and for assessing the severity of the accidents so that forecasts can be made of the extent and duration of any impact on the environment. In principle such information can be approximate; indeed the conditions in which it is gathered may make its approximate character inevitable. However, the conclusions based on the information will have wide implications and will be transmitted to very high levels of local and national government. Any subsequent changes in these conclusions, even if they result from extensions rather than retractions of the original data, are apt to be sharply criticized and will cause significant loss of confidence. It is of importance that early releases of information should allow as far as practicable for the inevitable uncertainties in the early data and particularly that they should not be incautiously optimistic. The need to obtain information as a basis for early and authentic statements is a major factor in the planning of an emergency survey (61).

After the first few hours of survey work the quantity of information to be assessed and the effort to be deployed will both increase sharply. After the initial assessment of the situation and the application of countermeasures, decisions will be needed on the withdrawal of these countermeasures. In some situations the source of the hazard will have passed and the decision is easy. In others, especially those concerned with the contamination of food, the activity will be decreasing steadily and a quantitative criterion must be adopted. This will be closely related to the emergency reference levels and will depend on a balance between the inconvenience and costs of continued countermeasures and the risks of the residual dose.

Howells and Dunster (61) have pointed out that even more than in other forms of environmental monitoring, the important aspect of the design of an emergency survey is to ensure that the results are capable of interpretation in terms of dose or, more directly, in terms of the protective action guides or emergency reference levels derived from doses but expressed in practical units. Second, the design must be intimately related to the proposed countermeasures and to the necessary time-scale of their application. For these reasons the design of a survey must be based on an appreciation of the type and the likely scale of accident, and can be usefully discussed only in terms of specific hazards.

The emergency reference levels used by the United Kingdom Atomic Energy Authority are shown in Table 112. An emergency reference level is a dose or a derived quantity below which it is unlikely that countermeasures will be justified, unless they have an exceedingly low impact on the community.

Table 112. Emergency Reference Levels (UK) for External Exposure and for Internal Exposure from Major Radionuclides (61)

External Exposure

	Type of Exposure		
Exposure Group	Gamma Radiation	Beta and Gamma Radiation[a]	Skin Contamination[b]
Children up to 16 yr pregnant women	20 R in free air	75 rads to superficial tissue	75 rads to superficial tissue
Other persons	30 R in free air	150 rads to superficial tissue	150 rads to superficial tissue

[a]Subject to a limit of 15 rads gamma, corresponding to 20 R in free air, or 25 rads gamma corresponding to 30 R in free air.

[b]Limited to 1/10 of the body surface and in addition to (a).

[c]A further 30 R may be permitted for essential duties by a special category comprising adult males (preferably in the older age group) or females above reproductive age.

Iodine-131 and Cesium-137

		Iodine-131		Cesium-137	
Parameter	Units	6-Month Old-Child[a]	Adult	6-Month Old-Child[a]	Adult
Critical organ		Thyroid	Thyroid	Whole body	Whole body
ERL of dose to critical organ	rad	25	25	10	10
Dose per microcurie inhaled	rad/μCi	15	1.5	0.049	0.047
Dose per microcurie ingested	rad/μCi	20	1.9	0.066	0.062
ERL of cloud dosage	Ci s/m^3	0.024[b]	0.075	2.9	0.93[c]
ERL in milk[d]	μCi/l	0.18	2.5	6.7	9.8
ERL on pasture[e]	μCi/m^2	1.3	18	22	33

[a]The values for the 6-month old child can be taken as typical of children in the first year of life.

[b]Where there is a dose contribution from other iodine isotopes and tellurium-132, the values for iodine-131 should be reduced by a factor of 2 or, in the case of a release of short-lived fission products from a criticality accident, by a factor of 10.

[c]The adult is the limiting case due to the much shorter half-life of cesium-137 in children than in adults.

[d]The tabulated values are for the maximum levels reached after a single deposition.

[e]The levels on pasture are the initial activities of the total deposits.

Continued

Table 112, Continued

Strontium-89 and Strontium-90

Parameter	Units	Strontium-89		Strontium-90	
		6-Month Old-Child[b]	Adult	6-Month Old-Child[b]	Adult
Critical organ		Bone	Bone	Bone	Bone
ERL of dose to critical organ		15 rads	15 rads	1.5 rads/yr	1.5 rads/yr
ERL of cloud dosage	Ci s/m^3	0.079	0.36	0.00083	0.0036
ERL in milk	μCi/g Ca	0.2	0.2	0.002	0.002
ERL in pasture	μCi/m^2	10	10	0.1	0.1

[a]The basis for these emergency reference levels, i.e., the assumption that the criterion should be local dose to mineral bone, is being reconsidered and it is likely that the basis may be changed and the values of the reference levels increased.
[b]The values for the 6-month old child can be taken as typical of children in the first year of life.

For planning purposes a realistic basis for discussion is the relationship between the magnitude of a release of activity and the probability of its occurrence, as outlined in relation to reactors by Farmer (97). If an approximate, even grossly approximate, relationship can be established between the size of release and its probability, rational decisions can be reached on the extent to which emergency procedures should be planned. For example, it will be widely accepted that detailed procedures must be available for the size of accident that may well occur in the lifetime of a typical program of work on a site with a probability of about 10^{-2}/yr. For larger releases with a probability of 10^{-4}/yr or 10^{-5}/yr, no detailed plans are needed to extend the prearranged procedures, but the consequences should be discussed and some thought given to the sort of improvisation that may be necessary.

The type of action necessary in each case is indicated in Table 113, which provides for three immediate responses—temporary isolation, elementary therapy by the use of stable iodide to reduce the dose from radioiodine, and evacuation. In practice, evacuation is feasible only for small population groups; with proper reactor siting mass evacuation of large population centers should never be required and may often be more hazardous than not.

The radiological hazard to the public is determined by the magnitude and composition of radionuclides accidentally released, together with their way of release. Accidental releases of high-level liquid effluents can

Table 113. Recommended Emergency Action Levels Based on the Dose in the Open Air from External Irradiation and Inhalation of Radionuclides (96)

Range of Risk	Expected Dose in Open Air (rem)	Recommended Action		
		Warning to Stay in House and Close Doors and Windows[b]	Distribution of Iodide Tablets[c]	Evacuation[d]
Irradiation of Thyroid by Inhaled Iodines and ^{132}Te[a]				
I	< 25	Useful	Not necessary	No
II	25-500	Necessary	Useful; necessary when > 100 rem	Not necessary
III	> 500	Necessary until evacuation	Necessary even when evacuated	Useful; necessary when > 1000 rem
Irradiation of the Whole Body by External Radiation and Inhaled Radionuclides				
I	< 25	Useful	Not applicable	No
II	25-100	Necessary	Not applicable	Useful
III	> 100	Necessary until evacuation	Not applicable	Necessary

[a]Dose calculation based on children
[b]Expected dose reduction a factor of two in case of inhalation and a factor of five in case of external irradiation
[c]Expected dose reduction about a factor of five.
[d]Prior dose reduction by staying house, etc., or taking iodide tables.

usually be handled like the release of any toxic chemical in industrial effluents, with appropriate warnings to downstream water users.

For airborne releases the rapid determination of the inhalation hazard proves the most difficult problem in early emergency monitoring. A release of rare gases alone can be easily assessed by measuring external irradiation, i.e., the dose rate from the cloud. In the case of a release of iodines, however, the thyroid dose resulting from inhalation will be much more important than external irradiation from the cloud. In this event, the concentration of the iodines in the cloud must be determined. In the same way, determination of the inhalation hazard from other radionuclides, e.g., Te- (thyroid dose), Cs- (whole body dose), Ru- (lung dose) and Sr-isotopes (bone dose), would also depend on identifying and measuring the individual radionuclides in the cloud. From practical

considerations it follows that ^{131}I together with the other short-lived radioiodines will be the most important radionuclides to be considered. Only
in the case of a release from overheated fuel, a large contribution of
tellurium and cesium and possibly ruthenium isotopes as well as some
[89]Sr and ^{90}Sr may also be expected.

When measuring the concentration of iodines in the cloud, one must
take into account the presence of possibly much larger activities of the
rare gases, as a release of rare gases is likely to be much larger than a
release of iodines. After the accidental release, the activity of the short-
lived iodines will exceed the activity of ^{131}I by a factor of six in the
case of a release of a reactor-equilibrium mixture of the iodines, and by
a factor of 500 in the case of a release from a criticality accident. However, the contribution of the shorter-lived iodines to the thyroid dose
will be less than a factor of two in the case of a reactor-equilibrium
mixture and about a factor of ten in the case of a criticality mixture.

From this it follows that on the basis of measurements of activity a
release of the criticality mixture of iodines could be over-estimated with
regard to the resulting thyroid dose, by a factor of twelve if mistaken
for a release of reactor-equilibrium iodines, and less likely by a factor of
50 if mistaken for a release of ^{131}I alone. This error can be lessened
if information from the reactor is available on the type of iodine mixture
to be expected, or if the method of measurement could discriminate to
a greater extent the shorter-lived iodines (96).

Howell and Dunster (61) recommend the following emergency program:

First Phase (Time 0-23 hr)

1. Plot, from meteorological data, plume limits and delineate possible
   affected downward area.
2. Direct two survey vehicles to preselected monitoring points within
   this area.
3. Measure existing air contamination concentration. Measure ground
   deposit by gamma dose rate meters and use these data to estimate
   integral air concentration.
4. Collect grass samples from affected area for gamma spectrometry
   to find quality of deposition.

Second Phase (Time: 24 hr onwards)

This phase of the emergency monitoring scheme is not precisely planned,
since action at this stage is dictated by information obtained in the first
phase. It may include the following:

1. Monitoring of milk samples, both by gamma spectrometry and by
   detailed radiochemical analysis.
2. Monitoring of other biological and food materials.

Table 114. Suggested Offsite Surveillance Program for Operating Light-Water-Cooled Nuclear Power Facilities (63)

Operation or Sample Type	Approximate Number of Samples and Their Locations	Collection Frequency	Analysis Type[a] and Frequency
Air particulates	1 sample from the 3 locations of the highest offsite ground level concentrations	Continuous collection—filter change as required	Gross long-lived β at filter change[b]
	1 sample from 1-3 communities within a 10-mile radius of the facility		Composite for gamma isotopic analysis and radiostrontium analysis[c] quarterly
	1 sample from a location greater than a 20-mile radius in the least prevalent annual wind direction[d]		
Air iodine	Same sites as for air particulates	Continuous collection—canister changes as required	Analyze weekly unless absence of radioiodine can be demonstrated
Direct radiation	2 or more dosimeters placed at each of the locations of the air particulate samples which are located at the 3 highest offsite ground level concentrations	Quarterly	Gamma dose quarterly
	2 or more dosimeters placed at each of 3 other locations for which the highest annual offsite dose at ground level is predicted[e]		
	2 or more dosimeters placed at each of 1-3 communities within a 10-mile radius of the facility[f]		
	2 or more dosimeters placed at a location greater than a 20-mile radius in the least prevalent annual wind direction[d]		
Surface water[g]	1 upstream	Monthly (Record status of discharge operations at time of sampling)	Gross β, gamma isotopic analysis[h] monthly. Composite for tritium and radiostrontium analysis[c] quarterly
	1 downstream after dilution (e.g., 1 mile)		
Ground water	1 or 2 from sources most likely to be affected	Quarterly	Gross β, gamma isotopic analysis[h] and tritium quarterly

Table 114, Continued

Operation or Sample Type	Approximate Number of Samples and Their Locations	Collection Frequency	Analysis Type[a] amd Frequency
Drinking water	Any supplies obtained within 10 miles of the facility which could be affected by its discharges or the first supply within 100 miles if none exists within 10 miles	Continuous proportional samples[i]	Gross β, gamma isotopic analysis[h] monthly. Composite for tritium and radiostrontium analysis quarterly[c]
Sediment, benthic organisms and aquatic plants	1 directly downstream of outfall[j] 1 upstream of outfall[j] 1 at dam site downstream or in impoundments[j]	Semiannually	Gamma isotopic analysis semiannually
Milk	1 sample at nearest offsite dairy farm in the prevailing downwind direction 1 sample of milk from local dairy representative of milkshed for the area	Monthly	Gamma isotopic analysis and radiostrontium analysis monthly[c]
Fish and shellfish	1 of each of principal edible types from vicinity of outfall 1 of each of the sample types from area not influenced by the discharges	Semiannually	Gamma isotopic analysis semiannually on edible portions
Fruits and vegetables	1 each of principal food products grown near the point of maximum predicted annual ground concentration from stack releases and from any area which is irrigated by water in which liquid plant wastes have been discharged 1 each of the same foods grown at greater than 20 miles distance in the least prevalent wind direction	Annually (at harvest)	Gamma isotopic analysis annually on edible portions
Meat and poultry	Meat, poultry, and eggs from animals fed on crops grown within 10 miles of the facility at the prevailing downwind direction or where drinking water is supplied from a downstream source	Annually during or immediately following grazing season	Gamma isotopic analysis annually on edible portions
Quality control[k]	Samples as required for accurate sampling and analysis		Minimum frequency– annually

aGamma isotopic analysis means identification of gamma emitters plus quantitative results for radionuclides that may be attributable to the facility.

bParticulate sample filters should be analyzed for gross beta after at least 24 hours to allow for radon and thoron daughter decay.

cRadiostrontium analysis is to be done only if gamma isotopic analysis indicates presence of cesium-137 associated with nuclear power facility discharges.

dThe purpose of this sample is to obtain background information. If it is not practical to locate a site in accordance with the criterion, another site which provides valid background data should be used.

eThese sites based on estimated dose levels, as opposed to ground level concentrations where the dose may be affected by sky shine, high plumes, or direct radiation from the facility being monitored.

fThese locations will normally coincide with the air particulate samplers used in the monitored communities.

gFor facilities not located on a stream, the upstream sample should be a sample taken at a distance beyond significant influence of the discharges. The downstream sample should be taken in an area beyond the outfall which would allow for mixing and dilution. Upstream samples taken in a tidal area must be taken far enough upstream to be beyond the plant influence when the effluent is actually flowing upstream during incoming tides.

hIf gross beta exceed 30 pCi/liter.

iDrinking water samples should be taken continuously at the surface water intake to municipal water supplies. Alternatively, if a reservoir is used, drinking water samples should be taken from the reservoir monthly. If the holding time for the reservoir is less than 1 month, then the sampling frequency should equal this holdup time. Increases in concentration of activation and/or fission products at these sources necessitate the analysis of tap water for the purpose of dose calculations. Additional analyses of tap water may be necessary to satisfy public demand.

jFor facilities located on large bodies of water, sampling sites should be located at the discharge point and in both directions along the shoreline.

kThe Analytical Quality Control Service of the Surveillance and Inspection Division of EPA provides low-level radiochemical standards and interlaboratory services to state and local health departments, federal and international agencies, and nuclear power facilities and their contractors. The service operates several types of cross-check programs for the analysis of radionuclide in environmental media, such as milk, food, water, air and soil. The samples are submitted on a routine schedule designed to fit the needs of each laboratory. Technical experiments are undertaken to permit detailed analyses of the accuracy and precision obtained by participating laboratories. In addition, low-level radioactivity standards are provided to the agencies participating in the various programs. Primary and secondary standardization is also performed as needed on those radionuclides not used on a routine basis.

Obviously, equipment and trained personnel must be on call at all times to provide the monitoring services and adequate dosimetry to serve as a basis for sound decisions on any subsequent course of action.

CONCLUSION

It has been stated that at the present state of technology, dose levels down to any arbitrary level can be determined given unlimited funds and ample time. In practice an extensive surveillance program can be quite costly and it is important to ensure its effectiveness by sound planning and by concentrating on measurements of real significance. Sample selection and proper sampling techniques are at least as important as the employment of sophisticated detection equipment and evaluation procedures. Table 114, by way of summary, presents one suggested program for off-site surveillance around operating light-water cooled reactor facilities (63). Actual programs at specific sites may well diverge from that program on the basis of demographic and geographical features peculiar to the site.

In all cases one must bear in mind that environmental surveillance meets an important social and legal obligation and its conduct requires intelligent and responsible performance at all times.

REFERENCES

1. Foster, R. F. "Sources and Inventory of Radioactivity in the Aquatic Environment," Paper, Health Phys. Soc. Meeting, Miami (1973).
2. *Radioactivity in the Marine Environment.* (Washington, D.C.: National Academy of Sciences, 1971).
3. "Radioactive Contamination of the Marine Environment," *Proceedings Seattle Symposium, 1972* (Vienna: International Atomic Energy Agency, 1973).
4. Miles, M. E., G. L. Sjoblom and R. D. Burke. "Environmental Monitoring and Disposal of Radioactive Wastes from U.S. Naval Nuclear-Powered Ships and their Support Facilities 1971," *Radn. Data Repts.* **13**, 469 (1972).
5. Clarke, R. H. "The WEERIE Program for Assessing the Radiological Consequences of Airborne Effluents from Nuclear Installations," *Health Phys.* **25**, 267 (1973).
6. "Environmental Analysis of the Uranium Fuel Cycle, Pt. II—Nuclear Power Reactors," (Washington, D.C.: U.S. Environmental Protection Agency, 1973).
7. Environmental Statement. Watts Bar Nuclear Plant, Units 1 and 2 TVA Report OHES-EIS-72-9 (Chattanooga, Tennessee: Tennessee Valley Authority, 1972).

8. Report of Committee II on Permissible Dose for Internal Radiation (1959), ICRP Publication 2 (Oxford: Pergamon Press, 1960). Supplement ICRP Publication 6 (New York: Macmillan Co., 1964).
9. Rowe, W. D., Ed. "Environmental Radiation Dose Commitment: An Application to the Nuclear Power Industry," Report EPA-520/4-73-002 (Washington, D.C.: U.S. Environmental Protection Agency, 1974).
10. "Reactor Safety Study, App. VI Calculation of Reactor Accident Consequences," Report WASH-1400 (Draft) (Washington, D.C.: U.S. Atomic Energy Commission, 1974).
11. Freke, A. M. "A Model for the Approximate Calculation of Safe Rates of Discharge of Radioactive Wastes into Marine Environments," Health Phys. 13, 743 (1967).
12. Reichle, D. E., P. B. Dunaway and D. J. Nelson. "Turnover and Concentration of Radionuclides in Food Chains," Nucl. Safety 11, 43 (1970).
13. "Numerical Guides for Design Objectives and Limiting Conditions to Meet the Criterion 'As Low As Practicable,' " Report WASH-1258 (Washington, D.C.: U.S. Atomic Energy Commission, 1973).
14. Soldat, J. K. "Modeling of Environmental Pathways and Radiation Doses from Nuclear Facilities," Report BNWL-SA-3939, (Richland, Washington: Battelle Pacific Northwest Laboratories (1971).
15. "Draft Regulatory Guides for Implementation—Attachment to Concluding Statement," Docket No. RM-50-2 (Washington, D.C.: U.S. Atomic Energy Commission, 1974).
16. Slade, D. H., Ed. Meteorology and Atomic Energy—1968. (Oak Ridge, Tennessee: U.S. Atomic Energy Commission, 1968).
17. Ritzman, R. L., P. C. Owzarski, A. K. Postma, D. L. Lessor, et al. "Release of Radioactivity in Reactor Accidents," Appendix VII, Report WASH-1400 (Draft). (Washington, D.C.: U.S. Atomic Energy Commission, 1974).
18. Martin, J. A., C. B. Nelson and P. A. Cuny. "AIREM Program Manual—A Computer Code for Calculating Doses and Depositions Due to Emissions of Radionuclides," Report EPA 520/1-74-004 (Washington, D.C.: U.S. Environmental Protection Agency, 1974).
19. Moore, R. E. "AIRDOS—A Computer Code for Estimating Population and Individual Doses Resulting from Atmospheric Releases of Radionuclides from Nuclear Facilities," Report ORNL-TM-4687 (Oak Ridge National Laboratory, 1975).
20. Killough, G. G., P. S. Rohwer and W. D. Turner. "INREM—A FORTRAN Code which Implements ICRP-2 Models of Internal Radiation Dose to Man," Report ORNL-5003 (Oak Ridge National Laboratory, 1975).
21. Baetsle, L. H. "Computational Methods for the Prediction of Underground Movement of Radionuclides," Nucl. Safety 8, 576 (1967).
22. Aberg, B. and F. P. Hungate, Eds. Radioecological Concentration Processes (Oxford: Pergamon Press, 1967).
23. Comar, C. L. "Movement of Fallout Radionuclides through the Biosphere and Man," Ann. Rev. Nucl. Sci. 15, 175 (1965).

24. "Environmental Contamination by Radioactive Materials," *Proceedings Vienna Seminar* (Vienna: International Atomic Energy Agency, 1969).
25. Fowler, E. B., Ed. *Radioactive Fallout, Soils, Plants, Foods, Man.* (Amsterdam: Elsevier, 1965).
26. Jinks, S. M. and M. Eisenbud. "Concentration Factors in the Aquatic Environment," *Radn. Data Reports* **13**, 243 (1972).
27. Polikarpov, G. G. *Radioecology of Aquatic Organisms* (New York: Reinhold Book Div., 1966).
28. "Environmental Behavior of Radionuclides Released in the Nuclear Industry," *Proceedings Aix Symposium 1973* (Vienna: International Atomic Energy Agency, 1973).
29. Garrett, A. R., S. L. Cummings and J. E. Regnier. "Accumulation of ^{137}Cs and ^{85}Sr by Florida Forages in a Uniform Environment," *Health Phys.* **21**, 67 (1971).
30. Jordan, C. F., J. R. Kline and D. S. Sasscer. "A Simple Model of Strontium and Manganese Dynamics in a Tropical Rain Forest," *Health Phys.* **24**, 477 (1973).
31. Borak, T. B., M. Awschalom, W. Fairman, F. Iwami and J. Sedlet. "The Underground Migration of Radionuclides Produced in Soil near High-Energy Proton Accelerators," *Health Phys.* **23**, 679 (1972).
32. Champlin, J. B. F. "The Transport of Radioisotopes by Fine Particulate Matter in Aquifers," Thesis, Georgia Institute of Technology, Atlanta (1969) (unpublished).
33. Champlin, J. B. F. and G. G. Eichholz. "The Movement of Radioactive Sodium and Ruthenium Through a Simulated Aquifer," *Water Resources Res.* **4**, 147 (1968).
34. Lomenick, T. F. "Movement of Ruthenium in the Bed of White Oak Lake," *Health Phys.* **9**, 835 (1963).
35. *Radionuclides in Foods.* (Washington, D.C.: National Academy of Sciences, 1973).
36. Eisenbud, M. *Environmental Radioactivity,* 2nd ed. (New York: McGraw-Hill, 1973).
37. BEIR Committee. "The Effects on Populations of Exposure to Low Levels of Ionizing Radiations," (Washington, D.C.: National Academy of Sciences/National Research Council, 1972).
38. UNSCEAR. *Ionizing Radiations: Levels and Effects.* (New York: United Nations, 1972).
39. Preston, A. "A United Kingdom Approach to the Application of ICRP Standards to the Controlled Disposal of Radioactive Waste Resulting from Nuclear Power Programs," in *Environmental Aspects of Nuclear Power Stations* (Vienna: International Atomic Energy Agency, 1971).
40. Inoue, Y. and S. Morisawa. "On the Selection of a Ground Disposal Site for Radioactive Wastes," *Health Phys.* **26**, 53 (1974).
41. Voilleque, P. G. and C. A. Pelletier. "Comparison of External Irradiation and Consumption of Cow's Milk as Critical Pathways for ^{137}Cs, ^{54}Mn and ^{144}Ce-Pr Released to the Atmosphere," *Health Phys.* **27**, 189 (1974).
42. Porter, C. R., C. R. Phillips, M. W. Carter and B. Kahn. "The Cause of Relatively High ^{137}Cs Concentrations in Tampa, Florida

Milk," in *Radioecological Processes Concentration Processes*, B. Aberg and F. P. Hungate, Eds. (Oxford: Pergamon Press, 1967).

43. Ekman, L., A. Eriksson, L. Fredrickson and U. Greitz. "Studies on the Relationship Between Iodine-131 Deposited on Pasture and its Concentration in Milk," *Health Phys.* **13**, 701 (1967).

44. "Draft Regulatory Guides: Implementation of Numerical Guides for Design Objectives and Limiting Conditions for Operation to Meet the Criterion as Low as Practicable," Docket RM-50-2 (Washington, D.C.: U.S. Atomic Energy Commission, 1974).

45. Cosslett, P. and R. E. Watts. "The Removal of Radioactive Iodine and Strontium from Milk by Ion Exchange," UK AEA Report AERE-R2881 (Harwell, 1959).

46. Tadmor, J. "Consideration of Stable Iodine in the Environment in the Evaluation of Maximum Permissible Concentrations for Iodine-129," *Radiol. Health Data Reports* **12**, 611 (1971).

47. Magno, P. J., T. C. Reavey and J. C. Apidianakis. "Iodine-129 in the Environment Around a Nuclear Fuel Reprocessing Plant," (Washington, D.C.: U.S. Environmental Protection Agency, 1972).

48. Woodhead, D. S. "The Radiation Dose Received by Plaice from the Waste Discharged . . . from the Fuel Reprocessing Plant at Windscale," *Health Phys.* **25**, 115 (1973).

49. Garner, R. J. *Transfer of Radioactive Materials from the Terrestrial Environment to Animals and Man.* (Cleveland: CRC Press, 1972).

50. Federal Radiation Council. Quoted in "Milk Surveillance, October 1972," *Radn. Data Repts.* **14**, 85 (1973).

51. "Assessment of Environmental Radioactivity in the Vicinity of Shippingport Atomic Power Station," Report EPA-520/5-73-005 (Washington, D.C.: Environmental Protection Agency, 1973).

52. Kahn, B., R. L. Blanchard, H. E. Kolde, H. L. Krieger, S. Gold, W. L. Brinck, W. J. Averett, D. B. Smith and A. Martin. "Radiological Surveillance Studies at a Pressurized Water Nuclear Power Reactor," EPA Report RD 71-1 (Cincinnati, Ohio: U.S. Environmental Protection Agency, 1971).

53. Kahn, B., R. L. Blanchard, H. L. Krieger, H. E. Kolde, *et al.* "Radiological Surveillance Studies at a Boiling Water Nuclear Power Reactor," Report BRH/DER 70-1 (Washington, D.C.: U.S. Public Health Service, 1970).

54. *Air Sampling Instruments for Evaluation of Atmospheric Contaminants.* 4th ed. (Cincinnati, Ohio: American Conf. of Governmental Industrial Hygienists, 1972).

55. Tennessee Valley Authority. "Environmental Radioactivity Levels— Browns Ferry Nuclear Plant, July-Dec. 1973," Report RH-74-BFI (1974).

56. "Routine Surveillance for Radionuclides in Air and Water," (Geneva: World Health Organization, 1968).

57. *Assessment of Airborne Radioactivity, Proceedings Vienna Symposium* (Vienna: International Atomic Energy Agency, 1967).

58. Godbold, B. C. and J. K. Jones, Eds. *Radiological Monitoring of the Environment.* (Oxford: Pergamon Press, 1965).

59. "Guide for Environmental Surveillance Around Nuclear Facilities,"

Report NF-67-8 (Rev. 1), Public Health Service, U.S. Department of Health, Education and Welfare (1967).

60. "La Radioprotection du Milieu devant le Developement des Utilisations Pacifiques de l'Energie Nucléaire," (Paris: Soc. Française de Radioprotection, 1968).

61. Reinig, W. C., Ed. *Environmental Surveillance in the Vicinity of Nuclear Facilities.* (Springfield, Illinois: C. C. Thomas, 1970).

62. Adams, J. A. S. and W. M. Lowder, Eds. *The Natural Radiation Environment* (Chicago: University of Chicago Press, 1964).

63. "Environmental Radioactivity Surveillance Guide," Report ORP/SID 72-2 (Washington, D.C.: U.S. Environmental Protection Agency, 1972).

64. Fitzgerald, J. J. *Applied Radiation Protection and Control.* (New York: Gordon and Breach, 1969).

65. Harley, J. H., Ed. "HASL Procedures Manual," Report HASL-300 (New York: U.S. Atomic Energy Commission, 1972), Suppl. 2 (1974).

66. "Radioassay Procedures for Environmental Samples," U.S. Public Health Service Publication No. 999-RH-27 (Rockville, Maryland: National Center for Radiological Health, 1967).

67. Brinck, W., K. Gross, G. Gels and J. Partridge. "Special Field Study at the Vermont Yankee Nuclear Power Station," preprint (1974).

68. "Draft Regulatory Guides for Implementation-Attachment to Concluding Statement of the Regulatory Staff (As Low As Practicable)," Docket No. RM-50-2 (Washington, D.C.: U.S. Atomic Energy Commission, 1974).

69. Budnitz, R. J. "Tritium Instrumentation for Environmental and Occupational Monitoring—A Review," *Health Phys.* 26, 165 (1974).

70. Freiling, E. C., Ed. *Radionuclides in the Environment.* Advances in Chemistry Series 93 (Washington, D.C.: American Chemical Society, 1970).

71. Matuszek, J. M., C. J. Paperiello, C. O. Kunz, J. A. Hutchinson and J. C. Daly. "Permanent Gas Measurements as Part of an Environmental Surveillance Program," in *Environmental Surveillance Around Nuclear Installations* (Vienna: International Atomic Energy Agency, 1973).

72. Fessler, H., L. A. König, K. Nester and M. Winter. "Preliminary Experience Gained in Monitoring Krypton-85 Received in the Neighborhood of the Karlsruhe Reprocessing Plant," in *Environmental Behavior of Radionuclides Released in the Nuclear Industry.* (Vienna: International Atomic Energy Agency, 1973).

73. Jarvis, A. N. and D. G. Easterly. "Measuring Radioactivity in the Environment—The Quality of the Data," *Nucl. Technol.* 24, 447 (1974).

74. Karches, G. J., H. E. Kolde, W. L. Brinck, R. L. Shearin and C. R. Phillips. "Field Determination of Dose from ^{133}Xe in the Plume from a Pressurized Water Reactor," in *Rapid Methods for Measuring Radioactivity in the Environment* (Vienna: International Atomic Energy Agency, 1970).

75. Kunz, C. O. "Separation Techniques of Reactor-Produced Noble Gases," Noble Gases Symposium, Las Vegas, Nevada (1973).

76. Cummings, S. L., R. L. Shearin and C. R. Porter. "A Rapid Method for Determining ^{85}Kr in Environmental Air Samples" in *Rapid Methods for Measuring Radioactivity in the Environment* (Vienna: International Atomic Energy Agency, 1971).

77. Cowser, K. E. "Current Practices in the Release and Monitoring of ^{131}I at NRTS, Hanford, Savannah River and ORNL," Report ORNL-NSTC-3, Oak Ridge National Laboratory (1964).

78. Goldstein, N. P., K. H. Sun, J. L. Gonzalez. "The Measurement of Extremely Low-Level Radioiodine in Air," *Nucl. Technol.* **23**, 328 (1974).

79. Eichholz, G. G., A. N. Galli and L. W. Elston. "Problems in Trace Element Analysis in Water," *Water Resources Res.* **2**, 561 (1966).

80. Eichholz, G. G., T. F. Craft and A. N. Galli. "Trace Element Fractionation by Suspended Matter in Water," *Geochim. Cosmochim. Acta* **31**, 737 (1967).

81. Theodorsson, P. "Improved Tritium Counting Through High Electrolytic Enrichment," *Internat. J. Appl. Rad. Isotopes* **25**, 97 (1974).

82. Terpilak, M. S. and B. L. Jorgensen. "Environmental Radiation Effects of Nuclear Facilities in New York State," *Radn. Data Repts.* **15**, 375 (1974).

83. Gabay, J. J., C. J. Paperiello, S. Goodyear, J. C. Daly and J. M. Matuszek. "A Method for Determining Iodine-129 in Milk and Water," *Health Phys.* **26**, 89 (1974).

84. Mitchell, N. T. "Monitoring of the Aquatic Environment of the United Kingdom and Its Application to Hazard Assessment," *Environmental Contamination by Reactor Materials* (Vienna: International Atomic Energy Association, 1969).

85. Porter, C. R., R. J. Augustine, J. M. Matusek and M. W. Carter, Eds. "Procedures for Determination of Stable Elements and Radionuclides in Environmental Samples," Publ. No. 999-RH-10 (Washington, D.C.: U.S. Public Health Service, 1965).

86. "Measurements of Radionuclides in the Environment-Sampling and Analysis of Plutonium in Soil," Regulatory Guide 4.5 (Washington, D.C.: U.S. Atomic Energy Commission, 1974).

87. Easterly, D. G., I. B. Brooks and J. K. Hasuike. "Development of Ion Exchange Processes for the Removal of Radionuclides from Milk," Report RO/EERL 71-7 (Rockville, Maryland: U.S. Environmental Protection Agency, 1971).

88. Hansen, W. G., J. E. Campbell, J. H. Fooks, H. C. Mitchell and C. H. Eller. "Farming Practices and Concentrations of Fission Products in Milk," Publ. No. 999-R-6 (Washington, D.C.: U.S. Public Health Service, 1964).

89. Csupka, S. and M. Petrashova. "^{90}Sr and ^{137}Cs Content in Soils, Cereals, Food Crops and Milk During 1963-1971," *Atomnaya Energiya* **36**, 226 (1974).

90. Daly, J. C., S. Goodyear, C. J. Paperiello and J. M. Matuszek. "Iodine-129 Levels in Milk and Water Near a Nuclear Fuel Reprocessing Plant," *Health Phys.* **26**, 333 (1974).

91. "Full-Scale System for Removal of Radiostrontium from Milk," U.S. Public Health Service Publication, No. 999-RH-28 (Rockville, Maryland: National Center for Radiological Health, 1967).

92. Kahn, B., G. K. Murthy, C. Porter, G. B. Hagee, G. J. Karches and A. S. Goldin. "Rapid Methods for Estimating Fission Product Concentrations in Milk," Publ. No. 999-R-2 (Washington, D.C.: U.S. Public Health Service, 1963).

93. Porter, C. R., M. W. Carter, B. Kahn and E. W. Pepper. "Rapid Field Method for the Collection of Radionuclides from Milk," in *Radiation Protection*, W. S. Snyder, Ed. (New York: Pergamon Press, 1968), pp. 339-346.

94. "Measurements of Radionuclides in the Environment—Analysis of I-131 in Milk," Regulatory Guide 4.3 (U.S. Atomic Energy Commission, 1973).

95. "Guide and Checklist for the Development and Evaluation of State Radiological Emergency Response Plans for Fixed Nuclear Facilities," Interim Guidance (Washington, D.C.: U.S. Atomic Energy Commission, 1973).

96. Schwibach, J. "Emergency Planning at Nuclear Installations as a Basis for Rapid Monitoring Programs," in *Rapid Methods for Measuring Radioactivity in the Environment* (Vienna: International Atomic Energy Agency, 1971).

97. Farmer, F. R. "Siting Criteria—A New Approach," in *Containment and Siting of Nuclear Power Plants* (Vienna: International Atomic Energy Agency, 1967).

ADDITIONAL REFERENCES

Alkaline Earth Metabolism in Adult Man, ICRP Publication 20 (Oxford: Pergamon Press, 1972).

Attix, F. H. and W. C. Roesch, Eds. *Radiation Dosimetry*, Vol. I (New York: Academic Press, 1968).

Bleher, G. L. and D. J. Holloway. "Selection of Monitors for Airborne Particulate Radioactivity," Paper presented at ANS National Meeting, Chicago (1973).

Bolch, W. E. and E. F. Gloyna. "Radioactivity Transport in Water—Behavior of Ruthenium in Algal Environments," *Env. Health Eng. Res. Lab. Technol. Report 4*, University of Texas (1963).

Bovard, P., J. Delmas, A. Grauby and P. Benard. "The Transfer of Radiostrontium and Radiocesium from the Soil to the Vine and to Wine" in *La Radioprotection du Milieu devant de Developpement des Utilisations Pacifiques de l'Energie Nucléaire* (Paris: Soc. Francaise de Radioprotection, 1968).

Brisbin, I. L., R. T. Beyers, R. W. Dapson, R. A. Geiger, J. B. Gentry, J. W. Gibbons, M. H. Smith and S. K. Woods. "Patterns of Radiocesium in the Sediments of a Stream Channel Contaminated by Production Reactor Effluents," *Health Phys.* 27, 19 (1974).

Budnitz, R. J. "Krypton-85: A Review of Instrumentation for Environmental Monitoring," Report LBL-1779, Lawrence Berkeley Lab., University of California, Berkeley (1973).

Clarke, R. H. "Physical Aspects of the Effects of Nuclear Reactors in Working and Public Environments," Berkeley Nuclear Laboratories, Central Electricity Generating Board, England (1973).

Currie, L. A. "The Measurement of Environmental Levels of Rare-Gas Nuclides and the Treatment of Very Low-Level Counting Data," *Trans. IEEE* NS-19, 119 (1971).

Dutton, J. W. R. and N. T. Mitchell. "Rapid Methods for Specific Radio-nuclide Analysis and their Application to Emergency Conditions," *Proc. Internat. Symp. Rapid Methods Measurement of Radioactivity Environment* (Vienna: International Atomic Energy Agency, 1971).

"Environmental Impact Monitoring of Nuclear Power Plants: Sourcebook of Monitoring Methods," AIF/NESP-004 (New York: Atomic Industrial Forum, 1975).

Holm, N. W. and R. J. Berry, Ed. *Manual on Radiation Dosimetry* (New York: Marcel Dekker, 1970).

"Interim Licensing Policy on As Low as Practicable for Gaseous Radio-iodine Releases from Light-Water-Cooled Nuclear Power Reactors," Regulatory Guide 1.42 (Rev. 1)(Washington, D.C.: U.S. Atomic Energy Commission, 1974).

Johihara, T. "Environmental Radiological Monitoring System at Nuclear Installations," *Health Phys.* 13, 549 (1967).

Jones, J. K., G. Lewis, H. C. Orchard, M. J. Owers and B. W. Skelcher. "The Experience of the Central Electricity Generating Board in Monitoring the Environment of its Nuclear Power Stations," *Health Phys.* 24, 619 (1973).

Kelleher, W. J. and H. R. Prins. "Significance of Stable Iodine-127 in Milk," *Radn. Data Repts.* 15, 567 (1974).

Krieger, H. L. and S. Gold. "Procedures for Radiochemical Analysis of Nuclear Reactor Aqueous Solutions," EPA Report EPA-R4-73-014 (Cincinnati, Ohio: U.S. Environmental Protection Agency, 1973).

Lieberman, J. A., E. D. Harward and C. L. Weaver. "Environmental Surveillance Around Nuclear Power Reactors," *Radiol. Health Data Repts.* 11, 325 (1970).

Lloyd, R. D. "Cesium-137 Half-Times in Humans," *Health Phys.* 25, 605 (1973).

Luetzelschwab, J. W. and C. Gignac. "Ability of an Ion Exchange Resin to Retain ^{137}Cs in the Presence of Ca^{2+} Ions at Fast Flow Rates," *Health Phys.* 27, 109 (1974).

Martin, J. A. and C. B. Nelson. "Calculations of Dose, Population Dose and Health Effects due to Boiling Water Nuclear Power Reactor Radio-nuclide Emissions in the United States during 1971," *Radn. Data Reports* 15, 309 (1974).

McLaughlin, S. and H. L. Beck. "Environmental Radiation Dosimetry for Nuclear Facilities and Problems," *IEEE Trans.* NS-20, 36 (1973).

"Measurement of Low-Level Radioactivity," ICRU Report 22 (Washington, D.C.: ICRU Publications, 1972).

"Measuring and Reporting of Radioactivity in the Environs of Nuclear Power Plants," U.S. Atomic Energy Commission Regulatory Guide 4.1 (1973).

Nuclear Power and the Environment (Vienna: International Atomic Energy Agency, 1973).

"Radiation Protection Procedures," Safety Series No. 38 (Vienna: International Atomic Energy Agency, 1973).

Recommendations of the ICRP, ICRP Publication 9 (London: Pergamon Press, 1966).

"Report on Releases of Radioactivity in Effluents and Solid Waste from Nuclear Power Plants for 1972," Directorate of Regulatory Operations (Washington, D.C.: U.S. Atomic Energy Commission, 1973).

"Results of Measurements of Iodine-131 in Air, Vegetation, and Milk at Three Operating Reactor Sites," Directorate of Regulatory Operations (Washington, D.C.: U.S. Atomic Energy Commission, 1973).

Reynolds, T. D. and E. F. Gloyna. "Radioactivity Transport in Water— Transport of Strontium and Cesium by Stream and Estuarine Sediments," *Env. Health Eng. Res. Lab. Tech. Report 1*, University of Texas (1963).

Rossano, A. T., Ed. *Air Pollution Control: Guidebook for Management* (Stanford, Connecticut: Environmental Sciences Service Division, ERA Inc., 1969).

Stover, B. J. and W. S. S. Jee, Eds. *Radiobiology of Plutonium* (Salt Lake City: The J. W. Press, 1972).

"Tentative Method for Analysis for Plutonium Content of Atmospheric Particulate Matter," *Health Lab. Sci.* 7, 141 (1970).

"Tentative Method of Analysis for Tritium Content of the Atmosphere," *Health Lab. Sci.* 8, 107 (1971).

Thompson, S. E., C. A. Burton, D. J. Quinn and Y. C. Ng. "Concentration Factors of Chemical Elements in Edible Aquatic Organisms," U.S. Atomic Energy Commission Report UCRL-50564, Rev. 1 (Livermore, California: Lawrence Livermore Laboratory, 1972).

Waite, D. A. "An Analytical Technique for Distributing Air Sampling Locations Around Nuclear Facilities," Report BNWL-SA-4534, BNWL-SA-4676 (Richland, Washington: Battelle-Northwest, 1973).

CHAPTER 9

POWER PLANT SITING

INTRODUCTION

The selection of a suitable site for the location of a nuclear power plant lies at the heart of any evaluation of its environmental impact, and a considerable methodology for such a selection process is being developed. Appropriate location of a plant may determine its economic viability and may contribute to the safety of its operation. Many of the factors involved in site selection are not peculiar to nuclear facilities but are associated with any power plant or, for that matter, with the location of any large industrial enterprise (1). In fact, the long-range demand for suitable sites is creating enough problems to prompt many governmental and local agencies to demand long-term planning and identification of such sites. It must also be recognized that detailed evaluation of alternative sites is a costly and time-consuming process.

In order to be recognized as suitable, a site must meet several criteria that can be categorized as technical, socioeconomic and environmental (2,3). The technical factors involve the following determinations:

1. need for power plant
2. proximity to load centers
3. size of generating system
4. type of power plant
5. number of steam supply systems
6. highway, rail and water access
7. lead time considerations
8. cooling water needs and availability
9. fuel supply
10. plant safety and security.

The physical and environmental factors include the following:

1. land requirements for the plant, fuel storage, waste disposal, switchyard and transmission lines

437

2. condenser cooling water, reservoir needs, water resources, water quality and thermal pollution
3. cooling tower effects and alternatives
4. foundations, which includes geology, water table, hydrology, seismology and soil mechanics
5. hydrology: interference with drainage patterns, flooding hazards and contamination
6. meteorology and air pollution, which include a study of normal weather patterns, storm and tornado frequency, and the site's relationship to built-up areas
7. radiological safety including land use and access restrictions
8. special site conditions encountered with ocean frontage, underground and offshore sites.
9. effects on fish and wildlife
10. effect of construction activities.

Socioeconomic factors include the following, not necessarily in order of importance:

1. effect on community planning
2. taxes and land values
3. amenities and services for employees
4. amenities and recreational considerations for the public
5. aesthetics, such as appearance of the plant and power lines
6. wildlife protection
7. preservation of archaeological and historical sites
8. present and projected population patterns
9. multipurpose plant siting
10. economic impact of air or water pollution
11. industrial development
12. transportation needs.

These factors must be considered for any industrial plant in order to evaluate the relative merits of several alternative sites, to balance economic, social and technical factors, to select the "best available" site, and to eliminate from consideration sites that are found undesirable for one or another significant reason. The best available site will rarely be ideal from all aspects, and in most cases the pros and cons will have to be balanced to arrive at a solution of greatest benefit to the public, at acceptable cost, and with the least environmental impact. Normally, initial decisions will be based on purely economic considerations. Considerable judgment is needed in applying the various criteria. In the case of nuclear power plants, additional considerations include plant safety and security, and possible population hazards following postulated accidental releases of radioactive materials.

Since the site selection is closely linked to any determination of environmental impact, the methodology of siting involves an iterative approach to test various alternatives as to location, type of power plant, cooling systems, land requirements, and population exposure from all causes. In general, a preliminary screening process can eliminate a number of superficially attractive locations. The remaining candidate sites must then be evaluated by applying weighting factors called variously "importance factors" (4), "weighting factors" (5), "site selection criteria" (3) or "site criteria" (6). Population dose in case of accidents would normally be considered a secondary criterion to be applied after all other criteria related to plant safety and economics have been met.

In the systematic site evaluation method described by Fischer and Ahmed (4), the environmental assessment proceeds to assign "importance factors" to all components in each category of selection factors. Table 115 shows such a table of importance or weighting factors proposed by Beer (3). Obviously, there is no absolute method for assigning importance

Table 115. Site Selection Criteria and Weighting Factors (3)

Criteria Number	Selection Criterion	Weighting Factor
1	Availability of land	3
2	Compatibility of land use	3
3	Availability of water	3
4	Cooling system development	2
5	Hydrology	2
6	Meteorology	1
7	Geotechnical	2
8	Seismology	2
9	Archaeology	2
10	Terrestrial biology	3
11	Aquatic biology	3
12	Population density	3
13	Exclusion radius consideration	3
14	Proximity to load center	1
15	Accessibility to transmission system	1
16	Accessibility to transportation system	1
17	Radiological impact	2
18	Health and safety	3
19	Social and economic implications and acceptance	3

to different environmental and technical factors; in fact, this constitutes the most obvious area of conflict between technologists and environmentalists. Fischer and Ahmed describe the "Delphi Method," an iterative questioning and answering procedure, as a means of reconciling differences in opinion among members of the siting team. In practice, this must simply reflect the best attainable compromise among all participants.

The positive or negative effect of the construction of a plant at a given site can then be expressed quantitatively by an "impact quotient," IQ, computed by (4):

$$IQ_j = \sum_{i=1}^{n} (EQ_0)_i \, W_i - \sum_{i=1}^{n} (EQ_1) \, W_i \tag{80}$$

$$i = 1, 2 \ldots n, \quad j = 1,2,3,4$$

where: EQ_0 = projected environmental quality without plant
$\quad\quad$ EQ_1 = projected environmental quality with plant
$\quad\quad$ W = importance factor
$\quad\quad$ n = number of components in each category j.

In order to arrive at a decision in site selection the attribute values, or the sum of IQ in each category are determined. To rank sites in the order of increasing environmental cost the utility function, U, which reflects the desirability of a particular alternative, is evaluated.

The utility function can be expressed as

$$U_j = P_1 \cdot u(j_1) + P_2 \cdot u(j_2) + P_3 \cdot u_{j_3} + P_4 \cdot u_{j_4} \tag{81}$$

where u = attribute value, and P = preference index subject to $0 < P_j < 1$, $\Sigma P_j = 1$, where the values for the preference values reflect an agreed weighting for the categories among the selection team members.

An alternative way to rank all impact factors involves assigning numerical values to site licensability and intercriteria weighting factors (3,8) and tabulating the results in an impact analysis matrix as a basis for decision. Table 116 shows another way of obtaining a numerical ranking.

No amount of mathematical formulation can disguise the fact that most of the input factors reflect subjective judgments by the participants. For that reason such computations are useful only for internal comparison of alternatives among a group of persons that are basically in agreement on the relative importance of the various factors involved. No site selection process can be considered acceptable that does not take into account all of the factors listed above and does not arrive at a publicly acceptable answer to any problem that might be anticipated in any of the categories. To this must be added the assessment of risk associated with any site (9-11), which will be discussed later.

For U.S. power plants, most of the selection factors are outlined in government regulations (2) and guidelines (12,13). Some of these factors will now be discussed.

Table 116. Example of Site Evaluation (Hypothetical Site) (5)

	Weighting Factor (WF)	Rating (R)	Evaluation Points (WF x R)
Capability of cooling system development	3	5	15
Proximity to load center	1	3	3
Land availability	3	5	15
Compatibility of land use	3	5	15
Resource consumption			
Water consumption	3	3	9
Land utilization, amount	1	1	1
Land utilization, critical environmental importance	2	5	10
Accessibility			
To rail transportation	1	5	5
To highway transportation	1	5	5
To water transportation	1	5	5
To a port	1	3	3
Suitable soil foundation conditions	1	3	3
Cost of transmission connections	1	5	5
Environmental impact			
Water quality impact	3	5	15
Terrestrial biological impact	3	3	9
Aquatic biological impact	3	2	6
Construction effects	1	5	5
Aesthetics	1	3	3
Air quality impact	2	5	10
Noise impact	1	5	5
Transmission system routing	1	5	5
Impact on fuel delivery corridors	2	3	6
Process water supply	1	5	5
Population density	2	5	10
Socioeconomic impact			
Community services	1	5	5
Area economy	1	3	3
System compatibility	3	1	3
Site Evaluation Quality (Total)			184

Key 1 PPSEI Weighting Factors		Key 2 PPSEI Rating Scale	
Weight	Criteria	Rating	Criteria
1	Least important	1	Very poor site
2	Moderately important	2	Poor site
3	Highly important	3	Fair site
		4	Good site
		5	Very good site

TECHNICAL FACTORS

It may be assumed that no utility will contemplate the enormous cost and commitment of personnel and resources involved in designing and building a nuclear plant without having established a projected need for the power to be generated and without due consideration of the cost and availability of alternative, conventional power sources.

For a nuclear steam supply, the type of reactor and its manufacturer will be determined by considerations of cost and availability, as well as by long-range projections regarding fuel availability, enrichment requirements and reprocessing needs. Environmental considerations will only affect these decisions marginally, except in regard to availability and need for cooling water, or the cost and security of spent fuel shipments that might be considered at this stage. Any up-to-date, state-of-the-art reactor design may be assumed to comprise all required engineered safeguards features and to generate effluents at an "as low as practicable" level.

Site selection then will concentrate on the suitability of a given site with respect to its capacity to accommodate one or more nuclear power plants, its proximity to load centers by direct transmission or as part of a balanced grid, its access to adequate cooling water by direct use of natural stream flow or with construction of any necessary reservoir or lake, access to adequate road, rail or water transportation links, and, finally availability of sufficient land to accommodate the power plant, any cooling towers, any cooling water reservoir, the switchyard, and enough surrounding ground to meet requirements for an exclusion area, at acceptable cost. Satisfaction of all of these requirements will rapidly shorten the list of potential sites in any preliminary screening. Of these factors the need for adequate land to accommodate future plants obviously depends on other considerations regarding future growth. Road connection by heavy-duty highway is essential both for construction activities and for waste shipments; rail or water transportation may be desirable for shipping in some of the more massive components of the plant, but is not essential and in many cases is impractical to attain. All the other factors mentioned obviously must be satisfied.

PHYSICAL AND ENVIRONMENTAL FACTORS

Land Requirements

In addition to sheer availability of sufficient land, the site must also be evaluated in terms of the suitability of the land as a construction site, the changes in the ecology resulting from its projected use, the economic and aesthetic impact of changing the previous land use, and the effects on surrounding areas with regard to drainage, ecology, interference with crops, interference with existing activities, traffic patterns or river access.

The environmental impact of power plant construction on land use and ecology was discussed in Chapter 2, which also described the environmental consequences of transmission line construction. As experience is gained in the operation of nuclear power plants and in the effectiveness of the various engineered safeguards, the total land requirements to meet exclusion zone criteria (see below) will probably be reduced, leaving transmission line right-of-way as perhaps the largest item of land to be leased or purchased. The plant site proper will probably continue to require 600-1000 acres (240-400 hectares) for a 1200 MWe plant in the United States, where land is relatively more plentiful than in other industrialized countries.

In view of the cost of the site selection and licensing process and the relatively minor increase in plant area associated with locating more than one nuclear plant at any one location, there is a strong incentive to use a given site for multiple reactor plants. Such a trend has been evident for some years, culminating in power complexes like Pickering or Browns Ferry that house three or four high-power nuclear generating plants. Additional land requirements for cooling ponds or cooling towers can usually be accommodated within the exclusion area for a single plant.

Cooling Water

Adequate cooling water must be available and the site must be suitable for dissipating the heat in an environmentally acceptable fashion. The various methods available for this purpose have been discussed in Chapter 6. The particular site may determine the type of cooling tower usable for the existing conditions of humidity, wind velocity, temperature inversion frequency, ultimate heat sink, and water economy. Within a given geographical area access to cooling water may often be a dominant selection criterion, but the type of cooling used will usually be determined by the site selected, not the other way around. Unusually restrictive water quality criteria may make certain areas less desirable than others, especially close upstream to intakes to municipal water treatment plants.

Ready access to cooling water and an essentially unlimited heat sink are usually advanced as the main arguments in favor of shore sites or offshore island locations. This will be discussed in a later section.

Geology and Soils

As for any other major construction job, it is essential that the soil be uniform, competent, free of cavities or landslide potential and meet all normal requirements for supporting the appreciable weight associated with such massive structures as are normally found in reactor buildings, fuel storage buildings, cooling towers, and associated structures. Sites with unfractured bedrock generally have suitable foundation conditions. If bedrock sites are not available, sites in areas having low liquefaction potential should be sought. This requires extensive geological and engineering investigations, including drilling at depth to determine the static and dynamic properties of the material underlying the site (2).

Ground motion from any cause, such as settling, ground water effects and seismic events, is of significance. Construction experience extending over many centuries has shown that it is possible to design buildings capable of long-term stability in swampy ground, subsidence locations and earthquake-prone regions, but such special construction is costly and worthwhile only in regions such as Andean countries, where no alternative site may be feasible.

Earthquakes can cause soil instability due to ground disruption, such as fissuring, liquefaction, differential consolidation, and cratering which is not directly related to surface faulting. The same effects can occur in many areas even without the triggering mechanism of an earthquake. These areas of potential surface disruption may result from natural features such as underground solution cavities or human activities such as mining or heavy withdrawal of petroleum or ground water (14).

Karst terrain is characterized by sinkholes, caves, and similar solution openings. Surface streams are few and typically flow into sinkholes to join underground drainage systems. The bedrock is characteristically limestone, dolomite, or marble. Areas of karst topography have notably irregular soil-bedrock interfaces. Thick soil overburden will fill enlarged solution joints, but thin soil often covers only the intervening rock. The active solution of rocks is common and can result in continued development of joints. Sudden surface collapse sometimes occurs, posing very real danger in the use of such areas. Regions of karst terrain often have drainage systems that may be locally difficult to define and that may allow surface contaminants to migrate rapidly into the main zone of saturation.

Other types of subsidence may occur from one or more of several causes including withdrawal of fluids, application of water to moisture-deficient deposits above the water table, drainage of wetlands, and mining operations. In the United States, particularly the West, subsidence of appreciable magnitude and area has resulted from withdrawal of ground water and oil, including drainage of wetlands, and by application of water to dry lands. For example, drainage of wet delta deposits of the Sacramento-San Joaquin River, begun in 1850, has caused a subsidence of more than 15 feet. Adding surface water to an area, as by irrigation, can also cause subsidence by a phenomenon known as hydrocompaction. Hydrocompaction has been reported in California, Missouri, Montana, Washington and Arizona, where subsidence of as much as 6 feet after wetting has occurred. Subsidence due to withdrawal of fluids is by far the most common type of man-made regional subsidence. In the Houston, Texas area, 5 feet of subsidence has occurred over a 4000-square-mile area due to heavy ground water withdrawal and oil-field development. This type of subsidence has also been documented in Arizona, Nevada, and particularly California, where subsidence of over 20 feet has occurred near Fresno. These phenomena may be easy to identify where they are currently active, but the planner must also consider the potential for future subsidence in areas with developed ground water supplies (14).

The instability of natural surface slopes will not generally pose any direct hazard to a nuclear power plant that has been well engineered to the environment. An exception might be the case of the foothill areas at the base of mountainous terrain. For example, on the west side of Mount Rainier in Washington, several geologically recent mudflows have traveled over 30 miles and carried up to 800 million cubic yards of material. The areas covered by the mudflows are currently inhabited by over 30,000 people. Although large mudflows are rare, disruption of access routes, transmission lines, hydraulic structures such as dams, intake-outfall structures, and cooling ponds can easily occur from quite minor slope instability and should be considered in site planning.

Early consideration should be given to the amount of excavation required for placement of the plant and, of no less importance, to the placement and stabilization of the excavated spoil materials. Disposal of quantities of a few thousand cubic yards at some sites may be relatively easy. However, large excavations may require the establishment of agreements and/or considerable expense along with potentially adverse environmental impact. In some coastal areas select materials could be dumped at sea but in other cases all such disposal may be prohibited.

In some cases ground instability has been discovered after site selection and construction commenced. Remedial solutions in that case may involve backfilling of soil, pile driving, cement grouting of porous base material, and soil stabilization, all standard methods but potentially costly.

Seismic Characteristics

Natural disasters such as earthquakes, volcanic action, landslides, flooding, and tsunamis are potentially so catastrophic that their possible occurrence at any site could be considered as sufficient cause to exclude the site from further consideration. A site at which a geologic event has either induced a great disaster in the past or at which a disaster might be induced in the future, or a site at which geologic hazards might increase the scale of a disaster, could preclude public acceptance. All areas of potential geologic hazard, however, cannot be avoided. The probability of occurrence of an earthquake or other geologic hazard and its potential consequences on a nuclear facility must be taken into account in the selection of any site.

Microseismic events occur continually at all points on the earth's surface, with no visible effect on any well-built structure. Other locations, especially those close to volcanic areas and active crustal faults, have a history of earthquake activity. Although engineering experience in countries such as Japan and Mexico has shown that it is possible to construct earthquake-proof concrete buildings capable of withstanding the ground shock associated with severe earthquakes (15-17), prudence dictates that, wherever possible, any sites including capable faults should be ruled out as sites for nuclear power stations. Sites within about 5 miles (8 km) of a surface capable fault, greater than 1000 ft (300 m) in length, are generally not suitable for a nuclear power station.

Two types of investigations are required: a compilation of all recorded earthquakes in the region to a radius of up to 100 miles, and, where appropriate, a geological and seismological study of surface and bedrock to locate the presence or absence of any geologically recent faulting. For any site the combination of the most severe motions from any earthquakes that may affect the site is defined as the *design earthquake*, which is then used to test the sufficiency of the proposed plant design from this aspect (16-22).

Several empirical scales have been used to describe the relative violence of ground motion during earthquakes. These scales, limited to qualitative assessment of the shock because they are not based upon recorded values, correlate the observed effects of ground motion with ranges of accelerations. The oldest widely used scale was the Rossi-Forel Scale, developed in 1883. The more recent Mercalli Intensity Scale, modified by Wood and Newman in 1931, is still used to evaluate earthquakes. A shortened version of the Modified Mercalli Intensity Scale is shown in Table 117.

In 1935 Richter devised an arbitrary magnitude scale for earthquakes based upon the maximum amplitude of a standard seismometer record

Table 117. The Modified Mercalli Intensity Scale

	Modified Mercalli Intensity	Ground Acceleration
II	Felt by a few persons at rest, especially on upper floors; suspended objects may swing.	
III	Felt noticeably indoors, but not always recognized as a quake; standing autos rock slightly.	0.005 g
IV	Felt indoors by many, outdoors by a few; at night some awaken; dishes, windows, doors disturbed.	0.01 g
V	Felt by most people; some breakage of dishes, windows, and plaster; tall objects disturbed.	
VI	Felt by all; many frightened and run outdoors; falling plaster and chimneys; damage small.	0.05 g
VII	Everybody runs outdoors; damage to buildings varies depending on quality of construction.	0.1 g
VIII	Panel walls thrown out of frames; fall of walls, monuments; sand and mud ejected; drivers disturbed.	
IX	Buildings shifted off foundations, cracked, out of plumb; ground cracked, underground pipes broken.	0.5 g
X	Most masonry and frame structures destroyed; ground cracked; rails bent; landslides.	1 g
XI	New structures standing; bridges destroyed; fissures in ground; pipes broken; landslides.	
XII	Damage total; waves seen in ground surfaces; lines of sight distorted; objects thrown into air.	5 g

taken 100 km from the earthquake epicenter. This Richter scale is defined as

$$\text{``Magnitude''} \; M = \log a_0 + 1.66 \log \Delta + 2.0 \qquad (82)$$

where a_0 = maximum ground amplitude for surface waves of 20 sec
period

Δ = distance from epicenter to station (in degrees of arc
along great circle over Earth's surface).

The total energy, E, released in seismic waves in ergs is related to the
Richter magnitude M by

$$\log E = 11.4 + 1.5M \tag{83}$$

Magnitude scales do not supersede intensity scales, but are complementary
to them.

For an earthquake whose intensity I on the Modified Mercalli Scale is
known at a given location, the local ground acceleration a, in cm/sec^2,
can be found from

$$I = 3 \log a + 1.5 \tag{84}$$

Table 118 shows a conversion table for seismic scales. Figure 86 (23)
shows the relationship between ground acceleration and earthquake
intensity.

Two consequences flow from such studies: the proposed reactor plant
and containment structure must be studied to establish that they can
readily withstand the acceleration associated with the design earthquake.

Table 118. Conversion Table for Seismic Scales (18)

Seismic Scale MSK 1964	Scale of the Inst. of Physics of the Earth, Soviet Acad. of Sciences 1952	American Modified Mercalli Scale (MM) 1931	Japanese Scale 1950	Rossi-Forel Scale 1873	European Mercalli-Cancani-Sieberg Scale 1917
I	1	I	0	I	I
II	2	II	1	II	II
III	3	III	2	III	III
IV	4	IV	2,3	IV	IV
V	5	V	3	V-VI	V
VI	6	VI	4	VII	VI
VII	7	VII	4,5	VIII	VII
VIII	8	VIII	5	IX	VIII
IX	9	IX	6	X	IX
X	10	X	6	X	X
XI	11	XI	7	X	XI
XII	12	XII	7	X	XII

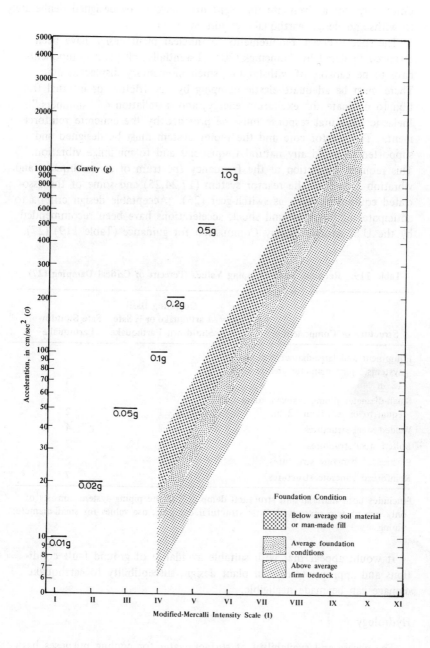

Figure 86. Relation between seismic intensity and ground acceleration (23).

Conversely, for a given site the plant may have to be designed deliberately to withstand design earthquake conditions (18). The special design requirements for nuclear plant design have been reviewed in detail by Lomenick (20). Essentially, all plant components have to be capable of withstanding small momentary displacements. There must be adequate elastic damping by dry friction or internal friction to dissipate the excitation energy, and installation of "snubbers." Inelastic structural response must be provided by the concrete reinforcement. The reactor core and the piping system must be designed and supported to avoid any natural frequencies and to minimize vibration. This requires prediction of the frequency spectrum of an earthquake and vibration testing of the reactor system (17,24,25) and some of the associated equipment, such as switch-gear (26). Acceptable design criteria to anticipate minimum ground shock accelerations have been recommended by the U.S. Atomic Energy Commission for guidance (Table 119) (27).

Table 119. Recommended Damping Values (Percent of Critical Damping)(27)

Structure or Component	Operating Basis Earthquake or ½ Safe Shutdown Earthquake	Safe Shutdown Earthquake
Equipment and large-diameter piping systems,[a] pipe diameter greater than 12 in.	2	3
Small-diameter piping systems, diameter equal to or less than 12 in.	1	2
Welded steel structures	2	4
Bolted steel structures	4	7
Prestressed concrete structures	2	5
Reinforced concrete structures	4	7

[a]Includes both material and structural damping. If the piping system consists of only one or two spans with little structural damping, use values for small-diameter piping.

It would appear that with suitable avoidance of ground fault conditions and appropriate care in plant design, susceptibility to earthquake damage can be made negligible.

Hydrology

The access and availability of surface water for cooling purposes has already been referred to, as has the need for an adequate receiving body for all liquid effluents, including blowdown water. Such a body of water must serve as the ultimate heat sink in closed-loop coolant systems or under emergency conditions, if the existing cooling systems fail; it may also be needed as a source of water in case of fire.

Hydrological assessment of the site is mainly concerned with such aspects as potential flooding or icing of the site in case of severe winter weather, excessive snow melts, or dam failures. Such problems can usually be circumvented by setting the plant grade sufficiently high, hardening the plant against ice or river flotsam, and by auxiliary pumps and shutdown procedures (28).

Ground water conditions must be such that the plant site is well drained, located well above the water table, and not subject to excessive drainage from higher ground. Ground water intrusion into excavation and storage areas is obviously undesirable and must be considered in site selection.

A safety-related water supply must be assured for normal or emergency shutdown and cooldown (13). The assurance of an adequate, long-term water supply for a nuclear power plant is dependent upon the water allocation policies of the state or political jurisdiction. In states that base their water rights on the riparian doctrine (common law), rights in surface water generally go with land rights. While upstream (higher) riparian landowners have precedence in water use, they must respect the uses that downstream riparian owners have made with the water, both traditionally and over time. If one of the downstream uses happens to be irrigation, reduction in the upstream volume flow by consumptive use of water by the nuclear power plant could leave a significant impact on land use. On the other hand, there is a possibility that upstream land use could change over the design life of the nuclear power plant and cause a reduction in the volume flow at the plant site. Such land use changes might involve industrial, commercial, and residential growth in upstream areas. Changes in the character of the watershed such as reforestation or afforestation of agricultural or pasture lands might result in a decreased water supply. These water use conventions and land use factors must be evaluated in site selection.

In states where the appropriation doctrine forms the basis of water law, land use could also influence the adequacy of the water supply. Appropriations are arranged in a hierarchy based on "first in time, first in right." This means that all the water in a stream can be appropriated and that the first priority in a given quantity goes to the one who made the earliest appropriation. During the periods of low flow, all of the water may be appropriated by those at the top of the list, leaving others without water. Location of a nuclear power plant in an arid area where appropriations already have been made for agriculture, mining, or other uses could reduce the assurance of an adequate water supply.

Utilization of ground water for nuclear power plants should not exceed the sustained yield of the ground water system or adversely affect higher priority uses. Other ground water uses to be considered are agricultural (domestic, irrigation, and livestock), industrial or municipal.

Depending upon hydrogeologic characteristics of the ground water system and the location of wells for the power plant, the ground water supply of other present and projected users in combination with the nuclear power plant may be affected adversely. Withdrawals can affect both upgradient and downgradient users by lowering the water table in confined aquifers and reducing the piezometric head in unconfined aquifers (14).

Inadvertent seepage from cooling ponds and canals or direct disposal of concentrated cooling tower effluent to the ground can result in ground water table rise. A consequence could be surface flooding of lowland areas. Depending upon the present or potential land use plans for these areas, corrective measures, such as construction of drain fields, may be required. Potential contamination of the ground water system can also result from seepage. Under less than ideal situations, engineering solutions are available to preclude inadvertent interactions with the ground water system, such as canal and pond linings, cut-off walls, slurry trenches, and bulkheads. Direct disposal of concentrated liquid wastes into the ground would not be acceptable in many locations.

Elevation of the water table in the plant vicinity may result in foundation problems because of changes in soil-bearing capacities. A rising water table may eventually intercept plant foundations, resulting in direct communication for contamination of the aquifer. Furthermore, seepage faces may be created on steep slopes, within or away from the plant vicinity, creating a potential for landslides. In the selection of sites on rivers routinely involved with ice jams, such as the Missouri and Upper Mississippi, the potential short-term reduction of capacity that may be imposed as a result of drastic lowering of available water flow must be carefully weighed.

A considerable amount of operating experience is available to justify concern over icing problems. Most nuclear power plant sites in northern latitudes are designed to use partial recirculation of heated effluent as an ice control practice. Siting on ice-prone waters requires consideration of the maximum allowable temperature gradient and the size of mixing zones imposed by state standards. Since flow reductions normally occur during icing conditions, the combination of effluent temperature limitations and mixing-zone size needs careful consideration in preliminary and final design stages. An important additional factor is the lessened surface heat transfer during winter conditions. All of these factors involve important operational considerations that may limit plant availability during critical load periods in northern latitudes.

A special case exists in regard to intake systems using submerged slots or infiltration beds. These systems are often considered for application on stream shorelines that are potentially subject to periodic ice accumulation. Since site feasibility is related to the potential for interruption of needed water supply, most careful consideration needs to be given to this aspect of the hydrology of nuclear power plant sites (14).

Installation of a sparging system that mitigates silt deposition may be necessary at sites where sedimentation is foreseen as a serious problem. Such systems agitate the water by use of jets or other means, keeping silts suspended, and hence prevent large build-up of sediment in the intake forebays, especially in idle bays. Particular attention should be given to the design of the intake structure with respect to the direction of river flow and shoreline geometry. For instance, the shape of the upstream river bank may cause a natural eddy to persist in front of the intake structure, a condition favorable to a high siltation rate. Sand bars and dikes that can result in eddy formation and otherwise impede the flow along the face of the intake trash racks should be avoided. In general, free river flow along the face of the intake system will not eliminate the suspended sediments problem, but the free flow condition will nevertheless minimize siltation (14).

Atmospheric Factors

For inland locations, meteorological conditions will rarely differ sufficiently among alternative sites to constitute a significant selection criterion. The principal requirement is that meteorological conditions at a site should provide sufficient dispersion of radioactive materials released during a postulated accident to reduce ground-level radiation exposures at the exclusion area boundaries. If the dispersion characteristics at a proposed site are unfavorable, the exclusion area may have to be unusually large to satisfy criteria on population exposure and the effluent treatment may have to be more thorough. Deep valleys and canyons may have characteristics such as low wind speeds and limited mixing depths that can retard atmospheric emissions; such locations should be avoided if unfavorable conditions may occur often enough to cause unacceptable dispersion of plant discharges to the atmosphere.

Special problems may arise if a plant is placed on a shoreline site. These arise from marked preferred wind directions that may change diurnally from onshore to offshore and restrict vertical dispersion of plumes in stable atmospheres. Vertical and horizontal dispersion of an airborne plume over a cold water surface is less than over a warm land surface because the water surface is smoother than the adjacent land, resulting in less mechanical turbulence. On the other hand, air moving from water to land will pass through a transition region where atmospheric stability decreases and atmospheric mixing increases.

Sea breezes or lake breezes are accompanied by turbulent transition-zone fumigations. These cold-water to warm-land breezes are initiated during the morning hours and may persist into the evening. Their depth rarely exceeds 1500 feet, and at their maximum strength they do not extend inland over 30 miles. A pure sea breeze is rarely observed, since a synoptic scale airflow is usually present. The airflow that accompanies the large-scale weather systems will frequently be stronger than the sea breeze and therefore can negate the sea breeze when the large-scale wind is blowing off the land. When winds from large-scale systems blow toward land, the trajectories are nearly straight lines unless channeled into deep canyons or diverted by mountains. In the absence of wind channeling conditions, measurements of the wind at a single point will reasonably represent the straight-line wind flow for long distances. This is more pronounced where the coastline is uniform, such as along the Florida coast.

The entrainment of drier air (unsaturated) will act to dissipate visible plumes produced by cooling towers. The degree of dissipation depends upon the degree of mixing with drier air and is thus dependent on the stability of the atmosphere. The greater the instability (mechanical or convective), the greater the mixing and the faster the visible plume will be dissipated by the entrainment of drier air.

At sites where the prevailing atmospheric conditions are less favorable for the dissipation of visible water droplet plumes, visibility hazards to transportation and navigation may result. In particular, environmental hazards may occur where water droplet plumes from cooling towers or ponds result in fog formation over corridors of land, sea, or air transportation. Additional hazards caused by icing may arise in areas under the influence of cooling towers if ambient air or surface temperatures below freezing are prevalent.

Sites should be avoided where the operation of plants having cooling towers, cooling ponds, or spray ponds would result in unacceptable fogging and icing.

Proximity of Industrial, Military and Transportation Facilities

In choosing a plant site it is common sense to avoid locations in the approaches of civil or military airfields, in the vicinity of factories making explosives or toxic chemicals, or near any other activity that may potentially affect the plant in any hazardous fashion. The take-off and landing patterns of aircraft, including training flights, must be considered in choosing a plant site, particularly where tall structures, such as natural-draft cooling towers or exhaust stacks are concerned. A specific analysis of such factors as frequency and type of aircraft movement, flight patterns, local meteorology, and topography should be performed for (1) sites located within 5 miles of any existing or projected commercial or military airport, (2) sites located between 5 and 10 miles from an existing or projected commercial or military airport with more than 500 d^2 (where d is distance in miles) aircraft movements per year, and (3) sites located at distances greater than 10 miles from an airport with more than approximately 1000 d^2 aircraft movements per year (13,29). Such an analysis must be based on estimates of the likelihood of plane crashes near airports. Table 120 presents some probability factors for fatal crashes near airports from U.S. statistics (23,30). It is necessary to show that the probability of such a crash affecting the plant in such a way as to cause the release of radioactive materials in excess of 10 CFR 100 (2) guidelines is less than 10^{-7} per year. If the probability is higher,

Table 120.   Fatal Accident Probabilities Near U.S. Airports (23)

Distance from End of Runway (miles)	Probability (x 10^8) of a Fatal Crash per Square Mile for Aircraft Movements			
	U.S. Air Carrier[a]	General Aviation	USN/USMC[a]	USAF[a]
0-1	16.7	84	8.3	5.7
1-2	4.0	15	1.1	2.3
2-3	0.96	6.2	0.33	1.1
3-4	0.68	3.8	0.31	0.42
4-5	0.27	1.2	0.20	0.40
5-6	0	NA[b]	NA	NA
6-7	0	NA	NA	NA
7-8	0	NA	NA	NA
8-9	0.14	NA	NA	NA
9-10	0.12	NA	NA	NA

[a]Reference 30.
[b]NA indicates that data was not available for this distance.

another site should be considered or design of the plant must be hardened against aircraft impact (13).

Other potentially hazardous facilities and activities within 5 miles of a proposed site must be identified. If a preliminary evaluation of potential accidents of these facilities indicates that the potential hazards from shock waves and missiles approach or exceed those of the design-basis tornado for the region, or potential hazards such as flammable vapor clouds, toxic chemicals or incendiary fragments exist, the suitability of the site can only be determined by detailed evaluation of the potential hazard. Such an evaluation should include pipeline accidents leading to fires, explosions or releases of flammable vapors, transportation accidents involving hazardous chemicals or high explosives, and tank farms or storage areas involving such materials. For example, Table 121 shows the maximum allowable chlorine inventory in a single container at various distances from the plant, for different ventilation conditions (23). Similar relationships may have to be developed for other contemplated hazards.

Table 121. Maximum Allowable Chlorine Inventory in a Single Container for Various Reactor Distances and Control Room Conditions (23)

Control Room Characteristics		Maximum Weight of Chlorine Container (1000 lb) vs. Distance from Control Room	
Isolation Time sec	Normal Air Exchange Rate hr^{-1}	200 m (660 ft)	2000 m (6560 ft)
10	1	2	1200
4	1	5	3400
10	0.3	6	2700
4	0.3	20	32000

Radiation Safety

The question of possible radiation exposure of people living in the vicinity of nuclear power plants is, of course, the aspect that has attracted the greatest concern and is the one most responsible for opposition to certain locations. As a consequence, U.S. Atomic Energy Commission reactor site criteria (2) deal predominantly with the relationship between the reactor site and the surrounding population. The general philosophy assumes "that reactors will reflect through their design, construction and operation an extremely low probability for accidents that could result in release of significant quantities of radioactive fission products. In addition, the site location and the engineered features included as safeguards against the hazardous consequences of an accident, should one occur, should insure a low risk of public exposure" (2). Implied in this approach is the presumption that the population exposure from routine as-low-as-practicable releases of radioactive effluents results in a population dose well below natural background in almost any location that incorporates the above criteria.

Exclusion Area Criterion

To ensure adequate isolation of the plant from any population concentration, U.S. NRC site criteria deal solely with means of ensuring low population exposure following the occurrence of that rare "design-basis accident" discussed previously. For this purpose it is required that the nuclear plant must be located inside an *exclusion area*, which is defined as "that area surrounding the reactor in which the reactor licensee has the authority to determine all activities including exclusion or removal of personnel or property from that area." In practice, this usually means a fenced area leased or owned by the plant operator and under his sole control. This area may be traversed by a highway, railroad, or waterway provided these are not so close to the facility as to interfere with normal operations of the facility and provided appropriate and effective arrangements are made to control traffic on the highway, railroad or waterway in case of emergency, to protect the public health and safety. Residence within the exclusion area shall normally be prohibited. In any event, residents shall be subject to ready removal in case of necessity. Activities unrelated to operation of the reactor may be permitted in an exclusion area under appropriate limitations, provided that no significant hazards to the public health and safety will result (2).

Immediately surrounding the exclusion area there must be a *low population zone* in which the population is sufficiently sparse that there is a reasonable probability that appropriate measures could be taken in their behalf in the event of a serious accident; that is, their exposure could be monitored and they could be readily evacuated if necessary. This definition does not specify a permissible population density or total population within this zone because the situation may vary from case to case. Whether a specific number of people can, for example, be evacuated from a specific area, or instructed to take shelter on a timely basis will depend on many factors such as location, number and size of highways, scope and extent of advance planning, and actual distribution of residents within the area. Low-density farming areas, woodlands and national parks may qualify as low-population zones. Essentially, though, such a zone implies the absence of any sizable town or other concentration of persons, such as military camps, large prisons or large educational institutions. The nearest *population center*, defined as any densely populated area containing more than about 25,000 residents, should be at least 30% further than the distance to the outer boundary of the low-population zone in that direction.

The areas and boundaries of the exclusion zone and the low-population zone are, of course, purely theoretical and are best described by an overlay on a site map. The legal exclusion zone may coincide with the plant fence line, but in many cases it may lie well within it. The exclusion area is determined as being "an area of such size that an individual located at any point on its boundary for two hours immediately following onset of the postulated fission product release would not receive a total radiation dose to the whole body in excess of 25 rem or a total radiation dose in excess of 300 rem to the thyroid from iodine exposure." The low-population zone is defined to be "of such size that an individual located at any point on its outer boundary who is exposed to the radioactive cloud resulting from the postulated fission product release (during the entire period of its passage) would not receive a total radiation dose to the whole body in excess of 25 rem or a total radiation dose in excess of 300 rem to the thyroid from iodine exposure." (2).

To determine the extent of the exclusion area and the low-population zone, the computations described in the preceding chapter must be performed to estimate the ground-level dose distribution following the design-basis accident and the associated accidental escape of radioactive materials from the containment building. The calculation can take the form of computing the dose at the proposed 'fence line in all directions, for any assumed "worst" meteorological conditions, and verifying that the calculated doses for wholebody exposure and thyroid dose do lie within the specified limits. Alternatively, isodose contours corresponding to those limits can be obtained and shown to lie within the plant site perimeter; this may free some of the outlying plant property for other purposes, such as hunting, agriculture or recreational uses that may involve only transitory public access. In general, neither the exclusion zone contour nor the postulated low-population zone boundary will be circular or concentric with the plant. Similarly, in applying the guide regarding the population-center distance of at least 1.30 times the distance from the reactor to the "boundary" of the low-population zone, due consideration should be given to the population distribution within the population center. Where very large cities are involved, a greater distance may be necessary because of total integrated population dose considerations.

For sites for multiple reactor facilities consideration should be given to the following:/

(1) If the reactors are independent to the extent that an accident in one reactor would not initiate an accident in another, the size of the exclusion area, low population zone and population center distance shall be fulfilled with respect to each reactor individually. The envelopes of the plan-overlay of the areas so calculated shall then be taken as their respective boundaries. (In recent U.S. reactor designs such independence is now a license requirement.)

(2) If the reactors are interconnected to the extent that an accident in one reactor could affect the safety of operation of any other, the size of the exclusion area, low population zone and population center distance shall be based upon the assumption that all interconnected reactors emit their postulated fission product releases simultaneously. This requirement may be reduced in relation to the degree of coupling between reactors, the probability of concomitant accidents and the probability that an individual would not be exposed to the radiation effects from simultaneous releases. To reduce coupling, current design practice completely isolates reactors and their control rooms in most new multiple reactor facilities.

Various estimates regarding the needed size of an exclusion area have depended strongly on the assumptions made regarding the composition and fraction of the radioactive inventory in a reactor released in case of a severe accident and the extent to which the radwaste treatment system may be assumed to fail as well. Undoubtedly, earlier estimates both of release levels and of the exclusion zone areas required were overly pessimistic (31) and have led to excessive demands for plant isolation. In practice the thyroid dose from radioiodine has proved to be the limiting factor in determining exclusion area dimensions, and the final value depends sensitively on the degree of internal scavenging of airborne iodine allowed for. Table 122 lists the exclusion radii and low population zone (LPZ) radii for some U.S. reactors. Whether or not current siting practice still results in excessive transmission distances can only be answered after more years of operating experience and more detailed verification of actual dispersion patterns (32).

Note particularly, that this site criterion does not involve any specific consideration of population dose. It is merely assumed that the population density postulated will minimize excessive exposure of individuals following a serious accident. General population dose is more effectively controlled by minimizing routine, continuous releases of radioactive effluents.

Table 122. Dimensions of Some U.S. Reactor Exclusion Areas and
Low-Population Zones (11)

Reactor	Type	Thermal Power MWth	Exclusion Radius, Miles (meters)	LPZ Radius, Miles (meters)[a]
Connecticut Yankee	PWR	1473	0.25 (400)	13.4 (21,500)
Dresden	BWR	700	0.50 (800)	8.1 (13,000)
Indian Point	PWR	615	0.30 (485)	7.3 (11,700)
Nine Mile Point	BWR	1538	0.76 (1220)	13.7 (22,000)
Yankee	PWR	600	0.50 (800)	7.3 (11,600)
Crystal River	PWR	2560	0.83 (1330)	20.0 (32,000)
Typical 1000-MWe	PWR	3200	1.6 (2600)[a]	23.0 (37,000)

[a]Radius based upon criteria of Reference 32.

For the population density criteria to be effective, the growth pattern of the neighboring areas must be projected for the expected lifetime of the plant. Since the availability of large blocks of electric power may tend to attract power-intensive industry, effective zoning controls are essential to avoid any major vitiation of the low population density assumed.

Risk Criterion

By using the limiting dose due to the most severe accident postulated to define exclusion area dimensions, the U.S. Atomic Energy Commission criteria tend to require very large exclusion areas.

An alternative method for evaluating the suitability of a potential reactor site with respect to radiological exposure has been advocated by Farmer (9,10). In this approach the site criterion is not based on limiting dose to individuals and the general public in the region surrounding the reactor site, but on the overall *risk* of exposure, principally to radioiodine accumulated in the thyroid, taking into account the probability of *any* given accident and any consequent massive release of activity actually happening. Once the relationships of radiation dose *vs.* probability have been determined for the distances of interest, expressions proposed in References 11 or 9 to correlate dose and mortality risk can be used to obtain mortality risk as a function of accident probability for various distances. The method is not restricted to use of the linear risk-dose assumption used here:

$$M_{t/wb}(P,s) = D_{t/wb}(P,s) \, m_{t/wb} \tag{85}$$

where: $M_{t/wb}(P,s)$ = mortality probability (thyroid or whole-body), a function of accident probability, P, and distance, s, from the reactor

$D_{t/wb}(P,s)$ = dose (thyroid/whole-body) in rads, a function of accident probability, P, and a distance, s, from the reactor

$m_{t/wb}$ = mortality probability (thyroid or whole-body) per rad of radiation.

The total risk to an individual is found by integrating the mortality risk of Equation 85 over all accident probabilities.

$$R(s)_{t/wb} = \int_{P_1}^{P_2} M_{t/wb}(P,s) \, dP \qquad (86)$$

where: $R(s)_{t/wb}$ = individual mortality risk per year (thyroid or whole-body), a function of distance, s.

This yields the total risk to an individual as a function of a specific distance from the reactor; this risk is then used as the criterion for determining exclusion-radius requirements so that risk can be set at any desired value.

The total risk to the population around the reactor site in terms of deaths per year or total deaths during the projected reactor lifetime can be found by integrating the individual risk (weighted by the number of individuals at risk) over distance. The only additional datum needed is the population distribution as a function of area covered by the fission-product cloud. We may assume that the cloud occupies a 30° arc at 100 m and decreases linearly with distance to a 15° arc at 10,000 m. The population density is conservatively assumed constant over this distance. The total risk is given by

$$R_t = \int_{s_1}^{s_2} R(s) \, P(s) \, ds \qquad (87)$$

where: R_t = total yearly death risk to the population in the affected area

$R(s)$ = $R_t(s) + R_{wb}(s)$ as defined by Equation 86

$P(s)$ = population distribution within the affected area as a function of radial distance from the reactor.

The limits of integration of Equation 87 are normally from the exclusion radius to infinity, or to a point where doses are below the threshold (11).

Figure 87 indicates the form the criterion takes. By plotting the equivalent ground release of radioiodine against the number of reactor-years elapsing between iodine releases of given magnitude, one obtains a

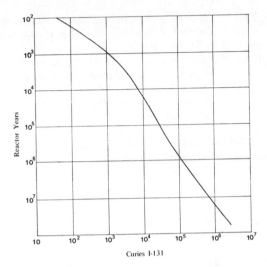

Figure 87. Release criterion, relating ground-level activity of I-131 with the time interval between release of such activity (9).

number of contours like the line shown, which represent conditions of comparable risk. For Great Britain for a "standard site" there was assumed to be a uniform population density of 13,000 per square mile in all directions, from 0.5 to 10 miles from the site. The line drawn in Figure 88 then represents the condition that the aggregate risk of casualties from the operation of a GCR reactor under those conditions on an urban site is of the order of 0.01 per year, or an individual risk of 1×10^{-7} per year, several orders of magnitude lower than the risks of incurring leukemia or accidental death from other causes.

Similar risk criteria can be derived for other sites using fault-tree methodology of safety analysis (34). Otway (11,33) has developed the risk approach and analyzed the contribution of design factors, site-related factors, and of assumptions regarding the effects of single exposures to high radiation fields on the risk calculation. He concluded that for a 1000 MWe PWR an exclusion radius of 350 m (1000 ft) would keep the individual mortality risk at the site boundary below a value of 10^{-7} per person per year, and the total number of deaths expected from a 30-yr operating life would be 0.003. This was based on an assumed mortality risk from I-131 irradiation of the thyroid of 1×10^{-6} cancer cases per person per rad, with a 1-rad threshold. Any reduction in the release consequences of a design-base accident by improvement in engineered safeguards would progressively shrink the required site area or further reduce the mortality risk at the site boundary.

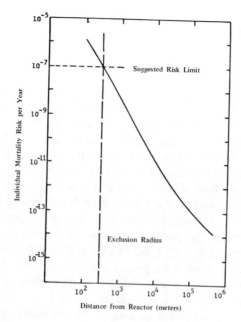

Figure 88. Derived risk criterion: relation between mortality risk and distance from reactor to be exposed to such risk (11,33).

Ecology

Chapter 2 dealt with the impact of any power plant on the local ecology, principally fish and wildlife. When a site is evaluated a detailed survey must be made of all vegetation and wildlife factors. Any unique or endangered species should be identified and the probable consequences of plant construction indicated. As mentioned previously, the removal of the exclusion area from public access may be beneficial to wildlife, but some of the construction work and the clearing of the right-of-way for transmission lines may change the habitat for many species.

Unless there are very specialized conditions involving rare or endangered species or certain types of commercial fishing, the location of a nuclear power plant will not cause any widespread change in ecological conditions, though precautions will have to be taken to minimize disturbance of the environment. The evaluation of environmental data has been discussed in the preceding chapter and its methodology has been described in Reference 35. Again, predictive modeling is required and a ranking in importance among various, possibly contradictory factors, to permit use of a relation of the type of Equation 81. All these considerations will have to be detailed in the environmental impact statement filed to obtain the construction license.

SOCIOECONOMIC FACTORS

Many of the socioeconomic aspects of large power installations were discussed in Chapter 2. The construction and staffing of a large plant will provide obvious employment opportunities and will benefit local trade, especially during construction when there will be an additional skilled workforce that has to be housed, fed and entertained. There may be need for additional school rooms, sewerage and water supplies; this will usually be more than offset by the additional tax base provided by the plant. The change in land use from agricultural will also affect tax assessments as will the construction or improvement of roads and highways in the vicinity of the plant. Any large local economic dislocation resulting from preemptive use of productive land should be avoided.

Careful consideration must be given to the benefits and drawbacks of changes in zoning and community planning as they relate to the attraction to industry presented by the ready availability of large blocks of power and by the need to avoid major changes in the population density in the low-population zone. The construction of water reservoirs and lakes may remove agricultural land from production, but may in turn provide recreational facilities and generate recreational business. All of these factors must be considered in site evaluation.

Areas valued for their historic, scenic, cultural, or natural significance may be affected by construction of the plant, its associated water reservoirs, road system, or power lines. Power companies are often accused of insensitivity in this regard. Recognized historic or cultural sites that conflict with power plant location will, in general, take precedence. Interaction with archaeologically interesting locations may be almost inevitable in some parts of the world and some effort should be made for their preservation or salvage, particularly since the plant buildings themselves may occupy only a small portion of the site area.

Considerations of plant appearance will rarely affect site selection since the plant already will normally be in a fairly remote location; the cooling towers may be the dominant visual impact feature. Available corridors for transmission lines have to be evaluated at an early stage as to their effect on land cost, land use and impact on value of adjoining properties. Finally, consideration must be given to the relative impact of waste truck shipments on local traffic conditions, even though they are minor in comparison with a comparable coal-fired plant.

From a practical point of view, site selection will also be affected by the public attitude toward any intended construction. At a fairly early stage a survey should be made of local opinion with regard to any proposed plant (35). In cases where several alternative sites, otherwise comparable in technical desirability, are considered, a favorable climate of opinion at any one locality will clearly decide the issue. Care should, of course, be taken to ensure that the opinion survey is based on availability of all relevant facts and of all factors related to plant location in that particular region.

SPECIAL SITES

Most of the present generation of nuclear power plants have been planned or constructed at sites on major freshwater lakes or streams, which supply the needed cooling water. Considerations of cooling water availability or of minimizing land requirements have prompted the location of some reactors on ocean shores and, more recently, the planning of plants on artificial islands or floating barges. Another special case is the location of reactors underground or in caves. Selection of such locations in competition with other land-surface sites involves the consideration of additional criteria and safety factors.

Ocean Shores

A fairly high proportion of nuclear power plants in Great Britain and the United States have been constructed on beach locations. The principal attraction of such sites is the practically unlimited availability of cooling water and an essentially boundless heat sink. Furthermore, exclusion zone and low-population criteria need only be applied on the land side of the plant, thus reducing the potential demographic and economic problems. Against this are the difficulties in finding suitable beach locations that are not already preempted for recreational, residential or commercial-fishing uses or that do not form areas of scenic or cultural significance. The use of salt water for cooling purposes imposes additional demands on water purification and demineralizer systems, and exposure to salt spray may cause serious corrosion and transmission line problems. The impact on marine biota from heat release and low-level radioactivity in effluents must also be evaluated carefully. As the experience at Turkey Point has shown, thermal pollution of marine areas can be a major problem; it can usually be solved only by cooling towers or by extensive diffuser systems (Figure 53). Long intake pipes or outfall diffusers also have to be considered in relation to local fishing activities, especially trawler fisheries.

A further possible reason for locating a nuclear power plant on beach sites is for dual-purpose plants that would generate both power and fresh water by desalination. The primary incentive for combining an electric power plant with a desalting plant is that the exhaust or extraction heat from a power plant can be effectively utilized in the desalination process, thus achieving an overall economic advantage. Savings also result from combined facilities operations and maintenance (36).

Offshore Sites

Offshore locations with power plants placed on floating barges or in submerged caissons have been seriously considered for some time. Their advantages arise from ready availability of cooling water, a large heat sink, the possibility of prefabricating standard units that are then towed to the final site, and no complications from hostile neighbors, drinking water quality standards, or exclusion zone considerations. The disadvantages are also evident: susceptibility to severe weather and oceanic storms, transmission lines under water to a shoresite switch yard of significant size, possibility of ship collision in poor-visibility weather, and trans-shipment needs of spent fuel and wastes.

It is being argued that population densities in some regions make it difficult to find land-based reactor sites and that the standardization in plant design achievable would offset the added expense of offshore installations. For this reason several nuclear power plants are under consideration for positioning off the coast of New Jersey and Florida. Figure 89 shows an artist's conception of the overall layout, and Table 123 sets forth some of the design parameters (37). To make such a site environmentally acceptable and to satisfy all safety requirements it is necessary to determine the effect of the physical existence of the plant on waves and currents and, therefore, on bottom and shoreline conditions. The effects of intake and exhaust of water and the heat release on marine biota must be evaluated (7,38). The design and cost of the buildings and the breakwater to withstand the most severe wind and water conditions, and the possible effects on any ship collisions, must be reviewed and assessed (37,39). For the proposed power station to be built 2.8 miles out in the ocean off New Jersey, the breakwater, believed to be the largest man-made structure ever placed in the ocean, is reported to cost in excess of $200 million (51). Problems of underwater transmission by means of a flexible cable from a floating plant or by a buried cable have to be considered, and its effect on bottom-dwelling organisms will also need evaluation (37,40).

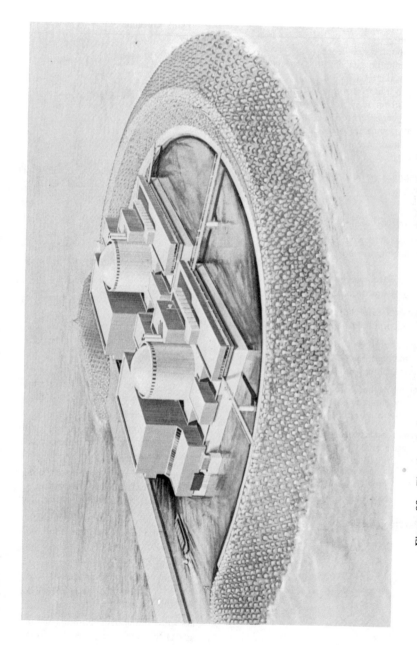

Figure 89. Floating nuclear power plant; artist's conception of breakwater design (37).

Table 123. Floating Nuclear Power Plant: Proposed Design Parameters (37)

Design Criteria	Plant Envelope Value
Maximum water depth minimum site depth at mean low water (MLW)	44 ft
maximum astronomical tide plus storm surge for one-in-a-hundred-year storm *or* maximum tsunami allowance for waves adjacent to vital structures	⎱ 32 ft ⎰
total	76 ft
Minimum water depth	31-ft draft of plant plus 13-ft allowance for extremely low tides and vertical motion for a total of 44 ft
Operating basis wind	180 mph
Design basis wind (tornado) maximum peripheral velocity	200 mph below elevation +64 MLW 300 mph above elevation +64 MLW
forward speed	20 to 60 mph
pressure drop at center	3.0 psi
One-half safe shutdown earthquake— maximum horizontal ground acceleration	0.15 g

Safe shutdown earthquake
  maximum horizontal ground acceleration
  maximum vertical ground acceleration
	0.30 g
	0.20 g

Operating basis storm— wave and wind plant motions (roll or pitch)	Accelerations due to 2-deg single amplitude motions at 13-sec periods must not be exceeded.
Design basis storm— wave and wind plant motions (roll or pitch)	Accelerations due to 3-deg single amplitude motions at 13-sec periods must not be exceeded.
Ship collision	The breakwater must prevent colliding ships or displaced portions of the breakwater from contacting the plant.
Hazardous cargoes	Probability of collision and explosion of munitions or LNG ship must be $<10^{-6}$/yr. Probability of collision and rupture of an anhydrous ammonia tanker must be $<10^{-6}$/yr, or provide emergency breathing apparatus. Fuel spills other than liquefied natural gas (LNG) must be prevented from approaching closer than 100 ft to the plant.

Table 123, Continued

Design Criteria	Plant Envelope Value
Aircraft collision	Probability of fixed-wing aircraft/plant collision must be $<10^{-6}$/yr.
Rainfall	Plant is designed for a maximum of 7 in./hr.
Air temperature	Plant is designed for a minimum of $-5°F$ near sea surface.
Water temperature	Plant is designed for a minimum of $30°F$ and a maximum of $85°F$.
Breakwater support	Seabed must support the breakwater under static and dynamic conditions.
Sunken plant support	Seabed must support a static load of 1600 lb/ft^2 (7800 kg/m^2).
Plant/breakwater contact	The mooring system must prevent plant/breakwater and plant/circulating-water discharge box contact.
Alignment of connections	Mooring system must sufficiently limit plant motion to maintain the integrity of the transmission lines.

Okrent and his group (41,42) have reviewed the environmental and technical aspects of floating nuclear power plants and recommend a floating buoy system, 1.5 miles in diameter, as a means of collision protection, and mooring towers with a buried cable as the best means of power transmission. Comparing various platform supports of the type familiar for oil drilling rigs with large barges as a method of support for the power plant, they decided that a barge-supported, heavily damped system was safer under conditions encountered in the Pacific off the California coast. Analysis of different site characteristics and cable routes showed that installation of the transmission system to shore is an appreciable cost item, considerably more than conventional overland systems (Table 124) (42).

Safety requirements of a floating barge station are unique in several respects. The design must be capable of surviving maximum seismic disturbances. Since disruption of power transmission lines by faulting or seismic vibrations from a severe earthquake may be difficult to avoid, on-site emergency power capacity of high reliability must be provided. Tsunamis and hurricanes will exert the greatest load on the mooring

Table 124.   Estimated Costs (1973) of a 2000 MWe, 175-Mile System for a
Floating Power Plant: 5 Cables, 3 Routes each Requiring 50% Suspended Cables (42)

Item	Total Length (miles)	Cost ($M/mile)	Total Cost
5 SF$_6$ 345 kV Systems	175	1.3	$230 M
Bottom pull installations	87.5	0.03	3
Cable suspensions	87.5	0.36	32
Towers		0.63	55
SF$_6$ terminations	–	–	0.1
			$320 M

system.   The plant must be designed so that sinking within its breakwater
would not cause undue risk to the health and safety of the public.   Ship
collision hazards, fire hazards, potential effects of commercial or military
aircraft crashes, and the effects of releases of radionuclides to the marine
environment all have to be evaluated and related to comparable hazards
affecting land-based plants.   At this time the economic and environmental
benefits of floating offshore power plants do not seem to be clearly
established (6,43).

Similar considerations apply to submerged offshore power plants (44).
Such a plant would be contained in a caisson floated to the site and
deposited on the sea floor at a depth of 200-250 ft (65-85 m).   It
would be less susceptible to wind and water effects, and presumably
collisions, but movement of fuel and supplies in and out would be more
complex.   The environmental consequences from heat exchange and release of liquid and gaseous radioactive wastes appear to be minimal;
however, severe technical problems associated with all offshore plants
remain, but are by no means insuperable.

Underground Location

Periodically one reads of some speaker advocating the location of
nuclear power plants in caves or generally underground as a safety
measure.   Implicit in this recommendation is the assumption that the
direct radiation from the reactor is a health hazard or that the reactor
is a disguised bomb liable to explode in case of an accident and, therefore, more safely contained under a heavy roof of rock and soil.

Even if one rejects these assumptions as false or misconceived, there
is a certain attraction to the idea of using a natural cave as a containment

structure and such underground plants have been built in Sweden and Norway. Jamne (45) has described the layout and construction of the Halden test reactor and the planning for other underground plants. The advantage of this approach, in a country like Norway with few open suitable areas that are not already densely populated, is that it permits location near load centers on inexpensive land with low population density. Unclad rock is used as containment with an epoxy lining near regions where cracking has occurred. The cost of underground construction is appreciably above that of surface construction, both because of the cost of blasting the cavity and clearing it and because of the added assembly and installation costs arising from difficult access. Rogers (46) in 1971 estimated the additional cost of underground location as $10 per kilowatt. Table 125 indicates the quantities involved. A recent review (47) arrived at comparable cost figures.

Table 125. Incremental Costs (1971) for Placing Two 1000 MWe Nuclear Units Underground (46)

		Cost in $1,000
Excavation		
Nuclear chamber	300,000[a]	$ 5,200
Shafts	40,000[a]	1,200
Water storage and passages	10,000[a]	400
Turbine-generator and auxiliary pit	1,000,000[a]	2,000
Concrete		
Nuclear chamber arch	10,000[a]	1,100
Shaft, water storage and passage lining	17,000[a]	1,600
Additional hoists, cranes and elevators	(lump sum)	1,000
Additional steam and feedwater lines	4,500[b]	4,500
Subtotal direct cost		$17,000
Contingencies and engineering		3,000
Total		$20,000

[a]cubic yards
[b]linear feet

Since rock falls are a major hazard, in most cases the cavity must have a concrete liner. Liquid and gaseous wastes must still be discharged to surface and problems of heat dissipation are accentuated by the extra pumping power required for coolant circulation. In case of a catastrophic accident the whole cavity can be flooded with water, with the assumption that the surrounding rock is free of cracks and fissures that would permit escape of the water. However, several independent assessments have indicated that, from an engineering viewpoint, underground location of 1000 MWe reactors is entirely feasible and that the additional cost is a moderate fraction of total project cost (47-50).

On balance it would appear that the advantages of underground locations would be largely psychological and would be achieved only at appreciably higher cost compared with contained surface installations.

REFERENCES

1. "Environmental Analysis of the Uranium Fuel Cycle, Pt. II—Nuclear Power Reactors," Report EPA-520/9-73-003-C. (Washington, D.C.: U.S. Environmental Protection Agency, 1973).
2. "Reactor Site Criteria," Title 10, Code of Federal Regulations, Part 100, as amended.
3. Beer, L. P. "Quantification of Siting Factors for Rational Selection of Alternate Sites," Trans. ANS 19, Suppl. 1, 12 (1974).
4. Fischer, J. A. and R. Ahmed. "A Systematic Approach to Evaluate Sites for Nuclear Power Plants," Trans. ANS 19, Suppl. 1, 7 (1974).
5. Jopling, D. G. "Plant Site Evaluation Using Numerical Ratings," Power Eng. 78(3), 56 (1974).
6. West, L. J. and K. Dyar. "The Methodology and Criteria Used for a Nuclear Power Plant Siting Study," Trans. ANS 19, Suppl. 1, 8 (1974).
7. Fischer, J. A. and F. L. Fox. "Siting Constraints for an Offshore Nuclear Power Plant," Dames and Moore Engineering Bulletin 42 (Los Angeles, California: Dames and Moore Inc., 1973).
8. Cordaro, M. C. and W. T. Malloy. "A Methodology for Power Plant Site Selection at the Reconnaissance Level," Nucl. Technol. 23, 233 (1974).
9. Farmer, F. R. "Siting Criteria—A New Approach," in Containment and Siting of Nuclear Power Plants (Vienna: International Atomic Energy Agency, 1967).
10. Farmer, F. R. "Reactor Safety and Siting: A Proposed Risk Criterion," Nucl. Safety 8, 539 (1967).
11. Otway, H. J. "The Application of Risk Allocation to Reactor Siting and Design," U.S. Atomic Energy Report LA-4316, Los Alamos Scientific Lab. (1969).
12. "Guide to the Preparation of Environmental Reports for Nuclear Power Plants," Regulatory Guide 4.2 (Washington, D.C.: U.S. Atomic Energy Commission, 1973).
13. "General Site Suitability Criteria for Nuclear Power Stations," Regulatory Guide 4.7 (Washington, D.C.: U.S. Atomic Energy Commission, 1974).
14. "General Environmental Siting Guides for Nuclear Power Plants," (Draft). (Washington, D.C.: U.S. Atomic Energy Commission, 1973).

15. Bell, C. G. (Conf. Report). "Earthquake-Resistant Design of Engineering Structures," *Nucl. Safety* **14**, 1 (1973).
16. Hansen, R. J., Ed. *Seismic Design for Nuclear Power Plants.* (Cambridge, Massachusetts: MIT Press, 1970).
17. Udoguchi, T., Y. Ohsaki and H. Shibata. "The Aseismic Design of Nuclear Power Plants in Japan," *Proc. 4th Internat. Conf. Peaceful Uses of Atomic Energy* **3**, 297 (1971).
18. "Earthquake Guidelines for Reactor Siting," Techn. Reports Series No. 139 (Vienna: International Atomic Energy Agency, 1972).
19. Hentschel, G., W. Novak and F. Orth. "Earthquake Protection for Nuclear Reactors," *Proc. 4th Internat. Conf. Peaceful Uses of Atomic Energy* **3**, 309 (1971).
20. Lomenick, T. F., Ed. "Earthquakes and Nuclear Power Plant Design," U.S. Atomic Energy Report ORNL-NSIC-28; UC-80 (Oak Ridge, Tennessee: Nuclear Safety Information Center, 1970).
21. Newmark, N. M. "Earthquake Response Analysis of Reactor Structures," *Nucl. Eng. Design* **20**, 303 (1972).
22. Wiegel, R. L., Ed. *Earthquake Engineering.* (Englewood Cliffs, New Jersey: Prentice-Hall, 1970).
23. Grimes, B. K. "External Events Affecting Plant Design" (1974) (unpublished).
24. Newmark, N. M. "A Study of Vertical and Horizontal Earthquake Spectra," U.S. Atomic Energy Commission Report WASH-1255 (1973).
25. Smith, C. B. and R. B. Matthiesen. "Vibration Testing and Earthquake Response of Nuclear Reactors," *Nucl. Applic. Technol.* **7**, 6 (1969).
26. Boyle, M. V. "Meeting Seismic Requirements of Switchboards and Switchgear," *Power Eng.* **78** (3), 50 (1974).
27. "Damping Values for Seismic Design of Nuclear Power Plants," Regulatory Guide 1.61 (Washington, D.C.: U.S. Atomic Energy Commission, 1973).
28. Carpenter, P. J. and E. L. Meyer. "Flood Considerations in Evaluating Nuclear Power Plant Sites," Proc. 5th Midyear Symp. Health Physics Aspects of Nuclear Facility Siting, Health Physics Society (1970).
29. Hornyik, K. and J. E. Grund. "Evaluation of Air Traffic Hazards at Nuclear Plants," *Nucl. Technol.* **23**, 28 (1974).
30. Eisenhut, D. G. "Reactor Sitings in the Vicinity of Airfields," *Trans. ANS* **16**, 210 (1971).
31. "Theoretical Possibilities and Consequences of Major Accidents in Large Nuclear Power Plants," Report WASH-740 (Washington, D.C.: U.S. Atomic Energy Commission, 1957).
32. Di Nunno, J. J., F. D. Anderson, R. E. Baker and R. L. Waterfield. "Calculation of Distance Factors for Power and Test Reactor Sites," U.S. Atomic Energy Commission Report TID 14844 (1962).
33. Otway, H. J. and R. C. Erdmann. "Reactor Siting and Design from a Risk Viewpoint," (Los Angeles: University of California, 1973).
34. Ritzman, R. L., P. C. Owzarski, A. P. Postma, D. L. Lessor, *et al.* "Release of Radioactivity in Reactor Accidents," Appendix VII.

Report WASH-1400 (Draft) (Washington, D.C.: U.S. Atomic Energy Commission, 1974).

35. Committee on Power Plant Siting. "Engineering for Resolution of the Energy Environment Dilemma," (Washington, D.C.: National Academy of Engineering, 1972).

36. "Considerations Affecting Steam Power Plant Site Selection," (Washington, D.C.: Office of Science and Technology, 1969).

37. Ashworth, J. A. "Atlantic Generating Station," *Nucl. Technol.* 22, 170 (1974).

38. Wright, J. H. "Environmental Considerations of Offshore Nuclear Power Plants," Proc. AIF Washington Conf. Water Quality Considerations, Atomic Industrial Forum (1972).

39. Klepper, O. H. and T. D. Anderson. "Siting Considerations for Future Offshore Nuclear Energy Stations," *Nucl. Technol.* 22, 160 (1974).

40. Perry, E. R. and J. C. Cronin. "Underwater High Voltage Power Transmission Using SF_6 Insulated Cable," Proc. IEEE Underground Transmission Conf. (1972).

41. Cramer, E. N. "Evaluation of Floating Nuclear Plants off the Coast of California," *Nucl. News* 16(13), 52 (1973).

42. Okrent, D., I. Catton, *et al. Environmental, Technical, Legal and Safety Aspects Related to Floating Nuclear Power Plants off the Coast of California* (Los Angeles: School of Engineering & Applied Science, UCLA, 1973).

43. Yadigaroglu, G. and S. O. Andersen. "Novel Siting Solutions for Nuclear Power Plants," *Nucl. Safety* 15, 651 (1974).

44. General Dynamics Inc. "Potential Environmental Effects of an Offshore Submerged Nuclear Power Plant (2 vols)," Report 16130-GFI-06/71 (Washington, D.C.: U.S. Environmental Protection Agency, 1971).

45. Jamne, E. "Underground Siting of Nuclear Power Plants in Norway," *Proc. 4th Internat. Conf. Peaceful Uses of Atomic Energy* 3, 359 (1971).

46. Rogers, F. C. "Underground Nuclear Power Plants," *Bull. Atomic Scientists* 27(8), 38 (1971).

47. Crowley, J. H., P. L. Doan and D. R. McCreath. "Underground Nuclear Plant Siting: A Technical and Safety Assessment," *Nucl. Safety* 15, 519 (1974).

48. Doan, P. L., J. R. Huffman and M. H. Smith. "Safety and Environmental Assessment of Underground Nuclear Siting," *Trans. ANS* 18, 242 (1974).

49. "Population Distribution Around Nuclear Power Plant Sites," (Washington, D.C.: U.S. Atomic Energy Commission, 1974).

50. Willett, D. C. and D. R. McCreath. "Rock Caverns for Nuclear Power Plant Siting," *Trans. ANS* 18, 243 (1974).

51. *Nuclear News* 17(8), 99 (1974).

ADDITIONAL REFERENCES

Anderson, T. D. "Offshore Siting of Nuclear Energy Stations," *Nucl. Safety* **12**, 9 (1971).

Beranek, J., Z. Kriz, I. Chochlovsky, C. Raisigl and J. Sevc. "Siting Nuclear Power Plants in the Czechoslovak Socialist Republic," *Proc. 4th Internat. Conf. Peaceful Uses of Atomic Energy* **3**, 339 (1971).

Blume, J. A. and Assoc. "Recommendations for Shape of Earthquake Response Spectra," U.S. Atomic Energy Commission Report WASH-1254 (1973).

Brasier, R. I. and C. B. Mills. "Reactor Power Plants for Undersea Applications," *Nucl. Technol.* **22**, 224 (1974).

Carter, L. J. "Floating Nuclear Plants: Power from the Assembly Line," *Science* **183**, 1063 (1974).

"Containment and Siting of Nuclear Power Plants," Proc. Vienna Symp. (Vienna: International Atomic Energy Agency, 1967).

"Design Response Spectra for Seismic Design of Nuclear Power Plant," (Rev. 1), Regulatory Guide 1.60 (Washington, D.C.: U.S. Atomic Energy Commission, 1973).

Moscati, A. F. and R. C. Erdmann. "Possible Effects of Ionizing Radiation upon Marine Life and Some Implications of Postulated Accidental Releases of Radioactivity," *Nucl. Technol.* **22**, 184 (1974).

Otway, H. J., M. E. Battat, R. K. Lohrding, R. D. Turner and R. L. Cubitt. "A Risk Analysis of the Omega West Reactor," U.S. Atomic Energy Commission Report LA-4449, Los Alamos Scientific Laboratory (1970).

Perla, H. F. "Power Plant Siting Concepts for California," *Nucl. News* **16**(13), 47 (1973).

Pickel, T. W. "Evaluation of Nuclear System Requirements for Accommodating Seismic Effects," *Nucl. Eng. Design* **20**, 323 (1972).

Piper, H. B. and F. A. Heddleson. "Siting Practice and its Relation to Population," *Nucl. Safety* **14**, 576 (1973).

Sieving, K. W. "Nuclear Plant Siting and Reactor Safety," Stone and Webster Engineering Corp. (1969).

Smith, C. B. "Power-Plant Safety and Earthquakes," *Nucl. Safety* **14**, 586 (1973).

Thoma, R. E., M. Simon-Tov and H. A. Vanderploeg. "Floating Nuclear Power Plants: Environmental Issues and Licensing," *Trans. ANS* **18**, 57 (1974).

U.S. Nuclear Regulatory Commission. "Draft Environmental Statement: Manufacture of Floating Nuclear Power Plants," Report NUREG-75/113 (1975).

CHAPTER 10

TRANSPORTATION OF NUCLEAR MATERIALS

INTRODUCTION

The complexities of the nuclear fuel cycle entail the movement of appreciable amounts of nuclear materials at various stages. While the actual quantities involved are small in comparison with the enormous transportation requirements of coal-fired power stations (Table 126), which in fact account for a major environmental impact of such stations, the nature of the nuclear material tends to engender enough public concern to require a detailed environmental analysis. Figure 90 shows a schematic diagram of the transportation pathways for the LWR nuclear power industry (2). Considering these pathways, most stages are seen to

Table 126. Projected Comparative Annual Fuel Cycle Logistics for AD 2000; Nuclear Electric Energy Demand of 8 Trillion kWh per Year (1)

Materials at Various Stages	Nuclear[a]	Coal Equivalent[b]
Mining (million tons)		
Fuel	0.1 (U_3O_8)	3,000
Ore	50	3,500
Transportation (thousand carloads)		
Fuel	50-100	30,000
Waste	1-3	2,500 (ash)
		500 (sulfur)
Waste management (million cubic feet)	0.1-0.3	2,500 (ash)
		800 (sulfur)

[a]Reflect total nuclear economy logistics including U_3O_8, UF_6, enriched uranium fuel for LWR and HTGR, and low-level waste.
[b]Based on 10,000 Btu per pound, 9% ash, and 90% removal of 2% sulfur.

477

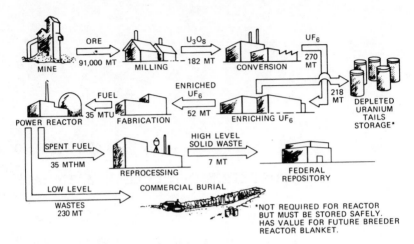

Figure 90. Average annual fuel materials requirements for a typical 1000 MWe light-water reactor (3).

involve the movement of fairly pure uranium compounds. Such materials are alpha-particle emitters of low specific activity and can be shipped in drums like any other chemical material of some value, subject to criticality considerations. For instance, UF_6 is shipped to enrichment plants in 48-inch, 10- and 14-ton capacity steel cylinders. Enriched UF_6 shipments are made in 30-inch cylinders holding about 2.5 tons of UF_6 containing up to 5 weight percent U-235.

The shipment of enriched uranium is subject also to nuclear safeguards considerations, which will be discussed later in this chapter. Fuel elements are shipped in protective packages or special "birdcages" to prevent mechanical damage in handling and to avoid criticality (4). Table 127 lists the various shipping stages in the fuel cycle, the method of packaging, and representative quantities (2). All projections of U.S. shipping requirements for the nuclear power industry show steep increases in the mid-1980s; Western European industry will require shipment of comparable quantities, although the logistics of shipments will vary for each country depending on the location of reprocessing facilities and waste disposal sites (1,3). At this time there are no plans to reprocess spent fuel from Canadian-type heavy-water reactors.

Whereas the fresh fuel sent to the reactor is of low activity, the spent fuel from the reactor, the intermediate-level wastes from the nuclear power plant and the high-level waste from a reprocessing plant represent substantial amounts of radioactivity. The movement of such material

Table 127. Summary of Transportation Parameters for the LWR Nuclear Power
Industry (2)

Material	Form	Mode	Annual Quantity Shipped to Facility	Quantity per Shipment
Reactor wastes	Packaged solids	Truck Rail	3,000 drums per reactor	50 drums 150 drums
Spent fuel	Spent fuel assemblies	Truck Rail (Barge)	1,500 tonne	0.5 tonne 3.0 tonne
Reprocessing low-level wastes w/ transuranics	Packaged solids and cladding	(a)	–	–
High level wastes w/ transuranics	Solid	Rail	500 casks (141.5 m^3)	10 casks (2.88 m^3)
Plutonium oxide	Non-dispersible solid	Truck	15,000 kg^b per chemical plant	30 kg^b
Fabrication wastes w/ transuranics	Packaged solids	Truck	(c)	–
Recycle fuel PuO$_2$ + UO$_2$	Fuel assemblies	Truck	–	–
Uranyl nitrate	Liquid	Truck	500 tonne	5 tonne
Uranium oxide from mine	Powder (yellow cake)	Truck Rail	15,000 tonne	15 tonne
UF$_6$ (natural)	Powder (natural U)	Truck	10,000 tonne	11 tonne (11 tonne/cask)
UF$_6$ (enriched)	Powder (enriched U)	Truck	750 tonne	11 tonne (2.2 tonne/cask)
Fresh fuel	Fuel assemblies	Truck	30 tonne	3 tonne

[a]Miscellaneous low and intermediate level wastes are currently stored onsite at the reprocessing plant. Decontamination of reprocessing sites upon decommissioning will probably result in shipment of these wastes to a repository.

[b]Very limited quantities of plutonium have been shipped to the mixed oxide fabrication plants. In addition, the quantity of plutonium permitted in shipping containers varies widely. Thus, it is not possible to estimate the quantities in this path with any accuracy at this time. Shipment of plutonium would range from 900 kg/yr for a 30 tonne mixed oxide plant to 4,500 kg/yr for a 150 tonne plant assuming a plutonium content of 3% in the fabricated assemblies.

[c]Insufficient information exists at this time to estimate these quantities. However, it is suspected volumes will be large with plant startups and then decrease gradually.

across the country occasions the most imminent chance of exposure to radiation of large segments of the population from routine operations in any portion of the nuclear fuel cycle. This consideration, which may account for as much as 20% of the total potential population dose from the operation of any nuclear power plant, however low it actually may be, is the reason why the transportation of nuclear materials must be included in any estimation of the environmental impact of nuclear power plants.

SHIPMENT OF SPENT FUEL

The spent fuel that is discharged from the reactor at the end of its scheduled burnup period is too radioactive to ship or reprocess immediately and is usually placed under many feet of water in a cooling basin. There the shorter-lived fission and activation products in the fuel are allowed to decay and the fission product heat in the fuel elements decreases proportionately. At the end of a storage period of the order of 4-6 months, the fuel is shipped to the reprocessing plant. At this time it still contains a large part of the uranium-235 initially present in the fresh fuel, some plutonium of varying isotopic composition (about 400 g Pu/MW/yr), most of the longer-lived fission products, including most of the krypton-85 as well as appreciable quantities of tritium, and residual amounts of the shorter-lived activation and fission products. Zirconium-95 with its daughter, niobium-95, will be the predominant gamma-ray emitter.

The material will still generate heat and must be shipped in restricted quantities and configurations because of criticality considerations. Table 128 lists the principal emitters that will constitute the source material for any computation of the radiation field surrounding a shipping container; additional data will be required on the relative intensities of the radiations emitted. High burnup fuel will also generate some fast neutrons from curium-244 and requires neutron shielding. Figure 91 shows the rate of heat generation per ton of spent fuel for various burnup conditions. At the time of shipment from an LWR the fuel may generate heat of the order of 18 kW/tonne.

Any shipping container for spent fuel, then, has to provide a capability to move safely a noncritical quantity of spent fuel in such a way that the integrity of the fuel elements is never endangered, that the heat generated is safely dissipated, and that the gamma-radiation and in some cases fission neutrons are so well absorbed that the radiation fields around the outside of the container permit its movement on public highways and railway lines. The physical characteristics and dimensions of some fuel assemblies are given in Table 129.

Table 128. Radioactive Emitters and Thermal Energy in Irradiated Fuel After Various Cooling Periods (5)

	Cooling Period (in days)			
	90	150	365	3650
Radioactivity of Irradiated Fuel (curies per metric ton of uranium)				
Fission Products	6.19×10^6	4.39×10^6	2.22×10^6	3.17×10^5
Actinides (Pu, Cm, Am, etc.)	1.42×10^5	1.36×10^5	1.24×10^5	
Total	6.33×10^6	4.53×10^6	2.34×10^6	
Predominant Fission Products in Gaseous Form Included in Radioactivity of Irradiated Fuel (curies per metric ton of uranium)				
Krypton-85	1.13×10^4	1.12×10^4	1.08×10^4	6.05×10^3
Xenon-131m	1.06×10^2	3.27	1.08×10^{-5}	
Iodine-131	3.81×10^2	2.17	1.98×10^{-8}	
Thermal Energy in Irradiated Fuel (watts per metric ton of uranium)				
Thermal Energy	2.71×10^4	2.01×10^4	1.04×10^4	1.06×10^3

To meet the needs of the nuclear power industry and the above requirements, specially designed shipping casks must be provided and these represent a substantial financial investment. U.S. NRC regulations (7,8) stipulate that shipping casks meet the following specifications:

1. The cask must be able to withstand an impact equivalent to a drop of 30 feet onto a nonyielding (concrete) platform without losing its integrity.

2. It must be able to withstand impact on an 8-inch steel spike, 6 inches in diameter without losing integrity.

3. Radiation levels when fully loaded must never exceed the dose levels in Table 130.

4. It must be easy to clean and decontaminate, and contamination levels must meet requirements in Table 130.

5. Either natural or forced air cooling must be adequate to prevent damage to the shielding materials at all times, with provision for accidental damage or interruption of a forced-circulation cooling system.

6. The size and weight of the cask and any carriage must meet all state regulations and load limits for movement by road or rail.

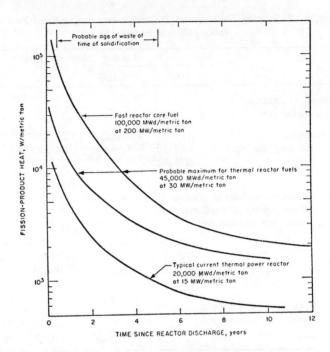

Figure 91. Heat content of spent nuclear fuel (6).

Table 129. Typical Spent Fuel Characteristics Affecting Fuel Shipment (1)

Characteristics	LWR (PWR)	HTGR	FBR
Average specific power (kWt/kgH)[a]	41	73	120-160 (core)
Average discharge exposure (MWD/kgH)	33	94	70-100 (core) 40-50 (fuel + blanket mixture)
Fuel type	UO_2 or $(Pu, U)O_2$ in Zr tubes	Th/U ceramic coated particles in graphite blocks	$(Pu, U)O_2/UO_2$ in stainless steel tubes
Fuel assembly geometry	204 rods, square array, 8.4 inch side, 12 feet long	210 fuel cavities, 108 coolant holes, hex graphite block 14.2 inch across flats, 31.2 inch high	217-331 rods, hex array, ~6 inch across 9-15 feet long
kgH/fuel assembly	~ 450	~ 10	~ 65 (core)
Fissile content, % Fabricated fuel Spent fuel	2.5-3.5 1-2 (open cycle) 2-3 (Pu recycle)	6-7 2-3	12-15 (av. core) 6-8 (core + blanket mixture)

[a]kgH = kilograms of "heavy" fuel material (U, Pu).

Table 130. Container Design Requirements (Maximum Permissible Levels)

	Normal Conditions	Accident Conditions
External Radiation Levels		
Surface	200 mR/hr	
3 ft from surface		1000 mR/hr
6 ft from surface	10 mR/hr	
Permitted Releases		
Noble gases	None	1000 Ci
Contaminated coolant	None	0.01 Ci alpha emitters
		0.5 Ci mixed fission products
		10 Ci iodine
Other	None	None
Contamination Levels		
Beta and gamma	2200 dpm/100 cm^2	
Alpha	220 dpm/100 cm^2	

In order to meet these specific requirements, which have also been categorized by the IAEA (9,10), a number of designs of specialized containers have evolved. Figure 92 shows a cut-away diagram of a water-cooled container of French design with an exchangeable "basket" to hold BWR or PWR fuel elements. It also contains a layer of neutron shielding to guard against spontaneous fission emissions. The shielding material in many cask designs is depleted uranium in order to economize on size and weight for a given required attenuation. It has to be supported by steel shells to provide the necessary mechanical strength. The fins and end plates serve to cushion impacts. Figure 93 shows the General Electric IF-300 shipping cask, one of the largest commercial units, and Table 131 presents its weight and dimensional parameters when loaded with either PWR or BWR fuel. Note the total weight of the order of 80 tons, and the skid length of 37.5 ft ($\sim$ 11.4 m). Clearly, the movement of such massive containers creates its own impact on the highways; for instance, the overall length of the transporter for such a cask shipment is 77-88 ft (23.5-27 m).

Shipment may be made by road, rail or barge, and the latter two permit movement of bigger containers, usually with a lower population exposure. However, in the United States only a small fraction of the reactor sites have a rail connection or can utilize barge traffic on a routine basis. In Europe, rail and barge movement may be more attractive, but this is limited by problems of bridge clearances and tunnel dimensions.

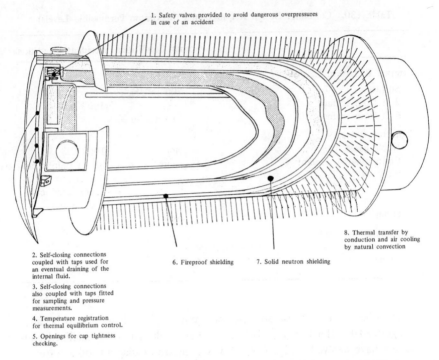

1. Safety valves provided to avoid dangerous overpressures in case of an accident

2. Self-closing connections coupled with taps used for an eventual draining of the internal fluid.

3. Self-closing connections also coupled with taps fitted for sampling and pressure measurements.

4. Temperature registration for thermal equilibrium control.

5. Openings for cap tightness checking.

6. Fireproof shielding

7. Solid neutron shielding

8. Thermal transfer by conduction and air cooling by natural convection

Figure 92. Cutaway view of spent fuel shipping cask (Genas 10-24).

Table 131. Capacity, Weight and Dimensional Parameters of IF-300 Cask

	BWR	PWR
Capacity		
Assemblies	18	7
Approximate loaded weight		
Cask only, lb	140,000	136,000
Skid, enclosure and cooling system, lb	35,000	35,000
Total system, lb	175,000	171,000
Cavity		
Length, inches	180	169
Diameter, inches	37.5	37.5
Cask Body		
Length, inches	208	198
Nominal diameter, inches	64	64
Skid		
Length, ft	37.5	37.5
Width, ft	8	8

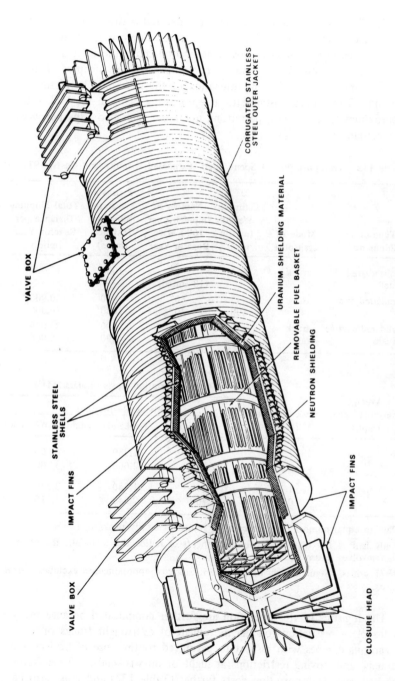

CORRUGATED STAINLESS STEEL OUTER JACKET

URANIUM SHIELDING MATERIAL

REMOVABLE FUEL BASKET

NEUTRON SHIELDING

VALVE BOX

STAINLESS STEEL SHELLS

IMPACT FINS

IMPACT FINS

VALVE BOX

CLOSURE HEAD

Figure 93. IF-300 spent fuel shipping cask (5).

Table 132, showing cask shipments for fuel and wastes (3), indicates the difference in weight possible for rail or truck shipments. With the projected expansion of nuclear power generation, both the number of shipments and the number of required containers will rise steeply. This will place a premium on efficient utilization of the casks and rapid turnaround. The average distance to a reprocessing plant in the U.S. will be approximately 500 miles, permitting not more than two trips per week per container.

Table 132. Transportation of Spent Fuel: Estimated Distance and Costs, 1971 (3)

Type of Shipment	Mode of Transport	Estimated Weight (metric tons)	Number of Shipments per 1000 MWe Reactor Year	Estimated Average Shipping Distance (miles)	Total Shipping Distance per Reactor Year (miles)
Unirradiated fuel	truck	24	6[a] (18 initial)[a]	1000	12,000[b]
Irradiated fuel	truck	35	60[a]	1000	120,000[b]
	rail	100	10[a]	1000	20,000[b]
Solid radioactive waste	truck	16	46	500	23,000
	rail	80	11	500	5,000

Average Transport Speed (mph)	Cask Capacity (MTU)	Transportation Costs (dollars/MTU)[c]		
		Driving Restrictions		
		None	Weekends	Nights & Weekends
25	1.35	1,550	1,980	2,960
	0.90	2,040	2,600	3,900
36	1.35	1,340	1,780	2,370
	0.90	1,760	2,340	3,110

[a]Plus an equal number of shipments for return of empty packagings.
[b]Only half of this distance involves shipments of radioactive material; the other half involves return of empty packagings.
[c]1971 prices. Standard trip of 500 miles; cask cost approximation excludes freight costs.

The logistics of this operation are further complicated because many states place restrictions on the movement of overweight trucks or hazardous cargoes with respect to permitted routing, use of bridges and tunnels, and driving restrictions at night or on weekends. These factors tend to raise transportation costs further (Table 132) and must also be considered in planning alternative shipping routes for the purpose of computing population exposure. Alternate routes must be indicated in environmental statements to establish still acceptable population doses in case the preferred route is closed or unavailable for any reason.

SHIPMENT OF SOLID RADIOACTIVE WASTES

As we have seen in Chapter 7, each nuclear power plant accumulates substantial quantities of low-level and intermediate-level radioactive wastes. The low-level wastes arise from laboratory glassware, contaminated lab coats, dust filters, and miscellaneous tools and supplies used in the handling of fuel elements, maintenance of exposed components and similar operations. At some plants no effort is made to identify and separate contaminated from uncontaminated wastes of this type, giving rise to the shipment of very large quantities of such "soft," dirty materials.

The intermediate-level wastes are composed of spent demineralizer resin, which contains the bulk of the activity dissolved in the primary and secondary coolant loops, evaporator "bottoms," contaminated HEPA filters and other filter materials, and dewatered sludges from settled contaminants from storage and holdup tanks. This material constitutes the bulk of any activity lost from the fuel by leakage in routine operations.

The low-level wastes may be shipped to government or commercial burial sites for burial in trenches in a licensed burial area. Intermediate level wastes may be combined with high-level wastes for ultimate disposal or may be stored to decay in subsurface storage vaults. Highly tritiated water from recirculated cooling water may have to be considered for shipment separately at infrequent intervals. Table 133 indicates estimated quantities of waste per reactor in these categories and the number of truckloads involved. It is evident that the number of these shipments is vastly greater than for spent fuel (but more material can be shipped per trip since no criticality problems exist) subject only to the limitation on the external radiation level, which must not exceed 10 mR/hr at a distance of 6 ft (2 m) from the container surfaces. To guard against any possibility of leakage from damp waste material, the drums should be double-walled with an absorbent packing, such as vermiculite. Still better, at most plants this waste now is mixed with cement or bitumen before being loaded into the drums (12,13).

High-Level Wastes

High-level wastes constitute the final product of reprocessing plants. They would be composed primarily of concentrates of cesium-137 and

**Table 133. Basic Estimates: Solid Radioactive Wastes from
a Nuclear Power Plant (5)**

Soft solid wastes compacted in 55-gallon drums

    100 drums produced per 1100 MWe reactor year

    1 curie of radioactivity per drum

    2 truck shipments per year; 50 drums per truckload

    1 rail car shipment per year; 100 drums per carload

1000 miles shipping distance from power plant to waste disposal site.
If the waste burned in an open fire, it is unlikely that much of the activity would be widely dispersed. Most of the activity, perhaps as much
as 99%, would remain in the ashes.

Resins, sludges, etc. dewatered and consolidated in 55-gallon drums

    3000 drums produced per 1100 MWe reactor year

    98% – Type A packages, limited to 20 curies/drum. About 3% low
        level compacted wastes and 95% average less than 0.3 curie
        per drum.

    2% – Type B packages, 100 Ci maximum estimated activity per
        package; average estimated about 20 curies per drum.

    60 truck shipments per year; 50 drums per truckload

    20 rail car shipments per year; 150 drums per carload

    500 miles shipping distance to waste disposal site.

Because of the form of the material, it is very unlikely that any significant
amount of the activity in material burned in an open fire would be released,
probably less than 10^{-5} of the activity in the contents.

strontium-90, with small admixtures of other long-lived fission products,
e.g., Ce-144, Ru-106 and Pm-147. In addition, trace concentrations of
the actinide elements, Np, Am, Pu and Cm, would be contained in them.
Current practice dictates storage of high-level wastes at the reprocessing
plant for periods of up to five years or more. The long-lived waste to
be shipped for burial will then be converted to solid form by one of the
calcination processes or by vitrification or cementation. The resulting
material will be in essentially nonleachable form in drums or metal containers that must meet all the mechanical heat dissipation and radiation
requirements outlined on page 481. There would not be many shipments
of this kind, and it is premature to indicate the necessary precautions or
the means of transport until the various problems associated with waste
disposal have been solved in a definitive fashion (Chapter 12).

CALCULATION OF RADIATION DOSE

Two population groups must be considered in calculating doses from exposure to radioactive waste shipments: (a) the drivers and handling crews occupationally exposed to the waste radiation, and (b) the general population living along the shipping route that is exposed momentarily to radiation from a passing truck, including the occupants of other vehicles on the highway. Two exposure conditions have to be evaluated: (a) routine shipments proceeding at a reasonable average speed, and (b) accidents involving the shipment, resulting in possible escape of activity or the close assembly of an abnormally large group of people for a protracted period.

Routine Shipments

For purposes of calculation it is assumed that the truck or rail car is loaded to the limit of external radiation permitted (10 mR/hr at 6 ft from the surface), which is about equivalent to 10 mR/hr at 13 ft (4 m) from the effective centerline of the shipment. The dose rates and doses are estimated by considering the source to be an isotropic point source located at the centerline of the shipping container. The average gamma-ray energy can be assumed to be about 1 MeV.

The dose rate as a function of distance from the shipping container is calculated by

$$D_{(r)} = \frac{I \, e^{-\mu r_1}}{r^2} \, B_{(E,Z,\, \mu r_1)} \tag{88}$$

where: I = source output, (mrem-ft^2/hr) or (mrem-m^2/hr)
r = source to receptor distance (ft or m)
r_1 = radial thickness of attenuating materials
μ = linear attenuation coefficient (ft^{-1} or m^{-1})
B = build-up factor

For air, neglecting self-absorption of the container, $\mu = 2.5 \times 10^{-3}$ ft^{-1} = 8.02×10^{-3} m^{-1}.

Since the source is assumed to move at constant speed, the dose received represents the contribution to dose at a distance r from the line of movement integrated over the time during which the shipment passed by. In practice it is convenient to divide the areas adjoining the railroad or highway into area bands over which the dose is assumed to be essentially uniform. The integrated dose received at a point within a given zone from a shipment moving at a given speed can then be calculated and related to the population density within that zone to yield the population exposure.   Figure 94 shows the assumed geometry of the situation, where other highway traffic is ignored.   For instance, the population dose within 1/2 mile of the route of travel can be calculated by considering the integrated dose at six intervals between 100 and 2500 ft from the centerline of the right of way.   Table 134 shows an example of the individual doses obtained for each zone for a fully loaded shipment moving at two different speeds.

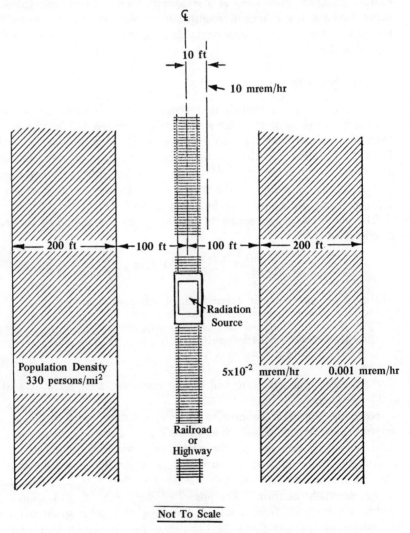

Figure 94.   Assumed population distribution along shipping route (5).

Table 134. Estimated Individual Dose at a Given Distance from the Apparent
Centerline of the Shipping Route for the Passage of One Shipment[a](2)

Distance (meters)	Dose at 320 km/day[b] (mrem)	Dose at 966 km/day[b] (mrem)
25	4.24×10^{-4}	1.4×10^{-4}
50	2.11×10^{-4}	6.97×10^{-5}
75	1.49×10^{-4}	4.59×10^{-5}
100	1.04×10^{-4}	3.41×10^{-5}
125	8.22×10^{-5}	2.7×10^{-5}
150	6.81×10^{-5}	2.25×10^{-5}
200	4.99×10^{-5}	1.65×10^{-5}
300	3.22×10^{-5}	1.06×10^{-5}
400	2.34×10^{-5}	7.76×10^{-6}
500	1.82×10^{-5}	5.99×10^{-6}
600	1.45×10^{-5}	4.79×10^{-6}
700	1.2×10^{-5}	3.97×10^{-6}
800	1.02×10^{-5}	3.36×10^{-6}
900	8.6×10^{-6}	2.88×10^{-6}
1000	7.5×10^{-6}	2.5×10^{-6}
Population dose (person-rem per km)	1.1×10^{-5}	3.7×10^{-6}

[a]Based on 10 mrem per hour at 13 feet ($\sim$4 m) from the apparent centerline of
the shipping route.
[b]Accurate to one significant figure.

The population dose then can be calculated for specific routes by taking
into account different patterns of settlement. By dividing the routes into
segments according to population densities, maximum population doses
for present distributions and projected population patterns can be calculated. For cities the population density can be averaged over the whole
affected area, whereas for country areas the population may be assumed
to be concentrated along a narrow band paralleling the highways, but not
expressways, within about 300 ft (100 m) from the road centerline.

Such calculations must be done for all types of shipment from a
nuclear plant taking into account the relative number of trips for spent
fuel and waste shipments. In addition, radiation exposures of truck
drivers, railroad personnel, and freight handlers must be calculated. The
truck driver and his mate will receive the most sustained radiation exposure during each trip, and the loading of the waste and the shielding of
the cab must be planned to ensure that they will not be exposed to doses
in excess of 300 mrem/yr each under any conditions unless they are treated

as radiation workers. Consideration must also be given to finding parking locations and exposure conditions at overnight stops that should permit as much isolation of the shipment as possible and some limitation on possible exposure of curious onlookers or overnight guests. Table 135, as an example, shows the result of such a calculation for an 1100 MWe reactor (5), assuming different modes of transportation and about 500 mile shipping distances. Note the small order of magnitude of the population doses actually obtained.

Table 135. Estimated Radiation Doses from Transportation under Normal Conditions per Reactor Year (5)

Unirradiated Fuel (by truck only)

	Man-rem	Number of People
Transport workers	0.01	40
General public: onlookers	0.0003	60
people along the route	0.001	3×10^5

Irradiated Fuel	Truck		Rail		Barge	
	Man-rem	No. People	Man-rem	No. People	Man-rem	No. People
Transport workers	1.2	4	0.05 $(2.6)^a$	100 $(22)^a$	0.04	10
General public:						
onlookers	0.8	600	0.1	100	–	–
people along the route	1	3×10^5	0.2	3×10^5	0.03	1×10^5

Solid Waste	Truck		Rail	
	Man-rem	No. People	Man-rem	No. People
Transport workers	1	4	0.05	100
General public:				
onlookers	0.6	500	0.1	100
people along the route	0.4	1.5×10^5	0.1	1.5×10^5

[a]For shipments transported by truck from the reactor site to a nearby railroad, transferred from truck to railroad car, and shipped by railroad car to the fuel recovery plant.

Accident Conditions

Since the shipments will take place over public rights-of-way, they are subject to the same risks of accidents due to collisions, overturns or fires as other freight shipments. Though size and weight of the shipping casks make it unlikely that they will become seriously damaged in any collisions, indirect escape of activity due to upsets or fires can be visualized, however remotely, and the consequent radiation exposure to the population must be estimated. Other injuries may be incurred just as for any other heavy shipment, and these can be estimated on the basis of official accident statistics (4). Table 136 shows U.S. traffic accident probabilities from common causes. If one includes all shipments to and from a nuclear plant, including cold fuel and empty fuel casks, the total number of truck miles would be about 155,000 per year, if all transport were by truck.

Table 136. U.S. Traffic Accident Statistics—Common Causes (5)

Mode	Data Year	Probability (Accidents/Vehicle-Mile)	Injuries per Accident	Fatalities per Accident
Truck	1969	1.7×10^{-6}	0.51	0.03
Rail	1969	1.4×10^{-7} [a]	2.7	0.2
Barge	1970	1.5×10^{-6}	0.06	0.0

[a]Single rail car

Based on the data in Table 136, this might cause 0.1 injuries and 0.01 fatalities per reactor year (5). Since the accident probability varies with speed as well as the accident severity and the chance of a fire in a collision, actuarial tables have been prepared to indicate such probabilities. Table 137 defines categories of severity associated with different impact velocities and indicates accident probabilities in each of these severity categories, with the assumed duration of a concomitant conflagration. The barge data are inserted in the corresponding severity category; all barges are assumed to move at less than 10 miles/hr.

Although a spent-fuel container is made of fire-retardant material and is assumed immune to direct impact, an accident in the "severe" or worse category may be assumed to result in damage to the irradiated fuel elements, with consequent release of some of the gaseous activity contained. Table 138 lists the fraction of the activity that is contained in fuel rod

Table 137.   Traffic Accident Probabilities (5)

Severity Category	Vehicle Speed (mph)	Fire Duration (hr)	Probability per Vehicle Mile		
			Rail	Truck	Barge[a]
Minor	0-30	<1/2	6×10^{-9}	6×10^{-9}	–
	0-30	0	4.7×10^{-7}	4×10^{-7}	1.6×10^{-6}
	30-50	0	2.6×10^{-7}	9×10^{-7}	1.4×10^{-7}
Total			7.3×10^{-7}	1.3×10^{-6}	1.7×10^{-6}
Moderate	0-30	1/2-1	9.3×10^{-10}	5×10^{-11}	–
	30-50	<1/2	3.3×10^{-9}	1×10^{-8}	8×10^{-9}
	50-70	<1/2	9.9×10^{-10}	5×10^{-9}	2×10^{-9}
	50-70	0	7.5×10^{-8}	3×10^{-7}	3.4×10^{-8}
Total			7.9×10^{-8}	3×10^{-7}	4.4×10^{-8}
Severe	0-30	>1	7.0×10^{-11}	5×10^{-12}	–
	30-50	>1	3.9×10^{-11}	1×10^{-11}	9.3×10^{-11}
	30-50	1/2-1	5.1×10^{-10}	1×10^{-10}	1.3×10^{-9}
	50-70	1/2-1	1.5×10^{-10}	6×10^{-12}	3.3×10^{-10}
	>70	<1/2	1×10^{-11}	1×10^{-10}	–
	>70	0	8×10^{-10}	8×10^{-9}	–
Total			1.5×10^{-9}	8×10^{-9}	1.6×10^{-9}
Extra Severe	50-70	>1	1.1×10^{-11}	6×10^{-13}	2.3×10^{-11}
	>70	1/2-1	1.6×10^{-12}	2×10^{-13}	–
Total			1.3×10^{-11}	8×10^{-13}	2.3×10^{-11}
Extreme	>70	>1	1.2×10^{-13}	2×10^{-14}	–
Total			1.2×10^{-13}	2×10^{-14}	–

[a]Barge accident probabilities are based on the duration of the fire and actuarial data on cargo damage. The impact velocities of all barge accidents were considered to be less than 10 mph, but for the purposes of this table, minor cargo damage is assumed to be equivalent to vehicle impact speeds of 0-30, moderate cargo damage 30-50 and severe cargo damage 50-70.

void space and hence may escape under accident conditions. In case of a fire or explosion the possibility of active particulates being scattered in the locality from damaged fuel or fractured waste shipments must be considered. Control of crowds and curious onlookers is obviously a prime prerequisite. Protection and training of the truck driver or railroad staff is of importance, as is prompt and informed cooperation of local law enforcement agencies. The main hazard may be assumed to arise from

Table 138. Estimated Radioactivity in Spent Fuel Shipments Available for Release
in the Event of Fuel Rod Cladding Rupture (14)

Type of Radioactive Material	Total Inventory 150 Days Cooling (Ci/tonne)	Percentage in Void Spaces of Fuel Rods and Percentage Leached by Coolant	Activity in Void Spaces (Ci/tonne)
Kr-85	1.12×10^4	30	3.4×10^3
I-131	2.17	2	4.3×10^{-2}
Other fission products	4.38×10^6	0.01[a]	400
Actinides (Pu, Am, Cm)	1.36×10^5	–	negligible[b]
Xe-131m	3.27	2	0.1[b]
I-129	2×10^{-3}	30	6×10^{-4}[b]
H-3	6.92×10^2	1	7[b]

[a]A conservative (high) value estimated on the basis of leaching the outer 1.2×10^{-5} inches from the surface of the uranium oxide fuel.
[b]Due to the small amounts present, the dose contribution from Xe, I-129, H-3, and the actinides may be neglected compared to the doses from the other radionuclides.

inhalation of any released activity, of which iodine may be considered most important.

Table 139 shows an example of the probability of people receiving a skin dose from Kr-85 or a thyroid dose from a given release of I-131, in case of a rail accident, resulting in substantial escape of activity from void space, in an area with a population density of 300 persons per square mile. The population dose from accidental releases of activity per year is given by

$$D_{(acc)} = \sum_i \text{Mileage} \times P_{acc} \times I_i \times L_i \times P_{exp} \times \rho \, \bar{t} \qquad (89)$$

where: P_{acc} = probability per mile of accident of appropriate severity
P_{exp} = individual exposure probability per isotope i
I_i = total inventory of isotope i in shipment
L_i = escape probability during accident
ρ = population density within 0.5 mile of route
t = estimated mean individual exposure time following accident.

Again, separate estimates must be made for transportation personnel.

Table 139. Calculated Doses from Severe Rail Accidents Involving Spent Fuel (5)

Organ		Centerline Dose[a] (rem)	Average Dose[a] (rem)
Kr-85	Skin	1.2	0.06
	Bone marrow, gonads, lens of the eye	0.02	8×10^{-4}
I-131	Thyroid	0.02	1×10^{-3}
Gross Fission Products	Bone	6	0.3
	Lung	8	0.4

[a]The radioactive material would be distributed downwind from the accident so that the isopleth (*i.e.*, boundary lines of equal doses) would be cigar-shaped. The centerline dose is the dose that might be received by a person on the centerline of that pattern at a distance of 50 meters from the accident.

In all cases published so far, the estimated conservative population exposures from postulated transportation accidents have been a small fraction of the routine-shipping population dose. Nevertheless, the possibility of finding oneself a part in an accident involving high-level radioactivity remains a major concern to the public. For this reason, maintenance of reasonable speeds, along routes that avoid dense population areas as much as possible, would be sensible and in accord with the desire to keep potential impact "as low as practicable." Training of personnel and the existence of well-understood and executed emergency procedures is essential, preferably with constant mobile radio contact. On balance, accident causes other than radiation events will tend to be the only significant ones (Table 140).

NUCLEAR SAFEGUARDS

A closely related subject, though not strictly of an environmental character, is that of nuclear safeguards. By this are meant the procedures that ensure the physical integrity and safety of any shipment involving actually or potentially fissile material, as well as procedures to provide for accurate record-keeping and precise inventory assays to prevent or readily detect any unauthorized diversion of fissile material (15-18). From the nature of the material, the size of the container, the high level of radioactivity, which can be readily traced, and the very special, remote-handling facilities needed to extract the valuable fissile contents,

Table 140. Comparative Environmental Impact of Transportation of
Fuel and Waste for one LWR (5)

Normal Conditions of Transport	
	Number of Shipments per Year
Unirradiated fuel and return of empty containers	12 truckloads
Irradiated fuel and return of empty containers	120 truckloads or 20 railcarloads or 10 barges
Solid radioactive wastes	46 truckloads or 11 railcarloads

	Environmental Impact
Heat, weight, and number of shipments	Negligible

Radiation Doses	Number of Persons Exposed	Estimated Dose Range to Exposed Individuals	Cumulative Dose to Exposed Population
Transport Workers	200	0.01 to 300 millirem/year	3 man-rem/year
General Public			
onlookers	1,100	0.003 to 1.3 millirem/year	2 man-rem/year
along route	600,000	0.0001 to 0.06 millirem/year	

Accidents in Transport	
	Environmental Risk
Radiological Effects	small
Common (nonradiological) causes	1 fatal injury in 100 years 1 non-fatal injury in 10 years $475 property damage per year

one may assume that diversion by theft or highjacking of spent fuel shipments is highly unlikely. Sabotage, of course, can never be fully eliminated and for that reason this operation, like all aspects of nuclear power, must be considered "hostages of civil law and order." (17).

Illegal diversion and consequent accidental dispersion of fissile materials may be both simpler, and on balance more profitable, by attacking shipments of fresh fuel, enriched fuel material or reprocessed fissile concentrates. Hence, these materials may require special safeguarding during transportation and appropriate precautions. It is to minimize the shipping distances and safeguarding the material in transit that the establishment of "nuclear parks" has been suggested, where as many of the related fuel cycle operations as possible would be located on adjoining sites. However, from the environmental impact point of view, such a concentration of nuclear activities may create additional problems.

REFERENCES

1. Golan, S. and R. Salmon. "Nuclear Fuel Logistics," *Nucl. News* **16** (2), 47 (1973).
2. U.S. Environmental Protection Agency. "Environmental Analysis of the Uranium Fuel Cycle; Pt. I—Fuel Supply," Report EPA-520/9-003B (1973).
3. "The Nuclear Industry, 1974," Report WASH 1174-74 (Washington, D.C.: U.S. Atomic Energy Commission, 1974).
4. Kelly, O. A. and W. C. Stoddart. "Highway Vehicle Impact Studies: Tests and Mathematical Analyses of Vehicle, Package and Tiedown Systems Capable of Carrying Radioactive Material," Report ORNL-NSIC-61, Oak Ridge National Laboratory (1970).
5. "Environmental Survey of Transportation of Radioactive Materials to and from Nuclear Power Plants," (Washington, D.C.: U.S. Atomic Energy Commission, 1972).
6. Schneider, K. T. "Solidification and Disposal of High-Level Radioactive Wastes in the United States," *Reactor Technol.* **13**, 387 (1970).
7. Conlon, B. and G. L. Pettigrew. "Summary of Federal Regulations for Packaging and Transportation of Radioactive Materials," U.S. Public Health Service Report BRH/DMRE 71-1, U.S. Dept. of Health, Education and Welfare (1971).
8. "Transportation of Nuclear Spent Fuel," Administrator's Guide No. 3, SINB, Atlanta (1972).
9. Gibson, R., Ed. *The Safe Transport of Radioactive Materials* (Oxford: Pergamon Press, 1966).
10. "Regulations for the Safe Transport of Radioactive Materials." Safety Series No. 6, (Vienna: International Atomic Energy Agency, 1967), Revised (1973).
11. Brobst, W. "Transportation of Nuclear Fuel and Waste," *Nucl. Technol.* **24**, 343 (1974).
12. Burns, R. H. "Solidification of Low- and Intermediate-Level Wastes," *Atomic Energy Rev.* **9**, 547 (1971).
13. Kibbey, A. H. and H. W. Godbee. "A Critical Review of Solid Radioactive Waste Practices at Nuclear Power Plants," Report ORNL-4924; Oak Ridge National Laboratory (1974).
14. "The Safety of Nuclear Power Reactors (Light-Water-Cooled) and Related Facilities," Report WASH-1250 (Washington, D.C.: U.S. Atomic Energy Commission, 1973).

15.   Leachman, R. B. and P. Althoff, Ed. *Preventing Nuclear Theft: Guidelines for Industry and Government* (New York: Praeger, 1972).
16.   "Safeguards Techniques," Proc. Karlsruhe Symposium, 1970 (Vienna: International Atomic Energy Agency, 1970).
17.   Weinberg, A. M. "Social Institutions and Nuclear Energy," *Science* **177**, 27 (1972).
18.   Willrich, M. and T. B. Taylor. *Nuclear Theft: Risks and Safeguards* (Cambridge, Massachusetts: Ballinger, 1974).

ADDITIONAL REFERENCES

"Advisory Material for the Application of the IAEA Transport Regulations," Safety Series No. 37 (Vienna: International Atomic Energy Agency, 1973).

Blackburn, R. W. "The Transportation of Radioactive Materials," (Ottawa, Canada: Atomic Energy Control Board, 1967).

Blum, P., H. Baatz and J. Mangusi. "Practical Experience in Spent Fuel Shipping and Relation to Cask Concepts," (New York: Transnuclear Inc., 1971).

Bresson, G., Y. Demoures, M. Lagorce and J. Hebert. "Aspects of the French Regulation in the Nuclear Field," *Proc. 4th International Conf. on Peaceful Uses of Atomic Energy, Geneva* **3**, 409 (1971) United Nations.

Brobst, W. A. "Transportation Accidents: How Probable?" *Nucl. News* **16** (7), 48 (1973).

Eynon, W. J. "Transporting Large NSSS Components," *Nucl. News* **14** (7) 35 (1971).

Groetch, D. J., Ed. "Spent Nuclear Fuel Transfer: Fuel Casks and Transfer Operations," Proc. 1971 AIME Symp., Am. Soc. Mech. Engrs. (1972).

"International Traffic of Radioactive Materials," U.S. Atomic Energy Commission Report WASH-2808 (Atlanta: Southern Interstate Nuclear Board, 1966).

Luten, D. B. "The Economic Geography of Energy," *Scient. Amer.* **224** (3), 165 (1971).

MacDonald, C. E. and R. H. Odegaarden. "Licensing of Packages for the Transport of Radioactive Materials," *Trans. ANS* **16**, 173 (1973).

"Packaging of Radioactive Material for Transport and Transportation of Radioactive Material Under Certain Conditions," Title 10 Code of Federal Regulations, Part 71 (10CFR 71 and 49 CFR 173).

Rose, D. J. "Nuclear Eclectic Power," *Science* **184**, 351 (1974).

Shappert, L. B., W. A. Brobst, L. W. Langhaar and J. A. Sisler. "Probability and Consequences of Transportation Accidents Involving Radioactive-Material Shipments in the Nuclear Fuel Cycle," *Nucl. Safety* **14**, 597 (1973).

Simens, H. G. and A. C. Cornish. *Shipping Radioactive Materials.* (San Francisco: Bechtel Corp., 1970).

Tarnuzzer, E. C. "Spent Fuel Shipping Experience—Yankee Atomic Electric Company," *Trans. ANS* **16**, 174 (1973).

Thompson, J. T. and J. M. Morgan. "Nuclear Transportation," *Nucl. Safety* 8, 443 (1967).
"Transportation of Nuclear Spent Fuel," Proc. Conf. CONF-7000207 (Atlanta: Southern Interstate Nuclear Board, 1970).

CHAPTER 11

THE FUEL CYCLE INDUSTRIES

INTRODUCTION

In assessing the environmental impact of nuclear power, attention naturally focuses on the nuclear power plants with their large core inventory of radioactive materials. However, in any determination of the overall effect of utilizing nuclear energy, to a growing extent one should consider the environmental effects of the supporting industries, both in their direct impact on their local environment as consumers of raw materials and energy and as sources of radioactive effluents. Most of the industries involved do not differ substantially in their impact from other chemical or manufacturing plants of comparable size and complexity. However, particular problems do arise in connection with the operation of uranium mines and mills and of reprocessing plants and these will be covered in more detail.

The fuel cycle operations supporting light-water-cooled and gas-cooled reactors naturally divide according to the type of activity involved. Those operations concerned with raw material preparation and fabrication up to the delivery of fuel assemblies to the power plant deal essentially only with natural or enriched uranium (or thorium), that is, long-lived alpha-emitters of low specific activity. In contrast, all "back-end" operations in the cycle following the discharge of spent reactor fuel, including fuel reprocessing and waste disposal, have to handle substantial quantities of highly radioactive beta-gamma emitters and are confronted by operational safety considerations of an entirely different character. Recalling the LWR fuel cycle (Figure 7) without consideration of plutonium recycle, one sees that the specific components comprising the LWR supporting fuel cycle include the following:

1) mining of uranium ore
2) milling and refining ore to produce uranium concentrates (U_3O_8)

501

3) production of uranium hexafluoride (UF_6) from uranium concentrates to provide feed for isotopic enrichment

4) isotopic enrichment of uranium hexafluoride to attain fuel enrichment requirements using the gaseous diffusion or competitive processes

5) fabrication of nuclear reactor fuel including: converting UF_6 to uranium dioxide (UO_2), pelletizing, encapsulating in rods, and assembling fuel elements

6) reprocessing irradiated fuel and converting uranium to UF_6 for recycle through the gaseous diffusion plant for reenrichment

7) radioactive waste management of high-level and lower-level wastes, including long-term storage of wastes

8) transportation activities associated with moving materials to and from each of the above operations.

The environmental considerations relating to the construction and operation of the individual nuclear power plants, including transportation of radioactive materials to and from the power plant, have been considered in the preceding chapters. Conventional manufacturing operations supporting the nuclear fuel cycle have been reviewed and have been judged to be relatively remote, indirect, and fairly insignificant contributors to the environmental impact of the nuclear fuel cycle; typical operations in this supporting category include production of commercial reagents such as sulfuric acid, hydrogen fluoride, ammonia, and nitric acid (1).

As the number of nuclear power plants increases, the number and capacity of plants in the supporting industries must increase too. Figure 95 shows a projection of this growth to the year 2000 for the U.S., based perhaps on rather sanguine expectations of increases in the power capacity. It is this projected growth rate that makes it imperative to ascertain that there are no unusual or threatening consequences associated with this anticipated expansion in the fuel cycle industries. As a matter of policy it appears unrealistic for the operator of each power plant to have to explain and account for the environmental impacts of all the ancillary industries. Instead it would seem more pragmatic to require each of them to keep their effluent levels as low as practicable, both in total quantity and in their content of radioactive materials. In assessing the cost-benefit relationships compared with alternative fuel sources, other considerations such as mine safety and spoil disposal have to be considered (Chapter 13) (3); overall impact effects might well be considered as a matter of national policy.

In reviewing the impact of the prereactor portion of the fuel cycle, one has to assess principally any health hazard arising from the handling of progressively purer uranium compounds and the presence of any natural daughter nuclides. It has usually been assumed that natural

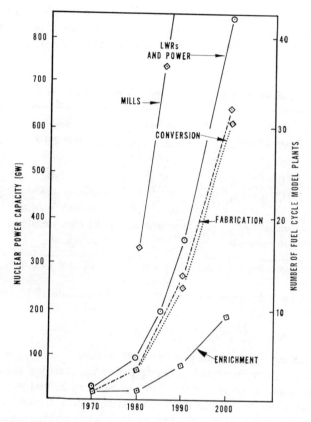

Figure 95. Projection (1973) of nuclear fuel cycle needs (2).

uranium, as U_3O_8 or any mineral form, is essentially only chemically toxic once the shorter-lived radioactive daughters are removed, because of the low specific activity associated with its long half-life and the appreciable self-shielding occurring in any purified bulk concentrate. The original protection guides for uranium grew out of early studies into its chemotoxic and radiotoxic effects. More recently, the significance of the difference in biological half-life associated with different compounds has been investigated. Scott (4) classifies materials in three classes of transportability: "highly transportable," biological half-life of days; "moderately transportable," weeks to months; and "slightly transportable," biological half-life of months to years, where *transportability* refers to the rate at which materials leave the critical organ, without regard for the route or mode of movement.

Table 141 lists some common uranium compounds falling into each of the three categories (4). Any change in isotopic distribution may also affect toxicity. For soluble materials with less than 5-8% U-235, the

Table 141. Classification of Uranium Compounds According to Transportability[a](4)

Highly Transportable	Moderately Transportable	Slightly Transportable
UF_6	UO_2[b]	UO_2[b]
UO_3[b]	U_3O_8[b]	U_3O_8[b]
$UO_2(NO_3)_2$	UO_3[b]	uranium oxides
UF_4[b]	UF_4[b]	uranium hydrides
uranium sulfates	uranium nitrates	uranium carbides
uranium carbonates		salvage ash

[a]Partially developed from data by Steckel and West (1966) and Task Group on Lung Dynamics (1966).

[b]Subjecting a particular uranium compound to higher temperatures tends to decrease rate of transportability.

chemical toxicity to the kidney is the limiting value.* For more highly enriched uranium or irradiated mixed fuel material, radiation hazards present the major toxicity factor. Thus, in handling natural uranium in the form of concentrate (yellow cake) or as natural or low-enrichment fuel material, external radiation exposure is of less consequence than ingestion or inhalation of dust or particles. Table 142 lists the maximum permissible concentrations for uranium dust in process area air. Dusty operations must be well-ventilated and the exhaust air filtered to reduce effluent levels to concentrations below those listed in Table 142 for air and water.

Basically the handling of uranium-containing compounds differs little from the handling of other valuable, high-purity chemical materials and does not require further discussion. There are, however, two operations that do need special consideration because they arise from the fate of the decay products of uranium; these are the mining of uranium ore and the disposal of mill tailings.

---

*Because of the toxicity of natural uranium, U-238, U-236 or U-235, ICRP Publication 6 limits short term exposures of soluble forms of any isotopic mix not to exceed inhalation of 2.5 milligrams of soluble uranium in one day, or ingestion averages over 2 days should not exceed 150 milligrams of soluble uranium.

Table 142. Air and Water Concentration Limits for Uranium Compounds (4)

	Transportability (μCi/cm^3)		
	high	moderate [b]	slight [b]
Process Area Air Limits [a]			
$< 5\text{-}8\%$ ^{235}U	7×10^{-11} kidney	10^{-10} lung	10^{-10} lung
$> 5\text{-}8\%\text{-}97\%$ ^{235}U	6×10^{-10} bone	10^{-10} lung	10^{-10} lung
^{233}U/^{232}U	5×10^{-10} bone	10^{-10} lung	10^{-10} lung
Plant Environs Air Limits [a]			
$< 5\text{-}8\%$ ^{235}U	3×10^{-12} kidney	5×10^{-12} lung	5×10^{-12} lung
$> 5\text{-}8\%\text{-}97\%$ ^{235}U	2×10^{-11} bone	4×10^{-12} lung	4×10^{-12} lung
^{233}U/^{232}U	2×10^{-11} bone	4×10^{-12} lung	4×10^{-12} lung
Plant Effluent Water Limits [a]			
$< 5\text{-}8\%$ ^{235}U	6×10^{-7}	4×10^{-5}	4×10^{-5}
$> 5\text{-}8\%\text{-}97\%$ ^{235}U	4×10^{-6}	3×10^{-5}	3×10^{-5}
^{233}U/^{232}U	4×10^{-6}	3×10^{-5}	3×10^{-5}

[a]International Commission on Radiological Protection, 1959
[b]Considered to be insoluble material.

URANIUM MINES

Uranium occurs in many places in the Earth's crust in the form of a variety of complex ores, such as uraninite $[(UO_2)(UO_3)_x]$, uranothorite $[(Th,U)SiO_4]$, thorianite $[(Th,U)O_2]$, carnotite $(K_2O \cdot 2UO_3 \cdot V_2O_5 \cdot 2H_2O)$, brannerite $[(UO, TiO, UO_2)TiO_3]$, uranophane $(CaO \cdot 2UO_3 \cdot 2SiO_2 \cdot 7H_2O)$ and various phosphate and carbonate minerals. To be economically mineable, the uranium content of the ore must exceed 0.05% U_3O_8 by weight, though lower grades have been treated under exceptional circumstances, such as the South African gold ores, and may be mined in future if scarcity of raw materials forces the market price of yellow cake to a high level (5-7).

Uranium-bearing minerals occur in many mines in conjunction with other ores, such as the historic silver mines at Jachymov (Joachimsthal) in Czechoslovakia, the gold mines of the Witwatersrand, and cobalt and silver occurrences at Port Radium in Canada's Northwest Territory. Several mines, such as the famous Shinkolobwe mine in Katanga (Zaire) have been open-pit operations, but the majority have been underground mines. Compared with coal mines, these mines are considerably safer;

they are hardrock mines, so that danger from roof falls or gas is greatly reduced; the quantity of ore mined is much smaller, and modern mining techniques are applied because these mines are relatively new, with attendant improvement in working conditions. Apart from the normal hazards inherent in any underground mine, the dominant health hazard in hardrock mines is the inhalation of dust, potentially leading to the endemic occupational disease of hardrock miners—silicosis—which affects the breathing ability of the workers. Dust problems, in theory, are mitigated by spraying the work area with water and by adequate ventilation, and an extensive program of lung surveillance has been in force in mining countries for many years. In uranium mines an additional hazard may arise from the inhalation of radon daughters, associated with the emanation of radon gas from exposed uranium-bearing minerals or from radon entrained in underground water through cracks or fissures in such rock. In contrast, the external dose from gamma emitters in uranium-bearing rock can be regarded as negligible.

Radon is a daughter element of uranium-238 and is a noble gas. As Table 143 shows, uranium-238 decays through several intervening daughters to radium-226, which decays to radon-222 with a half-life of 3.82 days before decaying in turn. In nature all the uranium daughters will build up in most minerals to their equilibrium concentration with their parent, in proportion to each half-life. Being a noble gas, radon may diffuse away during its existence, particularly in fractured rock, and may appear at its surface. Reference has been made already, in Chapter 4, to the role it plays in the natural radiation background. At normal breathing rates, very little of the radon would decay while in the body or be retained as such. However, a certain fraction of radon daughters will form in the air and, being relatively short-lived, may attain secular equilibrium concentrations with the radon. As Table 143 shows, the half-lives of the first four successive radon daughters are short; under static conditions in quiet air, radioactive equilibrium will develop in about three hours.

However, an equilibrium state is seldom found in an actively worked uranium mine where fresh air is continually brought into the mine. The amount of fresh air pumped into a mine affects the concentration of the radon daughter products more than it affects the concentration of radon. The radon daughter concentration in the air is reduced by dilution, by adherence to dust particles, and by preferential deposition on mine walls, whereas additional radon is diffusing into the air from the rock surfaces and from mine water. This attachment of the radon daughters on dust particles tends to affect their mobility in air and also the probability of their retention in the lung. The unattached daughter products of radon-222 exhibit a high rate of diffusion in air, quickly becoming attached to

Table 143. Radioactive Decay Series from U-238 to Pb-206; Main Sequence

Common Name or Symbol	Isotope	Half-Life	Principal Radiations	Alpha Energy (MeV)	Gamma-Ray Quanta per Disintegration	Average Gamma-Ray Energy (MeV)
Uranium I	^{238}U	4.49x10^9 yr	α	4.18		
Uranium X$_1$	^{234}Th	24.1 days	β			
Uranium X$_2$	^{234}Pa	1.17 min	β			
Uranium II	^{234}U	248,000 yr	α	4.76		
Ionium	^{230}Th	80,000 yr	α	4.68 (75%)		
				4.61 (25%)		
Radium	^{226}Ra	1,620 yr	α	4.78 (94.3%)		
				4.69 (5.7%)		
Radon	^{222}Rn	3.825 days	α	5.486		
Radium A	^{218}Po	3.05 min	α	5.998		
Radium B	^{214}Pb	26.8 min	β			
			γ		0.82	0.295
Radium C	^{214}Bi	19.7 min	β			
			γ		1.45	1.050
Radium C'	^{214}Po	164 μsec	α	7.68		
Radium D	^{210}Pb	22 yr	β			
			γ		1.0	0.047
Radium E	^{210}Bi	5.02 days	β			
Radium F	^{210}Po	138.2 days	α	5.298		
Radium G	^{206}Pb	Stable	Stable			

moisture or dust particles suspended in the air; the mean lifetime existence as free unattached ions is of the order of 10 to 50 seconds. The human respiratory system retains a substantial portion of radon daughters attached to moisture or dust particles and virtually all of the unattached portion.

Figure 96 shows the build-up and decay of alpha activity from individual initially isolated radon daughters; their combined effective half-life is of the order of 39 minutes. If they are lodged in lung tissue or the bronchi, they will lead to the deposit of the energy of four alpha particles in rapid succession and to residual activity from 22-year lead-210 ("radium-D") and its shorter-lived daughters.

The health effects of this inhalation of radon daughters have been studied in those countries were a stable mining population exists (8-12). Considerable arguments have persisted over the years regarding the incidence of lung cancer among uranium miners as compared with other

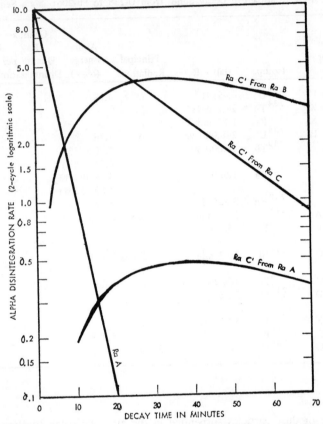

Figure 96. Build-up and decay of alpha activity of individual radon daughter isotopes, with an initial disintegration rate of 10 disintegr/minute (10).

hardrock miners, and the relative contribution to cancer deaths of radon daughters has been compared with cigarette smoking. Lundin *et al.* (10) appear to have established that there is an observable increase in cancer deaths among uranium miners that can be correlated with the exposure levels to radon daughters. Most of the respiratory cancer deaths have occurred ten or more years after the individual first mined uranium, and no excess deaths were observed in the first five years. It was estimated that smoking uranium miners experienced an excess of lung cancer ten times greater than nonsmoking miners. This apparent correlation of lung cancer and radon daughter concentrations prompted congressional hearings in the U.S. and subsequent legislation. The hearings record (13) constitutes the best available review of the subject up to 1967.

Since the concentration of radon daughters in mine air and hence lung exposures depends on the movement of air, the principal remedy for this problem lies in improved ventilation. In most mines active working areas already require high ventilation rates to control silicosis hazards; however, in unworked stopes and drifts and other side tunnels, radon-daughter levels may build up while ventilation is not supplied and any worker entering such areas could be exposed to high radon-daughter concentrations.

The radon daughter concentration is expressed in terms of a unit unfortunately called the *working level*. The working level (WL) is defined as any combination of radon daughters in 1 liter of air that will result in the ultimate emission of 1.3×10^5 MeV of potential alpha energy. The numerical value of the WL is derived from the alpha energy released by the total decay of the short-lived radon daughter products at radioactive equilibrium with 100 picocuries (pCi) radon-222 per liter of air. The 100 pCi of polonium-218 give 1.3×10^4 MeV from the total decay of the polonium-218 and the same number of terminal lead-214 atoms. The 100 pCi of lead-214 give 6.6×10^4 MeV from their decay leading to bismuth-214. The 100 pCi of bismuth-214 give 4.8×10^4 MeV from its decay to polonium-214. The resultant total is 1.27×10^5 MeV, which is rounded to 1.3×10^5 MeV.

A significant advantage in the concept of the WL is its practical application to field measurements of the radon daughter concentrations in mine air. The method of measuring the concentration of decay products in terms of total alpha particle emission is widely used for control and regulatory purposes. Exposure of an individual to radon daughters in air can be estimated from the length of time the individual breathes an atmosphere containing a stated burden of radon daughters. U.S. occupational exposures are usually expressed as Working Level Months (WLM), although other time periods are sometimes used. Inhalation of air with a concentration of 1 WL of radon daughters for 173 working hours results in an exposure of 1 WLM (13). One WLM has been calculated to be roughly equivalent to a dose to the bronchial epithelium of 5 rem (14), though Harley and Pasternack (15) recommend a value of 4.3 rads/yr per WLY. Sagan (3) has estimated that one man working for one year at 4 WL incurs an additional risk of lung cancer of 0.00053 case. As Figure 97 shows, there appears to be a clear correlation between cumulative exposure to radon daughters, expressed in WLM, and the observed annual mortality.

To evaluate probable exposure rates in U.S. mines, Table 144 presents a summary derived from records of radon daughter measurements in underground uranium mines during 1961-1967. The table shows the number of mines measured and the percentage of mines with radon daughter concentrations falling in various ranges of WL values. These

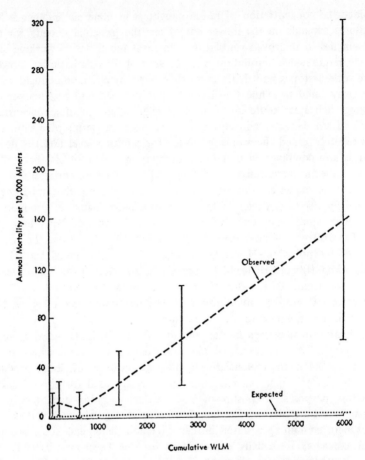

Figure 97. Observed and expected annual lung cancer mortality per 10,000 miners and 95% confidence limits in relation to exposure (13).

Table 144.   Radon-Daughter Levels at U.S. Uranium Mines (16)

Year	No. of Mines	No. of Samples	Percent of Mines in Each Working Level Range			
			<1	1-2.9	3-10	>10
1961	266	–	21	21.8	28.2	29
1962	327	1054	32	31	33	4
1963	244	1378	35	40	21	4
1964	250	1589	47	36	13	4
1965	229	1712	39	41	19	1
1966	231	1718	44	42	13	1
1967	217	2588	70	24	5	1

percentages are estimated to be about the same as the percentage of the work force whose annual average exposures fall within the WL ranges shown in the table and are considered to reflect the status of the whole industry during that period. It is evident that improvements in ventilation over the years have resulted in an appreciable reduction in working level values; alternatively, some of the worst-ventilated mines have been forced to close. U.S. legislation required reduction of exposures initially to less than 12 WLM per working year; since 1971, exposures have been limited to 4 WLM per year. To return to a uniform nomenclature, the U.S. Atomic Energy Commission (17) in 1971 proposed a return to expressing concentrations in $\mu Ci/ml$ and to a maximum permissible occupational exposure to alpha-emitting radon daughters of 7×10^{-8} $\mu Ci/ml$.

To limit exposures to such levels in mines producing high-grade ore requires either an excellent ventilation system and careful preparation of dead adits and abandoned stopes before entering or a reduction in the time a worker may spend in high-level areas. In view of the difference in transportability of different dust particle sizes, attention must also be paid to the particle size distribution and their removal by ventilation (18). The U.S. Bureau of Mines (19) has issued a very detailed guide to ventilation procedures and air sampling methods. The latter subject is also reviewed in a proposed American National Standard (20).

Some problems exist in the enforcement of any exposure standard since the radon daughter level is liable to vary from point to point within a mine and with differing ventilation conditions. For this reason it would be useful to develop a satisfactory method of personnel dosimetry, as well as a simple field method for underground air monitoring.

For most mines to obtain a concentration of radon daughters in air as low as 4 WLM/yr (0.3 WL) in active working areas requires a high degree of air movement, which results in appreciable disequilibrium among radon and its daughters. The method most generally used to determine the working-level value is the so-called Kusnetz Method (21). In this method a predetermined volume of air is passed through a filter; one then waits 40 minutes, measures the residual total α-activity, and extrapolates this measured activity to the WL-value at the time of sampling. This method is very approximate and tends to underestimate (22) since it does not take into account the disequilibrium among daughters nor the influence of the degree of build-up of activity during the sampling period. To some extent this may be overcome by a longer waiting period, at some loss in accuracy.

Alternative methods involve measurements of the beta emitters (23), and a method employing two alpha channels for Po-218 and Po-214 and one beta + gamma channel for Pb-214 and Bi-214 (24). The count rates of the different channels are multiplied electronically and combined to give the WL value. Another approach that lends itself to use as a personnel dosimeter employs plastic foil detectors, such as cellulose nitrate foil, in a portable, cylindrical air sampling unit (25). After exposure the foils must be removed for etch-development and track counting. Since both the radiation dose and the measurement methods depend on the degree of attachment, that is, the fraction of radon daughters adhering to dust particles, it would be useful to monitor this factor (24,26). Recent reviews on the various methods have been published by Budnitz (27) and Reiter (28).

Finally, studies have been undertaken on methods to reduce airborne radon and radon daughters under conditions where increasing the ventilation rates is impractical or impossible. Most of the methods are suitable for laboratories, but may be difficult to apply underground (29). One such method employs gas-scrubbing systems to remove radon in contact with BrF_3 and various solid complex fluorides (30). The cost and complexity of such systems would appear to make them more useful for analysis of radon in air, rather than continuous air cleaning. Reiter (28) advocates use of m-tricresyl phosphate as an air scrubber to reduce radon in abandoned drifts, as well as silica gel. Other suggestions to reduce radon emanation by coating drift walls or pressurizing the mine also seem impractical under mine conditions.

From a purely environmental point of view, exposure of miners to radon daughters is primarily an occupational health problem. However, it has been discussed in some detail here because it is a major component in the assessment of the human cost of the nuclear fuel cycle (3). The other, more conventional impact factors depend on the type of mine— whether open-pit or underground—and are related mainly to land use and disturbance, and the release of polluted effluents and mine drainage water (1). The consequences of open-pit mining can be reduced somewhat by careful site restoration and reclamation; the pit itself in some cases may be allowed to fill with water, forming a man-made lake with recreational possibilities. Gaseous wastes, mainly from fuel combustion, are rarely a pollution problem in remote locations. The wastewater should be cleaned, filtered and neutralized before release. Table 145 shows the composition of this water from surface and underground U.S. mines. No significant airborne alpha activity, above background, has been measured around mine sites.

Table 145. Composition of Some Uranium Mine Waters (1)

	Open Pit Mines			Underground Mines			
	Kerr-McGee Shirley Basin, Wyoming	Getty Oil KGS-JY-Mine Shirley Basin, Wyoming	Utah Intl. Shirley Basin, Wyoming	Cotter Corp. Schwartz-walder Golden, Colorado	Union Carbide Eula Belle Uravan, Colorado	Union Carbide Martha Belle Uravan, Colorado	Union Carbide Burro Slick Rock, Colorado
Flow rate, thousands gpd	460	1440	2880	72	69	47	25
pH	7.9	7.5	6.7-8.2	7.9	8.6	8.4	8.8
Alkalinity (as $CaCO_3$)	180	164	144-150	244	358	384	704
BOD 5-day	0	67	0-2	1	12	8	10.8
Chemical oxygen demand	2.4	0	0.8	10	<2	<2	11
Total solids	612	840	850-1275	1220	730	3103	1790
Total dissolved solids	411	627	750-825	1042	590	650	1780
Total suspended solids	163	49	40-420	178	140	2453	6
Total volatile solids	38	164	40-92	244	70.7	192	125
Ammonia (as N)	0.22	1.33	1.42-1.60	0.15	<0.10	<0.10	3.3
Kjeldahl nitrogen	0.22	1.33	1.42	0.55	145	0.3	21.8
Nitrate (as N)	<0.01	0.002	0-1.06	12.0	0.35	0.39	1.9
Phosphorus total (as P)	0.05	0.07	2.30	0.4	0.2	0.4	0.15
Alpha-total[b]	360	104	—	3.3			
Beta total[b]	168	77	—	1.05			
Gamma total[b]	N.D.	N.D.	—	3.3			
U_3O_8[b]	—	—	140-1100				

aComposition data given in mg/l unless otherwise specified.

b$\mu Ci/ml \times 10^{-9}$.

URANIUM MILLS

To extract the low concentrations of uranium from the mined ore and to concentrate the uranium content to marketable levels, it has to undergo a beneficiation and purification process at a mill. Such a process involves crushing and grinding of the rock to liberate the ore minerals from surrounding rock, their physical separation by gravity or flotation methods, dissolution and chemical leaching of the ore, and extraction of the uranium by ion exchange or solvent extraction methods. The concentrate is then further refined, precipitated and dried, and shipped as powdered sodium- or ammonium-diuranate (yellow cake), containing about 96% U_3O_8. Figure 98 shows a simplified flow diagram for one mill, employing acid leaching and solvent extraction (31).

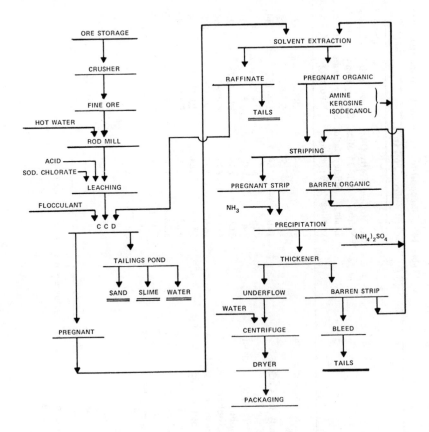

Figure 98. Flow diagram: Humble Oil and Refining Co., Highland Mill (31).

In terms of the radioactivity associated with the ore, the milling process separates the three (long-lived) uranium isotopes from the other decay products present in the ore, owing to their different chemical nature. The bulk of the radioactive uranium daughters remain in the uranium–depleted pulp (*tailings*)–that is pumped to the tailings retention system. These tailings consist of two types of materials: filter residues and slimes, and barren leach solutions. They are typically pumped to a tailings pond where the solids settle out, the shorter-lived activities decay away and the clarified liquid may be discharged to local streams after neutralization and filtering (32). Table 146 shows the analysis of the liquid tailings effluent for the Humeca mill (33) to indicate concentrations involved. Since uranium milling deals only with very low-level and

Table 146. Analysis of Plant Tailings Effluent (33)

Solution Analysis	$\mu Ci/ml$	Parts per Million
U_3O_8		6.8
Mn		0.01
Cu		0.01
Fe		1.0
Zn		0.6
SO_4		7500
CO_3		4000
HCO_3		1100
Th		2.0
Na		7100
pH		9.5
Ra-226	1×10^{-8} to 1.81×10^{-6}	
Th-230	1×10^{-10}	
U-238	3.6×10^{-6}	

Solids Analysis	Percent
U_3O_8	0.017
Mn	0.1
Cu	0.0028
Fe	1.36
Th	0.0005

dilute concentrations of radioactive materials, there are no operations or activities associated with the milling process itself that could result in a serious radiation exposure to either mill employees or members of the general public, even in the case of an accidental release of radioactive materials. External radiation levels associated with uranium milling activities are low, rarely exceeding a few mR/hr, at surfaces of process vessels. Liquid and solid wastes from the milling operations contain only low-level concentrations of radioactive materials. These wastes are retained and stored in the tailings retention systems on site. Concentrations of airborne radioactive materials escaping into the surrounding environs, especially from crushing and grinding operations, are not expected to be more than a few percent of limits. If the tailings pond is located appropriately, seepage into ground water or surrounding bodies of water can be controlled or minimized.

At a mill treating 2000 tons of ore per day, about 2000 tons per day of solid waste tailings slurried in 3000 tons of waste milling solutions will be generated. These wastes contain the unreacted portion of the leach acid and initially may have a pH of 1.5-2. Apart from various process reagents and the iron and silica from the dissolved gangue, they will contain radium-226 (2-5 x 10^{-7} $\mu Ci/ml$, depending on ore grade mined) and thorium-230 (1.5-3 x 10^{-5} $\mu Ci/ml$) as well as trace quantities of short-lived radon daughter products. Seepage of trace radioactivity can be reduced by raising the pH since $RaSO_4$ has rather low solubility in water. The solids average about 0.16 cubic yard per pound (0.27 m^3/kg) of recovered U_3O_8.

If the supernatant water is drained off or allowed to evaporate, the solid sediment is found to contain most of the radium and radium daughters. Radon emanation depends on exposed surface area and wind speeds. Schiager (34) has studied the radiation exposures on or near abandoned tailings piles. He showed that at any reasonable distance from the pile radon daughter inhalation would represent the principal source of potential radiation exposure, though in no case would it be expected to exceed measurably the natural background. Nevertheless stabilization of tailings piles by covering the tailings with plastic or asphalt sheets and earth or concrete can readily be achieved (35). The criteria for such tailings stabilization methods are the subject of an American National Standard, ANSI N313-1974 (36). Table 147 indicates the cost and effectiveness of proposed earth cover (2).

In this connection it is worth recalling one embarrassing incident of tailings disposal. In several mill locations crushed rock and mill tailings have been used for road material and land fill. At one time there was wide use of uranium tailings at Grand Junction, Colorado , as fill under and near homes, public buildings and sidewalks. Subsequent measurements in some homes and a school built on this material showed increases in external gamma-ray exposures to about three times background. Radon daughter concentrations measured in 1968 at about 100 locations,

Table 147. Control of Radiological Impact of Airborne Effluents from a
Model Uranium Mill Tailings Pile (2)

Controls	Source Term (Ci/yr)	Individual (mrem/yr, lung)	Total Health Effects per 30-yr Plant Operations	Present Worth (1970 dollars/facility)
None	20,000	1,300	0.25	0
Tailings pond[a] (2 ft water)	800	50	0.01	0
2 ft cover[b]	15,000	980	0.19	150,000
20 ft cover	2,000	130	0.025	1,400,000
100 ft cover[c]	0	0	0	Unknown

[a]Present level of control while operational for recent mills.
[b]Present level of control for recent mills, post operational.
[c]Tails used as backfill in strip mine; original layer of overburden is replaced. Ground water effects must be considered in this option.

identified as having used more than 20 cubic yards of tailings, showed that 42% of the residences and 62% of businesses sampled had radon levels exceeding 0.01 WL, the highest observed being 1.88 WL. The situation became the subject of congressional hearings (37) and it was decided that some remedial action was required. Complete excavation and removal of tailings from beneath the buildings would cost a minimum of $14 million (1971). Sealing of the basement floors with cement or epoxy resins reduces radon emanation but tends to build up gamma-ray fields from radon daughters accumulating below the surface. Excavation of high-concentration material in poorly ventilated but accessible areas seemed to be indicated, with asphalt or concrete sealing of most others, at an average estimated cost per residence of $3,220.

As a measure of potential environmental impact of tailings piles from uranium mills, the 100-year dose commitment has been calculated. This commitment estimates the additional, above-background health effects ascribable to the exposure of members of the general public to plant effluents for the 100 years following start of plant operations. The health effects committed for a single plant, 250-acre tailings pile for the eastern United States have been estimated as one health effect during plant operation and 120 following plant operation per year. Summed over a large population the health effects amount to 200 effects/facility for 30 years, even though the dose to any individual is extremely small (2).

Accidents from mill fires or tailings dike failures have been reviewed and in all cases the materials were rapidly deposited by sedimentation downstream and in no case was there any significant radiation exposure to the public (1).

FUEL PREPARATION

The environmental impact of the subsequent stages of the fuel cycle have been examined in details in References 1 and 2. These stages are UF_6 production, uranium enrichment, and fuel fabrication and assembly.

Uranium Conversion

The U_3O_8 concentrate extracted from the ore must be purified and converted to the volatile compound uranium hexafluoride (UF_6) for enrichment. Two processes are used for UF_6 production. The hydrofluor process consists of continuous successive reduction, hydrofluorination and fluorination of the ore concentrates followed by fractional distillation of the crude uranium hexafluoride to obtain a pure product. The second method employs a wet chemical solvent extraction step at the head end of the process to prepare a high-purity uranium feed prior to the reduction, hydrofluorination and fluorination steps. Roughly equal quantities of UF_6 feed to the enrichment plants are produced by each method.

The nature of the effluents from the two processes differs. The bulk of the impurities entering with the crude uranium feed are rejected from the hydrofluor process as solids; in the wet process, the bulk of the yellow-cake impurities is contained as dissolved solids in a raffinate stream. The "model plant" consists of a 5000 MTU plant, which is representative of about half of the currently operating industry, and is capable of supplying the annual fuel requirements of 27.5 model LWR's.

Table 148 gives the principal environmental considerations in the production of UF_6 related to the model LWR annual fuel requirement. The main environmental concern is related to the release of fluorides in gases and liquids. The off-gases are all scrubbed with KOH to minimize airborne effluents; analysis for fluoride in air in the vicinity of the wet process UF_6 plant indicated concentrations averaging less than 2 ppb (1). In liquid wastes from the dry process plant, increases in concentrations downstream under average conditions were 0.6% for fluorides and 4% for ammonia. The amount of fluoride in the waste is being further reduced (1). The highest radiation doses from these facilities are to the lung (30-70 mrem/yr), caused by insoluble uranium aerosols, to individuals

Table 148. Summary of Environmental Considerations for UF_6 Production Related to One Model 1000 MWe Reactor (1)[a]

	Total
Natural Resource Use	
Land (acres)	
Temporarily committed	2.5
Undisturbed area	2.3
Disturbed area	0.2
Permanently committed	0.02
Water (millions of gallons)	
Discharged to air	3.3
Discharged to water bodies	23.0
Total	26.3
Fossil Fuel	
Electrical energy (thousands of MW-hr)	1.7
Equivalent coal (thousands of tonnes)	0.62
Natural gas (millions of scf)	20
Effluents	
Chemical (tonnes)	
Gases	
SO_X	29
NO_X	10
Hydrocarbons	0.84
CO	0.2
F^-	0.11
Liquid[a]	
F^-	17.5
$SO_4^=$	4.5
NO_3^-	0.1
F^-	8.8
Cl^-	0.2
$Na^+ + K^+$	3.4
NH_3^+	1.6
Fe	0.04
Solids (tonnes)	40
Radiological (curies)	
Gases	
Uranium	0.00015
Liquids	
Ra-226	0.0034
Th-230	0.0015
Uranium	0.044
Solids (buried) other than high-level	0.86
Thermal (billions of Btu's)	20

[a]Assuming that half the output is produced by dry hydrofluor process, half by wet solvent extraction.
[b]Effluent gases from combustion of equivalent coal for power generation.
[c]From the combustion of coal and natural gas and process vents, hydrocarbons include 0.2 tonne/yr of hexane from wet process portion of model plant.

living within 1 km of the plants. It is believed that additional filtration of air streams can reduce this dose rate by at least a factor of 10 (2).

Enrichment Facilities

Since light-water-cooled and gas-cooled reactors employ enriched uranium fuel, extensive facilities are required to provide this material. For the current types of LWR's the enriched uranium-235 content is 2-4%, and about 116,000 kg of separative work units (SWU) are required to prepare enough uranium for the annual requirements of a 1000 MWe reactor. Gaseous-diffusion facilities are large in size because a large number of stages are required to attain the necessary enrichment, about 1700 stages to produce 4%-enriched UF_6 (1). Owing to the low stage–separation efficiency a large amount of power is required to drive the pumps and compressors associated with each stage.

Uranium enrichment requires approximately 98% of the electrical energy consumed in the entire LWR fuel cycle. About 310,000 MW-hr of electrical energy are utilized in the production of enriched UF_6 for the model LWR annual fuel requirement. The thermal-electric system providing power for the diffusion plants so far has been fossil-fueled and must burn about 113,000 tonnes of coal to meet this demand. In perspective, the electrical energy produced by the model 1000 MWe nuclear station annually (at 80% load factor) amounts to about 22-25 times the energy consumed to produce an annual fuel requirement.

Table 149 summarizes the environmental impact per 1000 MWe reactor of the enrichment process. The process cooling water requirements of the gaseous diffusion plant involve the evaporation of 84 million gallons of water per model LWR annual fuel requirement. This amounts to 54% of the evaporative use of water in the fuel cycle. An additional 6 million gallons of water per model LWR annual fuel requirement are discharged to water bodies at the enrichment plants. The additional water used for cooling by power stations supplying electrical energy to gaseous diffusion plants, assuming once-through cooling, is approximately 11 billion gallons discharged to water bodies per annual fuel requirement (1).

The primary source of environmental impact associated with the enrichment of uranium is related to the gaseous effluents from the coal-fired stations used to generate the required electric energy. Waste gas emissions, including particulates, of approximately 6600 tonnes are associated with the production of an annual fuel requirement. This is equivalent to the gaseous effluents released annually by a 45 MWe coal-fired plant.

Also related to the power requirements of the enrichment plant is the discharge of heat into the environment at the sites of individual electric

Table 149. Summary of Environmental Considerations for Uranium Enrichment by Gaseous Diffusion Related to One 1000 MWe Reactor (1)

	Total
Natural Resource Use[a]	
Land	
Temporarily committed	0.8 acres
Undisturbed area	0.6
Disturbed area	0.2
Permanently committed	0.0
Water	
Discharged to air (at GDP's)	84×10^6 gal
Discharged to water bodies (at GDP's)	6
Discharged to water bodies (at power plants)	11,000
Fossil Fuel	
Electrical Energy	310×10^3 MW-hr
Equivalent coal	113×10^3 tonnes
Effluents	
Chemical	
Gases (from coal-fired power plants)	
SO_x [b]	4,300 tonnes
NO_x [b]	1,130
Hydrocarbons[b]	11
CO[b]	28
F^-	0.5
Particulates[b]	1,130
Liquids (from GDP's)	
Ca^{++}	5.4
Cl^-	8.2
Na^+	8.2
$SO_4^=$	5.4
Fe	0.4
NO_3^-	2.7
Radiological[c]	
Gases	
Uranium	0.002 curies
Liquids	
Uranium	0.02 curies
Thermal (from coal-fired power plants & GDP's)[d]	3.200×10^9 Btu

[a]Based on 20-year life of gaseous diffusion plant (GDP).

[b]Estimated effluent gases based upon combustion of equivalent coal for power generation, assuming 100% load factor.

[c]Based on 4% isotopic enrichment.

[d]Approximately 67% of this heat is discharged by the electric generating plants servicing the model enrichment plant, assuming 100% load factor coal-fired plant.

generation plants. Since the power is drawn from the grids of large utility complexes, the environmental impact is difficult to evaluate. The heat rejection at the enrichment plant site is largely to the atmosphere, and although occasional misting and fogging result within the site from the operation of cooling towers, the thermal impact is insignificant.

Small quantities of airborne fluoride are generated at the diffusion plants. Measurements in unrestricted areas indicate concentrations that are below the range for which deleterious effects have been observed. In addition, oxides of nitrogen and sulfur are released at the diffusion plants. Conservative estimates of the offsite concentrations of these contaminants yield levels that are below EPA standards. Furthermore, the total quantity of these effluents is insignificant in comparison with the combustion products generated by the supporting electric power plants.

In summary, the highest radiation dose from the model enrichment facility is expected to be less than 3 mrem per year (bone) to the maximum exposed individual, delivered in about equal amounts through inhalation and drinking water. Approximately 0.1 health effects are to be expected from 30 years of plant operations under current levels of effluent controls. Less than 1 curie per year of uranium is discharged (2).

Fuel Fabrication

The processing technology used for fuel fabrication can be divided into three basic operations: chemical conversion of UF_6 to UO_2, mechanical processing including pellet production and fuel element fabrication, and recovery of uranium from scrap and off-specification material. The most significant potential environmental impact results from UF_6 to UO_2 conversion and chemical operations in scrap recovery.

The currently dominant method for UF_6 to UO_2 conversion is a wet process involving the use of ammonium hydroxide to form an intermediate ammonium diuranate (ADU) compound prior to final conversion to UO_2. Figure 99 shows a flow diagram for this process. Alternative dry processes to the conventional ADU process for conversion of UF_6 to UO_2 have been developed. These proprietary processes offer the potential of lower capital and operating costs as well as waste management advantages.

The model fuel fabrication plant used to estimate environmental effects has a capacity of 3 tonnes per day and operates 300 days per year. By today's standards, this is a large plant capable of producing 26 annual fuel requirements for the model LWR. The model plant lifetime was taken to be 20 years. The plant operations are based on currently dominant industrial practices using the conventional ADU process for

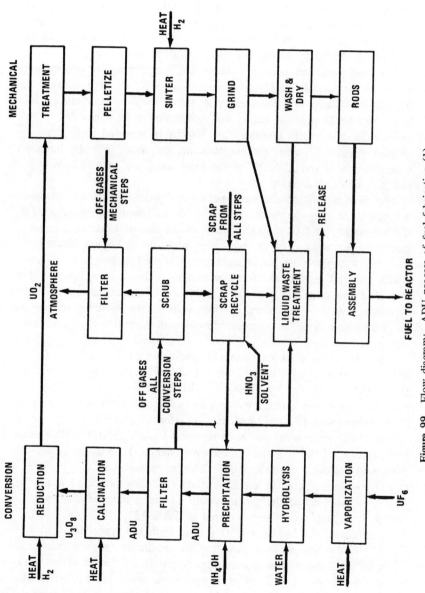

Figure 99. Flow diagram: ADU process of fuel fabrication (1).

conversion of UF_6 to UO_2. Of the processes currently in use or in development, the ADU process appears to create the greatest waste management problems. The most significant effluents from the standpoint of potential environmental impact are chemical in nature. Nearly all of the airborne chemical effluents result from the combustion of fossil fuels to produce electricity to operate the fabrication plant.

The only significant airborne chemical effluent from the process operations of the fabrication plant is fluorine in the form of fluorides. The fluorine introduced into the fuel cycle during the UF_6 production phase becomes a waste product during the production of UO_2 powder. The gaseous fluorine wastes generated are effectively removed from the air effluent streams by water scrubber systems and result in a site boundary concentration of roughly 20% of the most restrictive of a reference state's standard, 0.5 $\mu g/m^3$.

The most significant chemical species in liquid effluents are nitrogen compounds that are generated from the use of ammonium hydroxide in the production of UO_2 powder and from the use of nitric acid in scrap recovery operations. The nitrogen concentrations in liquids released from the waste holding ponds are about 420 mg/l in the form of ammonia and 280 mg/l in the form of nitrates. The limiting concentration is that of ammonia and is about three orders of magnitude above that of drinking water. Depending on the nature of the receiving stream and its downstream uses, the nitrogen releases could constitute a significant impact on the environment.

Water from the scrubber systems is combined with process liquid wastes and treated with lime to form a calcium fluoride (CaF_2) precipitate, which is removed by filtration. The 26 tonnes of CaF_2 filtered from the liquid per model LWR annual fuel requirement have a volume of 11 cubic yards (8.4 m^3) and may be retained onsite (1).

Table 150 summarizes the environmental impact per 1000 MWe model LWR reactor of the fuel fabrication component (1). Slightly higher discharges of enriched uranium, 0.005 Ci/yr in air and 0.5 Ci/yr in water, as UO_2, were predicted in the EPA study (2). The predicted health effects to members of the general public in the vicinity of the plant derived from that study are presented in Table 151.

In summary, the highest radiation dose from the model fabrication facility is 10 mrem per year (lung) to the maximum exposed individual living within 1 km of the plant. Approximately 0.1 health effect is to be expected from 30 years of plant operations under current levels of effluent controls. Less than 1 curie per year of uranium is discharged (2).

Table 150. Summary of Environmental Considerations for Fuel Fabrication Plants Related to One 1000 MWe Reactor (1)

	Total
Natural Resource Use	
Land	
Temporarily committed	0.2 acres
Undisturbed area	0.16
Disturbed area	0.04
Permanently committed	0
Water	
Discharged to water	5.2×10^6 gal
Fossil Fuel	
Electrical energy	1.7×10^3 MW-hr
Equivalent coal	0.62×10^3 tonnes
Natural gas	3.6×10^6 scf
Effluents	
Chemical	
Gases	
SO_x[a]	23 tonnes
NO_x[a]	6
Hydrocarbons[a]	0.06
CO	0.15
F^-	0.005
Liquids	
N as NH_3	8.4 ($\sim$10 tonne NH_3)
N as NO_3	5.3 ($\sim$23 tonne NO_3^-)
Fluoride	4.1
Solids	
CaF_2	26 ($\sim$13 tonne F^-)
Radiological	
Gases	
U	0.0002 curies
Liquids	
U	0.02
Th-234	0.01
Solids (buried)	
Uranium	0.23
Thermal	9×10^9 Btu

[a]Effluent gases from combustion of coal for power generation.

Table 151. Estimated Health Effects to Members of the General Population
from Operation of a Model Fuel Fabrication Plant (2)

| Pathway | Critical Organ | Predicted Health Effects/Facility-yr[a,b] | | |
		Mortality	Nonfatal Effects	Genetic Effects
Air	Lung	0.0002	0	0
Water	Bone (bone cancer)	0.0006	0.0006	0
	Bone (leukemia)	0.0004	0	0
Water	Soft tissue	0.0005	0.0005	0.0005
	Total	0.0016	0.0011	0.0005

[a]Listed health effects will result from each year of facility operations.
[b]Total health effects for 30 years of plant operations are 0.1.

REPROCESSING PLANTS

The spent fuel from a light-water-cooled or gas-cooled reactor is
shipped to a reprocessing plant to recover the remaining fissile materials:
uranium-235 and various plutonium isotopes in the case of LWR's, and
uranium-233 in the case of HTGR's. Of the various stages of the nuclear
fuel cycle, reprocessing is fraught with the highest potential risks from
radioactive contamination and release of radioactivity to the environment,
and for that reason exceptional precautions and safeguards must be included in plant design. On the other hand the number of reprocessing
plants is quite small and site selection is not subject to the same power
grid requirements as are power plants.

The principle of operation of reprocessing plants is to dissolve the
fuel elements, after stripping off the cladding in some cases, and to extract successively the uranium, plutonium, and the various fission products.
The fissile materials are refined to reactor grade purity and shipped to
the enrichment plant or fuel fabrication facility for recycle. The fission
products are concentrated for disposal, since insufficient industrial uses
as radiation sources have been found for such materials.

The spent fuel as received by the reprocessing plant will have undergone storage at the reactor site for 3-6 months and most of the shorter-
lived fission and activation products, such as iodine-131, xenon-133,
barium-lanthanum-140 and tellurium-132, will have decayed. Nevertheless, stripping and dissolving the fuel elements will now liberate the large
inventory of longer-lived volatile and gaseous fission products, notably
tritium, krypton-85 and ruthenium-103 and 106, which have to be

collected with very high efficiency from off-gas streams. Consequently, the radwaste treatment methods described in Chapter 7 must be applied to a much more elaborate extent at reprocessing plants to guard against excessive routine or accidental release of radioactive effluents. In selecting possible sites for new reprocessing facilities, computations involving a source term equivalent to the release of the radioactivity contained in plant inventory in process at the time of the postulated accident will determine exclusion zone dimensions and population doses. Particular attention must be paid to the plutonium and the higher actinides present in the plant, in solution form, in the final product, and in trace quantities in waste streams. To maintain some perspective one must remember that most of the reprocessing and plutonium extraction so far has been under military auspices in all the countries concerned, and the prospective quantities arising from commercial operations will not reach a comparable level for another decade.

A simplified flow diagram for the main process is shown in Figure 100. It is a modified Purex process, relying on successive solvent extraction to separate the uranium and plutonium; this process is used at the NFS and Barnwell (BNFP) plants. The initial form of the plutonium extraction product is plutonium nitrate; for safer storage or shipment this is usually converted to PuO_2. The principles of these extraction processes have been described by Long (38). In view of the high radiation levels of the reprocessing solutions the plants are designed for remote, highly automated operations. All process stages involving uranium and plutonium concentrates must also be designed to prevent any possibility of criticality occurrences. The fission product solutions can be further treated by solvent extraction and fractionation processes to concentrate the long-lived fission products, Sr-90 and Cs-134 + 137, in one stream and the shorter-lived waste activities, predominantly Zr-Nb-95, Co-60, Pm-147, Ce-144 and other rare earths in the other for storage, though this is not done in current commercial operations. Recently, it has been suggested that remaining actinides be separated out; however, improved U and Pu recovery should reduce the residual levels enough to make this impractical and uneconomic.

To reduce effluent activity levels it is desirable to minimize airborne activities that might otherwise overload the gaseous radwaste treatment system and to capture most of the tritium. This can be accomplished in the voloxidation process (39) in which the spent fuel elements are sheared into short segments and exposed to air or oxygen at high temperature. This treatment will convert the UO_2 fuel to other oxides of uranium and in so doing will tend to break up the fuel structure, releasing relatively high concentrations of the gaseous or volatile products

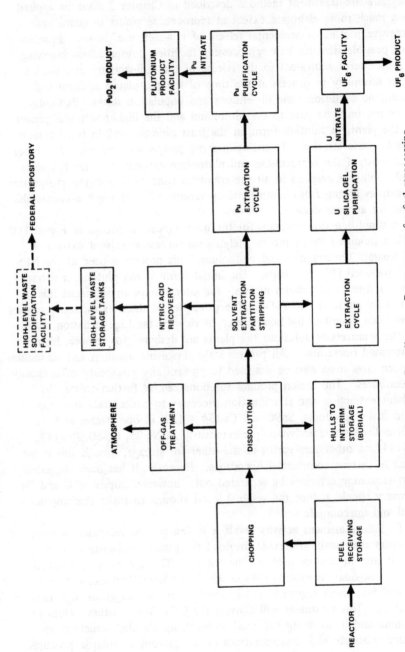

Figure 100. Simplified flow diagram: Purex process for fuel reprocessing.

not tightly incorporated into the fuel structure. This process drives off most of the tritium, which is then catalyzed over CuO to form HTO; about 5 gallons per day (20 l/day) of HTO may be produced, which is claimed to be easy to collect and store. Thirty to 90% of the noble gases and iodine are also volatilized under prolonged heating at 450°C; most of the remainder will be evolved during dissolution and can be removed from the dissolver off-gas. This early removal of iodine before solvent extraction is desirable because the iodine otherwise tends to spread through the process due to its solubility in both aqueous and organic solvents.

The source of all radioactivity in fuel reprocessing plants is the spent fuel shipped to the plant from a number of nuclear power plants. Table 152 presents typical amounts of radionuclides expected per metric tonne (MT) of uranium in the spent fuel. Since fuel reprocessing plant

Table 152. Estimated Annual Deliveries of Radionuclides to
U.S. Fuel Reprocessing Plants (46)

Nuclide	Typical Amount in Fuel[a] (Ci/MT)	Cumulative Annual Quantities Delivered (Curies)	
		300 MT/yr	1500 MT/yr
Fission products	4.4×10^6	1320×10^6	6600×10^6
Actinides	130,000	39×10^6	195×10^6
Krypton-85	11,000	3.3×10^6	16.5×10^6
Tritium	700	210,000	1.1×10^6
Iodine-131[b]	2.2	660	3300
Iodine-129	0.037	11	55
Total	$\sim 4.5 \times 10^6$	$\sim 1360 \times 10^6$	$\sim 6800 \times 10^6$

[a]PWR fuel with exposure of 33,000 MWd/tonne, average specific power of 30 MW/tonne and decay time of 150 days. Comparable BWR fuel would contain about two-thirds as much radioactivity per metric ton.

[b]Iodine-131 has an 8-day half-life, so that cumulative amounts indicate amounts delivered in fuel with 150 days decay. After 1 year, the 2.2 curies per tonne will have decayed to about 2×10^{-8} curies.

capacities will range from 300 MT to 1500 MT uranium per year in the foreseeable future, Table 152 also presents the total annual quantities of radioactivity that will pass through reprocessing plants of those capacities. After fuel reprocessing, essentially all of this radioactivity (less quantities that are released or decay) will reside in the high-level liquid waste tanks at the reprocessing plant until such time that the waste is converted to solid form and shipped to a storage facility.

Table 153 gives a breakdown by nuclide; note particularly the actinides represented and the dominant contributions of Ce-Pr-144, Ru-Rh-106, and Zr-Nb-95 to the total activity in the fuel inventory. (The discrepancy in numbers between Tables 152 and 153 arises from differences in burn-up assumed.) Since the Barnwell plant has a design capacity of 1500 MTU/yr it may have to collect about 16 MCi/yr of Kr-85 and treat about 350 MCi of transuranium isotopes and 9.4 GCi mixed fission products. In contrast to reactor fuel, most of this material will be present in the reprocessing circuits in dissolved or volatile forms. Even 0.01% of that activity appearing in plant effluents would dwarf any health effects from reactor effluents, making it obvious why a much higher level of performance and reliability must be expected of the rad-waste treatment system of a nuclear reprocessing plant. The treatment system thus should include effective recovery of noble gases, multiple zeolite filters for iodine removal, and ion exchange systems and evaporators to remove trace contaminants from final waste streams.

Table 153. Estimated Spent-Fuel Radionuclides Inventory (41)

Basis: 40,000 MWd/MTU, 50 MW/MTU

	Half-Life	Curies/tonne U
Tritium	(12 y)	490
Noble Gases		
Kr-85	(10.4 y)	10,750
Xe-131m	(12 d)	3
TOTAL		11,243
Iodine		
I-129	$(1.6 \times 10^7 \text{ y})$	0.04
I-131	(8.05 d)	1.73
TOTAL		1.77

continued

Table 153, continued

	Half-Life	Curies/tonne U
Other Fission Products (OFP)		
Sr-89	(50.4 d)	93,210
Sr-90	(28 y)	92,010
Y-90	(64.2 h)	91,040
Y-91	(59 d)	202,500
Zr-95	(65 d)	377,400
Nb-95m	(90 h)	7,549
Nb-95	(35 d)	712,700
Ru-103	(40 d)	132,900
Rh-103m	(57 m)	132,400
Ru-106	(1.0 y)	764,100
Rh-106	(30 s)	764,100
Ag-110m	(249 d)	2,689
Sn-119m	(250 d)	4,606
Sn-123m	(136 d)	3,293
Sb-125	(2.7 y)	9,466
Te-125m	(58 d)	2,840
Te-127m	(105 d)	5,099
Te-127	(9.3 h)	5,066
Te-129m	(33 d)	6,493
Te-129	(67 m)	6,503
Cs-134	(2.1 y)	203,100
Cs-136	(13 d)	23
Cs-137	(30 y)	132,900
Ba-137m	(2.6 m)	122,300
Ba-140	(12.8 d)	400
La-140	(40.2 h)	460
Ce-141	(32.5 d)	83,330
Ce-144	(285 d)	1,054,000
Pr-143	(13.7 d)	702
Pr-144	(17.3 m)	1,054,000
Nd-147	(11.1 d)	60
Pm-147	(2.7 y)	165,400
Pm-148	(5.4 d)	198
Sm-151	(90 y)	350
Eu-154	(16 y)	10,090
Eu-155	(1.7 y)	6,082
Tb-160	(73 d)	124
TOTAL OFP		6.25×10^6
Transuranium Isotopes		
Np-238	(2.1 d)	7
Pu-238	(86 y)	4,892
Pu-239	$(2.4 \times 10^4$ y)	338
Pu-240	$(6.6 \times 10^3$ y)	676
Pu-241	(13 y)	175,100
Pu-242	$(3.8 \times 10^5$ y)	4
Am-241	(458 y)	359
Am-242	(100 y)	7
Cm-242	(162 d)	46,310
Cm-243	(35 y)	35
Cm-244	(18 y)	5,663
TOTAL TRANSURANIUM ISOTOPES		2.33×10^5

For HTGR fuel a slightly different reprocessing system is projected, operating on a smaller scale in view of the smaller number of reactors of that type. In this process the spent fuel is taken through crushers that reduce the graphite hex blocks to 1/4-inch size without breaking the microsphere fuel particles. The particles then go into a fluid-bed combustor in which the graphite moderator and the pyrocarbon coats are burned away, leaving fertile oxide ash and SiC-coated fissile particles, which go to a sorter for separation of the fissile U material and fertile Th spheres. Subsequent recovery follows conventional processes (Figure 101). Carbon-14 as CO_2 will be the major radionuclide released in this process.

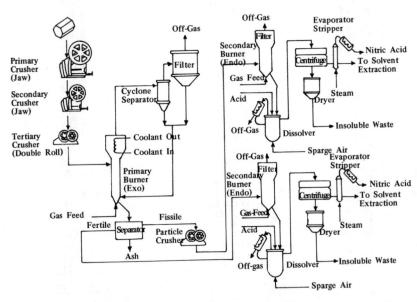

Figure 101. Flow diagram for HTGR fuel reprocessing: head-end portion (42).

The only U.S. experience in operating a full-scale, privately owned, commercial reprocessing plant has been at West Valley, New York, prior to 1971. The plant is currently closed down for reconstruction and modernization, and hence cannot be considered representative of the present state-of-the-art. The effluent treatment in use there would not be regarded as meeting the as-low-as-practicable criterion now, but this criterion was not in effect then. Whether the contamination observed around the plant is viewed as serious or relatively trivial clearly depends on the viewpoint of the observer (43,44), but in any case it would be expected to be greatly reduced, with improved radwaste treatment systems, there and elsewhere in future operations.

Estimates have been made (45) of the population exposure to radiation from radioactive material released from the stack; it is expected that future plants will have little, if any, radioactivity in the liquid effluents (46). For 33,000 MWd/T burn-up, 150-day decay for LWR fuel and 30-day decay for fast breeder (FBR) fuel, the fractional releases shown in Table 154 were assumed; no retention of krypton or tritium

Table 154. Estimated Fractional Release of Radionuclides and Accrued Annual Dose from 300 Tonnes/yr of Reprocessed Fuel from LWR's and Fast Breeder Reactors (45)

Estimated Fractional Release of Radionuclides Present at Time of Reprocessing

Radionuclides	LWR Fuel Reprocessing Plant	FBR Fuel Reprocessing Plant
^{85}Kr	1.0	1.0
^{133}Xe	0.1	0.1
Tritium	1.0	1.0
Halogens	0.001	10^{-7}
Particulates	1.2×10^{-8}	8.5×10^{-10}

Estimated Annual Dose Accrued from 300 Metric Tons of Fuel Reprocessed per Year

Body Organ	mrem/person at 3000 m		Accrued Dose (mrem/person/yr) within 100 km	
	LWR Fuel	FBR Fuel	LWR Fuel	FBR Fuel
Whole body	6.3	18	0.17	0.49
Skin[a,b]	60	72	1.6	1.9
Lung[b,c]	9.4	20	0.25	0.54
Bone[b,c]	7.7	18	0.21	0.49
Thyroid[b]	31	100	0.84	2.7
Respiratory lymph nodes[b,c]	420	150	11	4.0

[a]At 0.07 mm depth.

[b]Includes whole-body dose.

[c]Respiratory lymph node dose and a small portion of the bone and lung dose are received over the 50 years following exposure. Other doses are received within 1 year of exposure.

was allowed for. Under those conditions annual doses per person at a distance of 3 km and an individual dose to the population within a 100 km radius were obtained as shown in Table 154 per 300 MT of fuel reprocessed. Most of the skin dose is due to Kr-85, whereas the lymph node dose was ascribed to plutonium. It is evident that reduction in krypton and tritium release will substantially lower the skin and whole-body doses, while closer control of plutonium releases affects the critical dose in this framework. Actual observations at the NFS plant showed that much of the waste effluent was released in liquids rather than through the air and the above calculations are probably unduly high (44).

Table 155 compares the effectiveness and costs of some of the control options available to obtain very low levels of release in effluents from a reprocessing plant (47). In terms of indirect effects, observations around the Hanford site showed (1) a negligible effect on Columbia River water quality from wastes released to ground disposal sites and no detectable increase in offsite airborne radioactivity over background. Cs-137 and Sr-90 were detectable in local milk, but were shown to be attributable to atmospheric fallout from weapons tests. Fallout Sr-90, Cs-137, Ce-Pr-144, Zr-Nb-95 and Ru-106 activities were found to be higher in commercial foodstuffs than in local farm produce. Higher Zn-65 activities could be attributed to residual activity in irrigation water resulting from previous reactor operations (1). In summary, Table 156 collects postulated exposure estimates due to various nuclides for a "model," 900 MT/yr reprocessing plant. Table 157 shows the nonradioactive effluents expected from such a plant.

To calculate the overall radiological impact, exposures for a variety of postulated accidents have been computed (1). Table 158 summarizes the consequences of the more significant potential accidents. None of them should result in discernible effects on the environment and any contamination would be expected to be readily contained within the process buildings. Assuming accidental release outside the containment at ground level in the building wake, the estimated dose to the bone of an individual at the site boundary is approximately 10 mrem.

In respect to specific nuclides, I-129 has a very long half-life of 17 million years; iodine is reported not to concentrate to any appreciable extent in its transfer from soil to plants and to have a possible reaction with soil organic matter, making it largely unavailable. Iodine-129 accumulation in the soil is expected to require 50 years build-up before the dietary supply from this source would be comparable to the pasture-milk route. Careful surveillance for this isotope should be maintained at intervals (40). Cesium is known to be firmly bound by clay materials in many soils, and in sandy soils fixation may be complete in about three

Table 155. Cost and Performance Data for Effluent Control Systems for Fuel Reprocessing Plants (47)

Isotope	Control System	Reliability	System Decon Factor	Projected Release Rate	Costs (1972 estimates)	
					Capital/Plant	Operation/yr
Krypton-85	None	NA[a]	11	10^4 Ci/MTU	NA	NA
	Cryogenic distillation	Fair	10^3	10 Ci/MTU	$3x10^6$	$1x10^5$
	Cryogenic adsorption (charcoal)	Unproven	10^2	10^2 Ci/MTU	$3x10^6$	$1.5x10^5$
	Freon adsorption	Unproven	10^2	10^2 Ci/MTU	$1.5x10^6$	$1x10^5$
Tritium	None available	NA	1	800 Ci/MTU	NA	NA
	Voloxidation	Unproven	10^2	8 Ci/MTU	Unknown	Unknown
Iodine-131[b]	None	NA	1	2.0 Ci/MTU	NA	NA
	Scrubber + AgZ (zeolite)	Unproven	10^3	0.02 Ci/MTU	$1.2x10^6$	Unknown
Iodine-129	None	NA	1	0.04 Ci/MTU	NA	NA
	Scrubber + AgZ	Unproven	10^3	0.0004 Ci/MTU	$1.2x10^6$	Unknown
Actinides	None	NA	10^5-10^6	0.6-6 Ci/MTU	NA	NA
	Pre-filter + 2 HEPA's	Good	10^9	$6x10^{-4}$ Ci/MTU	$1.0x10^5$	$5.0x10^4$
	Pre-filter + HEPA + Sand filter	Excellent	10^9	$6x10^{-4}$ Ci/MTU	$3.5x10^5$	$7.0x10^4$

[a]NA indicates not applicable

[b]Iodine-131 content is estimated for fuel with 33,000 MWd/MTU burnup and 150 day cooling.

Table 156. Predicted Dose Levels from Gaseous Effluents from a
Model Reprocessing Plant (1)

Exposure Mode Isotope	Annual Release Ci/yr	Exposure, mrem/yr[a]				
		Whole Body	Thyroid	Bone	Exposed Skin	Lungs
Submersion						
H-3	430,000	0.76	0.76	0.76	0.76	0.76
Kr-85	9,000,000	0.16	0.16	0.16	6.75	0.32
Inhalation[b]				0.61		0.19
Dietary[c]						
H-3	430,000	1.44	1.44	1.44	1.44	1.44
Sr-90	0.18			0.94		
Ru-106	3.0	0.02		0.12	0.18	0.03
I-129	0.06		1.26			
I-131[d]	0.6		1.66			
Cs-134	1.2	0.09		0.07	0.09	0.02
Cs-137	0.6	0.03		0.06	0.03	0.01
		2.50	5.28	4.17	9.25	2.74

[a]Exposure normalized to 900 MT/yr; dose commitment in 50th year exposure at plant boundary.
[b]Dose from isotopes listed plus 25 Ci fission products (calculated as insoluble Ce-144) and 0.25 Ci of transuranium isotopes.
[c]30% of diet from local sources.
[d]1 liter/day milk intake by child.

Table 157. Nonradioactive Effluents from a Model Reprocessing Plant (48)

Constituent	MT/yr
Gaseous and Airborne	
SO_x	160
NO_x	185
Hydrocarbons	0.5
Fluoride	28
Liquid	
Na^+	137
Cl^-	6.3
$SO_4^=$	11.3
NO_3	21.6

Table 158. Estimated Radiation Exposure from Postulated Plant Accidents at Site Perimeter (1)

Accident	Organ of Greatest Exposure	Organ Dose (rem)
Criticality incident	Thyroid	0.04
Ion exchange resin fire	Lung	0.001
UF_6 loadout system leak	Bone	0.01
Storage pool cooler leak	Bone	0.01
Fuel assembly rupture	Thyroid	0.002

years after contamination. The general conclusion is that Cs-137 in plant materials has arisen from direct contamination or entrapment of solid particles, although Cs-137 may be taken up from plants grown in organic soils.

Other fission products released in small amounts include Zr/Nb, Ru, and Ce isotopes. Of these, Ru is reported to be slightly excluded (concentration ratio in plant/soil = 0.01-1), and Zr/Nb and Ce are strongly excluded (concentration ratio < 0.01). Radioisotopes such as strontium, cesium, and ruthenium can migrate via surface water/soil mechanisms and finally may reach dietary processes to cause human exposure (1). The dose from deposited radionuclides in the soil is expected to be almost completely due to tritium, which yields a mean dose of 0.6 mrem over much of the area within a 50-mile radius of the plant.

The annual exposure to the population within a 50-mile radius of the model reprocessing plant is estimated to be 167 man-rem, corresponding to 6 man-rem/annual fuel requirement. The average annual dose per person within the 50-mile radius is about 0.055 mrem/person (1).

The methodology for site selection of reprocessing plants is similar in essentials to that described for nuclear power plants in Chapter 9. Having taken steps to reduce all effluent levels to be as low as practicable, an exclusion area is determined on the basis of a design base accident. Special siting criteria applying in this case include (a) the need to make the plant location fairly central to its service area to hold down transportation costs for spent fuel shipments, and (b) the expected storage on-site of liquid, short-lived waste materials for periods of 1-100 years. The postulated leakage of one of these tanks would be assumed to be a major design accident to be considered both in examining the geology and hydrology of the site and in calculating the potential population dose due to plant operation.

Plutonium

Plutonium, as PuO_2, is one of the principal products of the reprocessing operation. Its isotopic composition will vary somewhat with irradiation conditions and, of course, its production is one of the main objectives of the breeder reactors. Its environmental effect is largely related to its high toxicity; while it is not nearly as toxic as many of the nerve gases or organic toxins, it is among the most dangerous among the heavy elements, both for its chemotoxicity and its radiotoxicity. This is reflected in the ICRP recommendations for maximum permissible concentrations of plutonium in air and water, which are among the lowest for any nuclide (6 x 10^{-13} μCi/cc in air; 5 x 10^{-5} μCi/ml in water). Both the chemical form and the isotope distribution are seen to affect the dose values. The chemical form may vary on exposure to natural phenomena, such as weathering or bacterial action, and thus affect the rate of uptake and transportability.

The isotope proportions depend on the type of reactor fuel and burn-up conditions, as shown in Table 159 (49). Stannard (4) has reviewed the uptake and mobility of plutonium in the biosphere and man. Fortunately it appears that there is a large discrimination factor against plutonium entering terrestrial food chains in quantity. However, any increase in the proportion of the shorter-lived isotopes will increase the effective dose. In addition, the growth of Am-241 during fuel or waste storage has a tremendous effect on the dose rate. For instance, for the unshielded Yankee sample in Table 159 only 3.3% of the total dose came from Pu-241 at the time of chemical separation. After one year this isobaric chain increases to 34% contribution and to 47% after two years, due to the growth of the two daughters of Pu-241, Am-241, and U-237.

It was pointed out in the preceding section that improved recovery of plutonium is the most effective means of reducing population dose from effluents; this also should include the other actinides, notably Am-241 and the curium isotopes shown in Table 153. Under those conditions actinide discharges could be held well below the levels predicted in Reference 50. This is particularly important if the quantity of plutonium recovered rises as steeply as has been suggested in various projections (51).

It would be idle to pretend that such observations fully meet public apprehension regarding the extraction and recycle of plutonium. Despite all assurances, a fire at a military weapons plant at Rocky Flats, Colorado, in 1972 did result in the dispersion of plutonium in the form of fine powder over a fairly large area. Actually that incident involved plutonium

Table 159. Plutonium Isotopic Distributions for Various Reactors and Burn-Up Conditions (49)

Source	MWd/MT	^{238}Pu wt %	^{239}Pu wt %	^{240}Pu wt %	^{241}Pu wt %	^{242}Pu wt %	^{236}Pu wt %	Dose Rate[a] Rad/hr
Yankee	34,000	1.92	63.3	19.2	11.7	3.9	4.5×10^{-6}	4.0
Dresden	30,000	1.44	52.9	27.7	12.1	5.8	—	3.3
Shippingport	27,000	1.28	46.2	30.5	11.0	11.0	1.4×10^{-6}	3.4
Yankee	24,000	1.0	67.7	18.8	10.0	2.5	3.4×10^{-6}	3.0
Dresden	20,000	0.36	69.5	21.5	6.9	1.74	3.0×10^{-6}	1.8
Shippingport	20,000	1.06	51.2	28.9	11.0	7.8	1.0×10^{-6}	2.8
Graphite moderated		0.07	72.4	22.7	3.9	0.97	2.9×10^{-7}	
Yankee	13,000	0.34	80.7	12.7	5.6	0.67	1.0×10^{-6}	1.5
Dresden	13,000	0.29	75.3	18.2	5.1	1.2	—	1.3
Shippingport	10,000	0.53	65.8	22.6	7.8	3.3	0.8×10^{-6}	1.8
Graphite moderated		0.04	86.0	12.0	1.8	0.16		
PWR ~ 1976	28,000	1.5	59.5	21.6	12.5	4.9	0.8×10^{-6}	

[a]Dose rate through about 33 mils of plastic bag material.

metal, none of which is used in the LWR fuel cycle. Although the Pu concentrations on the ground were fairly low, numerically, the plutonium dust did get carried along by the wind (52) and resuspension occurred (15). The plant involved clearly did not incorporate adequate engineering safeguards or containment of the type required for commercial reprocessing plants; one can only presume that such events will not recur in future with the more stringent design requirements for civilian plants.

The other concern regarding plutonium recycle is one for the physical safety of the material to keep it out of the hands of terrorists and blackmailers. Recent books on the potential risks of diversion of plutonium into unauthorized hands (53,54) have focused attention on this problem. The production of plutonium at a reprocessing plant makes it the obvious target for such activities, a situation which has resulted in proposals for counteracting this possible hazard. The "GESMO" Report (55,56), one of these proposals, would impose such a heavy economic and technical burden on plutonium extraction as to make plutonium recycle and even the operation of breeder reactors increasingly unattractive. Such proposals, if enforced, might well tip the balance towards natural uranium reactors, such as the CANDU type, that do not at present incorporate any need for reprocessing or plutonium recycle for commercial power.

If the recycle of plutonium is delayed or forbidden, the value of the uranium-235 recovered in reprocessing by itself may not pay for the cost of reprocessing and all attendant activities. In that case, economic considerations may dictate a "throwaway cycle," that is, the spent fuel is considered essentially just a high-level radioactive waste to be disposed of without further treatment. For light-water reactors this approach would represent a significant waste of fuel resources; for a breeder reactor it would be impossible, since the whole concept of the breeder hinges on recovery and recycle of the plutonium, or uranium-233, bred in the fertile material.

Colby (57) analyzed the economics of reprocessing and the throwaway cycle and arrived at a reprocessing cost, at 1975 prices, of the order of $180 per kg, including allowances for waste handling, safeguards and transportation. On the other hand, in the throwaway cycle the spent fuel is an economic liability with a range of $50-300 per kg U. Even in the absence of plutonium recycle, reprocessing would just barely break even; however, fuel conservation and waste treatment considerations seem to indicate that reprocessing is the more rational approach and consumption of plutonium as fuel in a reactor by far the preferred means of disposal.

Waste Storage

To provide for decay of most of the activation and fission products and to await decisions on ultimate disposal of the long-lived nuclides, the bulk of the waste solutions is stored on site as a dewatered slurry in large tanks. Such storage may have to be provided for 20-100 years and although the chemical industry has ample experience with the useful life of large storage tanks (58) little of that applies in this case. The tanks typically are large, double-walled carbon steel cylinders in an underground concrete "dish" or chamber to catch minor leaks; the Barnwell plant has provisions for stainless steel tanks at enormous cost. Figure 102 illustrates one type of storage tank.

In view of the high level of radioactivity of the slurries and the decay of the fission products, corrosion appears to be promoted by the synergistic effects of the direct radiation damage, the radiolytic decomposition of the nitric acid solvent, which promotes oxidation of the tank walls, and the thermal stresses set up near the liquid surface. To ensure adequate heat dissipation and to avoid gas bubble formation the slurry must be agitated or circulated. Adequate spare tank capacity must be provided to permit transfer of the slurry in case a leak develops, and adequate monitoring is essential to discover such leaks rapidly. Even so, it appears that some leaks did occur at Hanford in 1973 and remained undetected for some time. The small amount of activity escaping was retained by the soil below the tank; however, the increasing amounts of such waste kept in storage makes the possible consequences of leaks or catastrophic damage to these tanks a factor of prime environmental concern. By suitable site location, adequate concrete encasement, choice of soil foundations with high exchange capacity, and location well away from flooding hazards or any major aquifers in the ground, any potential environmental problems can be greatly minimized (59).

Table 160 indicates the experience with high-level waste tank storage at U.S. Atomic Energy Commission facilities. It can be seen that the stress-relieved carbon steel and stainless-steel tanks performed more reliably; some of the older tanks have since been heat-treated *in situ* for stress relief. Table 161 gives some orders of magnitude of waste solutions generated and the storage facilities required for plants of various capacities (46).

For the commercial reprocessing plants tank storage can be considered only as an interim operation. Permanent underground storage on the plant site will rarely be feasible or desirable, and most of the long-lived activity, including the actinides, should probably be shipped off-site regularly to minimize the fission product inventory in storage. For this reason, most reprocessing plants will incorporate a solidification system

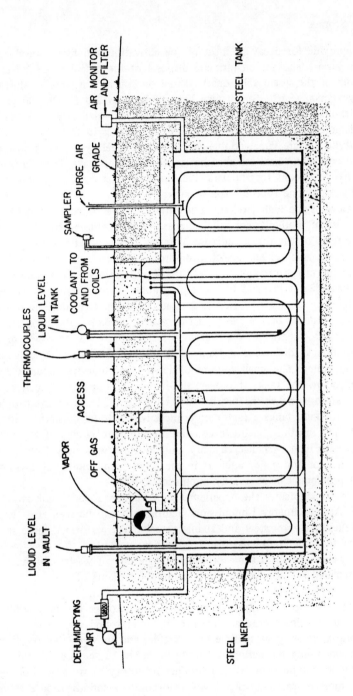

Figure 102. High-level waste storage tank (39).

Table 160. U.S. AEC High-Level Waste Tank Construction and
Failure Experience (59)

Type of Tank[a]	Construction Period	Number of Tanks (Leakers)	Tank Capacity m³	Method of Heat Removal
		Hanford		
A	1943-1944	16 (0)	204	None
A	1943-1944	48 (2)	2000	None
A	1946-1947	12 (1)	2000	None
A	1947-1949	30 (1)	2870	None
A	1950-1952	18 (2)	2870	None
A	1953-1954	5 (0)	3780	None
A	1953-1954	10 (7)	3780	Self-boiling
A	1954-1955	6 (1)	3780	Self-boiling
A	1963-1964	4 (1)	3780	Self-boiling
C	1968-1971	2 (0)	3780	Self-boiling
C	1970-1973	2 (0)	3780	Self-boiling
		Savannah River		
B	1951-1954	12 (3)	2840	Cooling coils
B	1953-1955	4 (3)	3780	Cooling coils
A	1957-1958	4 (0)	4920	None
A	1959-1962	4 (0)	4920	None
C	1967-1969	4 (0)	4920	Cooling coils
C	1969-1972	2 (0)	4920	Cooling coils
	National Reactor Testing Station			
D	1951-1952	1 (0)	1200	Cooling coils
D	1951-1952	1 (0)	1200	None
D	1954-1955	2 (0)	1130	Cooling coils
D	1954-1955	1 (0)	1130	None
E	1954-1955	4 (0)	113	Cooling coils
D	1957-1959	3 (0)	1130	Cooling coils
D	1957-1959	1 (0)	1130	None
D	1962-1964	2 (0)	1130	Cooling coils

continued

Table 160, continued

Leak Indicated	Type of Tank[a]	Initial Use	Approximate Amount Leaked to Ground Volume m³	kCi of ^{137}Cs	Postulated Failure Mechanism	Tank Service[b]
1956	A	1947	200	0.09	Structural stress[c]	Bismuth phosphate and TBP uranium-recovery wastes
1957	B	1955	0	0	Stress corrosion	Purex wastes and salt cake
1958	A[d]	1958	60	8	Structural stress[c]	Redox wastes
1959	A	1953	80	2	Pitting corrosion	TBP uranium-recovery wastes
1959	A	1946	110	23	Pitting corrosion	Bismuth phosphate and TBP uranium-recovery and redox wastes
1959	B	1956	0	0	Stress corrosion	Purex wastes and salt cake
1959	B	1957	0	0	Stress corrosion	Purex wastes
1959	B	1959	< 4	<0.1	Stress corrosion	HM (Hanford metal) wastes[e]
1960	A	1953	130	4	Pitting corrosion	TBP uranium-recovery wastes
1962	A[d]	1955	10	17	Structural stress[c]	Redox wastes
1963	A[d]	1963	Small		Structural stress	Purex wastes
1964	A[d]	1956	Small		Structural stress[c]	Purex wastes
1965	A[d]	1955	Small		Structural stress	Redox wastes
1965	A[d]	1958	190	40	Structural stress	Redox wastes
1969	A[d]	1956	110	45	Structural stress[c]	Redox wastes
1969	A	1948	260	51	Pitting corrosion	TBP uranium-recovery wastes
1969	B	1955	0	0	Stress corrosion	Purex waste and salt cake
1971	A	1954			Pitting corrosion	TBP uranium-recovery and redox wastes and salt cake
1972	B	1960	0	0	Stress corrosion	HM (Hanford metal) wastes[e]
1972	A[d]	1956	0	0	Pitting corrosion	Redox wastes and salt cake

[a]A, carbon-steel-lined concrete tank. B, carbon-steel tank inside partially steel-lined concrete vault. C, stress-relieved carbon-steel tank inside fully steel-lined concrete vault. D, stainless-steel tank inside concrete vault. E, direct buried stainless-steel tank on concrete pad.

[b]For alkaline wastes.

[c]Bulged liner.

[d]Self-boiling waste service.

[e]Wastes from processing (Purex) enriched uranium.

Table 161. Assumed Properties of Reprocessing Plants and
Waste Storage Facilities (46)

| | Fuel Processing Rate (metric tons/day)[a] | | | | | |
| | LWR Fuel | | | FBR Fuel | | |
	1	6	36	1	6	36
Processing plant						
Total dissolver solution, m³/day	3.33	20	120	10	60	360
No. independent lines	1	1	2	1	1	4
Relative processing rate/line	1	1	3	1	3	3
Max. head-end vessel capacity, m³	3.33	20	20	10	20	30
Total cell capacity/line, m³	2333	14,000	14,000	7000	14,000	21,000
No. cells/line	7	14	14	7	14	14
Cell size, m³	333	1000	1000	1000	1000	1500
Cell ventilation rate, m³/min	66.7	300	200	200	200	300
Total ventilation rate/line, m³/min	700	4200	4200	2100	4200	6300
Ventilation train[b]	F,M	F,M	F,M	F,A,M	F,A,M	F,A,M
Total off-gas flow rate	28	85	255	2.0	4.0	24
Off-gas train[b]	S,T,F	S,T,F	S,T,F	S,I,F	S,I,F	S,I,F
Interim[c] liquid waste (acid) storage facility						
Tank volume (80% filled), m³	812	3785	3785	990	3785	3785
No. tanks required for 2-yr accumulation	2	3	10	2	3	13
Off-gas flow rate/tank, m³/min	6.1	28	28	7.4	282	28
Off-gas train[b]	C,F	C,F	C,F	C,F	C,F	C,F
Vault ventilation rate, m³/min	6.1	56	224	7.4	56	280
Ventilation train[b]	C,F,M	C,F,M	C,F,M	C,F,M	C,F,M	C,F,M
Interim[c] waste solids storage canal						
Length for 14.6-m width, m	5.8	35	210	7.1	42	250
Ventilation rate, m³/min	170	1000	6100	210	1200	7300
Ventilation train[b]	C,F	C,F	C,F	C,F	C,F	C,F

[a]A 1.0 metric ton/day plant processes 260 metric tons of uranium + plutonium per year.
[b]S = caustic scrubber; 90% removal of iodine.
 T = silver tower; 99% removal of iodine
 A = activated charcoal filter; 99% removal of iodine.
 M = metal mesh or silica gel; 99.9% removal of Te, Cs and Ru.
 I = high-efficiency iodine removal units; iodine DF of 10^7.
 C = steam condenser; discharges air at 100°F and 100% relative humidity.
 F = either reliably-protected HEPA or deep bed filter.
    Normal effluent = 0.0012 mg/m³; accident effluent = 0.02 mg/m³.
[c]Two years.

to permit safe shipment and to reduce the need for subsequent handling. The off-gases from the waste evaporation and solidification may include some long-lived fission products, such as ruthenium, technetium, cesium, selenium and tellurium (46). These must be carefully recovered or included in any dose calculations. Table 162 indicates estimated ruthenium discharges and the consequent dose levels for three plants (50).

Table 162.   Estimated Ruthenium Discharges and Offsite Doses from
Reprocessing Plant Solidification Processes (50)

	NFS	Midwest	Barnwell
Fuel waste cooling time	5 yrs	150 days	5 yrs
Ruthenium present, Ci/MTU	1.7×10^4	5×10^5	1.7×10^4
Maximum average annual offsite dose, rem/yr	0.025	0.5	0.174
Percent of 1/10 ICRP MPC for Ru-106	1.67	33	11.6

One type of radioactive waste peculiar to fuel reprocessing plant operations is the fuel structural hardware and fuel hulls which, after hot nitric acid leaching processes, require disposal. This waste contains, typically, thousands of curies of induced activity consisting of comparatively short-lived radioisotopes and small amounts of fission products and plutonium that survived the acid leaching processes and remained affixed to the fuel hulls. At the present time, one might anticipate that these wastes will be disposed of in the same manner as high-level solidified wastes (i.e., transferred to a federal repository), or they may be buried in a licensed wasteburial facility.

SUMMARY

In reviewing the fuel cycle industries it is evident that there are only a few critical operations that may have significant environmental impact. These comprise the radium daughters from uranium mill tailings, the power consumption entailed in enrichment processes, the effluents—both gaseous and liquid—from reprocessing plants, and the potential impact from long-term storage of high-level radioactive wastes. Table 163, published by the U.S. Atomic Energy Commission in 1974 (48), was

Table 163. Summary of Environmental Considerations for the Uranium Fuel Cycle (48)

Natural Resource Use	Total	Maximum Effect Per Annual Fuel Requirement of Model 1000 MWe LWR
Land (acres)		
Temporarily committed	63	
Undisturbed area	45	
Disturbed area	18	Equivalent to 90 MWe coal-fired powerplant.
Permanently committed	4.6	Equivalent to 90 MWe coal-fired powerplant.
Overburden moved (millions of MT)	2.7	
Water (millions of gallons)		
Discharged to air	156	$\approx$ 2% model 1000 MWe LWR with cooling tower
Discharged to water bodies	11,040	
Discharged to ground	128	
Total	11,319	<4% of model 1000 MWe LWR with once-through cooling
Fossil fuel		
Electrical energy (thousands of MW-hour)	317	<5% of model 1000 MWe LWR output.
Equivalent coal (thousands of MT)	115	Equivalent to the consumption of a 45 MWe coal-fired powerplant.
Natural gas (millions of scf)	92	<0.2% of model 1000 MWe energy output
Effluents–chemical (MT)		
Gases (including entrainment)[a]		
SO_x	4,400	
NO_x[b]	1,177	
Hydrocarbons	13.5	Equivalent to emissions from 45 MWe coal-fired plant for a year
CO	28.7	
Particulates	1,156	
Other gases		
F^-	0.72	Principally from UF_6 production enrichment and reprocessing. Concentration within range of state standards–below level that has effects on human health.

Table 163, continued

Natural Resource Use	Total	Maximum Effect Per Annual Fuel Requirement of Model 1000 MWe LWR
Effluents—chemical (MT)		
Liquids		From enrichment, fuel fabrication, and reprocessing steps. Components that constitute a potential for adverse environmental effect are present in dilute concentrations and receive additional dilution by receiving bodies of water to levels below permissible standards. The constituents that require dilution and the flow of dilution water are:
SO_4^-	10.3	
NO_3^-	26.7	
Fluoride	12.9	
Ca^{++}	5.4	
Cl^-	8.6	NH_2—600 cfs. NO_2—20 cfs. Fluoride—70 cfs.
Na^+	16.9	
NH	11.5	
Fe^3	0.4	
Tailings solutions (thousands of MT)	240	From mills only—no significant effluents to environment.
Solids	91,000	Principally from mills—no significant effluents to environment.
Effluents—Radiological (curies)		
Gases (including entrainment)		
Rn-222	75	Principally from mills—maximum annual dose rate <4% of average natural background within 5 mi of mill. Results in 0.06 man-rem per annual fuel requirement.
Ra-226	0.02	
Th-230	0.02	
Uranium	0.032	
Tritium (thousand)	16.7	Principally from fuel reprocessing plants—whole body dose is 4.4 man-rem per annual fuel requirements for population within 50 mi radius. This is <0.005% of average natural background dose to this population. Release from Federal Waste Repository of 0.005 Ci/yr has been included in fission products and transuranics total.
Kr-85 (thousands)	350	
I-129	0.0024	
I-131	0.024	
Fission products and transuranics	1.01	

Liquids

Uranium and daughters	2.1	Principally from milling—included in tailings liquor and returned to ground—no effect on environment.
Ra-226	0.0034	From UF$_6$ production-concentration 5% of 10 CFR. 20 for total processing of 27.5 model LWR annual fuel requirements.
Th-230	0.0015	
Th-234	0.01	From fuel fabrication plants—concentration 10% of 10 CFR 20 for total processing 26 annual fuel requirements for model LWR.
Ru-106	0.15	From reprocessing plants—maximum concentration 4% of 10 CFR 20 for total reprocessing of 26 annual fuel requirements for model LWR.
Tritium (thousands)	2.5	

Solids (buried)

Other than high-level	601	All except 1 Ci comes from mills—included in tailings returned to ground—no significant effluent to the environment, 1 Ci from conversion and fuel fabrication is buried.
Thermal (billion Btu)	3,360	<7% of model 1000 MWe LWR.

Transportation (man-rem)

Exposure of workers and general public	0.334

[a]Estimated effluents based upon combustion of equivalent coal for power generation.
[b]1.2% from natural gas use and process.
[c]Cs-137 (0.075 Ci/AFR) and Sr-90 (0.004 Ci/AFR) are also emitted.

considered a realistic, if conservative, summary of the environmental impact that could be assigned to any representative 1000 MWe LWR through its portion of the nuclear fuel cycle operations.

REFERENCES

1. "Environmental Survey of the Uranium Fuel Cycle," Report WASH-1248 (Washington, D.C.: U.S. Atomic Energy Commission, 1974).
2. "Environmental Analysis of the Uranium Fuel Cycle. Part I, Fuel Supply," Report EPA-520/9-73-003B (Washington, D.C.: U.S. Environmental Protection Agency, 1973).
3. Sagan, L. A. "Human Costs of Nuclear Power," Science 177, 487 (1972).
4. Hodge, H. C., J. N. Stannard and J. B. Hursh, Eds. "Uranium–Plutonium–Transuranic Elements," Handb. Exp. Pharmac. XXXVI (New York: Springer Verlag, 1973).
5. Heinrich, E. W. Mineralogy and Geology of Radioactive Raw Materials (New York: McGraw-Hill, 1968).
6. Lang, A. H., J. W. Griffith, and H. R. Steacy. "Canadian Deposits of Uranium and Thorium," Econ. Geology Series No. 16, 2nd ed. (Ottawa, Canada: Geol. Survey of Canada, 1962).
7. Nininger, R. D. Minerals for Atomic Energy, 2nd ed. (New York: Van Nostrand Co., 1956).
8. Basson, J. K., C. H. Wyndham, A. J. A. Heyns, W. H. Kelley, C. P. S. Barnard, A. H. Munro and I. Webster. "Biostatistical Investigation of Lung Cancer Incidence in South African Gold Uranium Miners," Proc. Geneva Conf. Peaceful Uses of Atomic Energy 11, 13, United Nations (1971). Full Report South African Atomic Energy Board Report PEL-209, Pelindaba, Pretoria (1971).
9. Donalson, A. W. "The Epidemiology of Lung Cancer among Uranium Miners," Health Phys. 16, 563 (1969).
10. Lundin, F. E., J. W. Lloyd, E. M. Smith, V. E. Archer and D. A. Holaday. "Mortality of Uranium Miners in Relation to Radiation Exposure, Hardrock Mining and Cigarette Smoking–1950 through September 1967," Health Phys. 16, 571 (1969).
11. Mays, C. W. "Cancer Induction in Man from Internal Radioactivity," Health Phys. 25, 585 (1973).
12. Scott, L. M., K. W. Bahler, A. de la Garza and T. A. Lincoln. "Mortality Experience of Uranium and Nonuranium Workers," Health Phys. 23, 555 (1972).
13. "Radiation Exposure of Uranium Miners," Hearings, Joint Committee on Atomic Energy, 90th U.S. Congress, Washington, D.C. (1968).
14. International Labor Office. "Radiation Protection in the Mining and Milling of Radioactive Ores," IAEA Safety Series No. 26 (Geneva: International Labor Office, 1968).
15. Harley, N. H. and B. S. Pasternack. "Alpha Absorption Measurements Applied to Lung Dose from Radon Daughters," Health Phys. 23, 771 (1972).

16. Holaday, D. A. "History of the Exposure of Miners to Radon," *Health Phys.* **16**, 547 (1969).
17. U.S. Atomic Energy Commission. "Proposed Rules: Standards for Protection Against Radiation," *Federal Register* **39**, 22428 (1974).
18. Havlovic, V. and J. O. Snihs. "Behavior and Size Distribution of Natural Radioactive Ions," *Arch. Environ. Health* **24**, 432 (1972).
19. Rock, R. L., R. W. Dalzell and E. J. Harris. "Controlling Employee Exposure to Alpha Radiation in Underground Uranium Mines," (Washington, D.C.: U.S. Bureau of Mines, Dept. of Interior, 1971).
20. American National Standard N13.1-1971. "Radiation Protection in Uranium Mines," (New York: American National Standards Institute, 1971).
21. Kusnetz, H. L. "Radon Daughters in Mine Atmospheres; A Field Method for Determining Concentrations," *Am. Ind. Hyg. Assoc. Quart.* **17**, 85 (1956).
22. Groer, P. G. "The Accuracy and Precision of the Kusnetz Method for the Determination of the Working Level in Uranium Mines," *Health Phys.* **23**, 106 (1972).
23. Holmgren, R. M. "Working Levels of Radon Daughters in Air Determined from Measurements of RaA and RaB," *Health Phys.* **27**, 141 (1974).
24. Groer, P. G., R. D. Evans and D. A. Gordon. "An Instant Working Level Meter for Uranium Mines," *Health Phys.* **24**, 387 (1973).
25. Auxier, J. A., K. Becker, E. M. Robinson, D. R. Johnson, R. H. Boyett and C. H. Abner. "A New Radon Progeny Personnel Dosimeter," *Health Phys.* **21**, 126 (1971).
26. Chapuis, A., A. Lopez, J. Fontan, and G. J. Madelaine. "Determination de la fraction d'activité existant sous forme de Ra A non-attaché dans l'atmosphère d'une mine d'uranium," *Health Phys.* **25**, 59 (1973).
27. Budnitz, R. J. "Radon-222 and its Daughters—A Review of Instrumentation for Occupational and Environmental Monitoring," *Health Phys.* **26**, 145 (1974).
28. Reiter, R. "Equipment and Procedures for Protection Against Radon and its Daughters in Mine Atmospheres," *Nukleonik* **9**, 359 (1967).
29. Goodwin, A. "Problems and Techniques for Removal of Radon and Radon-Daughter Products from Mine Atmospheres," *Nucl. Safety* **14**, 643 (1973).
30. Stein, L. "Chemical Methods for Removing Radon and Radon Daughters from Air," *Science* **175**, 1463 (1972).
31. "Final Environmental Statement. Highland Uranium Mill (Exxon Co.)," (Washington, D.C.: U.S. Atomic Energy Commission, 1973)
32. "Radiological Health and Safety in Nuclear Materials Mining and Milling," *Proc. Vienna 1963 Symp.* (Vienna: International Atomic Energy Agency, 1964).
33. "Draft Environmental Statement. Humeca Uranium Mill (Rio Algom Corp.)," (Washington, D.C.: U.S. Atomic Energy Commission, 1972).

34. Schiager, K. J. "Analysis of Radiation Exposures on or near Uranium Mill Tailings Piles," *Radn. Data Repts.* **15**, 411 (1974).
35. Blomeke, J. O., C. W. Kee and J. P. Nichols. "Projections of Radioactive Wastes to be Generated by the Nuclear Power Industry," Report ORNL-TM-3965, Oak Ridge National Laboratory (1974).
36. "Stabilization of Uranium-Thorium Milling Waste Retention System," Regul. Guide 3.23 (Washington, D.C,: U.S. Atomic Energy Commission, 1974).
37. "Use of Uranium Mill Tailings for Construction Purposes," Hearings, Joint Committee on Atomic Energy, 92nd U.S. Congress (1971).
38. Long, J. T. *Engineering for Nuclear Fuel Reprocessing* (New York: Gordon and Breach, 1967).
39. "The Safety of Nuclear Power Reactors (Light-Water-Cooled) and Related Facilities," Report WASH-1250 (Washington, D.C.: U.S. Atomic Energy Commission, 1973).
40. Magno, P. J., T. C. Reavey and J. C. Apidianakis. "Iodine-129 in the Environment Around a Nuclear Fuel Reprocessing Plant," (Washington, D.C.: U.S. Environmental Protection Agency, 1972).
41. Allied Gulf Nuclear Services, Inc. "Environmental Report," (Barnwell, S.C.: Barnwell Nuclear Fuel Plant, 1971).
42. *Nucl. Ind.* **21** (6), 28 (1974).
43. Deuster, R. W. "Environmental Concerns in Fuel Reprocessing," *Trans. ANS* **19**, 210 (1974).
44. Magno, P., T. Reavey and J. Apidianakis. "Liquid Waste Effluents from a Nuclear Fuel Reprocessing Plant," Public Health Service Report BRH/NERHL 70-2 (Rockville, Maryland: Department of Health, Education and Welfare, 1970).
45. Klement, A. F., C. R. Miller, R. P. Minx, and B. Shleien, Eds. "Estimates of Ionizing Radiation Doses in the United States 1960-2000," Report ORP/CSD 72-1 (Rockville, Maryland: U.S. Environmental Protection Agency, 1972).
46. "Siting of Fuel Reprocessing Plants and Waste Management Facilities," U.S. Atomic Energy Commission Report ORNL-4451 (1970).
47. "Environmental Analysis of the Uranium Fuel Cycle: Pt. III—Nuclear Fuel Reprocessing," Report EPA-520/9-73-003-D (Washington, D.C.: U.S. Environmental Protection Agency, 1973).
48. U.S. Atomic Energy Commission. "Environmental Effects of the Uranium Fuel Cycle," *Fed. Register* **39**, 14188 (1974).
49. Smith, R. C., H. H. Van Tuyl and L. G. Faust. "The Effect of Radiation Levels from Plutonium on Fuel Fabrication Process Design," Report BNWL-SA-3139 (Richland, Washington: Battelle-Northwest, 1970).
50. Russell, J. L. and F. L. Galpin. "A Review of Measured and Estimated Offsite Doses at Fuel Reprocessing Plants," IAEA/ENEA Symp. Management of Radioactive Wastes from Fuel Reprocessing (Paris, 1972).
51. Golan, S. and R. Salmon. "Nuclear Fuel Logistics," *Nucl. News* **16** (2), 47 (1973).

52. Dunaway, P. D. and M. G. White, Eds. "The Dynamics of Plutonium in Desert Environments," Report NVO-142 (Las Vegas, Nevada: U.S. Atomic Energy Commission, 1974).
53. McPhee, J. *The Curve of Binding Energy.* (New York: Farrar, Straus and Giroux, 1974).
54. Willrich, M. and T. B. Taylor. *Nuclear Theft: Risks and Safeguards.* (Cambridge, Mass.: Ballinger Publ. Co., 1974).
55. Smiley, S. H. "Waste Management—Licensing and Criteria," *Nucl. Technol.* 24, 294 (1974).
56. "The Use of Mixed-Oxide Fuels in Light-Water Reactors. Draft Generic Environmental Statement," Report WASH-1327 (Washington, D.C.: U.S. Atomic Energy Commission, 1974).
57. Colby, L. J. "Fuel Reprocessing in the United States," *Nucl. News* 19 (1), 68 (1976).
58. Straub, C. P. "Low Level Radioactive Wastes: Treatment, Handling, Disposal," (Washington, D.C.: U.S. Atomic Energy Commission, 1964).
59. Lennemann, W. L. "Management of Radioactive Aqueous Wastes from AEC Fuel Reprocessing Operations," *Nucl. Safety* 14, 482 (1973).

ADDITIONAL REFERENCES

Archer, V. E., J. K. Wagoner and F. E. Lundin. "Lung Cancer Among Uranium Miners in the United States," *Health Phys.* 25, 351 (1973).
Auxier, J. A., W. H. Shinpaugh, G. D. Kerr and D. J. Christian. "Preliminary Studies on the Effects of Sealants on Radon Emanation from Concrete," *Health Phys.* 27, 390 (1974).
Bair, W. J. "Toxicology of Plutonium," *Adv. Radiation Biol.* 4, 255 (1974).
Blanchard, R. L., E. L. Kaufman and H. M. Ide. "Lead-210 Blood Concentration as a Measure of Uranium Miner Exposure," *Health Phys.* 25, 129 (1973).
Eichholz, G. G. "Notes on the Safe Handling of Uranium Alloys in Industry," Mines Branch Information Circular IC-125 (Ottawa, Canada: Canada Dept. of Mines and Technical Surveys, 1961).
"Evaluation of Radon-222 near Uranium Tailings Piles," Report DER 69-i (Rockville, Maryland: U.S. Public Health Service, HEW, 1969).
"Fuel Fabrication and Reprocessing," *Proc. 4th Internat. Conf. Peaceful Uses of Atomic Energy, Geneva* 8 (New York: United Nations, 1972).
Fusamura, N., R. Kurosawa and M. Maruyama. "Determination of f-Value in Uranium Mine Air," in *Assessment of Airborne Radioactivity* (Vienna: International Atomic Energy Agency, 1967).
Geesaman, D. P. "Plutonium and the Energy Decision," *Bull. Atomic Scient.* 27 (7), 33 (1971).
George, A. C. and L. Hinchcliffe. "Measurements of Uncombined Radon Daughters in Uranium Mines," *Health Phys.* 23, 791 (1972).
"Guidance for the Control of Radioactive Hazards in Uranium Mining," Federal Radiation Countil Report No. 8 (revised), Washington, D.C. (1967).

Hine, A. R. "Reprocessing in the United Kingdom," *Kerntechnik* 15, 157 (1973).

International Commission on Radiological Protection. "Implications of Commission Recommendations that Doses be Kept as Low as Readily Achievable," ICRP Publ. No. 22 (Oxford: Pergamon Press, 1973).

International Commission on Radiological Protection. "Report of Committee II on Permissible Dose for Internal Radiation," ICRP Publ. 2 (Oxford: Pergamon Press, 1960). Also, Supplement ICRP Publ. 6 (New York: Macmillan Co., 1964).

Lyon, W. S. "The Perils of Pu," *Radiochem. Radioanal. Letters* 18, 117 (1974).

Patton, F. S., J. M. Googin and W. L. Griffith. "Enriched Uranium Processing," (New York: Macmillan Co., 1963).

Raghavayya, M. and J. H. Jones. "A Wire Screen-Filter Paper Combination for the Measurement of Fractions of Unattached Radon Daughters in Uranium Mines," *Health Phys.* 26, 417 (1974).

Savignac, N. F. and K. J. Schiager. "Uranium Miner Bioassay Systems: Lead-210 in Whiskers," *Health Phys.* 26, 555 (1974).

Simpson, S. D., C. G. Stewart, G. R. Yourt and H. Bloy. "Canadian Experience in the Measurement and Control of Radiation Hazards in Uranium Mines and Mills," *Proc. Internat. Conf. Peaceful Uses of Atomic Energy* 23, 195 (1958).

Smith, R. C., L. G. Faust and L. W. Brackenbush. "Plutonium Fuel Technology. Part II: Radiation Exposure from Plutonium in LWR Fuel Manufacture," *Nucl. Technol.* 18, 97 (1973).

"Tentative Method of Analysis for Radon-222 Content of the Atmosphere," *Health Lab. Sci.* 6, 114 (1969).

Tsivoglou, E. C. "Research for the Control of Radioactive Pollutants," *J. Water Poll. Control Fed.* 35, 242 (1963).

Tsivoglou, E. C., H. E. Ayer and D. A. Holaday. "Occurrence of Nonequilibrium Atmospheric Mixtures of Radon and its Daughters," *Nucleonics* 14 (9), 40 (1953).

West, P. J. "Waste Management for Nuclear Power," *Bull. IAEA* 16, 78 (1974).

CHAPTER 12

RADIOACTIVE WASTE DISPOSAL

INTRODUCTION

The safe ultimate disposal of the highly radioactive wastes accumulating in the course of the nuclear fuel cycle represents a major challenge to the industry. It was shown in Chapter 7 that the low- and intermediate-level wastes generated at a reactor site could be stored temporarily, solidified, and shipped to a commercial disposal site for burial without undue trouble, although the number of drums shipped and the number of shipments to disposal each year may be appreciable.

High-level wastes are defined as "those aqueous wastes resulting from the operation of the first cycle solvent extraction system, or equivalent, and the concentrated wastes from subsequent extraction cycles, or equivalent, in a facility for reprocessing irradiated reactor fuels." These wastes contain virtually all of the nonvolatile fission products, several tenths of one percent of the uranium and plutonium originally in the spent fuels, and a large fraction of the other actinides formed by transmutation of the uranium and plutonium in reactors. They can be characterized generally by their very intense, penetrating radiation and their high heat-generation rates. U.S. Nuclear Regulatory Commission regulations call for these wastes to be solidified within five years after they are generated and for the resultant stable solids to be shipped to a federal repository within ten years after the liquids are generated. Table 164 lists the major fission products in wastes after 1 and 5 years' storage (1), including some stable ones.

Cladding wastes are the residual zircaloy and stainless steel cladding and structural components of the fuel assemblies that remain after the fuel cores have been dissolved. Although their radioactivity arises mainly from neutron-induced isotopes, the hulls are similar in some respects to high-level waste in that they may contain up to 0.1% of the plutonium

555

Table 164. Major Fission Products in High-Level Wastes (1)

Element	Average Atomic Weight	After 1 year[a]		After 5 years		After 1 year moles/liter (378 liters/metric ton)	
		Ci/metric ton	g/metric ton	Ci/metric ton	g/metric ton	45,000 MWd/metric ton	20,000 MWd/metric ton[a]
Rb	86	3.68×10^5	463	19.09×10^4	470	0.014	0.007
Cs	135	1.05×10^5	3,952	9.16×10^4	3,677	0.077	0.037
Sr	89		1,191		1,124	0.035	0.016
Ba	138		2,184		2,459	0.042	0.017
Y	89	1.12×10^5	622	9.16×10^4	621	0.018	0.009
La	139		1,729		1,729	0.033	0.015
Ce	141	4.41×10^5	3,440	1.26×10^4	3,305	0.064	0.028
Pr	141	4.41×10^5	1,611	1.26×10^4	1,611	0.030	0.013
Nd	145		5,531		5,665	0.101	0.043
Pm	147	1.03×10^5	111	3.58×10^4	39	0.003	0.002
Sm	150		1,039		1,111	0.018	0.008
Eu	153	0.16×10^5	270		256	0.004	0.001
Gd	156		127		144	0.002	
Zr	93	0.26×10^5	5,050		5,116	0.143	0.064
Nb	95	0.55×10^5					
Mo	98	2.84×10^5	4,856	1.80×10^4	4,859	0.131	0.058
Tc	99		1,153		1,153	0.031	0.014
Ru	102		3,118		3,040	0.081	0.033
Rh	103		520		520	0.013	0.007
Pd	106		1,736		1,814	0.043	0.016
Ag	109		68		68	0.002	0.001
Cd	111		103		103	0.003	0.001
Sn	123		78		78	0.002	0.001
Sb	123		22		16		
Te	130		705		711	0.014	0.006
Total		19.5×10^5	39,679	45.3×10^4	39,689	0.904	0.397

aExposure at a specific power of 15 MW/metric ton in a LWR.

originally in the spent fuel, require biological shielding equivalent to several inches of lead, and have heat-generation rates of 50 to 100 W/ft^3 (1750-3500 W/cm^3). To assist in the characterization of these wastes, reprocessing flowsheets were developed that avoided, wherever possible, the additions of chemicals that might be particularly troublesome in subsequent waste management operations (2).

The safe disposal of these high-level radioactive wastes produced at a reprocessing plant poses several problems that are technically quite soluble but may incur rather large long-term expenditures (3). There are several reasons for these problems. The first concerns the sheer magnitude of the activity involved, somewhere around 120 MCi/yr received for reprocessing per 1000 MWe reactor served and 200-400 MCi/yr to be discharged for burial as solidified wastes. The second relates to the projected growth in the quantity of wastes accumulated by the reprocessing of LWR and LMFBR fuels. Table 165 shows a 1971 projection of this growth for the U.S. (4). The actual rates will probably be lower, owing to delays in completion of power plants, but the table illustrates volumes of wastes generated, the total activity generated, and the contribution to this of some of the more significant nuclides. Note that the volumes of waste quoted are not particularly large, but one must realize that shielding requirements limit the quantity of solid waste that can be handled in any one shipment. The quantity of curium is also worth noting, since it may introduce some further disposal problems. More recent descriptions of the composition of the various waste products will be found in Reference 2.

The third problem arises from the lifetimes of the nuclides concerned. As reference to Table 158 will show, most of the fission and activation products are short-lived, with half-lives of less than one year; among these, Zr-Nb-95, Ce-Pr-144, Ru-103, Y-91 and Sr-89 are of particular importance. Materials in this range of half-lives can, if necessary, be stored, in liquid or solid form, in subsurface tanks or bunkers for 20-50 years to decay to low levels of activity. However, there are three other groups that are longer-lived, and they account for most of the philosophical and technical difficulties for disposal: These are

a. **Long-lived waste nuclides** ($t_{1/2}$ 1 - 100 yr)
   Ru-106 (1 yr), Eu-155 (1.7 yr), Pm-147 (2.7 yr), Sb-125 (2.7 yr), Cs-134 (2.1 yr), Co-60 (5.3 yr), Kr-85 (10.4 yr), H-3 (12 yr). Sr-Y-90 (28 yr), Cs-137 (30 yr) and Sm-151 (90 yr)

b. **Transuranium isotopes**
   Pu-238, 239, 240, 241, 242; Am-241-242; Cm-242, 243,244. All these have half lives ranging from 13-380,000 years.

Table 165. Projected High-Level and Alpha-Emitting Wastes (4)

	Calendar Year Ending		
	1980	1990	2000
Installed nuclear electric capacity, MW	150,000	450,000	940,000
Fuel reprocessed, metric tons/yr	3,000	9,000	19,000
Solidified high level waste[a]			
Annual volume, 10^3 ft^3	9.7	33	58
Accumulated volume, 10^3 ft^3	44	290	770
Total accumulated activity, MCi	19,000	110,000	270,000
Total thermal power, MW	80	410	1,040
Significant isotopes accumulated			
28.9-y Sr-90, MCi	960	5,700	12,000
30-y Cs-137, MCi	1,300	8,000	20,000
1.6 x 10^7-y, I-129, Ci	480	3,300	9,700
10.8-y Kr-85, MCi	120	690	1,500
12.3-y H-3, MCi	7.3	44	110
87.4-y Pu-238,[b] MCi	1.2	10	40
24,400-y Pu-239,[b] MCi	0.022	0.3	1.7
6600-y Pu-240,[b] MCi	0.041	0.5	2.4
14.3-y Pu-241,[b] MCi	6.6	58	240
433-y Am-241, MCi	2.3	28	150
18.1-y Cm-244, MCi	30	170	330
Number of shipments to repositories[c]	23	240	590
Alpha wastes			
Annual volume, 10^6 ft^3	0.36	0.92	2.5
Accumulated volume, 10^6 ft^3	4.6	10.4	27.0
Total activity, MCi	31	150	420
Total thermal power, MW	0.003	0.17	0.66
Significant isotopes accumulated, MCi			
87.4-y Pu-238	0.51	2.6	8.4
24,400-y Pu-239	0.11	0.58	2.0
6600-y Pu-240	0.16	0.83	2.8
14.3-y Pu-241	30	146	400
433-y Am-241	0.14	1.0	6.6
Number of shipments to repositories[d]	930	1,200	3,030

[a]Assumes 1 ft^3 of solidified waste per 10,000 MWd (th).

[b]Assumes 0.5% of plutonium in fuel is lost to waste.

[c]Each shipment consists of 57.6 ft^3 of waste in thirty-six 6-in.-diameter cylinders. Half of the waste is aged 5 years and half is aged 10 years at the time of its shipment.

[d]Each shipment contains 832 ft^3 of waste.

c. **Very long-lived waste nuclides**

These include I-129 (1.59 x 10^7 yr), Tc-99m (2.13 x 10^5 yr), Cs-135 (2.3 x 10^6 yr), Sn-126 (2 x 10^5 yr), Pd-107 (6.5 x 10^6 yr), Zr-93 (9.5 x 10^5 yr) and Se-79 (6.5 x 10^4 yr).

In the past the last group has been considered unimportant because of the low specific activities involved. In recent years particular attention has been paid to iodine-129 (5-7). The general conclusion seems to be that none of these materials will contribute significantly to population doses above natural background if they are properly contained; yet in the long run their production adds irreversibly to the global radioactive inventory.

In the first group, tritium and krypton-85 have been discussed already. Up to now they have been freely released to the environment and liberally dispersed in water and air, respectively. Current technology is moving towards containment and storage of these two, separate from other wastes at some extra cost. Of the remainder in this group, strontium-90 and cesium-137 have been the subject of the widest study, whereas the others have been considered sufficiently short-lived to be allowed to decay. Not so for Sr-90 and Cs-137; with half-lives of 28 and 30 years, their activity would reduce only to about 10% when stored for a century, not much reduction when we are dealing with hundreds of megacuries. That means that these materials must be stored safely for centuries or millennia to prevent their reappearance in human ecosystems. Such a requirement imposes unusual conditions, since few human activities are ever planned on such a time scale; in fact, in view of the inherent impermanence of human institutions on a historical scale, any "ultimate" disposal scheme of the long-lived and very long-lived materials must be designed to be independent of all vagaries of fortune.

The same applies in theory to the actinides, but actually, because of their very low concentrations in other wastes, there is some question whether it is feasible and worthwhile to separate and concentrate these elements for disposal or possible recycle. The actinides cause waste management difficulties at two distinct points in the nuclear fuel cycle. Some are carried over with the fission products during nuclear fuel reprocessing and also some highly dilute plutonium wastes will appear from fuel manufacturing plants. Thus, at the entry point to the waste facility we find a mix of many different transuranic actinides intimately combined with the shorter-lived and temporarily more hazardous fission products. The important points to note about this category of wastes are that (1) the offending actinides are relatively very toxic, and (2) although initially far less radioactive than the fission products, they become dominant after 500 years because of their much longer half–lives.

This disparity in hazard between the categories, graphically presented in Figure 103, suggests that it would be advantageous to separate them chemically and adopt different strategies for each kind (8). The *relative toxicity* of waste materials is found by comparing their MPC values (9). Alternatively,

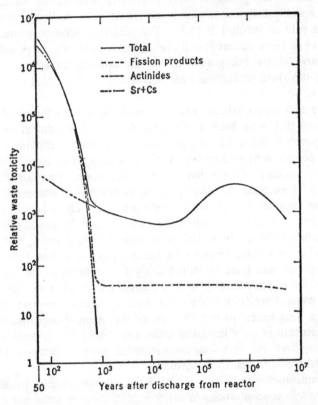

Figure 103. Decay plot for radioactive waste components, showing relative toxicity for an equilibrium fuel cycle with 99.5% removal of U and Pu and 30,000 MWd/T burnup (8).

toxicities may be expressed in terms of the *hazard index*, defined as the volume of air or water required to dilute a unit volume of waste or ore to its ingestion RCG (Radiation Concentration Guide recommendation). Figure 104 indicates an application of this concept to show the effect of secondary processing of PWR waste after primary extraction of the uranium and plutonium (10-12). It was found that a processing scheme that sends the individual actinides to the waste with equal hazard contributions is

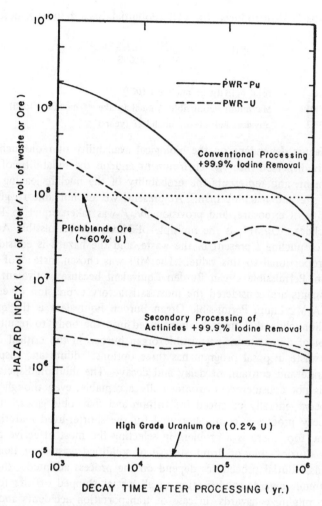

Figure 104. Hazard index of PWR waste as a function of decay time (10).

not practical. Since uranium and neptunium have little effect in the first few thousand years but can control after 200,000 years, the general strategy adopted was to select a high value of uranium removal (the easiest separation to perform) and determine a value for neptunium removal that shifts control to 1000 years or so when the americium and plutonium will control the hazard index (10).

To compare the potential hazards for mankind resulting from the presence in high-level waste of several nuclides having long half-lives, Gera and Jacobs (13) introduced the PHI (Potential Hazard Index), defined as

$$PHI_i = P_i \frac{Q_i}{MPI_i} \cdot \frac{\tau_i}{0.693} \tag{90}$$

where: Q_i = total activity of nuclide i (μCi)

MPI_i = Maximum Permissible Annual Intake of nuclide i (μCi)

τ_i = physical half-life of nuclide i (years)

P_i is a factor dependent on the biological availability of radionuclide i once it is dispersed into the environment and on the reliability of waste containment, and represents the probability of the nuclide leaving the disposal site and reaching man. At present, we are not able to give the probability of exposure, and provisionally, P was taken equal to 1 for all radionuclides. Q_i/MPI_i is the multiple of Maximum Permissible Annual Intakes of nuclide i present in the waste, and the hazard is considered to be proportional to this value. The MPI was chosen instead of the Maximum Permissible Organ Burden Equivalent because equivalent dose commitments are considered the most satisfactory expression of equivalent risks. (A Maximum Permissible Organ Burden Equivalent is the quantity of a radionuclide that must be introduced into the body to result in the retention of a maximum permissible organ burden in the critical organ.)

Any waste disposal program has three options: dilute and disperse, concentrate and contain, or delay and decay. The dilution-dispersion option is not considered environmentally acceptable, even though till now it has been generally practiced for tritium and the noble gases. The delay-decay process is commonly used for the shorter-lived materials, but for them, too, there is a problem in selecting the most effective process, whether slurries, bin or drum storage of solidified wastes, or tank storage of liquids. Burial procedures depend on the process adopted; there is a strong trend towards solidification of all wastes shipped off-site for disposal to minimize hazards in case of transportation accidents and to reduce the volume of material to be handled. In fact, the degree of volume reduction is an important selection criterion for any solidification process.

At this time there is some uncertainty regarding the actual permanence of some of the proposed disposal methods. Pending resolution of this question, retrievable storage sites have been proposed as an intermediate measure (2,14,15). It has also been proposed that transuranics content in disposed waste be limited to 10 nanocuries per gram. It is also increasingly obvious that the waste solidification plant must be an integral part of any reprocessing facility to avoid shipment of potentially hazardous solutions, as well as to minimize environmental impact problems.

Tank Storage

It was pointed out in the previous chapter that tank storage of high-level wastes has been practiced generally at U.S. government-owned facilities, where they could be monitored easily and had relatively large areas of land at their disposal. Future commercial wastes can probably be treated in this form only for brief times, as an intermediate measure, and to permit decay of the shorter-lived activities (2,14,16). The solutions are nitrate wastes raised in some cases to pH values around 14. Table 166 shows representative compositions of these liquids. Since tank space is expensive, the liquid volume is usually reduced by partial evaporation. The resultant slurry must be well-ventilated to remove off-gases and decay heat, as a means of minimizing corrosive action. The slurry must remain fluid enough to permit pumping into other tanks, both to vary its composition and in case of suspected tank leakage. Tanks must be stress-relieved, double-walled, and capable of withstanding earthquakes and tornadoes and of dissipating all decay heat in case of failure of the cooling or circulating systems. In Great Britain and France great care is taken during processing to keep the first-cycle wastes free of inert salts; consequently, volume reductions of 10 to 15 gal per metric ton (38-57 liters/T) of fuel (and proportionately higher volumetric heat generation rates) are achieved routinely.

The wastes are stored as acid solutions in stainless steel tanks, or they are neutralized and stored in carbon-steel tanks. In the United States, these tanks, which range in capacity from 0.33 to 1.3 million gallons, are encased in concrete and buried underground. Decay heat is removed during storage either by allowing the neutralized wastes to self-heat, condensing the vapors, and returning the condensate to the tanks, or by use of water-cooling coils submerged in the waste. The smaller volumes of more-concentrated fission product solutions in Europe are stored at environmental temperatures in stainless steel tanks of 15,000-20,000-gal (60-80 kl) capacity. These tanks are equipped with water-cooling coils and are housed in concrete vaults that are enclosed in industrial-type buildings.

The tanks are equipped with devices for measuring temperatures and liquid levels, detecting leaks, and agitating the contents; they are also equipped with emergency facilities to maintain cooling and other essential services. Costs depend on tank size, materials of construction, and the degree of cooling and secondary containment required. In the United States, capital costs range from about $0.50 per gallon of storage capacity for 1.3×10^6-gal carbon-steel tanks without cooling facilities to $5.40 per gallon for 30,000-gal stainless steel tanks equipped with cooling coils (17).

Table 166. Representative Composition of High-Level Liquid Wastes (16)

	Material[b]	Grams/MT from Reactor Type[a]		
		LWR[c]	HTGR[d]	LMFBR[e]
Reprocessing Chemicals	Hydrogen	400	3,800	1,300
	Iron	1,100	1,500	26,200
	Nickel	100	400	3,300
	Chromium	200	300	6,900
	Silicon	—	200	—
	Lithium	—	200	—
	Boron	—	1,000	—
	Molybdenum	—	40	—
	Aluminum	—	6,400	—
	Copper	—	40	—
	Borate	—	—	98,000
	Nitrate	65,800	435,000	244,000
	Phosphate	900	—	—
	Sulfate	—	1,100	—
	Fluoride	—	1,900	—
	Sub-total	68,500	452,000	380,000
Fuel Product Losses[f,g]	Uranium	4,800	250	4,300
	Thorium	—	4,200	—
	Plutonium	40	1,000	500
	Sub-total	4,840	5,450	4,800
Transuranic Elements[g]	Neptunium	480	1,400	260
	Americium	140	30	1,250
	Curium	40	10	50
	Sub-total	660	1,440	1,560
Other Actinides[g]		< 0.001	20	< 0.001
Total Fission Products[h]		28,800	79,400	33,000
	Total	103,000	538,000	419,000

[a]Water content is not shown; all quantities are rounded.

[b]Most constituents are present in soluble, ionic form.

[c]U-235 enriched PWR, using 378 liters of aqueous waste per metric ton, 33000 MWd/MT exposure. (Integrated reactor power is expressed in megawatt-days [MWd] per unit of fuel in metric tons [MT].)

[d]Combined waste from separate reprocessing of "fresh" fuel and fertile particles, using 3,785 liters of aqueous waste per metric ton, 94,200 MWd/MT exposure.

[e]Mixed core and blanket, with boron as soluble poison, 10% of cladding dissolved, 1,249 liters per metric ton, 37,100 MWd/MT average exposure.

[f]0.5% product loss to waste.

[g]At time of reprocessing.

[h]Volatile fission products (tritium, noble gases, iodine and bromine) excluded.

Since the high-level waste gives off considerable heat, gases and aerosols are released and the waste may boil. The off-gases must be decontaminated, and a sensitive monitoring system must be included in the design of the system. The underground tanks should be designed to withstand not only the internal pressures from boiling but also external pressures from earthquakes, soil loads, and rises in the water table. The storage tank should be located in an area where the release of activity from the tank would not result in an environmental problem. The storage-tank system should contain a pump to remove any liquid that may enter the containment unit since the introduction of water into the underground vaults containing the tanks may markedly reduce the 100-year lifetime of the tanks. In general, the materials in the tanks may be greater than 70% solids; consequently large quantities of the material would not be expected to be released to the environs upon rupture of the tanks. Figure 105 shows this type of tank installation (18).

The experience with tank storage over the past 20 years has not been uniformly good. A total of more than 80 million gallons of waste is currently being stored in about 200 underground tanks in the United States. Fifteen tank failures, all in carbon-steel systems, had been reported up to 1971 (Table 160), and others have occurred since then. Although the causes of the tank failures are believed to be understood and appropriate corrective measures have been incorporated in the designs of new tanks now under construction, a general lack of confidence in the long-term integrity of these systems seems to be merited, particularly as they pertain to the civilian nuclear power program (17). The requirement that the waste be initially maintained in a condition amenable to rapid, efficient transfer and the economics of future solidification would dictate that the fission products be stored as relatively pure acid solutions under nonboiling conditions.

Power-reactor wastes derived from fuels of high nuclear burn-up will contain much larger quantities of fission products than do current wastes. In designing tanks and cooling systems to remove decay heat, consideration must be given to: (1) the age and concentration of the fission products at the time they are added to the tank, (2) the thermal characteristics of the waste, as determined largely by the physical states and concentrations of the inert chemicals present, and (3) the rate at which the tank is filled. Careful review of Hanford and Savannah River operating experience with existing tank farms has indicated that 120-to-150-days old waste solutions with concentrations of inert salts comparable to present Purex production wastes could be stored at a volume of about 100 gal (379 l) per 10^4 MWd (thermal) of fuel exposure. If these wastes are neutralized, considerations of heat removal from the precipitated solids

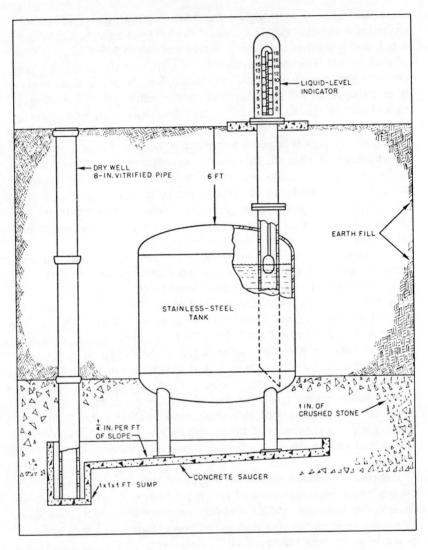

Figure 105. Dry-well and waste collection tank installation (18).

indicate that storage at about 600 gal (2271 l) per 10^4 MWd (thermal) of fuel exposure would be practical. On this basis, the radioactivity level of the acid wastes 150 days after removal of the fuel from the reactor is about 13,700 curies/gal, which is equivalent to 200 Btu hr^{-1} gal^{-1}; in the alkaline case, it is about 2300 curies/gal, or 30 Btu hr^{-1} gal^{-1} (27 W/l/hr).

The radiolysis of water or aqueous solutions results in the production of hydrogen and oxygen. In the case of nitrate solutions, the hydrogen yield, $G(H_2)$, defined as the number of molecules formed per 100 eV of absorbed energy, has been shown to be dependent on the nitrate ion concentration. It is estimated that $G(H_2)$ values for the acid wastes and the alkaline wastes are 0.03 and 0.10, respectively. These values are equivalent to about 3 ft^3 of H_2 (STP) per 10^6 Btu of fission-product heat in acid waste, and 10 $ft^3/10^6$ Btu in the alkaline case. If the waste is not stored under self-boiling conditions, provisions must be made to sweep the hydrogen from the vapor space above the waste and prevent its accumulation in the tank.

Although the general corrosion rate for the carbon steel used to store alkaline waste is only about 0.5 microns/year, some pitting has been observed; also, stress corrosion that occurs at the weld-affected areas has indicated the need for heat treatment of the tanks in place before use. Overall corrosion rates of types 304L and 347 stainless steel during storage of acid wastes at about 140°F (60°C) are a few tenths of a micron per month, with grain-boundary but no intergranular attack. However, the rate of corrosion of stainless steel by acid wastes is accelerated to 0.5-1 mm/year at temperatures near boiling (17).

SOLIDIFICATION OF WASTE SOLUTIONS

To overcome some of the problems of secure storage of high-level wastes for prolonged periods of time, storage in solid form is generally preferred. Solids can be transported more easily over great distances, with less risk of spillage in case of accidents or fires, and greatly reduced chances of seepage into the water table after long storage on surface or underground. Any treatment process for the conversion of liquid wastes into solids ideally should meet the following criteria:

a. substantial reduction in volume
b. process should be relatively simple
c. product should be chemically stable in all expected environments
d. no deterioration from the effects of self heating
e. product should be nonhygroscopic and dense
f. process should be adaptable to remote operation and maintenance
g. process should be relatively inexpensive
h. product should be in a form that is easily transported
i. final product should have enough mechanical strength to survive drops and other accidental impacts
j. activity losses can be kept low by careful design or by use of efficient trapping methods.

No one process can meet all these criteria equally well and every major establishment seems to favor its own procedure over all others. While most processes have assumed that a high degree of volume reduction is desirable on economic and safety grounds, Puechl (19) recently has argued in favor of the contrary approach of deliberately reducing the specific activity of the solidified waste. Most solidification processes suggested so far consist of two stages:

1. conversion of liquid to stable chemical form, usually an oxide
2. incorporation in a stable, inert matrix or container.

The choice of method adopted depends largely on cost and past experience. The various fixation methods have been described in detail by Amphlett (20) and Schneider (1). Most of them involve high-temperature oxidation of the waste slurry; they differ in the final stable matrix material.

Three of the inert chemical constituents that may be in high-level wastes are sufficiently troublesome in solidification processing to merit a special effort to keep them out of high-level wastes. These constituents are sulfate, fluoride, and mercury ions.

The sulfate ion is generally unstable chemically at the higher range of temperatures reached in solidification (700°C and higher) and tends to volatilize. Retention of sulfate in the solidified waste at temperatures above 700°C requires chemical additives (usually calcium). Retention above 850-900°C is very difficult, particularly in melts. This volatilization can result in added corrosion problems in the off-gas system, recycle of sulfates, and increased sulfate concentrations in the liquid waste for cases of partial volatilization, or it can produce another medium-to-high-level waste stream that requires special treatment and disposal for cases of complete volatilization.

Fluoride is retained with difficulty (by using calcium) in solidified-waste processing up to temperature of about 600°C. At higher temperatures a significant fraction of fluoride may volatilize and must be recycled or disposed of by other means.

Mercury cannot be retained in the solidified waste when processed at temperatures above 400-500°C. When volatilized, the mercury and its oxides condense at temperatures of about 350°C and present relatively serious potential plugging problems. However, mercury is rarely encountered in high-level wastes.

Ruthenium, one of the ever-present fission products, is just as troublesome in waste solidification as it is in fuel reprocessing. Of the ruthenium 1-80% will usually oxidize and volatilize during solidification, and it is more difficult to remove from the off-gas stream than are the entrained nonvolatile materials. Chemical additions to minimize the oxidizing potential during solidification are sometimes required to avoid oxidation to the volatile RuO_4 form. Even then, volatilization of at least 1% is usually encountered. Volatilized ruthenium can be recycled to the solidification step or be removed by filters and selective chemical treatment devices.

All processes for solidification of high-level waste generate additional waste streams that contain intermediate levels of radioactivity. These are the vapor or condensate streams from the solidifier that have radionuclide contents less than the aqueous high-level wastes by factors of 10 to 1000. These primary effluents from solidification should be decontaminated by additional factors up to 1×10^{10}, which is comparable to those of the system for handling high-level liquid waste in a fuel-reprocessing plant. Processing of these vapor streams would logically be done by recycle routing to the existing high-level liquid waste concentration and processing equipment. Only a modest increase in capacity (on the order of 10%) of the liquid-waste processing capacity of the reprocessing plant would be required (1).

The most important conversion and fixation processes are:

    a. bituminization
    b. cementation and cement block forming
    c. pot calcination
    d. fluidized-bed calcination
    e. spray calcination
    f. vitrification
    g. conversion to clay sinters

Bituminization and cementation were described in Chapter 7. They lend themselves particularly well to the fixation of low- and intermediate-level wastes because they are simple and fairly inexpensive processes (21). They are less suitable for high-level wastes where self-heating and gas evolution may seriously impair the integrity of the matrix. The bitumen compound may decompose in time, particularly if hot, or develop cracks if frozen. Cement blocks by themselves are too friable and porous to serve over the long periods of time that must be contemplated, though they may be quite adequate for intermediate storage in concrete underground bins.

The four primary solidification processes developed in various countries are pot calcination, spray solidification, phosphate-glass formation and fluidized-bed calcination (1).

Pot Calcination

The principal processing vessel in pot calcination, the pot, is also the final container for the solidified waste. In this process, liquid waste is added to a pot that is heated in a multiple-zone heating and cooling furnace. The waste is concentrated at a constant volume to the point that a scale of calcine cake forms on the walls of the pot. As calcination continues, the scale grows in thickness and reduces the heat transfer from the pot wall to the boiling sludge; therefore the feed rate must be reduced proportionately. When the feed rate is reduced to an unprofitable rate (about 5 liters/hr), the feed is shut off. At this point the scale has grown inward from the pot wall and upward from the bottom of the pot to fill the pot, except for a thin cone-shaped liquid-containing space in the upper 2 to 3 ft of the salt cake. Heating is then continued until the liquid is boiled to dryness and all the waste in the pot has been calcined and has reached the temperature of 850-900°C, where essentially all of the volatile constituents have been released. The pot is then cooled in the furnace, removed, sealed, and taken to storage.

Because the pots serve as the processing vessels, they must be made of corrosion-resistant material to withstand the severe corrosion conditions during calcination. Corrosion of 304L stainless steel is negligible during processing (less than 8 μ/day). Pots must have liquid-level and temperature-measurement devices. Liquid level may be measured either with a standard gas-purged dip tube or with an internal temperature sensor near the top of the pot.

Internal heat from decay of radioactive constituents requires slight modifications of operating techniques. With the presence of internal heat, cooling of the pot wall is necessary before the material at the center of the pot has reached its final maximum temperature or the center temperature will exceed that desired. (Higher temperatures can result in severe corrosion and undesirable volatilization of some constituents.) Typical time requirements for each of the calcination steps range from 10-30 hours, for a total batch time of 70 hr for a 12 in. (30 cm) diameter pot.

During pot calcination of Purex wastes, ruthenium is volatilized to the extent of about 5% and 10-30% for low-sulfate wastes and high-sulfate wastes, respectively, although other variables may also affect volatilization. Lower volatilization can be effected by the addition of chemical reductants, such as nitric oxide gas or phosphites. Volatilization of cesium and rubidium, which are always present as fission products, can be virtually eliminated by adding enough sulfate or phosphate ions to the feed to be chemically equivalent to the total amount of alkali metals present. Entrainment from the pot calciner, when operated on a reasonably conservative capacity basis, is approximately 0.4% of the total feed (1).

Pot calcination has the advantage of being a simple process that is adaptable to a wide variety of feed compositions. Its disadvantages are the necessity of using a stainless-steel pot, the low thermal conductivity of the calcine, the necessity of increasing the capacity of a system by multiple pot lines, and the leachability of solidified waste is greater in water than for glassy solids (17).

Spray Solidification

In the spray calciner (Figure 106), liquid waste (which contains some or all of the melt-making additives) is fed through a pneumatic atomizing nozzle into the top of a heated cylindrical tower. The atomized waste is sequentially evaporated, dried, and calcined to a powder as it falls into a continuous melter below the calciner tower where it is melted at temperatures of 800 to 1200°C. Process gases from calcination flow

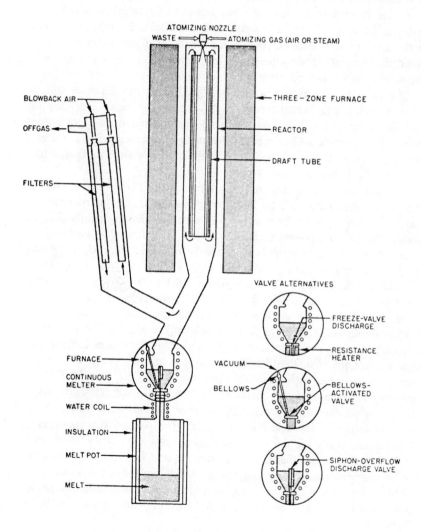

Figure 106. Spray calciner and continuous melter.

into the adjacent filter chamber, carrying much of the calcined powder as dust. The dust collects on the porous metal filters as the gas passes through. Dust deposits are periodically blown off the filters by sudden pulses of high-pressure steam or air directed backward through the filters by small nozzles. The dislodged dust falls into the melter with the main powder stream. The molten calcine flows through an overflow weir or a freeze valve into the receiver-storage pot below. After the pot is filled, it is cooled in the furnace, sealed, and taken to storage.

In the continuous melter the small amount of residual nitrate and water is driven off from the calcine, and the calcine is melted. Platinum is the only reliable metallic construction material found to date to withstand the corrosive environment and high temperature with phosphate melts, although other materials appear promising.

The capacity of a platinum melter 10-in. (25 cm) in diameter with a 14-in. (35-cm) high heated section is 1.7 liters/hr of melt. Discharge of melt from the melter has been adequately demonstrated on a continuous basis using overflow weirs or batchwise using straight-tube freeze valves wherein a plug of melt about 2 in. (5 cm) long is melted or frozen to provide on-off flow control.

The pot for receiving the molten waste may be made of mild steel if the pot is filled with melt by large, rapid batchwise dumps from the melter or if the pot is filled with melts with low melting points (less than about 700°C). These limitations are caused by the need to heat the pots under slow-filling conditions to the point where the melt will slump to assure complete filling of the pots without formation of stalagmites or voids. (Mild-steel pots can acceptably resist oxidation by air at temperatures up to about 650°C for periods of several days.)

Most of the flow sheets used in spray-solidification processing at Battelle-Northwest produce alkali metal-phosphate solids. These compositions are used primarily because they offer a relatively large latitude in chemical composition, they have generally low melting points (700-900°C), they produce melts with reasonably low viscosities (less than 50 P) at operating temperatures, the chemically adjusted feed solutions are relatively easy to handle, they generally produce homogeneous melts, and the final solid has desirable characteristics. The primary disadvantages of phosphate melts are the higher corrosion rates and somewhat higher leach rates than other melts, such as silicates and borosilicates. With the typical phosphate melts, microcrystalline solids are formed in spray solidification by enough phosphate addition to approach orthophosphate melts. Sufficient alkali metals are added to reduce the melting point to 700-900°C.

Up to 75% of the ruthenium can be volatilized during the spray-calcination step (not during melting) with the phosphate flow sheets. This volatilization can be reduced by completely eliminating melt-making flux from the feed and by adding all melt-making flux to the melter, by reducing the oxidizing potential in the calciner, by reducing the nitrate and acid concentrations in the waste, and by using borosilicate fluxes. Volatilization of the fission products cesium and rubidium has not occurred significantly in spray solidification flow sheets.

The spray solidification process has several advantages: it is a continuous process with low holdup volumes, it is adaptable to a moderately wide variety of feed compositions, and it produces a variety of good-quality solids. Its disadvantages are that it is a moderately complicated system, it requires good flow control of sometimes difficult-to-handle feed solutions, its performance requires high-quality atomization, and at present it requires the use of a relatively expensive platinum melter. Melting the calcined powder in the receiver pot, rather than using an expensive platinum melter, may eliminate one of the disadvantages (1,17).

Fluidized-Bed Solidification

Fluidized-bed solidification is a continuous process that has been developed for use with aluminum nitrate and zirconium fluoride-aluminum nitrate wastes. In fluidized-bed solidification, liquid waste is continuously converted to granular solids by being heated in a fluidized bed of the solids, and the solids are continuously withdrawn from the calciner to storage bins (or the solids may be further converted to monolithic forms). The liquid waste is injected through pneumatic atomizing nozzles into the side of a heated (400-600°C) bed of granular solids. This bed is continuously agitated (fluidized) by sparging gas upward through the fluidized-bed reactor. Contact of the waste with the hot, granular bed results in evaporation and calcination of the feed as coatings of the bed particles. The calcine that is entrained with the process gases from the calciner is removed from the gas stream by cyclone separators or filters, and is then returned to the main stream of particles. The main stream of particles is continuously removed from the reactor and transported to storage bins (Figure 107).

The product from fluidized-bed solidification is granular, with a mean particle diameter of about 500 μm. The granules may be composed of crystals or amorphous solids. They are generally spherically shaped, moderately soft, and friable. The thermal conductivity of the bulk calcine is relatively low.

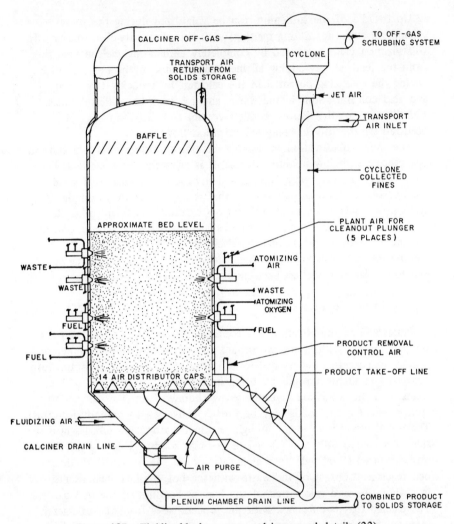

Figure 107. Fluidized-bed process: calciner vessel details (22).

The heat for calcining must be provided in such a manner that the maximum temperature of the heat-transfer surface is less than the sintering point of the calcine, and the heat must be distributed in such a manner that it can be absorbed by the needs of the reactor. Experimental systems have been built in which the calcining chamber was a rotary-ball kiln or the atomized feed solution was heated by radiant heat. However, a fluid bed of the type shown in Figure 107 with in-bed combustion of kerosene seems to produce the densest material with the highest thermal conductivity (22).

For high-level wastes with high rates of self-heat generation, the fluidized-bed system requires a means for cooling the contents of the bed or for dumping the bed during shutdown periods. Such provisions will eliminate the potential for self-overheating of the bed when flow of feed to the bed has stopped.

Containers for fluidized-bed calcine may be individual pots or large slab or annular containers. The latter geometry provides for heat removal by air or water circulating around the outsides of the concentric storage bins. Thus far, containers for storing fluidized-bed calcine have been made of stainless steel; however, mild steel could be used if air cooling were provided.

The volatility of ruthenium from aluminum nitrate wastes varies from less than 1% at 550°C to greater than 90% at 350°C, and averages 40% in the calciner during operation at 400°C. The addition of chemical reductants greatly reduces the volatility of ruthenium. For example, the volatility from Purex wastes at 500°C was about 70%, but was reduced to about 1% when sugar (a chemical reductant) was added to the feed.

The advantages of the fluidized-bed process are that it is continuous, it has a relatively high capacity for a given equipment size, its scale-up technology is relatively well-known, and its solidified waste products are readily transportable by pneumatic means. The disadvantages of the process are that it is a moderately complicated system and its solidified waste products have relatively poor thermal conductivities (1). A detailed discussion of this method and the different approaches studied has been provided by Dickey et al. (22).

Vitrification

The incorporation of waste material into glass has several inherent attractions and hence has been studied rather widely. Glass is a well-known substance of low leachability and a high-degree of stability up to reasonably high temperatures. Depending on the glass formulation, moderate proportions of waste oxide, up to 25%, can be incorporated in borosilicate glass (23). In the United States, phosphate-glass solidification has been developed extensively. It has the advantage of being a continuous process producing a good-quality glass product. Its disadvantages are that it is a moderately complicated system, it requires operation with slurries that are difficult to handle, it cannot retain sulfate in the final solid, and at present it requires the use of a relatively expensive platinum melter.

The feldspar nepheline, (Na, K) $AlSi_3O_8$, has been extensively investigated at Chalk River (Canada), as a natural source of alumina and silica; extensive deposits of considerable purity exist on the northern shore of Lake Ontario, and its cost is only 2 cents/lb (1961). On adding acid, a highly porous gel is formed, from which water and gases escape readily on heating; the time of gelation varies from 30 min in 3N HNO_3 to 1 min in 7N acid, and a suitable composition for incorporating acid wastes is 1 g of nepheline + 1 ml of acid (> 2N). After drying the gel, a hard glass is formed on fusion at 1350°C. Addition of a flux lowers the viscosity of the melt but not the fusion temperature, and gases escape readily to give a bubble-free glass. Lithia is the most effective flux but is too expensive for general use, and sodium hydroxide gives foamy glasses unless fired in graphite crucibles. Lime was finally chosen in sufficient amount to neutralize the acid (14% in the lime-nepheline mix for 5N HNO_3 solutions).

Under these conditions up to 3% of aluminum or uranium can be incorporated in the wastes to give clear, dense glasses. Only cesium and ruthenium are appreciably volatilized during fusion; losses of cesium are probably due to decomposition of Cs_2O before the cesium is incorporated into the glass, since they are much smaller with lower melting-point borosilicate glasses. Ruthenium losses occur in two stages; below 900°C they are due to the action of nitric acid, giving volatile RuO_4, while at higher temperatures oxidation by air of RuO_2 takes place (20). Volatilized ruthenium may be removed on a bed of firebrick coated with oxides of iron (prepared by soaking in a ferric salt solution and firing); a 10-cm bed removes more than 98% of ruthenium from a gas stream of linear velocity 30 cm/sec at 800°C. Uncoated firebrick is without effect. Nepheline glasses are highly resistant to distilled water leaching, much more so than lead borosilicate glasses prepared from commercial frits (20).

The borosilicate process has been adopted widely in Europe. Figure 108 shows simplified flowsheets of the British and Russian pilot plants, as presented at the 1964 Geneva Conference. Hine (23) has presented a more detailed flow sheet with the compositions of the glass product and off-gases. The feed to the process is envisaged as being composed of 43% wt of silica, 32% wt of borax ($Na_2B_4O_2$) and 25% wt of waste oxides. In certain cases heat output considerations may limit the waste oxide content to a lower value. The glassy material still has to be canned for shipment and handling to minimize ultimate leach effects (23).

In phosphate glass solidification, liquid waste containing all the melt-making additives is first fed to the evaporator where it is concentrated and denitrated by factors of 2 to 10 to a thick, syrupy, aqueous phosphate

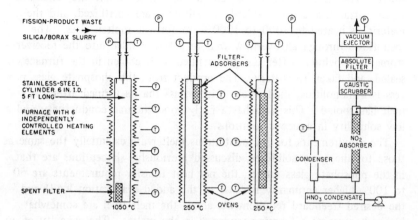

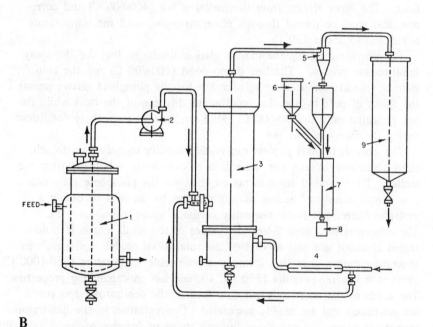

Figure 108. Schematic diagrams of vitrification processes for waste solidification
A. Harwell (England) process; B. Russian pilot plant (18).
(1) Feed tank, (2) Feed pump, (3) Fluidized bed, (4) Steam superheater, (5) Cyclone,
(6) Flux-feed machine, (7) Melting tank, (8) Vessel for the melt, (9) Offgas treatment.

slurry. The slurry is fed to the continuous melter where the remaining water, nitrates, and other volatile constituents are volatilized, and the material is heated to 1000 to 1200°C to form a molten glass. The melt flows through an overflow weir or a freeze valve into the receiver-storage pot below. After the pot is filled, it is cooled in the furnace, sealed, and taken to storage. The product from the phosphate glass process is a monolithic, moderately brittle glass that is formed after the melt has cooled. This glass has a fairly good thermal conductivity and a low solubility in aqueous solutions.

The requirements for the continuous melter are essentially the same as those for the spray solidifier, discussed previously. Exceptions are that, in the phosphate glass melter, the net heat transfer requirements are 50 to 100% higher (primarily because of the added evaporation load) and the desired freeboard requirements above the melt level are somewhat higher because of a foaming tendency in the melter. The capacity of a platinum melter that is 25 cm in diameter and has a 36 cm-high heated section is 1.2 liters of glass per hour, or about 3 liters of slurry feed per hour. The vapor stream from the melter is hot (400-600°C) and corrosive, and must be routed through platinum piping until the temperature is reduced to about 120°C.

The pot for receiving the molten glass is similar to that for the spray solidification process. The low slump point (600-700°C) and the continuous viscosity-temperature relationship for the phosphate glasses permit the filling of pots by the slow continuous dripping of the melt while the pot is heated only to 500-600°C. Mild steel can tolerate these conditions during the filling of one pot.

The phosphate glass process can readily solidify high-level waste solutions containing sulfate, but the sulfate is completely volatilized from the melter. The chemical adjustments required for the phosphate glass process consist mainly of adding phosphoric acid to the feed to obtain a metaphosphate melt (total normality of metal ions/phosphorus = 1). The concentration (mole %) of the oxides of the alkali metals is maintained at about one-half of that of the total metal oxides in the melt in order to obtain a glass that forms at a reasonable temperature (850-1000°C), melts at a low temperature (650-700°C), and has good handling properties. The solids in the chemically adjusted feed to the denitrator-evaporator are gelatinous and are readily suspended. Concentration in the denitrator-evaporator sometimes progresses through stages of foaming of heavy crystalline crystalline deposits at lower than, as well as higher than, normal concentration factors. These conditions must be defined for each flowsheet (17).

Ceramic Systems

Most ceramic systems utilize the ion exchange properties of clay minerals such as kaolinite or montmorillonite to adsorb radioactive cations from solution with high efficiency. On heating to several hundred degrees both the water content and the exchange capacity are reduced; above $900°C$ the structure changes to that of an aluminosilicate lattice within which the cations are firmly fixed. They are leached out in minute amounts except by reagents that attack the lattice chemically (20).

Clay mineral columns, packed in large cement pipes have been used satisfactorily for the immobilization of intermediate-level wastes. For high-level materials laboratory tests of fired clays on which radioactive materials have been fixed show a satisfactory resistance to leaching by distilled water and seawater. Fixation is improved as the firing temperature is raised from $900°C$ to $1200°C$.

Despite successful operation on a pilot plant scale there are limitations which render large-scale use unlikely except possibly in certain specialized cases:

1. The pH of the solution being treated is limited by the breakdown of clays in both acid and highly alkaline solutions. A range of pH 4-11 is preferable.

2. The loading of active material is limited by the exchange capacity of the clay; since the latter does not discriminate between active and inactive ions, the uptake of the former is reduced in the presence of inactive carriers or impurities. The latter predominate greatly in most processing waste solutions and include both hydrogen ions and polyvalent ions (Fe^{3+}, Al^{3+}), which are much more strongly absorbed than ions such as Cs^+ and Sr^{2+}. The method is therefore very inefficient except for solutions from which the interfering elements have first been removed.

3. Ruthenium in waste solutions is poorly removed because of its chemical complexity, and supplementary treatment is necessary.

4. The method is limited to solutions, and a more versatile technique applicable also to sludges and solids would be useful (20).

For these reasons vitrification processes are favored in general over ceramic processes.

Characteristics of the Solidified Wastes

The solidification process selected must meet the rather demanding criteria mentioned at the beginning of this section, the basic criterion of which is that radioactive material beyond safe limits not be permitted to enter the human environment. The desired characteristics of solidified waste of primary importance are: (a) high thermal conductivity, (b) low leachability by water (or possibly air), (c) good chemical stability and radiation resistance, (d) mechanical ruggedness, (e) noncorrosiveness to container, (f) minimum volume, and (g) minimum cost. The major characteristics of the principal processes are collected in Table 167. All of the materials have to be canned in a mild steel or stainless steel container, even the pot calciner where the pot forms the primary storage canister, but needs protection during handling.

Low leachability of the solidified products is desired in order to minimize the amount of contamination resulting in any water that might contact a breached container of solidified waste. In the case of the best solidified waste materials produced to date, less than one-millionth of the radionuclides are leached per unit specific surface per day. On the first contact of melt-solidified waste with water, the leachability is relatively high; then, over a period of 10 to 15 days, it decreases by about a factor of 10 to a relatively steady rate.

The chemical stability and radiation resistance of solidified waste are important for two reasons. First, they ensure that gases that may significantly affect the integrity of the product (or container, if present) are not generated during storage. Second, they ensure that the basic structure and properties of the solidified waste are known. Experience to date indicates that the formation of gas from solidified waste in enclosed containers is generally not significant if the storage temperature does not approach processing temperature. However, a few exceptions have been indicated for calcine prepared from feeds with a high sodium nitrate content (nitrogen oxide volatility) and for some phosphate-sulfate melts (sulfur oxide volatility).

The basic structure and chemical properties of solidified waste will change with time because about 15% of the fission products present after six months out of the reactor will eventually decay to other chemical elements. For calcines, this 15% represents up to 10% of the oxides present in the total waste: for melts, it represents up to 5% of the oxides present. A clear definition of these changes with regard to properties and their effects is not well known. Some glasses will devitrify to microcyrstalline structures if held at 400-800°C for days or weeks; calcined alumina granules change from amorphous to crystalline form; some volatile constituents migrate from thermally hot locations and condense at cooler locations, and phosphates and other glasses sometimes exude liquids.

Mechanical ruggedness of the solidified waste package is desirable, primarily during transportation. In the event that the container is breached, the ruggedness of the solidified waste is important in terms of its tendency to be dispersed. A waste that has low leachability but is

Table 167.   Characteristics of Solidified High-Level Waste (16)

	Pot Calcine	Spray Phosphate Ceramic	Phosphate Glass	Borosilicate Glass[a]	Fluidized Bed Calcine
Form	Scale	Monolithic	Monolithic	Monolithic	Granular
Description	Calcine cake, friable	Ceramic hard, tough	Glass hard, brittle	Glass hard, brittle	Calcine, mean particle diameter 100 to 500 μ
Bulk density, g/cc	1.2 to 1.4	2.7 to 3.3	2.7 to 3.0	3.0 to 3.5	1.0 to 1.7
Wt % fission product oxides (maximum)	90	30	25	50	50
Thermal conductivity, W/(m^2) ($^\circ$C/m)	0.3 to 0.4	1.0 to 1.4	0.8 to 1.2	1.0 to 1.4	0.2 to 0.4
Leachability in cold water, water, g/cm^2-day	1 to 10^{-1}	10^{-3} to 10^{-5}	10^{-4} to 10^{-6}b	10^{-5} to 10^{-7}	1.0 to 10^{-1}

[a]Produced by either spray or fluidized bed calcining followed by melting, or by in-canister vitrification processing.

[b]Devitrified phosphate glass exhibits increased leachability (leach rates = 10^{-2} to 10^{-3} g/cm^2-day).

very brittle or easily scattered may contaminate the environs to the same degree that a physically rugged waste with a higher leachability would (17). The corrosiveness of the solidified waste to the container determines, in part, the life of the container. Corrosion of containers by solidified wastes has indicated no problem areas in limited measurements to date; however, very long-term effects have not been evaluated. The useful life of the containers is expected to be much longer than the 15 to 40 years for containers for liquid wastes, though for salt mine storage lifetimes of the order of a few years only can be accepted.

The minimum volume of the solidified waste is important primarily for economic reasons. In general, reducing the volume will reduce the size and cost of containers, container storage areas, shipping equipment, and land to be used for storage areas. Minimizing cost, without affecting quality, is an obvious merit.

On balance, the vitrified waste products seem to exhibit the best combination of properties by being less friable than the calcined powders, having higher thermal conductivity, and being more resistant to leaching. For comparison with leach rates shown in Table 167, the leach rate in Pyrex glass is about 5×10^{-7} g/cm^2-day.

The cost of the various conversion processes appears to be comparable; it seems difficult to obtain recent estimates of the cost of full-scale conversion systems of each type, arrived at on an equivalent basis. Table 168 presents some 1971 cost data (17) for an overall waste management program, using pot calcination conversion, based on the assumption shown at the bottom of the table. The cost range of 0.037 to 0.045 mill/kWh is equivalent to $9500-$14,500 per tonne of fuel processed of which roughly 25-50% can be assigned to the cost of conversion (1).

WASTE UTILIZATION

In view of the large amount of heat generated by the high-level wastes and the enormous amount of radioactivity they represent, it is clearly attractive to explore possible methods of utilization of this material to conserve this energy for constructive purposes and to recoup some of the costs expended in concentration, fixation and encapsulation. As a heat source, the material might provide heat for low-grade steam generation or for space heating. As a radiation source, separated cesium-137 sources can find applications for sterilization of food, treatment of industrial effluents (24) or sterilization of municipal sewage (25-28). One 1000 MWe reactor generates about 2.9 MCi of cesium-137 per year, and total Cs-137 production in the U.S. annually may be of the order of 400 MCi by 1980. For a city of 600,000 a source of 23 MCi would be sufficient for

Table 168. Cost Range (1971) for Management of High-Level Waste from
LWR Fuels (1)[a]

	Probable Time Range, Years	Cost for Time Range, 10^{-3} mill/kWh	
		Min.	Max.
Interim liquid storage	For 0 to 5		22
Pot calcination[b]	At 0 to 5	8	22
Interim solid storage	For 0 to 10		10
Transportation of solids	At 3 to 10	1	3
Burial in salt	At 5 to 10	3	11
Comparison of total costs for specific cases			
Management of high-level wastes		37	45
Perpetual storage of liquid wastes		30	35

[a]33,000 MWd/metric ton of exposure.

[b]Range for a volume of 1.0 ft^3/10,000 MWd(t).

Based on: Reprocessing plants with capacities of about 6 metric tons/day; LWR fuels exposed to 33,000 MWd/metric ton at 30 MW/metric ton; 20-year plant life with straight-line depreciation; cost for private financing; 16% return on equity, 30% capital investment financed by bonds, 8% bond interest, escrow fund established with 5% interest, 48% federal income tax rate; pot calcination in cylindrical pots 6 to 24 in. in diameter to restrict maximum calcine centerline temperature to 900°C; 1.0 to 1.5 ft^3 of solid calcine for each 10,000 MWd(t); 2000-mile round-trip shipment in casks weighing 50 to 90 tons; long-term storage in salt mines.

tertiary sewage treatment (16); by 1981 this application alone could conceivably utilize all of the long-lived cesium-137 generated, which otherwise merely represents a costly liability.

Irradiation of sewage is a fairly inefficient process owing to the low concentrations of organic materials; treatment of more highly concentrated industrial wastes is more attractive economically, though reported cost figures are rather controversial. Katsumura (27) suggested construction and operating costs for a plant of 50,000 gal/day capacity to be 55 and 30 cents per 1000 gallons, respectively, for 1972 Japanese conditions, with substantial reductions with greater plant capacity. Figure 109 shows a projection of the relation between processing cost and plant capacity for various source costs. Current costs for both cobalt-60 and cesium-137 are of the order of $0.5-1.0 per encapsulated curie in large quantities. For many irradiation plants the cost of the source will represent the bulk of the capital investment (29); hence the availability of low-cost waste

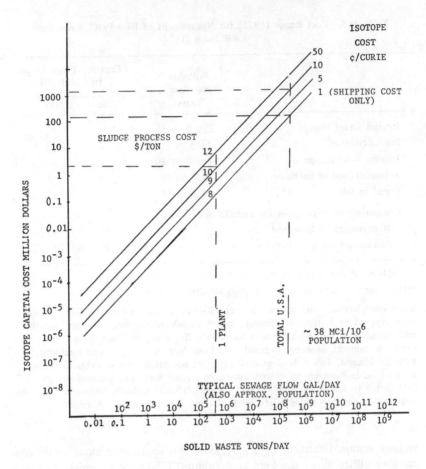

Figure 109. Cost of radioisotopes for sewage treatment (16).

may make them more attractive, even though the cost of source encapsulation remains the controlling factor.

Even if waste sludge and sewage irradiation processes become economically competitive, there remains a general problem of safety. The size of
radiation sources contemplated for many of these plants is in the range
of 1-30 MCi, much larger than those employed in industrial process
irradiation or for medical therapy. These sources would be under water
at all times and access to them could be virtually impossible. However,
there remains a finite possibility of source leakage, calling for extensive
monitoring, and some question of the 'desirability of having very large
essentially unattended radiation sources distributed in urban locations all

over the country. Technically, neither point is insurmountable, but in terms of potential population exposure and system security the public will expect substantial reassurance.

Similar considerations apply to suggestions for using such sources to keep ports and locks ice-free in major waterways, such as the Welland Canal or the Vistula River.

TRANSMUTATION OF WASTE NUCLIDES

A superficially attractive way of getting rid of long-lived radioactive waste nuclides is by transforming them into shorter-lived materials by neutron bombardment. In this way 30-year strontium-90 would be transmuted to its 9.7-hour isotope strontium-91, which would decay rapidly to stable zirconium-91, via its 58-day daughter yttrium-91. Similarly, cesium-137 could be changed to cesium-138, which decays to stable barium-138 with a 32-min half-life (30). Even more important, if some of the waste actinides could be transformed to stable or very short-lived nuclides, one of the most persistent long-term hazards could conceivably be removed.

Claiborne (31) has analyzed the requirements for reintroducing such waste fission products and actinides into power reactors or special high-flux reactors. Table 169 lists the properties of some of these nuclides and the time required to obtain significant inventory reduction. It is evident that high neutron fluxes are required in most cases, higher than are normally available in conventional power reactors and at substantial additional risk caused by introducing highly concentrated high-level radioactive material into the reactor system. Special burner reactors with neutron fluxes of the order of 10^{17} n/cm^2-sec would be required, but one could hardly justify the expense of operating a reactor solely for this purpose. Spallation reactors and fusion reactors can be visualized for this use, but are beyond near-term technical feasibility.

Table 170 analyzes the power cost and effects on inventory of the various suggested schemes for the transmutation treatment of Sr-90. It is seen that the reduction in inventory by conventional means is hardly worth the trouble, whereas burn-up in a high-flux reactor involves prohibitive costs.

Transmutation of the actinides is not really too different from other plutonium recycle schemes. Significant reductions in the cumulative toxicity of the actinides are achieved, though the transformation of long-lived Am-241 (458 yr) to the more hazardous Pu-238 (86 yr) via two decays may be of dubious benefit. Since many of the actinides are fissile, their recycle is clearly of advantage if it can be obtained through

Table 169. Reduction of Fission Product Wastes by Decay and Neutron Transmutation (31)

	^{90}Sr	^{137}Cs	^{85}Kr	^{3}H	^{129}I
Half-life, years	28.9	30.2	10.74	12.33	1.6×10^7
Burnout cross section, barns[a]	1.2	0.17	1.8	nil	35
Curies/metric ton in spent fuel[b]	77,600	108,000	11,400	708	0.0367
Relative hazard in spent fuel[c]					
m^3 air at RCG/metric ton	2.6×10^{15}	2.1×10^{14}	3.8×10^{10}	3.5×10^9	1.8×10^9
m^3 water at RCG/metric ton	2.6×10^{11}	5.4×10^9	–	2.3	6.1×10^5
Time required for 99.9% decay and burnout, years[d]					
Decay only	288	300.7	107	123	1.6×10^8
$\Phi = 10^{14}$ n/cm$^2 \cdot$ sec[e]	249	295	106	123	63
$\Phi = 10^{15}$ n/cm$^2 \cdot$ sec[e]	112	245	98	123	6.3
$\Phi = 10^{16}$ n/cm$^2 \cdot$ sec[e]	17	91	57	123	0.63
$\Phi = 10^{17}$ n/cm$^2 \cdot$ sec[e]	1.8	12	11	123	0.06

[a]Effective thermal cross section in typical spectrum of a PWR having average thermal flux of 2.91×10^{13} n/cm$^2 \cdot$ sec.

[b]Per metric ton of uranium charged to a PWR having average specific power of 30 MW/metric ton and burnup of 33,000 MWd/metric ton.

[c]Volume of air and water potentially contaminated to RCG (p. 560) by the content of a metric ton of spent fuel.

[d]Indicated times are doubled and tripled for reduction of inventory by factors of 10^6 and 10^9, respectively.

[e]Average thermal flux assuming spectrum typical of that in a PWR.

Table 170. Estimated Costs Associated with Various Schemes of Transmutation of Strontium-90 (31)

	Sr-90 Inventory (megacuries) per 1000 MW(e) of Capacity with Plant Factor of 0.8			Estimated Incremental Cost[a] (mills/kWh)
	In Reactors	Outside Reactors	Total	
PWR - conventional operation[b]	3.17	88.0	91.2	0
PWR with complete Sr-90 recycle[b]	64.6	21.5	86.1	0.1
LMFBR - conventional operation[c]	1.39	38.6	39.9	0
LMFBR with complete Sr-90 recycle[c]	28.9	9.6	38.5	0.1
High flux isotope reactor - conventional [d]	0.11	132	132	24
High flux isotope reactor - Sr-90 recycle [d]	38.3	12.7	51.0	25
98% of power from LMFBRs plus [e] 2% of power from fusion burner	1.36 0.93	0.91 0.31	3.51	0
90% of power from LMFBRs plus[e] 10% of power from spallation	1.54 0.16	1.03 0.51	3.23	0.8

[a]In excess of the typical power generation cost of approximately 7 mills/kWh.
[b]Assumes thermal efficiency of 32.5%.
[c]Assumes thermal efficiency of 41.7%.
[d]Assumes thermal efficiency of 30%.
[e]One 1000-MW(e) spallation burner reactor associated with nine 1000-MW(e) LMFBRs. The electricity generated by the burner reactor is used internally to generate a 500-MW beam of 10-BeV protons.

more efficient partition during reprocessing. It is not a viable approach when applied to trace concentrations of the order of nanocuries per gram in final waste residues.

WASTE DISPOSAL

The final stage of the nuclear fuel cycle is the ultimate disposal of the highly radioactive wastes in as safe and economically reasonable a fashion as can be devised. This disposal can be permanent or interim ("retrievable") in nature and may involve wastes in liquid or solid form. Because of the long half-lives of some of the waste products, more than ordinary precautions must be taken to ensure secure disposal under conditions that will not pose any hazards to future generations. There is a great deal of public discussion of this issue at the present time; unfortunately, much of it is ill-informed and tends to exaggerate the technical problems involved. Some of the issues involved are political and philosophical rather than technical and may require a sociopolitical decision. Interim storage is used whenever the radionuclides contained in the waste material are sufficiently short-lived to make storage to decay practical, when some later use of the material may be contemplated, or when no decision has been reached regarding the optimum method of ultimate disposal. It is the latter condition that seems to affect U.S. waste disposal programs at this time.

Retrievable Storage

Tank Storage

Tank storage for high-level waste solutions has been described previously; Figures 101 and 105 illustrated two types of storage tanks. Although such tanks are suitable for storage of liquid wastes for up to 50 years, the possibility of crack formation and leakage is always present. Catch basins and choice of underlying soils with high ion-exchange capacity can greatly minimize any risk of dispersion into the ground. Nevertheless, it seems prudent to convert most wastes intended for indefinite storage into solid form, to provide a more stable, nonleachable form to the greatest extent possible. Figure 110 shows two examples of the estimated migration rates through Conasauga shale for Sr-90 and Am-241 assumed leached out from a solid-waste canister. It is evident that such movement will be slow and will affect only the immediate vicinity of the storage vault. However, there can be no question of the greater safety margin provided by solidification as compared to tank storage for given conditions of leakage (13).

The cost of perpetual tank storage depends on the chemical nature of the waste, the method of financing and the assumed tank life. For a plant generating 90,000 gal/yr of high-level wastes, stored in three 530,000-gal tanks equipped with submerged coils for heat removal, a cost of $3090 per ton of fuel (0.02 mill/kWh) was obtained (17).

Near-Surface Storage Vaults

Various types of storage vaults have been described in the literature (1,13,15,16,32-34). The general principle is to store the wastes in their canisters, which typically are 6-24 inches (16-60 cm) in diameter, 2-10 ft (60-300 cm) long and have a heat dissipation rate of about 5 kW (15),

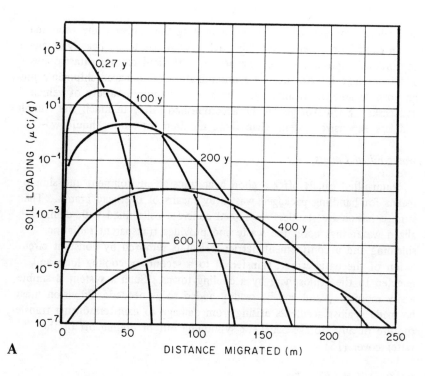

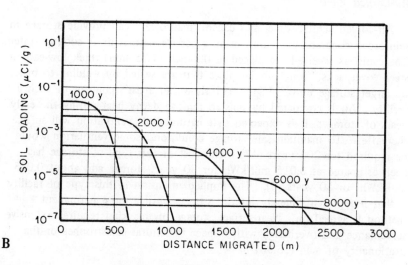

Figure 110. Estimated migration of solidified, leached waste material in soil (13).
A ^{90}Sr from leaching of pot calcine material
B ^{241}Am from diffusion-controlled leaching of glass

in shielded storage vaults. To avoid melting the canisters the vault may be air-cooled or water-cooled. In each case it has to be heavily shielded, continuously monitored and designed to withstand natural catastrophes and man-made interference. Decontamination of damaged shipments must be possible and the facility must be able to handle a variety of vehicles and casks. Such storage can be accomplished safely in much less complex and less expensive systems than those required for storing liquid wastes.

Water-Filled Canals

Water-filled canals offer a good heat transfer environment and simple means for handling packaged wastes by means of overhead cranes. The storage basins are stainless steel-lined concrete cells filled with demineralized water that cools the waste and provides transparent radiation shielding and a confinement barrier. Heat is removed by constant circulation of the basin water, transferred to a secondary cooling loop and rejected to the atmosphere by a cooling tower. Such a system is simple, easily monitored and can be readily added to as needed. Provision must be made against accidents arising from damage to canisters during transfer, rupture of any storage canal, or syphoning out or boiling off of the water cover (15).

Air-Cooled Vaults

Air-cooled vaults, sometimes called "mausolea," can provide storage in reinforced concrete structures, using a passive natural draft cooling system. The canisters received are placed in 0.5-inch (12.5 mm) thick low-carbon steel "overpacks," long tubular pipes that are sealed by welding to make up larger storage units. Figure 111 shows an artist's concept of such a facility. Air is circulated by fans to remove decay heat during the early years of storage; it is expected that natural-draft ventilation will suffice thereafter. To maintain safe storage conditions, some ventilation would be required for 200 years (17). The cost of this type of storage has been estimated as 0.015 mill/kWh for 30 years storage and about 0.025 mill/kWh for 50 years (8). The implied intention in this type of facility is that it represents merely interim storage until a more permanent method is agreed on. Near-surface storage of this kind requires extensive surveillance and overdesign with respect to natural catastrophes or the irrationality of man (35,36).

Sealed-Cask Storage

This concept differs from the water-basin and air-cooled vault concepts in that no vault is required. Each waste canister is stored in its own

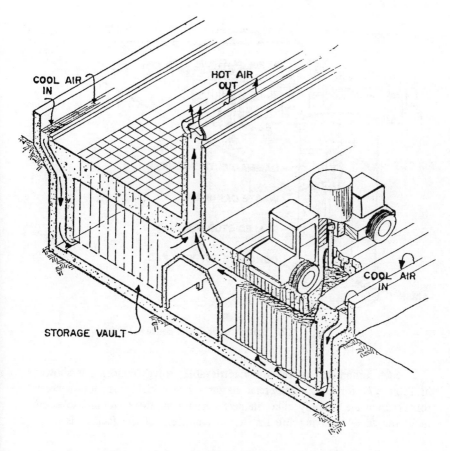

Figure 111. Artist's concept of air-cooled vault (15).

massive cask and concrete shield, which provide both structural ruggedness and radiation shielding. Radioactive decay heat is dissipated by natural-draft cooling.

Three canisters are sealed in one mild steel cask with a wall thickness of 2-16 in. (5-40 cm) surrounded by a gamma-neutron shield about 9 ft (2.70 cm) i.d., 15½ ft (4.65 m) high and a wall thickness of about 6 ft 5 in. (1.93 cm). These massive cylinders would be individually placed on pads on the ground, and cooling would be accomplished by convective circulation of air around the steel sleeve. Figure 112 shows a schematic drawing of a cemetery of this type with 3-ft (1 m) thick walled casks. Care must be taken that air inlets cannot be blocked by drifting sands or dirt (15,36).

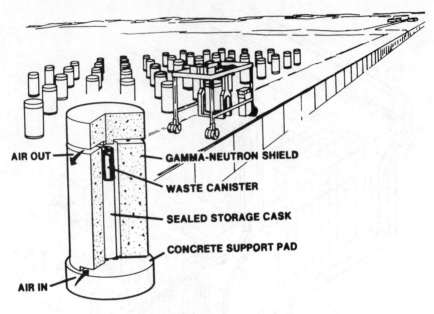

Figure 112. Sealed storage cask concept (32).

A cost comparison of the three retrievable storage concepts is shown in Table 171 for U.S. commercial wastes. The initial construction costs for all three are roughly alike, in part because all three would use essentially the same receiving and inspection facilities. If the facility is

Table 171. Cost Comparison of Retrievable Storage Concepts (millions of 1973 dollars) (15)

	Sealed Cask	Air-Cooled Vault	Water Basin
Cost of initial construction	45	49	43
Total operating through 2010[a]	960	320	170
Total program costs through 2010	1,100	620	450
Total program costs through 2110[b]	1,500	1,300	1,480
Cost in mills/kWh[c]	0.026	0.023	0.026

[a]Includes cost of steel overpacks and concrete shields, as required.

[b]Last waste shipment received in year 2010, stored with surveillance for 100 years. Does not include cost of recovery or shipment to disposal facility, or credits for salvageable materials.

[c]Based on estimated 57×10^{12} kWh generated from nuclear power through year 2000.

assumed to operate through the year 2010, the costs for the sealed cask storage concept become much higher because of the greater consumption of materials (some of which might be recoverable). If it is assumed that the last waste shipment is received in the year 2010 but that all waste is stored under surveillance for another 100 years, the total costs through 2110 for the three approaches become roughly the same, primarily because of the greater annual maintenance and operating costs for the water-basin, with its artificial cooling requirement. These costs would be roughly the same at any of the three leading alternative sites (15).

Tunnel Storage

Transitional between near-surface and underground storage is the concept of using a long tunnel, accessible from the surface, to provide interim storage for waste canisters. This method, which has been used widely, requires a remotely operated crawler or transporter that deposits the waste casks in double-walled burial holes, from which they can be removed at will at a later date. No cooling is provided so that the concentration of wastes must be controlled or previous decay must have taken place elsewhere. The principal accident for which provision must be made is partial collapse of the roof or damage to canisters during transit. In the worst cases the tunnel could be sealed off permanently; obviously site selection should avoid locations where flooding of the tunnel may be expected or where they are subject to seismic activity.

Transuranic Waste Disposal

The disposal of transuranic wastes—wastes containing Pu, Np, Am and Cm—lately has received much public attention. Much of this concern is misdirected in that it fails to distinguish between plutonium recovered for future use, which represents by far the largest fraction of transuranics handled at a reprocessing plant, and trace amounts of transuranics that may remain in waste streams or adhere to plant components and are impossible or impractical to recover. Bulk transuranic waste, consisting mainly of plutonium, does present a long-term environmental problem because of the long half-lives involved, the appreciable toxicity of all of the transuranics, and the gradual build-up of shorter-lived daughters. Even so the time scale for build-up is exceedingly slow and the transuranic oxides possess low mobility in the terrestrial environment.

Plutonium-238 is of particular interest since it decays with a half-life of 86.4 yr to U-234, and the specific activity of U-234 from this source is 18,000 times that found in natural uranium. The Th-230 daughter of U-234 has an 80,000 yr half-life and this tends to limit the rate at which

Ra-226 and its radon daughter build up (Figure 113). Because of the potential use of plutonium in reactor fuel, as mixed oxides or pure PuO_2, plutonium concentrates cannot be regarded as wastes and have to be stored in highly secure, absolutely noncritical configurations with all appropriate physical safeguards. Any actinides that can be extracted from reprocessing plant effluents should be extracted and concentrated (12). If the plutonium concentration is low, such actinide products, often referred to as "alpha wastes," pose a storage and disposal problem that has been the subject of intensive studies (37,38). The basic problem is to incorporate the transuranics in a stable matrix that is more resistant to abrasion or leaching than the more conventional solidification methods and to provide a degree of retrievability not envisaged for other waste products (38).

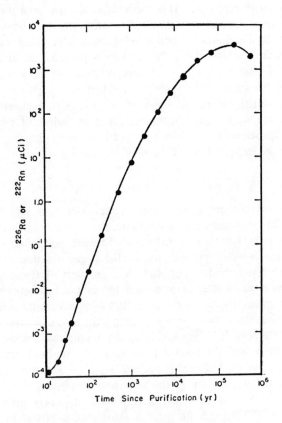

Figure 113. Time-dependent build-up of Ra-226 and Rn-222 from 1 gram of Pu-238 (37).

When applied to concentrations as low as 10 nCi/g, problems of precise measurement arise and it becomes evident that the cost for retrievability at such levels of actinide concentrations may be excessive when judged against the risks, considering the alternatives of disposal by deep burial. The practicability of recovery or separate storage of trace-level actinide wastes will have to be determined over the next few years. At this time retrievability seems to be largely a temporizing solution adopted possibly for political reasons. Considering the relative vulnerability of near-surface retrievable storage, the economic and environmental cost of retrievability for 10-500 mCi/g actinides in a nonleachable matrix may not be justifiable (8,15).

Underground Disposal

Most long-range solutions to the problem of ultimate disposal of long-lived high-level radioactive wastes involve the deposition of the material deep in the Earth's crust. The advantage of such deep disposal is that it removes this potentially hazardous material from all human access or contact in conditions that are unlikely to change, except on a geological time scale, and with reasonable assurance that it will long have decayed away before it could conceivably reach the surface again. This implies, of course, that the disposal site is in a geologically stable location, free of volcanic activity and far from any water-bearing stratum through which leached activity might return to surface. This requirement involves great care in site selection, depending on the method proposed for disposal, and careful evaluation of alternative options.

For the reasons given earlier, most disposal procedures envisage disposal of solids; however, some liquid disposal methods have been proposed and operated on a pilot scale, and these will be described briefly.

Deep-Well Disposal

In many central areas of the U.S. large quantities of chemical wastes are injected into wells up to several thousand feet in depth, which extend to sedimentary formations far below the water table. The same method is used to return oilfield brines to the formations from whence they came. The volumes so handled are considerably greater than any associated with foreseeable nuclear waste disposals, *e.g.*, in oilfield operations the volume injected daily is $\sim 10^8$ gallons, while one large oilfield alone in east Texas disposed of 4×10^9 gal/year into 58 7-in. wells. There are many known natural deep reservoirs of capacity greater than 10^8 gallons, and it has been calculated that the disposal of 4×10^9 gallons/year into a sandstone formation 100 ft (30 m) thick, with a porosity of 20%, would only cover 1 square mile/year.

In practice the method depends upon discharge to a deep stratum of limited permeability and low water transmissibility, preferably bounded above and below by relatively impermeable strata so that vertical movement of wastes towards higher, potable water-bearing formations is negligible. Such zones contain the connate or fossil waters trapped during the deposition of ancient marine sediments, the water being highly saline and therefore unfit for use. In the U.S. extensive zones of connate waters exist at depths from 2000 to 10,000 ft; the flow-rates vary from 10^3-10^4 gal/min in cavernous limestones and dolomites to <100 gal/min in dense, compacted sands, the latter value corresponding to a linear flow-rate of only a few feet per year (20).

Radioactive waste disposal in deep wells involves injection of the liquid waste into permeable reservoirs in rock that are below and isolated from freshwater-bearing formations. The permeable formation, which normally contains naturally occurring fluids such as briny water, oil or gas, should generally be at least 3000 ft deep and overlain with layers of thick and impermeable formations such as shale or salt deposits.

The concept is based on a slow, lateral movement of fluid in the permeable formation, typically no more than about 3 ft (1 m) per year. This movement, in combination with the width of suitable basins being tens to hundreds of miles in width, should provide containment in the formation for thousands of years (1). Studies on the applicability of deep-well injection for disposal of highly radioactive wastes have been limited to engineering evaluations.

The conditions that must be satisfied in deep well disposal include the following:

a) The wells must be cemented and sealed to depths below the potable water levels in order to avoid ingress of activity into drinking water supplies.

b) The area should contain no other wells that could become contaminated and so allow the escape of wastes to the surface, to potable water levels, or to mineral deposits. If, as has been suggested, abandoned oilfields were to be used, it would be necessary to line and cap all wells not being used for disposal.

c) The hydraulic behavior of solutions must be studied in the connate zone in question.

d) Suspended matter, bacteria and chemicals incompatible with the connate formation or its natural waters must be removed by pretreatment before disposal or else the porous formation may become blocked and useless.

Because of the extensive and expensive studies required to apply deep well disposal to highly radioactive wastes, because a reprocessing plant would have to be located near or over a suitable deep well formation, and because of the progress made in studies on solidification and storage of wastes in salt mines, major experimental studies have not been performed. Deep well disposal of the larger volumes of low- and intermediate-level radioactive wastes may become applicable since some of the potential problems with high-level wastes would not exist; however, its main limitation is that it accommodates only the soluble waste fraction.

Hydraulic Fracturing

The disposal process for intermediate-level liquid waste by hydraulic fracturing of shale consists of mixing the radioactive waste solution with a cement base mix and injecting the resultant grout into a 1000-ft-deep (300 m) impermeable shale formation that has been fractured previously by injection of water under high pressure. The injected radioactive grout forms a large, flat, nearly horizontal sheet several hundred feet in diameter and up to 1 in. (2.5 cm) thick, which quickly sets, fixing the radionuclides in a deep shale formation (see Figure 114).

Largely since World War II, the petroleum industry has developed the technique of hydraulic fracturing to increase oil recovery that is today almost universally used in reservoir rocks of low permeability. Single injections of 100,000 gal of oil containing 250,000 lb of sand are not unusual. Much the same equipment and procedure may be used for waste disposal, although there are certain significant differences. The disposal wells at Oak Ridge were drilled and cased to a depth of 1000 ft (300 m) in shale by using standard oilfield methods. After plugging the bottom, the casing is slotted near the bottom, and for a self-hardening mixture of waste, Portland cement, and clay is pumped down under high pressure to form a widespread, thin, horizontal fracture in the shale in which the waste sets up solid. After four injections into the same slot, totaling roughly 500,000 gal, the bottom of the well is plugged and a new slot cut a few meters higher up the well. Problems that had to be solved at Oak Ridge to make possible the operational disposal of medium-level radioactive waste were (1) determining the geometry of the fractures formed in the shale, (2) establishing the formula for a satisfactory waste-cement-clay mix, (3) designing and constructing the surface plant, and (4) formulating methods of monitoring the operation (39).

Once flow down the well is established, cement and clay and possibly other additives, in carefully predetermined amounts, are mixed into the flowing stream of water until the mixing process is stabilized. The liquid

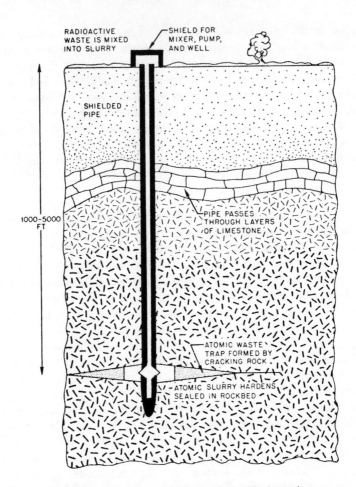

Figure 114. Waste disposal by hydraulic fracturing.

feed is then switched to the radioactive waste, and the waste, also mixed with the proper proportion of solids, is pumped down the well until the waste tanks are empty. The liquid feed is then switched back to water until the remaining solids are all used up, and finally a few hundred gallons of water are pumped down the well to flush it and the slot clear of cement. The well is then shut in until the cement has set and the waste-cement-clay mixture converted into a thin layer of artificial rock embedded in depth in the shale. Four 124,000-gal injections were made into a single slot, the wells plugged with neat cement, and a new slot cut a few meters above the old one. The injection cycle was then repeated. In this way each well segment could be used to dispose of 500,000 gal of slurry containing 320,000 gal of medium-level waste, or rather of medium-level waste concentrate.

If the well and the slot are flushed clear with water at the end of an injection, it is possible to make repeated injections into the same slot. However, when a sufficient volume has been injected into one slot to build up the maximum desirable local strains, the bottom of the well, including the slot, is plugged with cement, a new slot is cut at a slightly higher level, and the injections are continued. There is no precise way of determining the volume that may be injected into any one slot in a well or the total capacity of a well, so the operations are limited to volumes that theory suggests are much under the acceptable limits (39).

As each injection is made, the overlying red shale and the grey shale cover rock are uplifted; this is the mechanism by which space is provided for the material that is forced into the earth. Although the fractures extend out from the injection well only about 500 ft, the uplift, as measured on the surface by the very precise leveling of a net of bench marks, can be measured to some 2000 ft (660 m) from the well top.

The method is limited inherently by two factors: (1) finding a suitable shale formation in the vicinity of the reprocessing plant of adequate thickness to contain liquid wastes over the life of the plant and (2) the pumping power available to force the grout into the fractured formation. The total volume available for disposal is expected to be small and the cost of drilling the injection well and any monitoring wells is high. Hence it may be expected that this method will find only limited application.

Mine Storage in Bedrock

The concept of storing solidified radioactive wastes underground is a simple one and attractive on several grounds: the construction of underground storage vaults at depths of 2000 ft (650 m) or more involves no serious technological problems; shaft sinking and tunneling procedures for even greater depths are well established, and existing mines in hard rock in many areas can provide data on hydrology, seismic frequency and geology at little extra cost (40). Figure 115 illustrates the general concept. Particular attention has to be paid to isolating the mine from aquifers, careful installation of a shaft lining during shaft sinking and long-term stability of all underground workings. This implies continued operation of pumps and maintenance of ways over a very long period of time, though after a century or so collapse of the workings could be contemplated with equanimity provided that the storage area was well-sealed

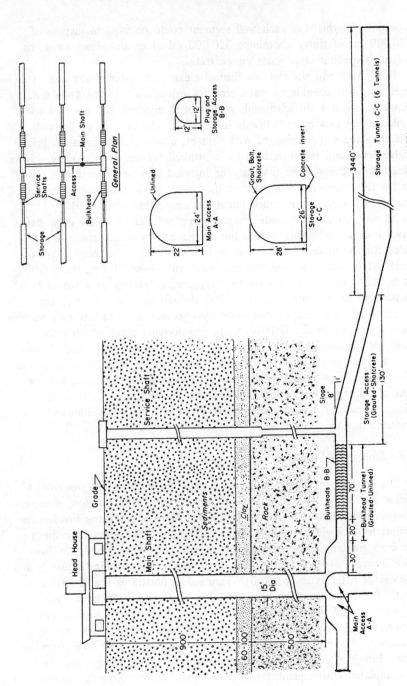

Figure 115. Cross section and dimensions of proposed bedrock waste storage facility (41).

from future access and water seepage. Storage of liquid wastes in tanks in such a location, though feasible, should probably be discouraged because of the greater seepage hazard (33,41,42).

Dry mine workings are probably not as commonplace as wet mines. However, thick and relatively undisturbed beds of limestone and shale (and even granite and other crystalline rocks) exist, many of which are essentially free of circulating water. For instance, in Barberton, Ohio, a dry 2000-ft-deep limestone mine is in operation. Also, mined caverns in chalk near Demopolis, Alabama, have been found to be relatively free of water. Excavations in thick shale beds in Illinois have remained dry since they were opened. It is reported that a mine in crystalline rocks in Ontario, Canada, has remained free of water even though the mine is situated directly beneath a large lake. Loess deposits offer another possibility for the disposal of high-level waste in some areas above the water table. Evaporite deposits other than rock salt (e.g., potash, trona, anhydrite and gypsum) may also be suitable.

For hard rock, such as limestone and granite, it is expected that mined cavities will remain stable under loads up to several thousand psi and temperatures up to a few hundred degrees centigrade. Model pillar tests on samples of dolomite from a Tennessee quarry showed that there are negligible amounts of deformation in the rock up to loads of 10,000 psi (70 MPa) and temperatures as high as 200°C. In comparison, it is of interest that, in similar pillar model tests for rock salt at temperatures of 200°C and 6000 psi (40 MPa) pillar deformation had exceeded 35% after only 1 hr. Thus it appears that the structural integrity of the excavated openings in these rocks, due to the superincumbent load, will not be of primary concern in the event that these rocks should be used as storage media; however, it is likely that such factors as ensuring the isolation of these excavations from migrating ground waters and the geographic location of suitable deposits and their vertical and lateral extents, along with possible radiation and heat effects on the rocks, would be critical (17).

In summary, it is apparent that dry openings that could be utilized for the storage of radioactive wastes can be excavated in rocks other than salt; however, investigations are needed to define more precisely such factors as the geohydrological and geotopographical conditions that determine the usefulness of local sites within the most desirable geographic regions and the effects of heat and radiation on the enclosed rock media.

Bedded Salt Formations

Some of the difficulties arising in bedrock disposal can be overcome by the use of bedded salt formations (40,43-47). Some of the advantages of natural salt formations as repositories for radioactive wastes are:

a) Salt is essentially impermeable due to its plastic properties.
b) Salt is widely distributed and abundant, underlying about 500,000 square miles in the United States and with known reserves greater than 6×10^{13} tons.
c) The cost of developing space is relatively low as compared with other rock types.
d) The heat-transfer properties of salt are good as compared with other rock types (k = 2.5 Btu hr^{-1} ft^{-1} °F^{-1} at 200°F).
e) Salt formations in many parts of the world are located in areas of low seismicity.
f) The compressive strength of salt is similar to that of concrete, or about 3000 psi (20 MPa).
g) The existence of a stable thick salt formation in itself provides evidence for the absence of leaching from ground water.

Both theoretical and experimental results indicate that rock salt is approximately equivalent to concrete as an absorber of gamma radiation. Approximately 5 ft (1.5 m) of solid salt or 7 1/2 ft (2.25 m) of crushed salt (assuming one-third to be composed of voids) will give adequate biological shielding to allow unlimited access to a salt mine room whose floor is filled with containers of the most radioactive waste of the future. The containers would be located in backfilled holes in the floor, with the tops of the containers at the proper depth and with container spacing based on heat dissipation calculations.

Field tests have indicated that the heat-transfer properties of salt are close enough to the values determined in the laboratory that confidence can be placed on theoretical heat-transfer calculations (17). Extensive field tests have been carried out at an abandoned salt mine at Lyons, Kansas (44,46), and at Asse, West Germany (47).

The most significant finding in the field tests regarding the effects of heat on salt behavior is that the insertion of heat sources in the floor of a mine room produces a thermal stress whose effects are instantaneously transmitted around the opening (to the pillars and roof). These stresses produce increased plastic flow rates in the salt, and could possibly cause mine stability problems if the roof of the room is very near a shale layer (a plane of weakness). In the demonstration area, such a shale layer existed about 2 ft above the ceiling; however, it was found that conventional roof-bolting techniques were adequate to handle the problem. In an actual disposal operation, it is anticipated that rooms would be filled with waste and then backfilled with crushed salt rapidly enough that roof bolts probably would not be required.

The combined field and laboratory tests have provided sufficient information on the deformation characteristics of the salt to allow the development of both general and specific empirical criteria for design of a disposal facility in almost any bedded salt deposit. Because of the hygroscopic nature of rocksalt any storage area would be expected to be very dry, so that container corrosion would be minimal. However, a typical bedded salt deposit might contain about 1/2 % water by volume. Calculations based on theoretical models and laboratory tests of the migration rates, as a function of temperature, were in reasonable agreement. Based on theoretical calculations, one might expect a total inflow of 2-10 liters of brine per waste container hole, which would take place over a period of 20-30 years after burial. The peak inflow rate of 200 ml to 1 liter per year per hole would occur about 1 year after burial. This brine inflow rate would be expected to taper off and approach zero after 20 to 30 years. Inflow rates similar to these were observed in the demonstration. Any resultant corrosion effects would be expected to be very slow, and container integrity should be maintained for an indefinite period of years (17).

Ventilation of the storage area may be a problem; however, it has been suggested that the hot waste containers may cause enough plastic flow in the surrounding salt that they may slowly sink into the ground, thus effectively sealing themselves. After backfilling, a room would be expected to be completely self-sealed in a few decades.

Figure 116 shows a drawing of the planned repository at Lyons; although it appears to have been abandoned for minor political and technical reasons, a similar repository may be developed in other locations. The deposition areas consist of a number of niches cut into the salt where the waste containers can be stored by remotely controlled transporters.

At the Asse mine, wastes in drums are lowered down the shaft and transported to rooms accessible via large bore holes (48,49). High-level wastes are in the form of vitrified fission products packed in stainless steel cylinders. To ensure sufficient dissipation of decay heat the individual boreholes are spaced at least 10 m apart (49).

Salt mines would appear to be suitable also for receiving transuranium wastes. Currently this concept is receiving the greatest development effort for the establishment of high-level waste repositories (32). The cost of disposal is expected to range between 0.045 and 0.055 mill/kWh for disposal of solidified high-level wastes, after 10 years of interim storage. That is about 0.5% of the cost of generating nuclear power, a price that may be acceptable (8). The long-term safety of the project depends on preventing the intrusion of water into the salt beds by any means. This could occur by natural means such as erosion, failure of overlying or underlying shale beds, boundary dissolution, and by man-induced means such as well borings. The Lyons site had several flaws, chiefly man-induced.

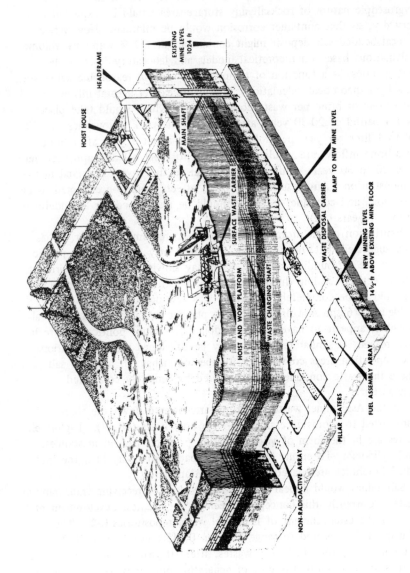

Figure 116. Artist's concept of salt mine depository (4,46λ.

The concept has some advantages: salt is easy to mine, it will in time flow plastically to seal the whole midden, and the very presence of the salt guarantees that no water was present in the geologic past. However, the demonstrated stability refers only to past time and not to the future, when conditions may conceivably be different (8).

Arguments like these, related to future uncertainty, now appear in the scientific and public literature and will not be stilled until some factual experience is available. However, mining is an ancient art and there is in fact a considerable body of experience to back up the geological evidence (13).

Other Ground-Disposal Concepts

Self-Burial by Rock Melting

Logan (50) has suggested that high-level radioactive wastes be allowed to dig their own holes by utilizing the decay heat of the wastes to melt crystalline rock and slowly move downwards under gravity, with the molten rock resolidifying in the wake of the container above it. The postulated descent rate in granite is assumed to be of the order of 1 m/day initially, in basalt 3-4 m/day, with ultimate depths ranging from 3-50 km depending on the waste content. Several practical difficulties arise, particularly with respect to the continued integrity of the waste container and the rate of heat loss from the capsule. However, high melting point capsule materials, such as those developed for space vehicles, could be visualized. The capsules were assumed to have radii of 50 and 100 cm and a wall thickness of 1.5 and 3 cm. If the capsules are released from a tunnel below ground, this concept may be workable as long as all volatile materials have been excluded from the wastes.

"Plowshare" Concept

In this method in-situ incorporation of liquid wastes in molten silicate rock is proposed (51). Initially a deep underground cavity would be blasted by means of a nuclear explosion in ground adjacent to the reprocessing plant. This would produce a deep chimney in silicate rock. Liquid wastes would be injected into the cavity and be allowed to self-boil. The steam could be utilized on surface, although this is not an essential feature. Ultimately the waste would boil dry and solidify. It is claimed that the remaining decay heat would melt the surrounding rock, which would dissolve and incorporate the waste. This would leave the waste in an "insoluble" rock matrix underground. Table 172 shows

Table 172.   Summary of Waste Disposal Options (8)

Option	Cost (mill/kWh)	Advantages	Disadvantages
Salt mine	0.045–0.050	Most technical work to date; plastic media with good thermal properties occur in seismically stable regions	Corrosive media; highly susceptible to water; normally associated with other valuable minerals; difficult to monitor and retrieve wastes
Granite	0.050–0.055	Crystalline rock; low porosity if sound; comparable to salt in thermal properties; retrievable wastes	Nonplastic media; presence of groundwater; difficult to monitor
Further chemical separation; recycle actinides	0.065–0.320	Reduced long-term toxicity; technology feasible; increases future options	Additional handling and processing; more toxic materials in fuel inventory; waste dilution due to processing; fission products remain
Further chemical separation; remove Sr and Cs; recycle actinides	0.140–1.100	Reduced long-term toxicity; reduced short-term thermal power; some reduction of fission product toxicity; increases future options	Additional handling and processing; more toxic materials in fuel inventory; waste dilution due to processing; storage and disposal of Sr and Cs extract; fission products remain
Melt *in situ*	0.011–0.016	*In situ* creation of insoluble rock-waste matrix; no transportation; reduced handling	Highly mobile wastes during 25-year boiling phase; presence of ground water; irretrievable wastes; proliferation of disposal sites; difficult to monitor
Melt *in situ*, central repository	0.031–0.036	*In situ* creation of insoluble rock-waste matrix; short boiling period; no proliferation of sites	Presence of ground water; irretrievable wastes; difficult to monitor
Antarctic rocks		Immobile water	Very narrow temperature limits; not a permanent geologic features; difficult environment
Continental ice sheets		Immobile water	Cannot dispose of actinides; limited amount of ice; not a permanent geologic feature; difficult environment;

the estimated costs for this process and others; it is claimed to be 0.027-0.052 mill/kWh less expensive than salt mine disposal (51).

The principal objection to this approach is the notion of detonating a nuclear explosive near the foundations of a reprocessing plant, when all other environmental considerations stress the desirability of stable, unfractured foundations with a low, undisturbed water table.

Deep Hole Concept

This concept is merely an extension of the deep-well approach. The proposal would be to drill holes 10 miles (16 km) into the earth, where storage caverns would be produced and used to store solid or liquid wastes. For high-level wastes, rock melting might occur, too, leading to an embedded waste-rock mixture (52). The cost of this concept is largely that of drilling and casing a very deep hole of 2-3-ft (60-100 cm) diameter.

Tectonic Trench Disposal

Current theories of crustal plate movements assume that the continents ride on tectonic plates that are slowly moving over the underlying structures. At continental edges, such as the Pacific coast, deep trenches exist where one plate is subducted under another plate and reabsorbed into the mantle. It has been suggested that waste containers be placed on locations of plate contact where they would be drawn deeply beneath the continents. Consideration of the slow rate of movement, less than 10 cm per year, quickly dispels this notion as a realistic disposal option.

Navigational problems and dispersal of the waste containers as they descend through several thousand meters of water make it unlikely that the wastes could be deposited on the floor of the trench in an array smaller than one km in diameter. Even if we assume the maximum rate of velocity between plates and that it will persist, the time needed to subduct the entire array would be 10,000 years, so that isolation of the materials would be deferred for a long time. Second, subduction zones are characterized by extreme tectonic instability. Evidence at some zones suggests that a portion of the sediments is scraped off the sinking plate and piled up at the margin of the continental slope. At other zones, the lighter, low-melting-point materials of the descending plate are vented through volcanoes. Therefore, removal of the wastes from the earth's crust may be neither complete nor permanent.

Ice Sheet Disposal

It has been suggested that radioactive wastes could be safely disposed of in the major ice sheets of the world, in Greenland and Antarctica (53). The advantages of ice caps are their remoteness from inhabited regions of the earth and the simplicity of allowing the waste containers to bury themselves by melting their way down through the ice. This concept suffers from the obvious practical difficulties of routine operations with highly hazardous materials in extremely harsh arctic climates. It also has certain philosophical deficiencies, principally that in the geologic sense ice caps are very transient features. The existence of vast accumulations of ice is probably due to the concurrence of unusual geographic and climatic conditions, with a distinct possibility for drastic changes in the present ice caps with only fairly minor changes in the world mean climate. Also, the conditions at the base of ice caps, which would be the final resting place for the wastes, are poorly known. In some cases, the ice may be floored with a layer of liquid water. Of course, it is known that even very slowly moving ice is efficient at grinding rocks (11).

The radioactive waste should be in solid form and encased in durable sealed canisters. The canisters could be placed in shallow drilled holes and allowed to melt their way downward, either to the bottom of the ice or to a specified depth within the ice. The melted ice would refreeze as the canister descended, sealing the wastes within the ice. A surface storage facility might also be used in which the waste was stored in a secure structure on the surface of the ice. This method offers the advantages of ready surveillance in addition to remoteness and isolation (53).

Both ethical and cost considerations seem to make this approach highly undesirable. Apart from that, ice-cap disposal has several drawbacks. First, wastes still containing actinides require extremely long periods of storage; for example, if the original concentration of ^{239}Pu is 10^6 times the permissible concentration in drinking water and no credit is allowed for insolubility, dilution, or adsorption, the required period of isolation is 500,000 years, and the ice may not be that permanent. Even if the actinides were removed, an area problem remains: to preclude appreciable heating at the ground-ice interface (and hence increased ice flow), the heat generated from the wastes must be a small fraction of that appearing via the geothermal gradient—1% would be 63 kW/km^2. Wastes from the United States aged 10 years before burial, if accumulated and spread out in Antarctica to give that heat load, would cover 10^6 km^2 by A.D. 2025 or 25% of the area that has ice with an anticipated lifetime exceeding 10,000 years. Finally, transportation and working conditions in the Antarctic are difficult and hazardous, and at present the Antarctic is kept free of nuclear wastes by international treaty (8).

Underground Disposal of Waste Gases

The underground storage of gases is practiced in the natural gas industry where three types of reservoirs have been developed: depleted natural gas reservoirs, depleted oil reservoirs, and water bearing sands or aquifers. The main geologic requirement for the underground disposal of ^{85}Kr would be to find a capping formation that is free of permeable vertical cracks or fractures, and of earthquakes. In the absence of vertical permeation to the ground surface (defined here as the flow through cracks or fractures due to pressure gradient), underground vertical movement would be controlled largely by molecular diffusion. In addition, there would be other mechanisms that would ensure the underground containment or decay of ^{85}Kr. These mechanisms might be trapping (containment) and adsorption. An evaluation was made of the delay required in the underground vertical movement of ^{85}Kr and of the geologic requirements for containment or sufficient retardation of ^{85}Kr by the aforementioned mechanisms (54).

Although very little experimental work has been done to evaluate the possibility of underground disposal of radioactive noble gases, it seems that, with given suitable geologic conditions, this method would be attractive for the disposal of ^{85}Kr from reprocessing plants. The advantages, as compared to the other methods developed or under development, would be its relative simplicity—the removal and storage stages are combined in one stage—its relative invulnerability to accidents, and its cost. However, a severe requirement would have to be imposed on the formation overlying the disposal layer because it must be essentially free of vertical channels.

The geologic requirements for the underground disposal of radioactive gases are not necessarily common features of sites that may be chosen for locating a reprocessing plant. However, the waste disposal criteria for either gaseous, liquid or solid waste should play an important role in establishing suitability of a site for the location of a fuel reprocessing plant. While the distance criterion for siting nuclear reactors, *e.g.*, distance to consumers and to large water supplies required for cooling, is of extreme economical importance, this criterion loses its importance in siting reprocessing plants. Once the irradiated fuel elements are removed from a reactor and loaded on a vehicle, additional transport to a suitable site within a couple of hundred of miles would not add appreciably to the reprocessing cost. Mudra and Schmalz (1965) evaluated the cost of a single underground disposal of 2.5 x 10^6 ft^3 (7 x 10^4 m^3) of radioactive gases within 24 hr (1740 cf/min; 50 m^3/min) to be about $210,000, consisting of:

exploration	$61,000
testing	15,000
development of wells	43,700
equipment	87,000
pipelines	3,500
Total	$209,200

Reist (1964) estimated the capital cost for a continuous underground disposal of ^{85}Kr at a rate of 35 ft^3/min (1 m^3/min) to be $137,000. The annual operating cost would be $13,600. The average cost for the underground ^{85}Kr disposal into a 90-ft-thick (30 m) sandstone formation 4500 ft (1350 m^3) below ground level was thus estimated to be $0.61/1000 ft^3 (54).

Site Selection

To a large extent the choice of location for a ground disposal site and the method of disposal are obviously interdependent. However, salt formations of suitable characteristics underlie many portions of the central United States and the North European plain and there are other sites available for bedrock disposal. On the other hand, liquid disposal methods probably have to take place at or near a reprocessing facility, and this requirement may determine the choice of a site for such a facility.

Because of the very long-term commitment a disposal plant represents by removing a given site from alternative uses indefinitely, site evaluation has to consider carefully all political and economic aspects and to project future growth potentials of the surroundings. In its direct impact, an underground disposal site will not differ much from a small mine in terms of access requirements, land occupation and soil disturbance. Since the number of repository sites will be very small for the foreseeable future, each site will have to be considered individually and will be subject to different constraints. Nevertheless, some generalized methodology has been developed, among others by Deonigi (55) and by Inoue and Morisawa (56) and others, all of whom attempt to quantify weighting factors and safety considerations (53).

Two types of computations are required to determine the safety of a given underground site, or conversely the projected population dose from it over the lifetime of the site, which has to be summed over centuries. These are: predictions of the escape rates, leaching rates, and possible pathways to surface for each type of waste deposited; and the population exposure resulting from an accident, such as rupture of a shipping cask during unloading at the site. Consideration of maximum accident conditions will impose an exclusion area concept on the selected site.

To be a satisfactory site, it must comply with certain geological criteria (57):

1. burial site devoid of surface water and stable geomorphically
2. ground water flow paths that do not lead to surface flow
3. predicted residence time of radionuclides in the order of hundreds of years (hydrologic system must be simple enough to make possible reliable residence-time predictions)
4. the highest water table several meters below the burial zone.

The basic data needed for site evaluation include the following:

1. depth to water table
2. location and distance to points of water use
3. minimum of 2 years precipitation and land pan evaporation records
4. water-table contour map for different seasons of the year
5. magnitude of annual water-table fluctuations
6. detailed stratigraphic and structural data to base of shallowest confining aquifer
7. base-flow data on nearby perennial streams
8. chemistry of water in aquifer, confining beds and of leachate from burial trenches
9. laboratory measurements of porosity, permeability, mineralogy, and ion exchange capacity of each lithology in saturated and unsaturated zones
10. a record of at least 2 years of moisture content and *in situ* soil moisture-tension in the upper 10 to 15 m of unsaturated zone at burial site
11. three-dimensional distribution of heat to base of shallowest confining aquifer
12. field test determination of storage coefficient and transmissivity
13. definition of recharge and discharge areas
14. field measurements of dispersion coefficient
15. laboratory and field determination of the distribution coefficient
16. rates of denudation and slope retreat.

Leach rates and ground water movement have been studied for various soil types and rock conditions for years. However, each site has its own conditions of ground water movement, salinity and soil conditions, stream erosion conditions, subsidences, faulting and volcanism. In view of the unusual long-term commitment even rather slow geologic processes must be considered (13). The possibilities and consequences of catastrophic events, such as meteoritic impact, volcanic activity and disastrous floods, should be evaluated. However, it would be unrealistic to be concerned about remote geological events involving time spans greatly in excess of

3000-10,000 years such as glaciation or major climatic changes, as long as basic stability criteria are met.

Finally the cost-benefit evaluation must make a realistic assessment of alternative sites and alternative disposal methods. Table 172 presents one summary of the various waste disposal options in terms of direct costs and relative safety considerations (8).

Added to this one would need an appraisal of the accident potential for each process; this normally would favor permanent, deep disposal over retrievable sites. This evaluation of the risk to the population presents great difficulty in view of the many imponderables; however, technically there appears to be little reason to assume that disposal locations cannot be found that are deep enough and stable enough to contain the radioactive waste forever. The costs of doing this are a different matter. Rowe and Holcomb (3) have pointed out the ever rising costs for waste disposal implicit in the growing number of nuclear power plants. They conclude that by the year 2000 the radioactive waste management costs, for all wastes from U.S. commercial power plants, will have risen to a total of $1.7 billion, compared with only $39 million in 1980. This corresponds to about 0.05 mill/kWh, or 0.3% of fuel cycle cost. For the more limited Canadian program, Lewis (58) arrived at a comparable cost level.

Oceanic Disposal

The oceans have served man as disposal sinks for many centuries, and for densely populated countries with access to the sea it is tempting to consider the oceans as an unlimited disposal site, subject to few legal barriers. Growing concern with the increasing contamination of the oceans by human activities through pesticides, leaks from oil tankers, dumping of waste chemicals and accumulation of detritus carried by polluted rivers has led to several international conferences aimed at protection of the seas. An international treaty has been signed by most developed countries banning future disposal of radioactive wastes in the seas. Consequently, any fresh considerations of the ocean areas for waste disposal contemplate drilling below the sea bed.

Originally, the seas were used to dilute low-level wastes from nuclear facilities, such as the Windscale plant on the Irish Sea. The vagaries of currents and tides made this type of disposal uncertain and hazardous. The next type of disposal practiced in the U.S. and Europe involved the dumping of drummed wastes in deep waters and off the continental shelf. Such drums have to be heavy enough to sink, sturdy enough to withstand bottom impact, capable of resisting the high pressures at depth or of equalizing pressures, and able to survive corrosive effects over a long enough period to minimize subsequent release by leaching of the solidified waste products (59,60). In addition, disposal areas have to be remote from beaches, tidal action or fishing grounds. All of these requirements dictated the use of locations 40-60 miles (65-100 km) off shore. For the type of vessel available, this meant that only two round trips per week, including loading and dumping of waste, were possible. Hence, this type of operation proved completely uneconomical and capable of handling only small waste quantities. The recent antidumping treaty has made such operations obsolete in any case.

Subseafloor disposal, illustrated in Figure 117 has been studied conceptually in References 16, 53 and 61. Holes would be drilled from a submersible drilling platform into bottom sediments at depths of 4-10 km,

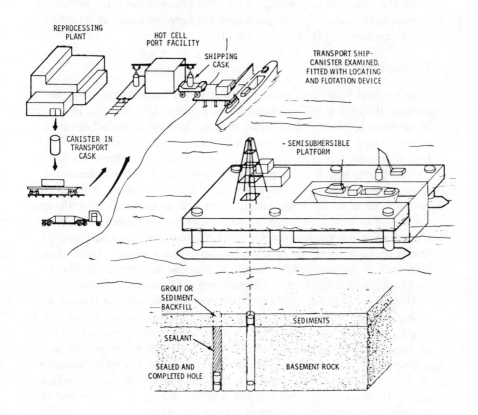

Figure 117. Conceptual engineering requirements for waste disposal in the seabed (61).

in abyssal plains or at the base of sea mounts. The stable deep areas, considered to be among the most stable geophysical features in the earth, have the potential for isolating the waste for necessary lengths of time if theoretical knowledge of the ocean floor is confirmed by future exploration. Other advantages of seabed concepts are the ability of sediments to retain the waste in case it is released by the bedrock, safety from surface storms and fishing activities, the availability of large volumes of water for dilution of any released wastes, and the opportunity for international cooperation.

Balancing the advantages, however, is the lack of knowledge of the ocean floor. Further, there are problems caused by the need for considerable transportation and precise placement of the wastes, the danger of transport of released waste by seawater, the possibility of biological accumulation of radionuclides by sea life, and international political uncertainties.

In any case such a proposal seems to contravene the spirit of the international treaty. The disposal cost has been estimated at $17,000/tonne processed, or 0.063 mill/kWh (16).

Extraterrestrial Disposal

In contrast to sinking waste in the seabed, extraterrestrial concepts envision shooting it off the Earth entirely—either into the sun, in a solar or earth orbit, or out of the solar system. Because of uncertainties in course control and orbit stability, solar system escape is considered the best alternative and is attainable with present-day technology (16,53). Assuming improved knowledge of deep space travel, the extraterrestrial route would completely remove the waste from earth, thus avoiding any possible release to the environment by natural or man-made disturbances. On the other hand, the waste would have to be partitioned to separate the actinides and transuranics for space disposal, while the shorter-lived fission products would be handled by conventional means. Other problems include the questionable safety of multiple launches, difficulty in monitoring and retrieval, and international political considerations.

Drumheller (62) has reviewed alternative destinations and relative advantages (Table 173). Solar system escape can be achieved directly by a single propulsion burn from the low-Earth orbit with all propulsion and guidance provided by the launch vehicle. Solar system escape can be achieved with somewhat less energy expenditure by a properly designed swingby of Jupiter, using a single propulsion phase (tug) from low-Earth orbit. However, either case requires multiple shuttles per waste package to supply the necessary sequential propulsion energy.

Table 173. Summary of Potential Space Destinations for Waste Removal (62)

Destination	Delta-V^a (km/sec)	Advantages	Disadvantages
High-Earth orbit	4.11	low Delta-V launch any day passive waste package can be retrieved	long-term container integrity required orbit lifetime not proven
Solar orbits via Single burn beyond Earth escape	3.65	low Delta-V launch any day passive waste package	long-term container integrity required Earth re-encounter possible (may not be able to prove otherwise) abort gap past Earth escape velocity ^{b}
Circular solar orbit	4.11	low Delta-V launch any day	long-term container integrity required orbit stability not proven abort gap past Earth escape velocity ^{b}
Venus or Mars swingby	4.11	low Delta-V	long-term container integrity required limited launch opportunity (3 to 4 mo every 19-24 mo) requires mid-course systems need space propulsion or have possibility of unplanned encounter
Solar system escape: direct	8.75	launch any day passive waste package removed from solar system	high Delta-V abort gap past Earth escape velocity ^{b}
via Jupiter swingby	7.01	removed from solar system	high Delta-V limited launch opportunity (2-3 mo every 13 mo) requires mid-course systems abort gap past Earth escape velocity ^{b}
Solar impact: direct	24.08	package destroyed launch any day passive waste package	extremely high Delta-V abort gap past Earth escape velocity ^{b}
via Jupiter swingby	7.62	package destroyed	high Delta-V limited launch opportunity (1-2 mo every 13 mo) requires mid-course systems abort gap past Earth escape velocity ^{b}

^{a}Delta-V is the incremental velocity required to leave a low-Earth orbit and is a direct indication of the size and propulsion energy of the rockets required.
^{b}An abort gap is a short time period wherein a controlled abort of the mission cannot be accomplished if the flight is off-course.

Direct solar impact with a single propulsion phase would require vehicles using advanced technology. Solar impact can be achieved by a swingby of Jupiter, using a single tug phase from low-Earth orbit. However, the complexities of course control in a swingby mission may make this mission impractical.

The destination considered most likely is direct solar system escape. About 190 kg of transuranic waste can be transported in each flight to direct solar system escape with the proposed space vehicles. This capacity provides for the disposal of the transuranics from ~280 tonnes of spent light-water-reactor fuel in each flight. For total system costs, however, the figure shoots far above that of other ideas: $40,000/tonne reprocessed (0.15 mills/kWh) for orbiting the waste, or $90,000/tonne reprocessed (0.34 mills/kWh) for solar system escape. In addition, disposal of nonactinides and transuranics would cost some $10,000/tonne reprocessed, or 0.04 mills/kWh.

The safety aspects for space disposal primarily include safety during launch and control of the extraterrestrial destination of the waste constituents. The potential for an abort that could cause a release of radionuclides during any one space launching is moderately high; however, relatively small amounts of waste constituents are associated with each launch, and package integrity is high even in an abort (62). Even so, it is unlikely that at the present stage of space vehicle technology the risk of an abort would be acceptable to the public, since potentially it might lead to high-impact disintegration of the packaged waste and its dispersion over a large area.

REFERENCES

1. Schneider, K. J. "Solidification and Disposal of High-Level Radioactive Wastes in the United States," *Reactor Technol.* 13, 387 (1970).
2. Blomeke, J. O., C. W. Kee and J. P. Nichols. "Projections of Radioactive Wastes to be Generated by the Nuclear Power Industry," Report ORNL-TM-3965, Oak Ridge National Laboratory (1974).
3. Rowe, W. D. and W. D. Holcomb. "The Hidden Commitment of Nuclear Wastes," *Nucl. Technol.* 24, 286 (1974).
4. Culler, F. L., J. O. Blomeke, and W. G. Belter. "Current Developments in Long-Term Radioactive Waste Management," *Proc. 4th Internat. Conf. Peaceful Uses of Atomic Energy* 11, 427, United Nations, New York (1972).
5. Daly, J. C., S. Goodyear, C. J. Paperiello and J. M. Matuszek. "Iodine-129 Levels in Milk and Water near a Nuclear Reprocessing Plant," *Health Phys.* 26, 333 (1974).
6. Klement, A. F., C. R. Miller, R. P. Minx and B. Shleien, Eds. "Estimates of Ionizing Radiation Doses in the United States

1960-2000," Report ORP/CSD 72-1 (Rockville, Maryland: U.S. Environmental Protection Agency, 1972).
7. Magno, P. J., T. C. Reavey and J. C. Apidianakis. "Iodine-129 in the Environment Around a Nuclear Fuel Reprocessing Plant," (Washington, D.C.: U.S. Environmental Protection Agency, 1972).
8. Kubo, A. S. and D. J. Rose. "Disposal of Nuclear Wastes," *Science* 182, 1205 (1973).
9. International Commission on Radiological Protection. *Rept. of Committee II on Permissible Dose for International Radiation (1959)*, ICRP Publ. 2 (Oxford: Pergamon Press, 1960), Supplement: ICRP Publ. 6 (New York: Macmillan, 1964).
10. Claiborne, H. C. "Effect of Actinide Removal on the Long-Term Hazard of High-Level Waste," *Trans. ANS* 18, 190 (1974); Report ORNL-TM-4724, Oak Ridge National Laboratory (1975).
11. Blomeke, J. O., J. P. Nichols and W. C. McClain. "Managing Radioactive Wastes," *Physics Today* 26 (8), 36 (1973).
12. Bond, W. D., H. C. Claiborne and R. E. Leuze. "Methods for Removal of Actinides from High-Level Wastes," *Nucl. Technol.* 24, 362 (1974).
13. Gera, F. and D. G. Jacobs. "Considerations in the Long-Term Management of High-Level Radioactive Wastes," U.S. Atomic Energy Commission Report ORNL-4672 (1972).
14. Lennemann, W. L. "Management of Radioactive Aqueous Wastes from AEC Fuel Reprocessing Operations," *Nucl. Safety* 14, 482 (1973).
15. "Management of Commercial High-Level and Transuranium Contaminated Radioactive Wastes," Draft Envir. Statement, Report WASH-1539, U.S. Atomic Energy Commission (1974).
16. "High-Level Radioactive Waste Management Alternatives," Report WASH-1297 (Washington, D.C.: U.S. Atomic Energy Commission, 1974).
17. ORNL Staff. "Siting of Fuel Reprocessing Plants and Waste Management Facilities," Report ORNL-4451, Oak Ridge National Laboratory (1970).
18. Fitzgerald, J. J. *Applied Radiation Protection and Control* (New York: Gordon and Breach, 1969).
19. Puechl, K. H. "The Nuclear Waste Problem in Perspective," *Nucl. Eng. Internat.* 20, 950 (1975).
20. Amphlett, C. B. *Treatment and Disposal of Radioactive Wastes* (New York: Pergamon Press, 1961).
21. Burns, R. H. "Solidification of Low and Intermediate Level Wastes," *Atomic Energy Rev.* 9, 547 (1971).
22. Dickey, B. R., B. R. Wheeler and J. A. Buckham. "High-Level Waste Solidification: Applicability of Fluidized-Bed Calcination to Commercial Wastes," *Nucl. Technol.* 24, 371 (1974).
23. Hine, A. R. "Reprocessing in the United Kingdom," *Kerntechnik* 15, 157 (1973).
24. Craft, T. F. and G. G. Eichholz. "Decoloration of Textile Dye Waste Solutions by Combined Irradiation and Chemical Oxidation," *Nucl. Technol.* 18, 46 (1973).

25. Ballantine, D. S. and L. A. Miller. "The Potential Role of Radiation in Wastewater Treatment," in *Chem. Eng. Progress Symp. Series* 66, No. 104 (New York: Am. Inst. Chem. Eng. 1970).

26. Compton, D. M. J., S. J. Black and W. L. Whittemore. "Treating Wastewater and Sewage Sludges with Radiation: A Critical Evaluation," *Nucl. News* 13 (9), 58 (1970).

27. Katsumura, T. "Application of Ionizing Radiation to the Treatment of Waste Waters and Sewage Sludge," *Nizu Shori Gijutsu* 14, 197 (1973); Abstract NSA 29:29125 (1974).

28. Reynolds, M. C., R. L. Hagengruber and A. C. Zuppero. "Thermoradiation Treatment of Sewage Sludge Using Reactor Waste Fission Products," Report SAND 74-0001 (Livermore, California: Sandia Laboratories, 1974).

29. Eichholz, G. G. *Radioisotope Engineering* (New York: Marcel Dekker, 1972).

30. Steinberg, M., G. Wotzak and B. Manowitz. "Neutron Burning of Long-Lived Fission Products for Waste Disposal," Report BNL-8558 (Upton, New York: Brookhaven National Laboratory, 1958).

31. Claiborne, H. C. "Neutron-Induced Transmutation of High-Level Radioactive Waste," Report ORNL-TM-3964, Oak Ridge National Laboratory (1972).

32. Pittman, F. K. "U.S. Atomic Energy Waste Management Programs and Objectives," *Nucl. Technol.* 24, 273 (1974).

33. "Radioactive Waste Management Practices in Western Europe," (Paris, France: European Nuclear Energy Agency, OECD, 1971).

34. "Environmental Survey of the Uranium Fuel Cycle," Report WASH-1248 (Washington, D.C.: U.S. Atomic Energy Commission, 1974).

35. Morgan, W. W. "The Management of Spent CANDU Fuel," *Nucl. Technol.* 24, 409 (1974).

36. Nelson, D. C. and D. D. Wodrich. "Retrievable Surface Storage Facility for Commercial High-Level Waste," *Nucl. Technol.* 24, 391 (1974).

37. Healy, J. W., T. J. Hirons, M. M. Thorpe, K. A. Pashman, G. R. Waterbury and E. B. Fowler. "Transuranic Waste Repository Studies," Report LA 5127-MS, Vol. 1, Los Alamos Scientific Laboratory (1972).

38. H-Division Staff. "Guidelines for the Interim Storage of AEC-Generated Solid Transuranic Wastes," Report LA-5645, Los Alamos Scientific Laboratory (1974).

39. de Laguna, W. "Radioactive Waste Disposal by Hydraulic Fracturing," *Nucl. Safety* 11, 391 (1970).

40. "Disposal of Radioactive Wastes into the Ground," Proc. Vienna Symp. (Vienna: International Atomic Energy Agency, 1967).

41. National Academy of Sciences. *An Evaluation of the Concept of Storing Radioactive Wastes in Bedrock below the Savannah River Plant Site* (Washington, D.C.: National Academy of Sciences, 1972).

42. Regan, W. H., Ed. "Solidification and Long-Term Storage of Highly Radioactive Wastes," Proc. Richland Symp (1966), U.S. Atomic Energy Commission Report CONF-660208 (1966).

43. Blomeke, J. O. and W. C. McClain. "A Salt Mine Repository for Radioactive Wastes," *Nucl. News* **14** (4), 35 (1971).
44. Bradshaw, R. L. and W. C. McClain. "Project Salt Vault: A Demonstration of the Disposal of High Activity Solidified Wastes in Underground Salt Mines," U.S. Atomic Energy Commission Report ORNL-4555, Oak Ridge, Tennessee (1970).
45. Krause, H., H. Ramdohr and M. C. Schuchardt. "Project for Storing Radioactive Wastes in a Salt Cavity," in *Disposal of Radioactive Wastes into the Ground.* (Vienna: International Atomic Energy Agency, 1967).
46. McClain, W. C. and A. L. Boch. "Disposal of Radioactive Waste in Bedded Salt Formations," *Nucl. Technol.* **24**, 398 (1974).
47. Schwibach, J. "Research on the Permanent Disposal of Radioactive Wastes in Salt Formations in the Federal Republic of Germany," in *Disposal of Radioactive Wastes into the Ground.* (Vienna: International Atomic Energy Agency, 1967).
48. Krause, H., H. Ramdohr, G. Böhme and E. Albrecht. "Experimental Storage of Radioactive Wastes in the Asse II Salt Mine," in *Disposal of Radioactive Wastes into the Ground.* (Vienna: International Atomic Energy Agency, 1967).
49. "Management of Low- and Intermediate-Level Radioactive Wastes," Proc. Aix-en-Provence Symp. (Vienna: International Atomic Energy Agency, 1970).
50. Logan, S. E. "Deep Self-Burial of Radioactive Wastes by Rock-Melting Capsules," *Nucl. Technol.* **21**, 111 (1974).
51. Cohen, J. J., A. E. Lewis and R. L. Braun. "*In situ* Incorporation of Nuclear Waste in Deep Molten Silicate Rock," *Nucl. Technol.* **14**, 76 (1972).
52. Rubin, J. H. *Nucl. News* **16** (1), 56 (1973).
53. "Overview of High-Level Radioactive Waste Management Studies," Report BNWL-1758, (Richland, Washington: Battelle Pacific Northwest Laboratories, 1973).
54. Tadmor, J. and K. E. Cowser. "Underground Disposal of Krypton-85 from Nuclear Fuel Reprocessing Plants," *Nucl. Eng. Design* **6**, 243 (1967).
55. Deonigi, D. E. "Evaluation Methodology of Waste Management Concepts," *Nucl. Technol.* **24**, 331 (1974).
56. Inoue, Y. and S. Morisawa. "On the Selection of a Ground Disposal Site for Radioactive Wastes: An Approach to its Safety Evaluation," *Health Phys.* **26**, 53 (1974).
57. deBuchananne, G. D. "Geohydrologic Considerations in the Management of Radioactive Waste," *Nucl. Technol.* **24**, 356 (1974).
58. Lewis, W. B. "Radioactive Waste Management in the Long Term," AECL Report DM-123/AECL-4268 (Chalk River, Ontario: Atomic Energy of Canada Ltd., 1972).
59. "Disposal of Radioactive Wastes into Seas, Oceans and Surface Waters," (Vienna: International Atomic Energy Agency, 1966).
60. Slansky, C. M. "Principles for Limiting the Introduction of Radioactive Waste into the Sea," *Atomic Energy Rev.* **9**, 853 (1971).

61. Bishop, W. P. and C. D. Hollister. "Seabed Disposal—Where to Look," *Nucl. Technol.* **24**, 425 (1974).
62. Drumheller, K. "Extraterrestrial Disposal of Nuclear Wastes," *Nucl. Technol.* **24**, 418 (1974).

ADDITIONAL REFERENCES

Cheverton, R. D. and W. D. Turner. "Thermal Considerations and Analysis for a Radioactive Waste Repository," *Nucl. Technol.* **19**, 21 (1973).
Fineman, P. "Progress in Research and Development on Waste Disposal," *Reactor Fuel Proc. Technol.* **11**, 159 (1968).
Hespe, E. D., Ed. "Leach Testing of Immobilized Radioactive Waste Solids," *Atom. Energy Rev.* **9**, 195 (1971).
Meyer, W. "An Argument for a Recoverable High-Level Waste Container," *Nucl. News* **14** (4), 38 (1971).
Morisawa, S. and Y. Inoue. "On the Selection of a Ground Disposal Site by Sensitivity Analysis," *Health Phys.* **26**, 251 (1974).
"Radiation Preservation of Food," Proc. Bombay Symposium (1972) (Vienna: International Atomic Energy Agency, 1973).
Robinson, B. P. "Ion-Exchange Minerals and Disposal of Radioactive Wastes," Geol. Survey Water-Supply Paper 1616 (Washington, D.C.: U.S. Govt. Printing Office, 1962).
Spitsyn, E. Ya. "Treatment and Disposal of Radioactive Wastes," Transl. *AEC*-tr-6881 (1968).
Straub, C. P. "Low Level Radioactive Wastes: Treatment, Handling, Disposal," (Washington, D.C.: U.S. Atomic Energy Commission, 1964).
West, P. J. "Waste Management for Nuclear Power," *Bull. IAEA* **16**, 78 (1974).

CHAPTER 13

TECHNOLOGICAL ASSESSMENT

Recent history has made it clear that the industrialized countries will have to rely on nuclear power to provide an increasing fraction of the large-scale base load capacity of their electric power generation. This development is an inescapable consequence of the steady growth in population, the diminishing supply of oil and gas at competitive prices, and the competing high-priority demand for petrochemicals as raw materials by such industries as textiles and plastics. Given the need for a further increase in size and sophistication of such power generating plants, the question must be answered as to how such plants can be fitted into our environment with the least disruption, the least insult to existing ecosystems, under conditions providing for the lowest possible addition to environmental radioactivity and dose commitments to future generations, and the best and most effective utilization of limited resources. Like many other decisions, the options involved here are rarely clear-cut but represent a deliberate compromise between conflicting requirements. This process of optimization, by weighing alternatives and attempting to attach value and priorities to all significant input factors, is often called technological assessment. It recognizes that nuclear energy is neither "good" nor "bad," but is a stage in the technological development of mankind that can provide great benefits to civilization if handled in a deliberate and responsible fashion.

The development of criteria to judge and weight the importance of the various factors involved can proceed by different routes. One such approach is the *cost-benefit* assessment, whereby an attempt is made to express all factors on a uniform scale, usually in monetary terms, to ensure that the final choice adopted shows the greatest positive benefit value compared with alternatives. The difficulty, of course, is to attach monetary values to essentially aesthetic or moral factors, such as the

621

preservation of a wildlife refuge, a scenic beach, a historical site, or the consequences, however small, of genetic damage.

Many large-scale enterprises involve a certain risk to life and limb. Society accepts many of these as a normal part of ordinary living; some of them are readily identified, and special inducements must be offered for people to accept the risks involved. Examples include highway driving, coal mining, high-wire balancing, aeroplane piloting, high rigging, scuba diving, or being a policeman. Such activities can be evaluated by a *risk-benefit* assessment, which is optimized by minimizing the risks and maximizing the benefits. This approach is complicated by the fact that public awareness of certain risks changes and not all risks are judged comparable in severity. Similarly, a distinction is made between voluntary risks and involuntary risks, that is between risks that the individual can choose to accept or not (to fly or drive), and those imposed on him by political decisions of the society or community in which he lives (the amount of air pollution or traffic risks). The acceptance of certain risks is not absolute either, but depends on circumstances, such as age and family status, convenience factors, and even the time of year or day.

Risk-benefit assessment is even more difficult to quantify than cost-benefit because it depends on a numerical rating of risk severity and on actuarial predictions of risk incidence for events that may not have an adequate statistical basis. This is particularly the case in nuclear technology where the probability of occurrence of certain events may have to be formulated on the basis of none or only a very few known previous instances. Since opponents of nuclear power keep stressing the risks and hazards it presents, it is obviously important to define the nature of these risks and to predict their probable incidence.

A separate issue, then, is to determine the "acceptability" to society of a given risk that may be expected to occur with a specified probability, especially since it is human nature to assume that even high-risk accidents will probably occur to other people first. Another ingredient to risk assessment that is peculiar to nuclear technology is the long time scale for some effects, such as potential long-range pollution by improperly stored radioactive wastes that may become effective centuries later, or the genetic effects of radiation exposure that may become apparent generations later. That type of risk, rightly, would be subjected to more stringent acceptability criteria than effects where erroneous predictions would become apparent rather quickly and would permit early correction.

In assessing the feasibility of constructing a nuclear power plant at any given location one can in theory use absolute or relative criteria. Actually, formulation or application of an absolute criterion of inherent merit or desirability is rarely possible; few industrial operations can be shown to be inherently or unquestionably unattractive or beautiful on all counts. Usually decisions have to be made on a relative basis, by comparing alternative options and evaluating differential costs and benefits over a common base cost. This type of relative cost-benefit evaluation may compare one type of power source with another, *e.g.*, in terms of relative air pollution or in terms of improved recreational opportunities. Table 174 compares air pollution effects of the principal fuel sources in terms of the major contributing factors for a base load plant (1-3).

Data of the type presented in Table 174 quickly show the difficulty of such comparisons since it is not immediately obvious if the fly ash from coal-fired plants is more objectionable than the smaller volume of radioactive waste or the greater thermal discharge from nuclear reactors. In other words, ideally to be useful the relative criteria would have to lead to comparisons in compatible units and permit the use of an optimization procedure capable of balancing otherwise incompatible quantities. Although many optimization models and procedures have been proposed, most cost-benefit analyses for nuclear power plants, as presented in environmental statements, list factors considered in qualitative terms, but they are unable to present a total final balance comparing alternative sites or design options (4).

COST-BENEFIT OPTIMIZATION

Cost-benefit evaluation is a basic business management function and has been essential in all long-range planning operations. To a more limited extent it has always been required for private and public works of any type, although here political and social considerations often introduce further intangible factors. These intangible factors assume even greater importance to the public in the case of nuclear facilities, and a problem arises as to how one would evaluate their "cost." Once the need for additional generating capacity has been established, the traditional criterion for new power plants has been the assumption that the primary design objective is to produce power at lowest cost to the consumer or at highest profit to the utility per kilowatt-hour sold. This approach remains the basis of the publicly regulated, but privately owned, U.S. power industry, though it is being eroded severely by considerations of public policy regarding fuel supply and environmental constraints. In other words, until recently power generation was the sole benefit considered and facility cost and operating costs the only costs.

The newer approach broadened the definition both of the costs and the benefits, with some consequent loss of clarity and purpose. The lack of compatible units for some of the quantities considered imposes an

Table 174. Volume of Air Required to Meet Concentration Standards and Average Site Boundary Concentrations for Various Fuels for Yearly Emission from a 1000 MWe Power Station (3)

Type Plant	Pollutant	Standard[a]	Discharge Quantity[b]	Dilution Volume 10^9 m^3	Site Boundary Concentration[c]
Coal	SO_2 (3.5% S)	0.03 ppm	3.66×10^8 lb	2.14×10^6	0.20 ppm
	NO_2	0.05 ppm	5.50×10^7 lb	2.49×10^5	0.04 ppm
	CO	9.0 ppm[d]	1.38×10^6 lb	63.5 0	0.02 ppm
	Hydrocarbons	0.24 ppm[e]	5.50×10^7 lb	156	0.001 ppm
	Particulates (97.5% removal)	75 $\mu g/m^3$	1.25×10^7 lb	75,500	18 $\mu g/m^3$
	Ra-226	2 pCi/m^3	0.0170 Ci	8.5	4.2×10^{-16} Ci/m^3
	Th-228	0.2 pCi/m^3	0.108 Ci	708	2.6×10^{-16} Ci/m^3
Oil	SO_2 (1.5% S)	0.03 ppm	1.23×10^8 lb	6.95×10^5	0.07 ppm
	NO_2	0.05 ppm	5.42×10^7 lb	2.45×10^5	0.04 ppm
	CO	9.0 ppm[d]	2.08×10^4 lb	0.95	2.61×10^{-4} ppm
	Hydrocarbons	0.24 ppm[e]	1.17×10^6 lb	4,720	0.004 ppm
	Particulates (97.5% removal)	75 $\mu g/m^3$	5.88×10^6 lb	35,400	8.4 $\mu g/m^3$
	Ra-226	2 pCi/m^3	6.0×10^{-4} Ci	0.3	1.5×10^{-18} pCi/m^3
	Th-228	0.2 pCi/m^3	1.3×10^{-3} Ci	6.7	3.2×10^{-18} pCi/m^3
Gas	SO_2	0.03 ppm	2.78×10^4 lb	157	1.5×10^{-5} ppm
	NO_2	0.05 ppm	2.71×10^7 lb	1.23×10^5	0.02 ppm
	Particulates (97.5% removal)	75 $\mu g/m^3$	1.04×10^6 lb	6,290	1.5 $\mu g/m^3$
Nuclear	Kr-85 & Xe-133	0.3 $\mu Ci/m^3$ PWR	9,650 Ci	33.2	2.3×10^{-11} Ci/m^3
	Short-lived radioactive gases	0.3 $\mu Ci/m^3$ BWR	1.66×10^6 Ci	5.54×10^4	4.0×10^{-9} Ci/m^3
	I-131 (inhalation)	1.0×10^{-10} Ci/m^3	0.2 Ci	2.0 PWR	4.2×10^{-16} Ci/m^3
			5.3 Ci	53.0 BWR	1.3×10^{-16} Ci/m^3
	I-131 (air-grass-milk)	1.4×10^{-13} Ci/m^3[f]	0.2 Ci	1,430 PWR	4.2×10^{-16} Ci/m^3
			5.3 Ci	37,800 BWR	1.3×10^{-14} Ci/m^3

[a] EPA "National Primary and Secondary Air Standards," (Federal Register, Vol. 36, No. 84, Part II, pp. 8186-87, 4/30/71), and AEC "Standards for Protection Against Radiation," 10CFR20.

[b] Discharges from PWR and BWR are derived from weighted average 1967-1972 release data as summarized by the Directorate of Regulatory Operations, U.S. Atomic Energy Commission.

[c] Based on average χ/Q at 500 m (for release height of 100 m) of 6.2×10^{-8} sec/m^3 for 25 operational or proposed nuclear power stations.

[d] Maximum eight-hour concentration, once per year. Yearly average not specified.

[e] Maximum three-hour concentration (6-9 a.m.) once per year.

[f] "Concentration factor" of 700 applied to inhalation standard for I-131.

arbitrary valuation procedure; however, this approach is not entirely novel since, for instance, the value of a human life or of a lost limb have long been determined in court in workman's compensation suits or medical negligence actions.

Any cost-benefit analysis is an iterative optimization process in which, first, certain fundamental options are evaluated; then, when the choice has been narrowed, certain refinements are imposed and examined in terms of the differential costs they introduce over and above the base cost. For power plants, the basic options are usually examined solely on the basis of cost per kilowatt installed and bus-bar cost per kilowatt hour delivered. These basic factors would include the following:

1. plant capacity
2. type of power plant (coal, gas, nuclear)
3. single or multiple site
4. fuel availability and projected cost
5. capital investment and amortization
6. construction costs
7. site-dependent factors, *e.g.*, foundation costs, road construction costs and water costs
8. transmission line requirements
9. land costs
10. labor costs
11. licensing costs (negligible for fossil plants)

All of these can usually be expressed in monetary terms at present worth with power generation being considered the only "benefit" at this stage.

The basic calculation will follow conventional cost estimate procedures. Once it is decided to go to nuclear power, quotations will be required to determine actual price ranges for the nuclear steam supply and major plant components and to establish fuel costs over the life of the plant, or at least the first two reloads. Generating equipment and switchgear will be essentially the same for any of the options, and so will land costs and transmission costs for a given site or district. Generating costs will depend critically on anticipated load conditions, since a nuclear plant with its higher overhead is most economically operated near full capacity, base loaded. This is illustrated in Table 175, which compares estimated operating costs and generating costs for one 1161 MWe power plant (5) for different fuels and capacity factors. Fuel alternatives and financial costs will, of course, vary from plant to plant and with the type of ownership.

Table 176 shows a 1973 breakdown of projected generating costs for U.S. conditions. Since then the cost of fossil fuels has risen more steeply than predicted and many of the cost comparisons must be reevaluated. Analysis shows that when oil is priced at greater than $4/barrel, oil-fired

Table 175. Typical Cost Comparison of Alternative Generating Systems: Comanche Peak Plant (5)

Alternative	1982 Estimated Operating Costs (mills/kWh)			Annual Operating, Maintenance, and Fuel Costs (millions of 1982 dollars) for Capacity Factor of		
	Fuel	Operating and Maintenance	Total	60%	70%	80%
Nuclear (1200 MWe PWR)	3.10	0.68	3.78	45.7	53.3	60.9
Lignite-fired (mine mouth)	2.33	1.40	3.73	45.1	52.6	60.1
Lignite-fired (Squaw Creek)	4.50	1.40	5.90	71.3	83.2	95.1
Coal-fired (Wyoming)	9.15	1.54	10.69	110.6	150.8	172.3

Alternative	1982 Present Worth			Annualized 1982 Dollars		
	Generating Costs[a] (millions of dollars) for Average Capacity Factor of			Generating Costs[a] (millions of dollars) for Average Capacity Factor of		
	60%	70%	80%	60%	70%	80%
Nuclear	1389 (1702)	1469 (1834)	1549 (1966)	132.2	139.8	147.4
Lignite-fired (mine mouth)	1368 (1676)	1447 (1807)	1525 (1937)	130.2	137.7	145.2
Lignite-fired (Squaw Creek)	1568 (2057)	1693 (2263)	1818 (2470)	149.3	161.1	173.1
Coal-fired (Wyoming)	1904 (2662)	2326 (3360)	2552 (3733)	181.2	221.4	242.9

[a]Using an operating period of 30 years, a discount rate of 8.75%, no increase beyond 1982 costs (and in parentheses an increase of 5% per year beyond 1982 in operating, maintenance, and fuel costs), and the staff-estimated capital costs.

Table 176. Estimated 1981 Generation Costs for 1000 MWe Steam-Electric Power Plant (1973 dollars) (6)

A. Light-Water Reactor Fuel Cost Breakdown

Cost Component	Cost (mills/kWh)
Mining & milling ($10/lb U_3O_8)	0.54
Conversion to UF_6	0.07
Enrichment ($42/kg SWU)	0.76
Reconversion & fabrication ($70/kg U)	0.33
Spent fuel shipping ($5/kg U)	0.02
Reprocessing ($35/kg U)	0.14
Waste management	0.04
Plutonium credit ($8/g)	(0.22)
Subtotal	1.68
Fuel inventory carrying charge (at 12%)	0.82
Total	2.50

B. Comparative Cost for Alternative Fuels (mills/kWh) incl. escalation to 1981

Cost Component	LWR	Coal	Oil
Capital	11.70	10.90	8.00
Fuel	2.50	5.50	24.60
O&M	1.00	1.60	0.80
Total	15.20	18.00	33.40

Assumptions:
1. Annual generation of 7.0 x 10^9 kWh (80% utilization)
2. Annual fixed charge of 15% used to translate unit capital costs to annual capital costs. All costs were escalated at 5% per year with the exception of nuclear fuel costs, which were assumed to remain constant because of improvements in technology and cost reductions as a result of increases in scale of manufacturing. For purposes of the analysis presented in the table an arbitrary price of $10/barrel (equivalent to $1.67/MBtu or 16.7 mills/kWh) was selected for oil.

power plants are uneconomical to operate compared with nuclear or coal-fired power plants. Furthermore, the prospective lack of availability of oil eliminates it from consideration as an expanding source of fuel for large central station power plants.

The price utilities pay for coal includes the selling price at the mine and transportation costs. The average as-burned cost of coal in the U.S.

in 1972 was $8.50/ton, 37 cents/million British thermal units (MBtu). Of this, mining amounted to $4.50/ton and transportation, $4.00/ton. It should be noted that the cost of coal at the mine varied from $2.00 to $10.00/ton depending on where the coal was mined, the thickness and depth of the coal, and whether it was produced by strip mining, underground mining, or auger mining methods. By 1974 the delivered price of coal in the southeastern U.S. had risen to $30-50/ton. Transportation costs are a function of the distance and mode of transportation. At a power plant heat rate of 10,000 MBtu/kWh, 37 cents/MBtu is equivalent to 3.7 mills/kWh.

The optimization process has to examine fuel alternatives, plant design alternatives and site alternatives. The costs are computed and benefits may be tabulated in appropriate units (7). Since the environmental enhancement terms can rarely be monetized, the concept of a cost-benefit optimization in purely monetary terms and a final dollar balance is usually impractical. Values assigned to recreational benefits, for instance, are frequently arbitrary and unrealistically optimistic. This has also long been the history of many proposed flood control and irrigation projects.

CONCEPT

At this time many utilities use a U.S. Atomic Energy Commission-produced computer program called CONCEPT to evaluate cost estimates (8); to a growing extent applicants are using this program also to provide a common basis for comparison of cost models. The use of the code provides a means of calculating future capital costs of a project with various assumed sets of economic and technical ground rules.

The procedures used in the CONCEPT code are based on the premise that any central station power plant involves approximately the same major cost components, regardless of location or date of initial operation. Therefore, if the trends of these major cost components can be established as a function of plant type and size, location, and interest and escalation rates, then a cost estimate for a reference case can be adjusted to fit the case of interest. The application of this approach requires a detailed "cost model" for each plant type at a reference condition and the determination of the cost trend relationships. The generation of these data has comprised a large effort in the development of the CONCEPT code. Detailed investment cost studies by an architect-engineering firm have provided basic cost model data for pressurized water reactor nuclear plants and coal-fired plants. These cost data have been revised to reflect plant design changes since the 1971 reference data of the initial estimates.

The cost model is based on a detailed cost estimate for a reference plant at a designated location and a specified date. This estimate includes a detailed breakdown of each cost account into costs for factory equipment, site materials, and site labor. A typical cost model consists of over a hundred individual cost accounts, each of which can be altered by input at the user's option. The AEC system of cost accounts is used in CONCEPT.

To generate a cost estimate under specific conditions, the user specifies the following input: plant type and location, net capacity, beginning date for design and construction, date of commercial operation, length of construction workweek, and rate of interest during construction. If the specified plant size is different from the reference plant size, the direct cost for each two-digit account is adjusted by using scaling functions that define the cost as a function of plant size. This initial step gives an estimate of the direct costs for a plant of the specified type and size at the base date and location.

The code has access to cost index data files for 20 key cities in the United States. These files contain data on cost of materials and wage rates for 13 construction crafts as reported by trade publications over the past 12 years. These data were used to determine historical trends of material costs, providing an estimate for the materials costs as of mid-1972. Cost escalation after mid-1972 must be based on labor rates, labor productivity, and labor and materials escalation rates estimated by the staff for the nearest metropolitan area.

This technique of separating the plant cost into individual components, applying appropriate scaling functions and location-dependent cost adjustments, and escalating to different dates is the heart of the computerized approach used in CONCEPT. The procedure is illustrated schematically in Figure 118. The code is a macroeconomic model and must be supplemented by purely local factors for some parameters.

External Costs

There are also external costs that should be estimated. For nuclear plants these involve the direct and indirect costs associated with the fuel cycle—cost and hazards of uranium mine operations, costs and energy demand of conversion and enrichment, fabrication costs, reprocessing and waste disposal costs, carrying charges on fuel inventory, and credit for recycled plutonium. Only some of these charges are directly reflected in the cost of fuel; the others may be largely environmental or social in character and their magnitude at this time a matter of contention. In addition there are maintenance costs of plant and property and insurance costs for the property and for liability contingency.

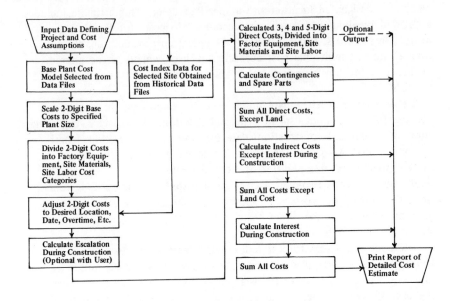

Figure 118. Use of the CONCEPT program for estimating capital costs.

After base costs are computed for plant and site, differential or enhancement costs must be determined for safety and environmental features.
Typically these will include comparison of alternative designs for plant
containment, waste heat dissipation and effluent treatment. These differential features must be evaluated in two respects: whether they enable
the plant to meet certain legal minimum standards, such as water quality
standards (ΔT) or as low as practicable radwaste emissions, and whether
the approach suggested is the most cost-effective in relation to the improvement obtained.

As was pointed out in Chapter 6, not all of the heat dissipation systems
necessarily improve the environmental impact, and merely reducing temperature rise at the outlet point may not always constitute the most
desirable solution environmentally. Table 177 exemplifies the type of
comparison between heat dissipation options. A question rarely answered
explicitly is the value to be attached to larval or adult fish mortality, as
related to the total fish population in the river, in reference to the
appreciable incremental cost of cooling towers.

Evaluation of alternative radwaste systems is done in two ways: (1) by
evaluating it in terms of its conforming with the state-of-the-art in meeting

Table 177. Differential Cost-Benefit Evaluations: Crystal River Unit No. 3 (9)

	Reference Case (Applicant's Design)	Closed Cycle Cooling (Unit 3 Only)			Supplemental Cooling for 7°F ΔT All Units				Supplemental Cooling for 11.5°F ΔT All Units				
					Differential Cost of Alternatives (From Reference Case)								
		Cooling Pond	Mechanical Draft (Freshwater)	Natural Draft (Saltwater)	Dilution	Hold-Up Pond	Spray Modules	Natural Draft Cooling Tower	Dilution	Hold-Up Pond	Spray Modules	Natural Draft Cooling Tower	Extension of Discharge Canal
Monetary Costs													
Capital Cost ($10⁶)	242	20	9	17	11	15	8	14	4	4	3	9	3
Annual Costs, Capitalized ($10⁶) Replacement Power Fuels Oper. Exp.	131	2	14	10	1	–	11	4	–	–	4	2	–
Total Present Worth	373	22	23	27	12	15	19	18	4	4	7	11	3
Environmental Considerations	Reference Case Impact				**Differential Environmental Impact (From Reference Case)**								
Area diverted to facility use (acres)	30	700	5	5	20	750	0	5	15	20	15	5	0
Heat added to gulf mW$_t$[a]	1718	-1718	-1718	-1718	0	-1462	-1462	-1462	0	-585	-585	-585	0
Aquatic area exposed to ¼°F ΔT (over full tidal cycle, acres)	1250	-1250	-1250	-1250	-250	-1300[b]	-1300[b]	-1300[b]	80	470	470	470	0
Fish kill intake screens (pounds)	36,000	-36,000	-36,000	-36,000	not significant	negligible	negligible	negligible	not significant	negligible	negligible	negligible	negligible
Chemicals Released to water	negligible	some chemicals in blowdown	some chemicals in blowdown	some chemicals in blowdown	negligible	negligible	negligible	some chemicals in blowdown	negligible	negligible	negligible	some chemicals in blowdown	negligible
Released to air	negligible	negligible	negligible	approx. 10 tons of salt per day	negligible	negligible	approx. 9 tons of salt per day	approx. 9 tons of salt per day	negligible	negligible	approx. 3 tons of salt per day	approx. 3 tons of salt per day	negligible
Rad wastes	negligible	negligible	negligible	negligible	negligible	negligible	negligible	negligible	negligible	negligible	negligible	negligible	negligible
Effect on terrestrial animal life	negligible	removal of 700 acres of habitat	removal of 5 acres of habitat	negligible	negligible	negligible	removal of 5 acres of habitat	removal of 5 acres of habitat	negligible	negligible	removal of 5 acres of habitat	removal of 5 acres of habitat	negligible
Effect on aquatic animal life	major localized	improved	improved	improved	improved	improved	improved	improved	improved	improved	improved	improved	improved
Effect on aquatic plant life	major localized	improved	improved	improved	improved	improved	improved	improved	improved	improved	improved	improved	improved

[a]Above that used by Crystal River Units 1 and 2 at full power.
[b]These cases would reduce the affected acreage of the reference case by 1250 acres and would reduce the present (Units 1 and 2) case by an additional 50 acres.

as-low-as-practicable criteria for effluents, except for noble gases and tritium, and (2) by evaluating—in terms of the reduction in population dose from routine releases per dollar spent. This may be expressed as $\Delta(dose)/\Delta\$$, and examples of this type of evaluation were given in Figures 73-76. The "cost" due to radiation effects is usually obtained from relative population dose in man-rems, and Table 178 gives an example of the costs and corresponding improvements obtainable for liquid and for gaseous radwaste systems in terms of dose rates to people and fish. By themselves these numbers tend to be meaningless; since the added costs are not insignificant, individual dose rates in particular should be related to natural background rates. Very detailed sample calculations for several options as to site, heat dissipation system and radwaste treatment have been presented in Reference 6. Such considerations of costs and expected effects show that for a BWR like the Browns Ferry plant, reductions in liquid effluent contamination are less significant than the dose due to airborne radionuclides, and hence additional liquid waste treatment may not be justifiable in that case.

For a true cost-benefit balance, the man-rem population dose must be expressed in monetary terms also. Even though one might argue about the linearity of radiation effects and hence the evaluation of the *man-rem* as a measure of risk, there are many precedents in legal practice for determining a monetary value of human life or bodily impairment, subject to variations in economic and social status of the individual. At the present time, estimates of the risk-cost per man-rem arrive at values of $30 to $1000 per man-rem, with a mean value around $200/man-rem (11,12). Using a value of $30/man-rem, Sagan (13) presented a comparison of the cost effectiveness for noble gas and tritium removal systems, shown in Table 179. On this basis few of the more elaborate radwaste systems can be considered to be cost-effective.

At an estimate of the cost of biological damage from 1 man-rem at $1000, the addition of any of that equipment could not be justified in economic terms. Nevertheless, TVA decided to add both the recombiners and six charcoal beds to the plant in order to meet AEC requirements. Ideally, in a society with unlimited resources, no expenditures would be spared to reduce risk regardless of cost; practically, however, radiation protection must compete with other pressing societal needs (13). Finally an attempt must be made to weigh the social and economic costs and benefits associated with the construction and operation of the nuclear facility and the alternatives proposed.

The primary internal costs are: (1) the capital costs of land acquisition and improvement, (2) the capital costs of facility construction, (3) the incremental capital costs of transmission and distribution facilities, (4) fuel costs including spent fuel disposition, (5) other operating and

Table 178. Cost and Dose Reduction from Alternative Radwaste Systems:
Browns Ferry Nuclear Plant (10)

Alternative Liquid Radwaste System	Filtration and Demineralization	Additional Demineralization	Filtration, Demineralization, and Evaporation
Incremental generating cost (thousands of dollars)	base	5,711	1,991
Dosage rates to people due to radionuclides discharged to water body—external contact			
(rem/yr)	2×10^{-6}	3×10^{-7}	2×10^{-8}
(man-rem/yr)	5×10^{-3}	6×10^{-4}	5×10^{-5}
Dosage rates to people due to ingestion			
(rem/yr)	1×10^{-3}	1×10^{-4}	1×10^{-5}
(man-rem/yr)	80	10	0.8
Dosage rate to fish			
(rad/yr)	5×10^{-2}	7×10^{-3}	5×10^{-4}
Dosage rates to people due to radionuclide contamination of ground water			
(rem/yr)	1×10^{-5}	2×10^{-6}	1×10^{-7}
(man-rem/yr)	0.1	1×10^{-2}	1×10^{-3}

Alternative Gaseous Radwaste System	30-Minute Holdup	Hydrogen Recombiners	Hydrogen Recombiner & 6 Charcoal Beds per Unit	Hydrogen Recombiner & 12 Charcoal Beds per Unit	ORGDP Systems	Cryogenic
Incremental generating cost (thousands of dollars)	base	6,000	9,000	10,500	not estimated	9,700
Dosage rates to people from external contact						
(rem/yr)	0.170	1.6×10^{-2}	1.6×10^{-3}	6×10^{-4}	5×10^{-4}	5×10^{-4}
(man-rem/yr)	890	260	23	6.8	1.6	1.6
Dosage rates to people from ingestion						
(rem/yr)	2×10^{-4}	1.8×10^{-4}	1.8×10^{-4}	1.8×10^{-4}	1.8×10^{-4}	1.8×10^{-4}
(man-rem/yr)	0.79	0.73	0.73	0.73	0.73	0.73
Dosage rate to plants and animals						
(rad/yr)	0.170	1.6×10^{-2}	1.6×10^{-3}	6×10^{-4}	5×10^{-4}	5×10^{-4}

Table 179. Cost Estimate of Dose Reduction Measures for a Large
BWR Plant (13)

Equipment	Reduction in External Dose (man-rem)	Cost ($)	Cost per Man-rem of Radiation Reduction ($)	Incremental Cost per Man-rem of Radiation Reduction ($)
Recombiners only	3,600	6,000,000	1,700	1,700
Recombiners and 6 charcoal beds	4,305	9,000,000	2,070	4,250
Recombiners and 12 charcoal beds	4,350	10,500,000	2,400	30,000

maintenance costs including license fees and taxes, (6) plant decommissioning costs, and (7) research and development costs associated with potential future improvements of the facility and its operation and maintenance (including backfitting).

There are also external costs. Their effects on the interests of people need to be examined. They can be divided into long-term (or continuing) costs and temporary costs, which are generally associated with the period of construction or the readjustment of the lives of persons whose jobs or homes will have been disrupted or displaced by the purchase of land at the proposed site. Examples of temporary external costs are:

- shortages of housing
- inflationary rentals or prices
- congestion of local streets and highways
- noise and temporary aesthetic disturbances
- overloading of water supply and sewage treatment facilities
- crowding of local schools, hospitals, or other public facilities
- overtaxing of community services
- the disruption of people's lives or the local community caused by acquisition of land for the proposed site.

Examples of long-term external costs are:

- impairment of recreational values (*e.g.,* reduced availability of desired species of wildlife and sport fish, restrictions of access to land or water areas preferred for recreational use);
- deterioration of aesthetic and scenic values
- restrictions on access to areas of scenic, historical or cultural interest
- degradation of areas having historic, cultural, natural or archaeological value

- removal of land from present or contemplated alternative uses
- creation of locally adverse meteorological conditions (*e.g.*, fog from cooling towers or cooling ponds)
- creation of noise, especially by mechanical-draft cooling towers
- reduction of regional product due to displacement of persons from the land proposed for the site
- lost income from recreation or tourism that may be impaired by environmental disturbances
- lost income of commercial fishermen attributable to environmental degradation
- decrease in real estate values in areas adjacent to the proposed facility
- increased costs to local governments for the services required by the permanently employed workers and their families

There are certain social and economic benefits that affect various political jurisdictions or interests, to a greater or lesser degree. Some of these reflect transfer payments or other values that may partially, if not fully, compensate for certain services, as well as external or environmental costs. This fact should be reflected in the designation of the benefit. A list of examples follows:

- tax revenues to be received by local, state and federal governments
- temporary and permanent new jobs created and payroll
- incremental increase in regional product (value-added concept)
- enhancement of recreational values through making available for public use any parks, artificially created cooling lakes or marinas
- enhancement of aesthetic values through any special design measures as applied to structures, artificial lakes or canals and parks
- environmental enhancement in support of the propagation or protection of wildlife and the improvement of wildlife habitats
- creation and improvement of local roads, waterways, or other transportation facilities
- increased knowledge of the environment as a consequence of ecological research and environmental monitoring activities associated with plant operation, and technological improvements from the applicant's research program
- creation of a source of heated discharge that may be used for beneficial purposes (*e.g.*, in aquaculture, in improving commercial and sport fishing, and other water sports)
- provision of public education facilities (*e.g.*, a visitors' center) (7).

RISK-BENEFIT ASSESSMENT

The cultural level of any human civilization perhaps may be measured by the value it puts on the individual lives of its citizens. No human activity can be tolerated for very long that wantonly endangers life and limb; yet, almost any human activity carries with it a finite risk of injury. We are accustomed to and accept such risks without question, although they may vary sharply with location or profession. Nuclear power plants have been attacked in public for carrying with them an intangible risk to the public both through a presumed inevitable contamination threat from plant operations and waste disposal, and through the alleged danger of catastrophic accidents leading to widespread destruction or gross contamination of air, soil and water. Attempts to lay at rest public fears of a catastrophic accident have been only partially successful in spite of the superior safety record of the industry, and it will probably require decades of peaceful operation of commercial plants to demonstrate their safety to everyone's satisfaction. At the same time, realistically one must expect some accidents to happen, including some fatalities because they are bound to occur in any industry of the size and complexity of the nuclear industry. However, it is incumbent on the designers and operators to ensure that the consequences of any accident remain localized and minimal.

This leaves as a major issue the population radiation dose, from inhalation and ingestion of radioactive trace contaminants that are discharged at as-low-as-practicable levels and enter the environment and any that are released from waste storage areas. Following the BEIR report and others (14-17), certain health effects can be ascribed to any large-scale population exposure on statistical grounds, however low the individual exposure. This results in projections for health effects (cancer, leukemia) due to operation of nuclear power plants of the type shown in Table 180 (16). Without arguing about specific numbers presented in that and other studies, it is important to assess the significance of such numbers. A blunt statement that operation of a certain industry will result in, say, 37 extra deaths due to radiation-induced injuries, would certainly justify questioning the right of the plant operators to exact such casualties. However, one can establish in most cases that predicted added casualties are in fact significantly less than the year-to-year normal fluctuations observed for many of these illnesses invoked, making such predictions difficult to prove or disprove.

In assessing the risk-benefit balance associated with the operation of a given nuclear power plant, two aspects must be examined: the risks and hazards posed by the nuclear facility as compared with those associated

Table 180. Projected Numbers of Health Effects Attributable to Release of Certain Long-Lived Radionuclides by Normal Plant Operation Under Current Release Practices (16)

| | Cumulative Number of Health Effects | | | | | | | |
| | Iodine-129 | | Tritium | | Krypton-85 | | Actinides | |
Year, t	Past-Present[a]	Future[b]	Past-Present[a]	Future[b]	Past-Present[a]	Future[b]	Past-Present[a]	Future[b]
1970	0	0	0	0	0	0	0	0
1975	0	0	2	0.5	0.3	5	0	0
1980	0	0	11	3	3	26	0	0.1
1985	0	0	35	8	14	79	0	0.4
1990	0	0.1	88	21	42	190	0.1	1
1995	0	0.2	190	43	110	410	0.2	2
2000	0.1	0.3	360	81	230	760	0.4	4
2005	0.2	0.5	630	140	460	1,300	0.7	7
2010	0.3	0.8	1,000	230	830	2,100	1.2	10
2015	0.5	1.2	1,600	340	1,400	3,200	2	15
2020	0.8	1.7	2,300	500	2,300	4,600	3	21
	One-fourth fatal		Two-thirds fatal		Two-thirds fatal		All fatal	

aThe number of health effects committed from doses received through year t.
bThe number of health effects committed from doses delivered after year t by radionuclide releases up through year t only.

with alternative power sources, and the magnitude of the actual risks involved relative to other risks that are normally assumed and judged acceptable by people in an industrialized society.

Table 181 lists the risks and benefits associated with the various energy sources used for the generation of electricity. Comparison of these plants indicates direct hazards arising predominantly in three areas: air pollution, mining accidents and the consequences of plant accidents. Of these, air pollution effects are the most obvious and the most directly evaluated hazards. Table 174 compared the pollutant emissions from fossil- and nuclear-powered plants. If the dilution volume required to meet emission standards is taken as a measure of toxicity, it is evident that fossil-fueled plants cause far more serious pollution than nuclear plants.

Table 181. Comparative Risks and Benefits from the Generation and Distribution of Electricity (1,2)

Type of Plant	Risks	Benefits
Hydroelectric	Alteration of stream flow; destruction of habitats and scenery, such as by reservoirs and long transmission lines	Energy; employment; flood control; recreation
Gas-fired	Destruction of scenery, such as by pipelines and plant stacks; air pollution with many substances; alteration of local ecology by thermal waste	Energy; employment; by-products
Oil-fired	Destruction of scenery, such as by pipelines, storage tanks, plant stacks, and ash-disposal areas; water pollution; air pollution with many substances; alteration of local ecology by thermal waste	Energy; employment; by-products
Coal-fired	Destruction of scenery, such as by strip mining, transport and storage facilities, plants, stacks, and ash-disposal areas; stream pollution (from mining refuse); air pollution with many substances; alteration of local ecology by thermal waste	Energy; employment; by-products
Nuclear	Destruction of scenery, such as by mining and processing facilities, plants, and stacks; minimal routine air and water pollution with radioactive ash; possible leakage during the long-term confinement of high-level radioactive wastes from fuel-reprocessing facilities; possible accidental release of significant quantities of radioactivity due to a reactor malfunction; alteration of local ecology by thermal waste	Energy; employment; by-products (*i.e.,* isotopes useful in medicine, industry, research, etc.)

Actual comparative mortality values for locations close to the plants also reflect stack effects and existence of low-population zones near nuclear plants. Figure 119 relates the low permissible level of pollutant emissions from nuclear plants in terms of government standards and medically perceivable effects (18) to other pollutants. In addition, the sheer volume of fossil fuel consumed (Table 174) causes a severe environmental impact through the volume of mining required to produce it, the number of trains or trucks that are needed to move the fuel and the fuel required in turn to power them.

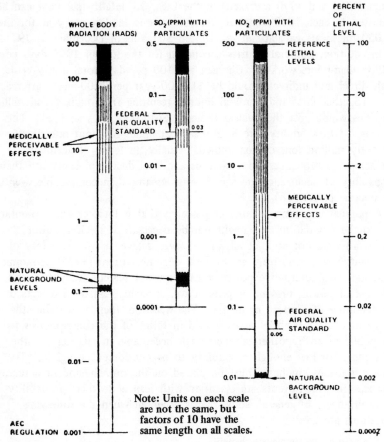

Figure 119. Physiological effects on humans of various pollutant plant emissions (18).

Coal mining is among the most hazardous of occupations, involving many deaths and injuries both in underground accidents and due to black-lung disease, and the accident rate is over twice that encountered in hardrock mines. The radon hazard in uranium mines can be reduced to a small fraction of the mechanical accident risk by adequate ventilation (Chapter 11). Averaging employment and production data for the three years 1967 to 1969, one finds that the mining and milling of 140 metric tons of U_3O_8 would require, at the rate of 2.3 metric tons per man, 62 man-years. Assuming that there are 1760 hours of employment per man per year, the total number of hours at risk per reactor per year would be 109,120. Since the rate of fatal accidents was 0.892 per million man-hours (1969 and 1970 averaged) in the U.S., 0.1 fatality per year can be allocated to each 1000-megawatt reactor, or one fatality per year for the 10,030 megawatt capacity in the United States as of December 1, 1971.

In addition, nonfatal injuries accounted for the loss of 1065 days per million man-hours worked. Charged at $100 per day, total injury costs, both direct and indirect, would be $11,700/year per 1000-megawatt reactor. Together fatal and nonfatal injuries resulting from mining and milling activities would cost the nuclear industry $417,000 per year (13). This contrasts with a fatality rate in British coal mines of about one death for every million tons of coal mined. Finally, in terms of accidents, the health consequences from oil fires and coal dust explosions and their probability of occurrence are significantly greater than comparable events for nuclear power (18).

A popular objection to nuclear power is that it imposes an involuntary risk to the population that might not be necessary if nuclear power plants (a) were not built at all, or (b) were delayed until the safety of the plants was conclusively established. As the current (1975) economic crisis indicates, alternative power sources are not readily available and a shortage of power, present or prospective, is having disastrous effects on the economy of Western countries. The question, then, is whether the safety of nuclear facilities as expressed in terms of the consequences to the public of an hypothetical accident is acceptable in relation to other risks and is of low enough probability to be considered negligible. To answer it means emphasis must be placed on the probabilistic or actuarial approach in which events are classified with respect to their probability of occurrence. In general, some risk is accepted when the following conditions are met:

    a. there is an immediate benefit
    b. some personal influence is possible
    c. analysis of risk is transparent
    d. someone who is not forced to enter the risk is convinced
    e. personal satisfaction
    f. familiarity
    g. known risk level and benefit.

Conversely, risk is not accepted if the following conditions prevail:

a. decision process is not transparent
b. no personal influence over risk
c. time for enjoying benefits is too remote
d. proponents raise suspicions
e. too little quantitative analysis
f. not enough own knowledge
g. the risk cannot be eliminated
h. residual risk exists whose consequences could be large and whose true probability is unknown.

Accident statistics are available to show the incidence of many common causes of injuries and deaths. Table 182 lists major causes of fatal accidents for the U.S. in 1967-72 and the probability of death per person per year from each cause. To these might be added about 1000 deaths

Table 182. Risk of Fatality by Various Causes (19)

Accident Type	Total Number	Individual Probability of Fatal Injury Per Year of Exposure
Motor vehicles	55,791	1 in 2.5 x 10^{-4}
Falls	17,827	1 in 10^{-4}
Fires and hot substances	7,451	1 in 4 x 10^{-5}
Drowning	6,181	1 in 3.3 x 10^{-5}
Firearms	2,309	1 in 10^{-5}
Air travel	1,778	1 in 10^{-5}
Falling objects	1,271	1 in 6.3 x 10^{-6}
Electrocution	1,148	1 in 6.3 x 10^{-6}
Lightning	160	1 in 5 x 10^{-7}
Tornadoes	91	1 in 4 x 10^{-7}
Hurricanes	93	1 in 4 x 10^{-7}
All accidents	111,992	1 in 6.3 x 10^{-4}
Nuclear reactor accidents (100 plants projected)	0	1 in 3.3 x 10^{-9}

per year from bee stings and snake bite or the estimated 800 deaths and 80,000 illnesses annually ascribed to pesticide poisoning. Such statistics in themselves are not very useful because they do not identify the exposed group among the general population, though it is obvious that the chance of exposure to bees or snakes is significantly lower in New York City than, say, in Georgia. This is taken into account in Figure 120, which presents historical trends for fatalities in transportation accidents.

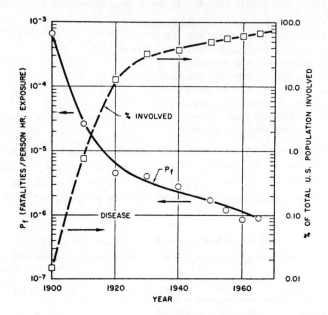

A. Risk and participation trends for motor vehicles

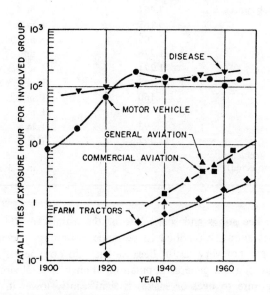

B. Risk and participation trends for certified air carriers.

Figure 120. Risk and participation trends for motor vehicles and air carriers and trends in group risks (20).

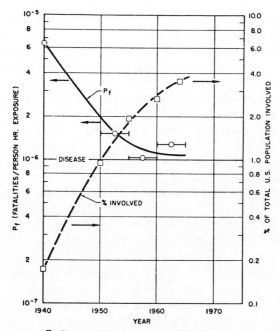

C. Group risk versus year, 1900-1960.

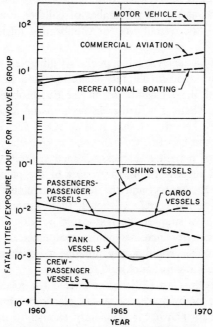

D. Group risk versus year, 1960-1970.

Figure 120, continued

Though the size of the exposed population groups has increased, the risk has decreased significantly (18,21).

In Great Britain and the United States the hourly risk of death for rail travel is an order of magnitude lower than for travel by private car. Thus, the percentage risk for a lifetime of rail travel would be about one-tenth of the risk for travel by motor car, provided the number of hours were equal for each form of transport. In Great Britain the mean number of hours traveled by rail is in fact lower than by motor car, being about 800 hours in 70 years, with a fatality risk of 50 deaths per 10^9 travel-hours, making the lifetime risk in that country $(50 \times 800)/10^9$, or about 0.004%.

The hourly risk of death for travel on scheduled airlines has been shown to be 2400 per 10^9 passenger-hours. On the assumption of a mean annual air travel time of one hour, the lifetime risk would therefore be $(2400 \times 70)/10^9$, or about 0.02%.

The total risk of death from pulmonary embolism or cerebral thrombosis among users of oral contraceptives in Great Britain was estimated to be 0.06% over the age span 20-44 years. For nonusers of the contraceptive pill the risk for the same age group was 0.008%. In this connection, it is worth noting that the total risk of death from pregnancy, delivery and the puerperium is 0.02% among women aged 20-34, and 0.06% for women aged 35-44. As most births are to mothers in the lower age group and as the mean number of children born to a mother is slightly over two, the overall risk for childbearing might be about 0.05% (22).

Starr (18,20,21) has analyzed the public's willingness to accept higher risks under voluntary than involuntary conditions and finds a significant gap, of the order of three magnitudes, in the risk probability that is judged acceptable. There is a further increase in the acceptability of risk as the perceived benefits of that activity rise. This is illustrated in Figure 121, where the benefits are expressed in monetary terms (Figure 121a) or in terms of benefit awareness, defined arbitrarily as the product of the relative level of advertising, the square of the percentage of population involved in the activity, and the relative usefulness (or importance) of the activity to the individual. Although these assumptions may be crude, Figure 121 does support the reasonable position that advertising the benefits of an activity increases public acceptance of a greater level of risk. This, of course, could subtly produce a fictitious benefit-risk ratio as may be the case for smoking (20).

In a similar analysis, Gast (23) has compared the relative safety and accident probability of hydroelectric power plants and nuclear plants. Major dam disasters, which occur frequently enough to permit statistical

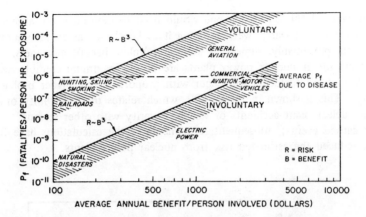

A. Risk versus benefit for voluntary and involuntary exposure.

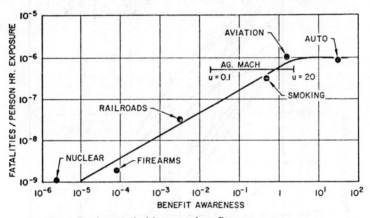

B. Accepted risk versus benefit awareness.

Figure 121. Risk acceptance versus benefit or benefit awareness (20).

analysis, are characterized by total failure to contain the impounded water. Comparing different studies for all types of dams, Gast concluded that the probability of major dam disasters is of the order of 10^{-4} per dam year, and for hydroelectric plants the fatality risk is 10^{-1} per plant year and damage risk $20,000 per plant year. The probability of major dam disasters is significantly greater than for hypothetical major nuclear plant disasters; the consequences of actual historical major dam disasters approach, but are somewhat smaller than, those of hypothetical nuclear plant disasters (23).

In view of this discussion one would need to determine the acceptable risk level for nuclear power, assuming it to be classed as an involuntary risk and, presumably, subject to increasing public benefit awareness. Not all accidents at nuclear power plants are of the maximum credible size; the less severe accidents will occur with proportionately higher probability. This is shown in Figure 122, which relates the probability of fatal nuclear plant accidents of varying severity with other serious man-caused events; subsequent refinements of the calculations have further reduced the estimated risk from nuclear power plants.

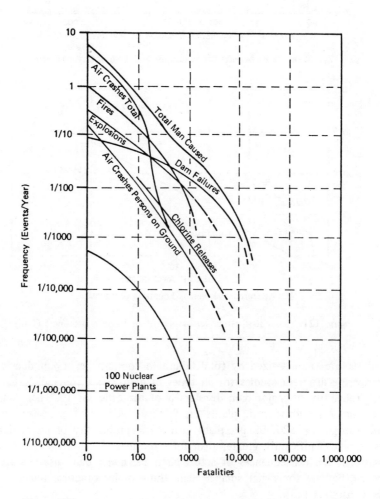

Figure 122. Frequency of fatalities due to man-caused events (19).

This plot is taken from the "Rasmussen Report" (19), which is the most detailed attempt to date to predict probabilistically the chances of occurrence of various reactor-related accidents and the associated risks. It indicates, for instance, that the predicted frequency and severity of nuclear plant accidents with 100 power plants operating is comparable to effects from falling meteors. Comparing major causes of fatalities to the population living in the vicinity of nuclear power plants, it can be shown that the added risk from nuclear reactor accidents makes a negligible contribution to the total (Table 183), with maximum consequences of property damage outside the plant itself on the order of $100,000 for a core melt accident (19). Even if one does not accept all of the assumptions of the Rasmussen Report or predicts more serious operator errors, it seems unlikely that the probability of occurrence for accidents of any severity would be in error by more than an order of magnitude. Even then the imputed hazard should lie well within the range of acceptable risks, particularly in comparison with the alternatives (24).

Table 183. Annual Fatalities and Injuries Expected Among the 15 Million People Living Within 20 Miles of U.S. Reactor Sites (19)

Accident Type	Fatalities	Injuries
Automobile	4,200	375,000
Falls	1,500	75,000
Fire	560	22,000
Electrocution	90	–
Lightning	8	–
Reactors (100 plants)	0.3	6

It would be desirable to quantify risks better in relation to plant reliability factors and escape probabilities for radioactive contaminants (25-27), but such refinements in methodology will remain largely academic in the absence of input data on actual plant and component failures. The ultimate test will lie in demonstrated plant safety, depending on a high level of engineering design and quality control. Some accidents are bound to occur, but should not cause significant delay or abandonment of further use of nuclear power (28). It also makes little sense to insist, in the case of nuclear power alone, on enforcement of risk criteria vastly more stringent than for other human activities.

ENERGY BALANCE ASSESSMENT

A third, independent method of technological assessment is through the energy balance approach. This type of assessment has long been advocated by Odum (29), who has emphasized that power generation invariably implies power consumption and that there is no environmental or society gain if the power consumed equals or approaches the power generated. It was shown in Chapter 11 that although uranium enrichment is the main power consumer in the fuel cycle, appreciable amounts of power are consumed in mining and milling of ores and in providing cooling for the condenser water at the power plant. As long as high-grade uranium ore is available, the inherently higher energy content of uranium compared with coal will maintain a favorable balance in support of nuclear power. On the other hand if shortage of ore forces the mining of low-grade minerals such as phosphate shales, the energy usage per ton of uranium produced may rise steeply and shift this balance. A fairly detailed analysis of the energy investment in fossil and nuclear power generation has been published by Rombaugh and Koen (30).

Similar considerations apply even more strongly to some advanced reactor types. From energy-balance considerations, the breeder reactor, which requires no or little enriched fuel, is clearly ahead of the thermal neutron reactor. In the case of fusion devices, demonstrated energy break-even between input power and power generated represents a vital first stage of development.

ENVIRONMENTAL IMPACT STATEMENTS

The need for available electric power will generate increasing pressure to build larger nuclear generating plants and at a faster rate. This pressure may encroach on the just demands for preservation of the environment and conservation of resources. As a protection against rash, potentially harmful design decisions, U.S. regulations require the submission of an environmental statement, as well as a safety analysis report (4,6,7,31-33). Such a statement, which is presented in a fairly standardized format, should be a balanced account of all of the environmentally significant factors inherent in the overall plant design and location. As such it should analyze the proposed plant in terms of alternative plant types, alternative sites and alternative methods of effluent treatment. A draft report is made available to all interested government agencies and private groups for review and comments. These are then incorporated in a final statement that forms the basis for public hearings prior to issuance of a construction license.

The compilation of an environmental statement represents an appreciable investment in time and money and must be planned on a careful schedule. At least one year's, but usually more than two years', observations on meteorological, hydrological and wildlife factors are required to provide adequate baseline data. Borehole drilling is needed for geological data, and background investigations on economic and historical factors require specialized study. Typically, the time between issue of the draft and final statements has been six to twelve months. Special or unusual site conditions may extend this period and the time for licensing action considerably. On the other hand, increasing standardization of nuclear steam supply systems may permit blanket licensing of certain reactor types and auxiliary features. Figure 123 illustrates the steps involved in providing "environmental accountability " (34).

Finally, a cost-benefit analysis must be presented for the principal alternatives. This is rarely a meaningful process since the operator probably has done such an analysis already at a more fundamental level for his own benefit, and the alternatives presented, on heat dissipation and radwaste treatment, often seem mainly destined for public consumption.

The last step in the analysis includes a consideration of the uranium fuel cycle exclusive of the reactor. To simplify that part of the statement the U.S. Atomic Energy Commission has provided a standard summary of environmental considerations for the uranium fuel cycle that will meet the intent of the law. This was presented in Table 163 (33).

Before and after filing an application for construction of the power plant it is up to the prospective plant operator to convince affected members of the public that the small risks from construction and operation of the plant are heavily outweighed by its benefits, that the consequent availability of electric power is desirable both on a local and a national scale, and that the engineering design and all measures taken to ensure environmental protection are soundly conceived, properly instituted and will be faithfully adhered to. Much has been written on the favorable and adverse effects on public opinion of political-style campaigns and of well-conceived public information programs relating to nuclear power (35-38). Unquestionably many of the doubts held by the public concerning the safety aspects and hazards associated with nuclear facilities will be allayed in time by more extended experience and operation of power plants in all parts of the world. Familiarity should never breed contempt, and more pressing demands for more power should never justify rape of the environment.

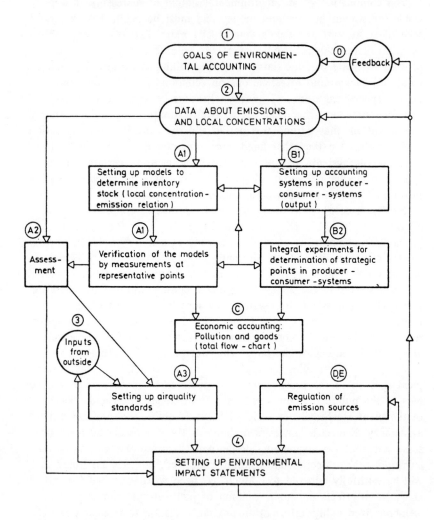

Figure 123. Environmental accountability relations (34).

REFERENCES

1. Buchanan, J. R., Ed. "Nuclear Power and Radiation in Perspective," Report ORNL-NSIC-100, Oak Ridge National Laboratory (1974).
2. Hull, A. P. "Reactor Effluents: As Low as Practicable or as Low as Reasonable?" *Nucl. News* **15** (11), 53 (1972).

3.   Hull, A. P. "Comparing Effluent Releases from Nuclear- and
      Fossil-Fueled Power Plants," *Nucl. News* **17**(5), 51 (1974).
4.   Hill, R. M., R. H. Bryan and B. L. Nichols. "Benefit-Cost Analyses
      in Licensing of Nuclear Power Reactors," *Nucl. Safety* **15**, 665
      (1974).
5.   "Final Environmental Statement: Comanche Peak Units 1 and 2,"
      U.S. Atomic Energy Commission (1974).
6.   "Numerical Guides for Design Objectives and Limiting Conditions
      for Operation to Meet the Criterion 'As Low As Practicable',"
      Report WASH-1258 (Washington, D.C.:   U.S. Atomic Energy Com-
      mission, 1973).
7.   "Preparation of Environmental Reports for Nuclear Power Plants,"
      Regulatory Guide 4.2, U.S. Atomic Energy Commission (1973).
8.   DeLozier, R. C., L. D. Reynolds and H. I. Bowers. "CONCEPT:
      Computerized Conceptual Cost Estimates of Steam-Electric Power
      Plants," User's Manual, Phase I, Report ORNL-TM-3276; Phase II,
      Report ORNL-4809, Oak Ridge National Laboratory (1971, 1973).
9.   "Final Environmental Statement: Crystal River Unit 3," (Washington,
      D.C.: U.S. Atomic Energy Commission, 1973).
10.   Environmental Statement. "Browns Ferry Nuclear Plant Units 1, 2
      and 3," Report TVA-OHES-EIS-72-6 (Chattanooga, Tennessee:
      Tennessee Valley Authority, 1972).
11.   Cohen, J. J. "On Determining the Cost of Radiation Exposure to
      Populations for Purposes of Cost-Benefit Analyses," *Health Phys.*
      25, 527 (1973).
12.   "Implications of Commission Recommendations that Doses be Kept
      as Low as Readily Achievable," International Commission on Radio-
      logical Protection Publication No. 22 (Oxford: Pergamon Press,
      1973).
13.   Sagan, L. A. "Human Costs of Nuclear Power," *Science* **177**, 487
      (1972).
14.   BEIR Committee. "The Effects on Populations of Exposure to
      Low Levels of Ionizing Radiation," (Washington, D.C.:   National
      Academy of Sciences/National Research Council, 1972).
15.   Klement, A. F., C. R. Miller, R. P. Minx and B. Shleien. "Estimates
      of Ionizing Radiation Doses in the United States, 1960-2000,"
      Report ORP/CSD 72-1 (Rockville, Maryland: U.S. Environmental
      Protection Agency, 1972).
16.   Rowe, W. D., Ed. "Environmental Radiation Dose Commitment:
      An Application to the Nuclear Power Industry," Report EPA-520/4-
      73-002 (Washington, D.C.:   U.S. Environmental Protection Agency,
      1974).
17.   UNSCEAR. "Ionizing Radiations: Levels and Effects," (New York:
      United Nations, 1972).
18.   Starr, C., M. A. Greenfield, and D. F. Hausknecht. "A Comparison
      of Public Health Risks: Nuclear vs. Oil-Fired Power Plants," *Nucl.
      News* **15**(10), 37 (1972).
19.   "Reactor Safety Study: An Assessment of Accident Risks in U.S.
      Commercial Nuclear Power Plants," Report WASH-1400 (Draft)
      U.S. Atomic Energy Commission (1974).

20. Starr, C. "Benefit-Cost Relationships in Social Technical Systems," in *Environmental Aspects of Nuclear Power Stations* (Vienna: International Atomic Energy Agency, 1971).
21. Starr, C. "Social Benefit Versus Technological Risk," *Science* 165 1232 (1969).
22. Sowby, F. D. "Some Risks of Modern Life," in *Environmental Aspects of Nuclear Power Stations* (Vienna: International Atomic Energy Agency, 1971), pp. 919-924.
23. Gast, P. F. "Divergent Public Attitudes toward Nuclear and Hydroelectric Plant Safety," Paper, ANS Annual Meeting, Chicago (1973).
24. Forbes, I. A., M. W. Goldsmith, J. P. Kearney, A. C. Kadak, J. C. Turnage, and G. J. Brown. "The Nuclear Debate—A Call to Reason," Boston, Massachusetts (1974).
25. Beattie, J. R., G. D. Bell and J. E. Edwards. "Methods for the Evaluation of Risk," U.K. Atomic Energy Agency Report AHSB(S)R 159 (Risley, Lancashire, 1969).
26. Weisman, J. "Uncertainty and Risk in Nuclear Power Plant Design," *Nucl. Eng. Design* 21, 396 (1972).
27. Wilson, R. "Man-rem, Economics and Risk in the Nuclear Power Industry," *Nucl. News* 15(2), 28 (1972).
28. Otway, H. J. "Nuclear Power Plant Safety," *Bull. IAEA* 16, 72 (1974).
29. Odum, H. T. *Environment, Power and Society.* (New York: Wiley-Interscience, 1971).
30. Rombaugh, C. T. and B. V. Koen. "Total Energy Investment in Nuclear Power Plants," *Nucl. Technol.* 26, 5 (1975).
31. Gustafson, P. F. "On the Preparation of Environmental Impact Statements," *IEEE Trans.* NS-20, 33 (1973).
32. "Standard Format and Content of Safety Analysis Reports for Nuclear Power Plants (Rev. 1)," U.S. Atomic Energy Commission (1972).
33. U. S. Atomic Energy Commission. "Environmental Effects of the Uranium Fuel Cycle," Title 10 Code of Federal Regulations, Part 50, *Federal Register* 39, 14188 (1974).
34. Häfele, W. "Energy Systems," *Bull. IAEA* 16, 3 (1974).
35. Bross, I. D. J. "Adversary Science in Aliquippa," *Health Phys.* 26 581 (1974).
36. Hess, D. N. "Nuclear Power in Perspective: The Plight of the Benign Giant," *Nucl. Safety* 12, 283 (1971).
37. Kathren, R. L. "Nuclear Power and Public Opinion," *Health Physics* 26, 483 (1974).
38. Weinberg, A. M. "Social Institutions and Nuclear Energy," *Science* 177, 27 (1972).

ADDITIONAL REFERENCES

Farmer, F. R. "Reactor Safety and Siting: A Proposed Risk Criterion," *Nucl. Safety* 8, 539 (1967).
Farmer, F. R. "Development of Adequate Risk Standards," Proc. Jülich

Symp. on Principles and Standards of Reactor Safety (Vienna: International Atomic Energy Agency, 1973).

Green, H. P. "Nuclear Safety and the Public Interest," *Nucl. News* **15** (9), 75 (1972).

Hausknecht, D. F., D. M. O'Connor and C. F. Smith. "Comparative Risks from Nuclear and Fossil-Fueled Power Plant Accidents," *Trans. ANS* **15**, 75 (1972).

Jordan, W. H. "Nuclear Energy: Benefit Versus Risks," *Physics Today* **23**(5), 32 (1970).

Morrison, D. L., R. H. Males, K. M. Duke, V. L. Sharp and R. L. Ritzman. "Environmental Benefit-Cost Analysis for Nuclear Power Generation," *Nucl. News* **15**(6), 50 (1972).

"Perspectives on Benefit-Risk Decision Making," Proc. 1971 Colloquia (Washington, D.C.: National Academy of Engineering, 1972).

Reiquam, H. "Establishing Priorities Among Environmental Stresses," in *Indicators of Environmental Quality*, W. A. Thomas, Ed. (New York: Plenum Press, 1972).

Rose, D. J. "Energy Policy in the U.S.," *Scient. Amer.* **230**(1), 20 (1974).

"The Safety of Nuclear Power Reactors (Light-Water-Cooled) and Related Facilities (Final Draft)," U.S. Atomic Energy Commission Report WASH-1250 (1973).

Shapar, H. K. "Trends in Federal and State Legislation in the USA for the Protection of the Environment and the Regulation of Nuclear Power Plants," *Proc. 4th Internat. Conf. Peaceful Uses of Atomic Energy* **3**, 432 (New York: United Nations, 1971).

Solomon, K. A., R. C. Erdmann and D. Okrent. "Estimate of the Hazards to a Nuclear Reactor from the Random Impact of Meteorites," *Nucl. Technol.* **25**, 68 (1975).

Tattersall, J. O., D. M. Simpson and R. A. Reynolds. "A Discussion of Nuclear Plant Safety with Reference to Other Hazards Experienced by the Community," *Proc. 4th Internat. Conf. Peaceful Uses of Atomic Energy* **11**, 487 (1971).

U.S. Atomic Energy Commission. "The Nuclear Industry—1974," Report WASH 1174-74 (Washington, D.C.: U.S. Government Printing Office, 1974).

Waldbott, G. L. *Health Effects of Environmental Pollutants.* (St. Louis: C. V. Mosley Co., 1973).

White, I. L. "Energy Policy-Making: Limitations of a Conceptual Model," *Bull. Atomic Scient.* **27**(8), 20 (1971).

CHAPTER 14

CONCLUSION

Never before has any human activity been the subject of so much public scrutiny, such soul-searching, so many studies and investigations as nuclear power. Never before has such a gigantic enterprise been conducted, all over the world, with so little loss of life or limb, with such a rapid rate of technical evolution.

We are faced with a change in life style as great or greater than that accompanying the demise of the horse as the principal means of transportation, around the turn of this century. With this change there comes a reordering of priorities, including those involving our use of energy, our consumption of nonrenewable resources and our concern with the preservation of the open countryside as we find it and as we wish to pass it on to future generations. The adoption of nuclear energy as a major power source is an integral part of this transition.

The purpose of studying the environmental aspects of nuclear power is to establish the magnitudes of any direct or indirect effects ascribable to the construction and operation of nuclear power plants, primarily as they involve changes in habitat and population of the existing flora and fauna and to the extent that they result in significant changes in the ambient radiation background. The secondary purpose is to document and measure these effects to relate the operation of nuclear facilities to existing legal requirements, some of which are not well defined at this time. A further purpose, of course, is also to confirm or lay at rest many of the public objections that have long been voiced in regard to nuclear power. In this connection it must be recognized that much of the public agitation concerning environmental questions is at heart merely a vehicle for those whom no argument will ever convince that nuclear power is not a hidden bomb and a menace to humanity.

At this stage nuclear power has become an essential energy source in most developed countries. The magnitude and ever-growing commitment of financial resources to this energy source by governments and industry in so many countries attest to their inherent confidence in the reliability and safety of nuclear power. Those who wish to abolish or delay any further use of nuclear power ignore the real need for ever more electricity to meet the demands of a steadily growing world population. There seems to be no ready solution to this population growth, which in itself embodies the most severe attack on the world's environment at this time. In comparison the local effects of any number of nuclear power stations may be regarded as trivial.

By now the safety of nuclear power stations is well established. This does not mean that there will not be some accidents due to human error, shoddy workmanship, poor engineering design, or a combination of circumstances; considering the size and complexity of these plants it would be surprising if there were none. The main consideration at all times has been to design the plant with multiple safeguards to confine any damage to the interior of the plant and to minimize absolutely any possible effects to the surroundings. Again there are practical limits to the number and extent of precautions that can be justified realistically and economically. At some point we must be willing to accept a finite, if highly improbable risk. This is nothing new for we do so all the time if the benefits seem worthwhile; we have not closed down automobile manufacture because of the carnage on the highways, mines are not shut down despite their poor safety record, and people keep flying. In fact, only the grave is the absolutely safe place.

The reason for accepting this fairly small risk of nuclear power is that modern society would collapse without adequate energy, provided, of course, that this low risk assessment can be maintained. In a growing world population, countries without adequate access to raw materials, food, energy and employment are facing disaster. There can be no reliance on a faster consumption of coal at this time, depriving future generations of a valuable raw material and imposing barely acceptable environmental penalties. Alternative energy sources are neither sufficiently developed nor economically practicable to take the place of nuclear energy for at least another generation.

This does not mean that current nuclear facilities are perfect and environmentally desirable. To future generations present plants undoubtedly will seem crude and primitive. Much remains to be done to improve their efficiency, their fuel design and inherent reliability. From an environmental point of view, constructive uses have to be found for the large quantities of waste heat. At present no entirely satisfactory solution exists for the reduction of tritium and noble gases in plant effluents and their ultimate disposal. Ultimate waste disposal of long-lived fission products and actinide wastes awaits a definitive solution. Transmission of power other than by high-tension wires is certainly conceivable and would do much to reduce environmental impact. Alternative fuel cycle approaches may be practical and deserve study.

Overall then, one may conclude that a knowledge of environmental factors is important in preserving our quality of life. Their detailed study over the past few years has contributed much to other fields of knowledge and deserves further effort. In looking at the information available and the experience accumulating from the operation of nuclear plants, it seems fair to state that much is known and much has been learned. We can look to the future with some confidence.

APPENDIX

LIST OF ELEMENTS

Atomic Number	Symbol	Name	Atomic Weight	Atomic Number	Symbol	Name	Atomic Weight
0	n	neutron	–	52	Te	tellurium	127.60
1	H	hydrogen	1.0079	53	I	iodine	126.9045
2	He	helium	4.00260	54	Xe	xenon	131.30
3	Li	lithium	6.941	55	Cs	cesium	132.9054
4	Be	beryllium	9.01218	56	Ba	barium	137.34
5	B	boron	10.81	57	La	lanthanum	138,9055
6	C	carbon	12.011	58	Ce	cerium	140.12
7	N	nitrogen	14.0067	59	Pr	praseodymium	140.9077
8	O	oxygen	15.9994	60	Nd	neodymium	144.24
9	F	fluorine	18.99840	61	Pm	promethium	–
10	Ne	neon	20.179	62	Sm	samarium	150.4
11	Na	sodium	22.98977	63	Eu	europium	151.96
12	Mg	magnesium	24.305	64	Gd	gadolinium	157.25
13	Al	aluminum	26.98154	65	Tb	terbium	158.9254
14	Si	silicon	28.086	66	Dy	dysprosium	162.50
15	P	phosphorus	30.97376	67	Ho	holmium	164.9304
16	S	sulfur	32.06	68	Er	erbium	167.26
17	Cl	chlorine	35.453	69	Tm	thulium	168.9342
18	Ar	argon	39.948	70	Yb	ytterbium	173.04
19	K	potassium	39,098	71	Lu	lutetium	174.97
20	Ca	calcium	40.08	72	Hf	hafnium	178.49
21	Sc	scandium	44.9559	73	Ta	tantalum	180.9479
22	Ti	titanium	47.90	74	W	tungsten	183.85
23	V	vanadium	50.9414	75	Re	rhenium	186.2
24	Cr	chromium	51.996	76	Os	osmium	190.2
25	Mn	manganese	54.9380	77	Ir	iridium	192.22
26	Fe	iron	55.847	78	Pt	platinum	195.09
27	Co	cobalt	59.9332	79	Au	gold	196.9665
28	Ni	nickel	58.71	80	Hg	mercury	200.59
29	Cu	copper	63.546	81	Tl	thallium	204.37
30	Zn	zinc	65.38	82	Pb	lead	207.2
31	Ga	gallium	69.72	83	Bi	bismuth	208.9804
32	Ge	germanium	72.59	84	Po	polonium	–
33	As	arsenic	74.9216	85	At	astatine	–
34	Se	selenium	78.96	86	Rn	radon	–
35	Br	bromine	79.904	87	Fr	francium	–
36	Kr	krypton	83.80	88	Ra	radium	–
37	Rb	rubidium	85.4678	89	Ac	actinium	–
38	Sr	strontium	87.62	90	Th	thorium	232.0381
39	Y	yttrium	88.9059	91	Pa	protactinium	–
40	Zr	zirconium	91.22	92	U	uranium	238.029
41	Nb	niobium	92.9064	93	Np	neptunium	–
42	Mo	molybdenum	95.94	94	Pu	plutonium	–
43	Tc	technetium	–	95	Am	americium	–
44	Ru	ruthenium	101.07	96	Cm	curium	–
45	Rh	rhodium	102.9055	97	Bk	berkelium	–
46	Pd	palladium	106.4	98	Cf	californium	–
47	Ag	silver	107.868	99	Es	einsteinium	–
48	Cd	cadmium	112.40	100	Fm	fermium	–
49	In	indium	114.82	101	Md	mendelevium	–
50	Sn	tin	118.69	102	No	nobelium	–
51	Sb	antimony	121.75	103	Lr	lawrencium	–

INTERNATIONAL NUMERICAL MULTIPLE
AND SUBMULTIPLE PREFIXES

Multiples and Submultiples	Prefixes	Symbols
10^{12}	tera	T
10^9	giga	G
10^6	mega	M
10^3	kilo	k
10^2	hecto	h
10	deka	da
10^{-1}	deci	d
10^{-2}	centi	c
10^{-3}	milli	m
10^{-6}	micro	μ
10^{-9}	nano	n
10^{-12}	pico	p
10^{-15}	femto	f
10^{-18}	atto	a

SYMBOLS, UNITS AND EQUIVALENTS

Symbol	Unit	Equivalent
Å	angstrom	10^{-10} meter
A	ampere(s)	
a	annum, year	
BeV	billion electron volts	GeV
Ci	curie	3.7×10^{10} dps-2.22×10^{12} dpm= 3.7×10^{10} becquerel
cpm	counts per minute	
dpm	disintegrations per minute	
dps	disintegrations per second	
eV	electron volt	1.6×10^{-12} ergs
g	gram(s)	3.527×10^{-2} ounces = 2.205×10^{-3} pounds
Hz	hertz	cycle per second
kVp	kilovolt peak	
m	meter(s)	39.4 inches - 3.28 feet
m^3	cubic meter(s)	
mCi/mi^2	millicuries per square mile	0.386 nCi/m^2 (mCi/km^2)
mi	mile(s)	
ml	milliliter(s)	
nCi/m^2	nanocuries per square meter	2.59 mCi/mi^2
R	roentgen	
rad	unit of absorbed radiation dose	100 ergs/g = 0.01 gray
r/min	revolutions per minute	
s	second	
yr	year	

FREQUENTLY USED ABBREVIATIONS

ALAP	As Low As Practicable
ALARA	As Low as Readily Achievable
BEIR Committee	Advisory Committee on the Biological Effects of Ionizing Radiation
BWR	Boiling Water Reactor
CVCS	Chemical Volume Control System (coolant purification system in water-cooled reactors)
EPA	U.S. Environmental Protection Agency
ERDA	U.S. Energy Research and Development Agency
FBR	Fast Breeder Reactor
GCR	Gas-Cooled Reactor
HTGR	High-Temperature Gas-Cooled Reactor
HWR	Heavy-Water-Cooled and Moderated Reactor
IAEA	International Atomic Energy Agency
ICRP	International Commission on Radiological Protection
kWh	kilowatt hour
LET	Linear Energy Transfer
LMFBR	Liquid Metal Fast Breeder Reactor
LOCA	Loss-of-Coolant Accident
LWR	Light-Water-Cooled Reactor
MPC	Maximum Permissible Concentration
MT	Metric ton (1 tonne = 1000 kg)
NCRP	U.S. National Council for Radiation Protection
NRC	U.S. Nuclear Regulatory Commission
PWR	Pressurized Water Reactor
RBE	Relative Biological Effectiveness
RPG	Radiation Protection Guide
SGHWR	Steam-Generating Heavy-Water Reactor
UNSCEAR	United Nations Scientific Committee on the Effects of Atomic Radiation
US AEC	U.S. Atomic Energy Commission (to January 1975)
US NRC	U.S. Nuclear Regulatory Commission

CONVERSION FACTORS

Multiply	by	to obtain
Acre	0.405	hectare
	4047	m^2
	43560	sq ft
Acre-feet	3.26×10^5	gallon
Atmospheres	33.90	ft of water
	1013.2	millibar
	760	Torr
Becquerel (Bq)	2.70×10^{-11}	Curie
Btu	1.055×10^{10}	erg
	252	g-cal
	1055	joule
	2.93×10^{-4}	kWh
	6.59×10^{15}	MeV
Btu/sec	1.055×10^3	watt
cm	0.394	inch
cubic ft	0.0283	cubic meter
cubic meter (m^3)	1000	liter
	264.2	gallon (U.S.)
	35.31	cu ft
curie (Ci)	3.70×10^{10}	disint/sec (Bq)
	2.22×10^{12}	disint/min
erg	9.49×10^{-11}	Btu
	2.78×10^{-14}	kWh
foot	1.894×10^{-4}	mile (statute)
	0.3048	meter
gallon (U.S.)	3785	cu cm
	0.1337	cu ft
	3.785	liter
	0.8327	gallon (Imperial)
gram	2.205×10^{-3}	lb
g-calorie	3.969×10^{-3}	Btu
	2.61×10^{13}	MeV
	1.163×10^{-6}	kWh
hectare	2.471	acre
	0.01	km^2
hour	4.167×10^{-2}	day
inch	2.540	cm
	1.578×10^{-5}	mile
inch of water	2.458×10^{-3}	atm
	1.866	Torr
joule	9.48×10^{-4}	Btu
	2.778×10^{-4}	watt-hr
kg	2.250	lb
	9.842×10^{-4}	ton (long)
	1.102×10^{-3}	ton (short)

CONVERSION FACTORS, cont.

Multiply	by	to obtain
km	0.6214	mile
	1094	yard
kWh	1.341	horsepower-hr
	3413	Btu
	3.6×10^6	joule
liter	0.0353	cu ft
	0.2642	gallon (U.S.)
	3.281	ft
	39.37	inch
	6.214×10^{-4}	mile
mm	0.03937	inch
m/sec	3.281	ft/sec
	2.237	mile/hr
micron (μm)	10^{-3}	mm
mile (statute)	1.61	km
MeV	1.517×10^{-16}	Btu
	1.602×10^{-6}	erg
	3.83×10^{-14}	g-cal
pounds/sq in.	0.068	atm
	6894.7	N/m^2 (Pa)
	144	lb/sq ft
Q-unit	1.06×10^{21}	joule
rad	100	erg/g tissue
rem	0.01	joule/kg
sq km	247.1	acres
sq ft	929	cm^2
	0.111	sq yard
	3.59×10^{-8}	sq mile
sq mile	640	acres
	2.590	sq km
ton (long)	1016	kg
	2240	lb
	1.120	ton (short)
ton (short)	907.2	kg
	2000	lb
Tonne (MT)	1000	kg
	2205	lb
watt	3.413	Btu/hr
	6.24×10^{12}	MeV/sec

GLOSSARY

absorbed dose: The energy imparted to matter by ionizing radiation per unit mass of irradiated material at the place of interest. The unit of absorbed dose is the rad. (The SI unit is the gray.)

acute exposure: Radiation exposure of short duration.

approach: ($°F$, $°C$) The temperature difference between cold water leaving a cooling tower and wet-bulb temperature of the surrounding air.

background radiation: The radiation in man's natural environment, including cosmic rays and radiation from the naturally radioactive elements, both outside and inside the bodies of men and animals. It is also called natural radiation. The term may also mean radiation that is unrelated to a specific measurement.

beta particle: [β(beta)] An elementary particle emitted from a nucleus during radioactive decay, with a single electrical charge and a mass equal to that of an electron.

biological half-life: The time required for a biological system, such as a man or an animal, to eliminate by natural processes half the amount of a substance (such as a radioactive material) that has entered it.

blowdown: The continuous or intermittent discharge to waste of a small portion of circulating water from the cooling tower basin. It is usually expressed as a percentage of the circulating water rate. It prevents build-up of dissolved solids left behind during evaporation.

body burden: The amount of radioactive material present in the body of a man or an animal.

boiling water reactor: A reactor in which water, used as both coolant and moderator, is allowed to boil in the core. The resulting steam can be used directly to drive a turbine.

bremsstrahlung: Secondary photon radiation produced by deceleration of charged particles passing through matter.

burnable poison: A neutron absorber (or poison), such as boron, that gradually "burns up" (is changed into nonabsorbing material) under neutron irradiation when purposely incorporated in the fuel or fuel cladding of a nuclear reactor. This process compensates for the loss of reactivity that occurs as fuel is consumed and fission-product poisons accumulate, and keeps the overall characteristics of the reactor nearly constant during its use.

burn-up: A measure of nuclear fuel consumption in a reactor. It can be expressed as the percentage of fuel atoms that have undergone fission or the total amount of heat released per unit weight of fuel. In the latter case it is usually expressed as Megawatt-days per tonne (MWd/t).

chemical shim: Chemicals, such as boric acid, that are placed in a reactor coolant to control the reactor by absorbing neutrons.

chemical waste stream: Normally liquids that contain relatively high concentrations of decontamination, regeneration, or other chemical compounds other than detergents. These liquids originate primarily from resin regenerant and laboratory wastes.

chronic exposure: Radiation exposure of long duration by fractionation or protraction.

cladding: The outer jacket of nuclear fuel elements. It prevents corrosion of the fuel and the release of fission products into the coolant. Aluminum or its alloys, stainless steel and zirconium alloys are common cladding materials.

clean waste stream: Normally tritiated, nonaerated, low-conductivity liquids consisting primarily of liquid waste collected from equipment leaks and drains and certain valve and pump seal leakoffs.

containment: The provision of a gastight shell or other enclosure around a reactor to confine fission products that otherwise might be released to the atmosphere in the event of an accident.

coolant: A substance circulated through a nuclear reactor to remove or transfer heat. Common coolants are water, air, carbon dioxide, liquid sodium and sodium-potassium alloy (NaK).

cooling range: ($^\circ$F, $^\circ$C) The temperature difference between hot water entering a cooling tower and cold water leaving.

core: The central portion of a nuclear reactor containing the fuel elements and usually the moderator, but not the reflector.

curie (Ci): The physical unit of radioactivity, as distinct from a measure of its biological significance. It is equal to 3.7×10^{10} disintegrations per second. A curie is also a quantity of any nuclide having 1 curie of radioactivity.

daughter: A nuclide formed by the radioactive decay of another nuclide, which in this context is called the parent.

decay, radioactive: The spontaneous transformation of one nuclide into a different nuclide or into a different energy state of the same nuclide. The process results in a decrease, with time, of the number of the original radioactive atoms in a sample.

decontamination factor (DF):   The ratio of the initial amount (specified) in terms of concentration or activity of radioactive material to the final amount resulting from a process.

dirty waste stream (floor drains):   Normally nontritiated, aerated, high-conductivity, nonprimary-coolant quality liquids collected from building sumps and floor and sample station drains. These liquids are not readily amenable for reuse as primary coolant makeup water.

dose commitment: Dose equivalent, in rem, per microcurie body intake calculated for a 50-year period.

dose equivalent (DE):   Quantity that expresses all radiations on a common scale for calculating the effective absorbed dose. It is defined as the product of the absorbed dose in rads and certain modifying factors. The unit of DE is the rem.

dose-rate: The radiation dose delivered per unit time, *e.g.*, rads per hour.

doubling dose: The amount of radiation needed to double the natural incidence of a genetic or somatic anomaly.

drift, carry-over, or windage loss: Water carried out of a cooling tower in mist or small droplet form. Usually expressed as a percentage of the circulating water rate.

electron volt (eV):   Unit of energy equivalent to the energy gained by an electron in passing through a potential difference of one volt. $(1 \text{ eV} = 1.6 \times 10^{-12} \text{ erg})$.

exclusion area: An area immediately surrounding a nuclear reactor where human habitation is prohibited to assure safety in the event of accident.

exposure: A measure of the ionization produced in air by X- or gamma radiation. The special unit of exposure is the roentgen.

external radiation: Radiation from a source outside the body.

fallout: Airborne particles containing radioactive material that fall to the ground following a nuclear explosion.

fast neutron: A neutron with energy greater than approximately 100,000 electron volts.

fill: Material placed in a cooling tower over which water flows. It increases the air-water surface area and time of contact and maintains uniform air and water flow distribution.

fission product: A nuclide produced either by fission or by subsequent radioactive decay of the nuclides formed in the fission process.

food chain: The pathways by which any material (such as radioactive material from fallout) passes from the first absorbing organism through plants and animals to man.

fuel: Fissionable material used or usable to produce energy in
                    a reactor. Also applied to a mixture, such as natural
                    uranium, in which only part of the atoms are readily
                    fissionable, if the mixture can be made to sustain a chain
                    reaction.

fuel cycle: The series of steps involved in supplying fuel for nuclear
                    power reactors. It includes mining, refining, the original
                    fabrication of fuel elements, their use in a reactor, chem-
                    ical processing to recover the fissionable material remain-
                    ing in the spent fuel, reenrichment of the fuel material,
                    and refabrication into new fuel elements.

fuel element: A rod, tube, plate, or other mechanical shape or form
                    into which nuclear fuel is fabricated for use in a reactor.
                    (Not to be confused with element.)

fuel reprocessing: The processing of reactor fuel to recover the unused
                    fissionable material.

gamma rays: [γ(gamma)]   High-energy, short-wavelength electromag-
                    netic radiation. Gamma radiation frequently accompanies
                    alpha and beta emissions and always accompanies fission.
                    Gamma rays are essentially similar to X-rays, but are
                    usually more energetic and are nuclear in origin.

gaseous effluent stream: Processed gaseous wastes containing radioactive
                    materials resulting from the operation of a nuclear power
                    reactor.

genetic effects of radiation: Radiation effects that can be transferred
                    from parent to offspring. Any radiation-caused changes
                    in the genetic material of sex cells.

half-life, biological: *See* biological half-life.

half-life, effective: The time required for a radionuclide contained in a
                    biological system, such as a man or an animal, to reduce
                    its activity by half as a combined result of radioactive
                    decay and biological elimination.

half-life, radioactive: The time required for a radioactive substance to
                    lose 50% of its activity by decay. Each radionuclide
                    has a unique half-life.

heat load: (Btu/hr, Btu/min) The amount of heat dissipated in a
                    cooling tower per unit time. It equals water circulation
                    rate multiplied by cooling range.

heavy water: [D_2O] Water containing significantly more than the
                    natural proportion (one in 6500) of heavy hydrogen
                    (deuterium) atoms to ordinary hydrogen atoms.

internal radiation: Radiation from a source within the body (as a result
                    of deposition of radionuclides in body tissue).

ion exchange: A chemical process involving the reversible interchange
                    of various ions between a solution and a solid material,

670 ENVIRONMENTAL ASPECTS OF NUCLEAR POWER

usually a plastic or a resin. It is used to separate and purify chemicals, such as fission products and rare earths, in solutions.

ionization chamber: An instrument that detects and measures ionizing radiation by measuring the electrical current that flows when radiation ionizes gas in a chamber, making the gas a conductor of the electricity.

ionizing radiation: Any radiation displacing electrons from atoms or molecules, thereby producing ions. Examples: alpha, beta, gamma radiation, short-wave ultraviolet light.

isotope: Nuclides with the same atomic number (*i.e.*, the same chemical element) but different atomic weights.

linear energy transfer (LET): The average amount of energy lost per unit of particle spur-track length.

linear hypothesis: The assumption that a dose-effect curve derived from data in the high-dose and high-dose-rate ranges may be extrapolated through the low-dose and low-dose range to zero, implying that, theoretically, any amount of radiation will cause some damage.

liquid effluent stream: Processed liquid wastes containing radioactive materials resulting from the operation of a nuclear power reactor.

load factor: The ratio of average load carried by an electric power plant or system during a specific period to its peak load during that period.

makeup: Water required to replace normal system losses from evaporation, drift, blowdown, and small leaks.

man-rems: The product of the average individual dose in a population times the number of individuals in the population.

maximum permissible concentration (MPC): The amount of radioactive material in air, water, or food that might be expected to result in a maximum permissible dose to persons consuming them at a standard rate of intake; an obsolescent term. MPC values are expressed in microcuries per milliliter and represent concentrations, which, if continuously maintained, would result in the maximum permissible doses to the critical organs specified by ICRP.

median lethal dose (MLD): Dose of radiation required to kill, within a specified period, 50% of the individuals in a large group of animals or organisms. Also called LD_{50}.

megawatt-day per ton: A unit used for expressing the burn-up of fuel in a reactor; specifically, the number of megawatt-days of heat output per metric ton of fuel in the reactor.

nuclide: A general term applicable to all atomic forms of the elements. The term is often erroneously used as a synonym for "isotope," which properly has a more limited definition. Nuclides are distinguished by their atomic number, atomic mass, and energy state.

packing: *See* fill.

partition coefficient (PC):   The ratio of the concentration of a nuclide in the gas phase to the concentration of a nuclide in the liquid phase when the liquid and gas are at equilibrium.

partition factor (PF):   The ratio of the quantity of a nuclide in the gas phase to the total quantity in both the liquid and gas phases when the liquid and gas are at equilibrium.

personnel monitoring: Determination by either physical or biological measurement of the amount of ionizing radiation to which an individual has been exposed, such as by measuring the darkening of a film badge or performing a radon breath analysis.

plant factor: The ratio of the average net power load of an electric power plant to its rated capacity. Sometimes called capacity factor.

pressure vessel: A strong-walled container housing the core of most types of power reactors; it usually also contains moderator, reflector, thermal shield, and control rods.

pressurized water reactor: A power reactor in which heat is transferred from the core to a heat exchanger by water kept under high pressure to achieve high temperature without boiling in the primary system. Steam is generated in a secondary circuit. Many reactors producing electric power are pressurized water reactors.

primary coolant (reactor coolant):   The fluid circulated through the reactor to remove heat. In a PWR, the fluid is kept under sufficient pressure to keep it from boiling.

protective action guide (PAG):   The absorbed dose of ionizing radiation to individuals in the general population which would warrant protective action following a contaminating event, such as a nuclear explosion.

Q: A unit used to express very large energy resource figures. One Q equals 10^{18} Btu. Present world energy consumption runs at 0.1 Q/year.

quality factor (QF):   The linear-energy-transfer-dependent factor by which absorbed doses are multiplied to obtain (for radiation protection purposes) a quantity that expresses, on a common scale for all ionizing radiations, the effectiveness of the absorbed dose.

rad: The unit of absorbed dose, equal to 0.01 J/kg in any medium.

radiation monitoring: Continuous or periodic determination of the amount of radiation present in a given area.

radiation protection guide: The officially determined radiation doses that should not be exceeded without careful consideration of the reasons for doing so. These standards are equivalent to what was formerly called the maximum permissible dose or maximum permissible exposure.

radioactive release rate (Q'): The average quantity of radioactive material released to the environment from a nuclear power reactor during normal operational period and anticipated operational occurrences.

radionuclide: Any radioactive species of atom that exists for a measurable length of time. Individual radionuclides are distinguished by their atomic weight and atomic number.

radiosensitivity: Relative susceptibility of cells, tissues, organs, organisms, or any living substance to the injurious action of radiation.

recovery rate: The rate at which recovery takes place after radiation injury. It may proceed at different rates for different tissues. Among tissues recovering at different rates, those having slower rates will ultimately suffer greater damage from a series of successive irradiations.

relative biological effectiveness (RBE): A factor used to compare the biological effectiveness of different types of ionizing radiation. It is the inverse ratio of the amount of absorbed radiation, required to produce a given effect, to a standard (or reference) radiation required to produce the same effect.

relative humidity (%): The ratio of the amount of water vapor actually present in the air to the greatest amount it could hold if saturated at that temperature and pressure.

relative risk: The ratio of the risk in those exposed to the risk to those not exposed (incidence in exposed population to incidence in control population).

rem: Unit of dose equivalent. The dose equivalent in rems is numerically equal to the absorbed dose in rads multiplied by the quality factor, the distribution factor, and any other necessary modifying factors. The rem represents that quantity of radiation that is equivalent—in biological damage of a specified sort—to 1 rad of 250 kVP X-rays.

roentgen (R): Unit of exposure to ionizing radiation. It is that amount of gamma or X-rays required to produce ions carrying 1 electrostatic unit of electrical charge (either positive or negative) in 1 cubic centimeter of dry air under standard conditions. One roentgen equals 2.58×10^{-4} coulomb per kilogram of air.

secondary coolant: A separate stream of coolant that is converted to steam by the primary coolant in a heat exchanger (steam generator) to power the turbine.

somatic effects of radiation: Effects of radiation limited to the exposed individual, as distinguished from genetic effects (which also affect subsequent unexposed generations).

source term: The calculated average quantity of radioactive material released to the environment from a nuclear power reactor during normal operation including anticipated operational occurrences.   The source term is the isotopic distribution of radioactive materials used in evaluating the impact of radioactive releases on the environment.

specific activity: The radioactivity per unit weight or volume of a radioactive substance, often expressed in curies per gram (Ci/g) or curies per cubic meter (Ci/m^3).   It is used to describe the radioactivity of both pure radionuclides and other materials (*e.g.*, wastes) into which radionuclides have been incorporated.

spent fuel: Nuclear reactor fuel that has been irradiated (used) to the extent that it can no longer effectively sustain a chain reaction.

steam generator blowdown: Liquid removed from a steam generator needed to maintain proper water chemistry.

survey meter: Any portable radiation detection instrument especially adapted for surveying or inspecting an area to establish the existence and amount of radioactive material present.

thermal efficiency: The ratio of the electric power produced by a power plant to the amount of heat produced by the fuel; a measure of the efficiency with which the plant converts thermal to electrical energy.

thermal (slow) neutron: A neutron in thermal equilibrium with its surrounding medium.   Thermal neutrons are those that have been slowed down by a moderator to an average speed of about 2200 meters per second (at room temperature).

transuranic element (isotope):   An element above uranium in the periodic table, that is, with an atomic number greater than 92. All transuranic elements are produced artificially and are radioactive.

threshold hypothesis: The assumption that no radiation injury occurs below a specified dose level.

tritium: A radioactive isotope of hydrogen with two neutrons and one proton in the nucleus.   It is man-made and is heavier than deuterium (heavy hydrogen).

wet-bulb temperature ($^\circ$F, $^\circ$C):   Obtained by covering the bulb of an ordinary thermometer with wetted gauze and reading in moving air.   It is the theoretical limit to which water can be cooled through evaporation.

working level (WL): Any combination of short-lived radon daughters in 1 liter of air that will result in the ultimate emission of 1.3×10^5 MeV of potential alpha energy.

working level month (WLM): Inhalation of air with a concentration of 1 WL of radon daughters for 173 working hours results in an exposure of 1 WLM.

MISCELLANEOUS PROBLEMS

All problems require making reasonable assumptions and presume access to standard nuclear reference works.

1. Calculate the mass, carrier-free, of 1 kCi of cesium-137 and 1 kCi of ruthenium-106.

2. Calculate the volume of 10 kCi of argon-41 and of krypton-85, at normal temperature and pressure, and in liquefied form.

3. Calculate the amount of Pu-239 produced in a 1000-MWe BWR after a burn-up of 30,000 MWd/t. Neglect secondary reactions.

4. Calculate the radiation field at 1 meter from the surface of a 25-cm-thick lead shield containing 300 kCi of Zr-Nb-95.

5. Calculate the population dose received when the above container is moved along the middle of a highway through a populated area with an average population density of 5000 per square kilometer at maximum legal speed.

6. Design a shielded container for two PWR spent fuel elements that had a burn-up of 25,000 MWd/t and a storage time of three months. Identify the major gamma-ray emitters.

7. Select a suitable site for a 1000 MWe nuclear power plant within a 60 km radius from the center of your city. Justify your choice.

8. In the previous problem, assuming typical stack release rates, calculate the population dose in the 22 1/2° sector along the prevailing wind direction. State all your assumptions.

9. Estimate, with simplified but specified assumptions, the fence line thyroid dose due to iodine-131 in case of a loss-of-coolant accident of a 1000 MWe BWR, if it occurs after 6 months' operation of that reactor core. Exclusion area radius 1 km.

10. Calculate the tritium dose due to ingestion from a stream receiving 50,000 Ci/yr. The average stream flow is 1400 cfs (39640 l/s). its water is used as the sole source of drinking water by a downstream community.

11. Estimate the costs and benefits of installing a cooling tower, instead of once-through cooling, at a 2000 MWe power plant, deriving its cooling water from a river with an average flow rate of 900 cfs.

12. Making reasonable assumptions regarding sample size and detector type and sensitivity, estimate the minimum concentrations of (a) strontium-90, (b) iodine-131, (c) iodine-129, and (d) tritium that can be detected with acceptable accuracy in water samples, without or with radiochemical concentration.

13. A waste mixture containing 60% Cs-134, 20% Cs-137, 8% Pm-147, 2% Pu-238 and 0.2% Am-241 is stored for disposal. Calculate the proportions of this mixture after storage of 10 years, 30 years, 150 years and 6000 years.

14. The exhaust stack of a 1000 MWe BWR power plant is raised from 200 m to 300 m height. What effect will this have on the size of the exclusion area?

15. A 600 MWe power plant uses a 300-acre cooling pond, 10 ft deep on the average. Calculate the effect on the pond's surface temperature if the wind speed varies from 5 to 25 miles per hour and the air temperature from 40 to $80°F$.

16. Assuming 50% escape of the noble gas inventory in case of a serious reactor accident, account for the contribution of all gaseous nuclides to the ground concentration 1 mile (1.6 km) downwind under Pasquill category E conditions.

17. A certain power plant releases an average of 0.5 Ci of iodine-131 per year from its 500-ft-high (150 m) stack. During winter daytime, wind conditions are type C, with a mean wind speed of 6 m/sec, at night E, with 3 m/sec wind speeds. Calculate ground concentrations at a point 2 miles (3.2 km) downwind and the resultant thyroid dose to a person residing there.

18. A certain BWR type sends 17.18×10^6 lb/hr of steam at an outlet pressure of 1070 psia to the turbine generators. If the core volume constitutes 5% of system volume and average core neutron flux is 10^{13} n/cm^2 sec, estimate the equilibrium concentration of nitrogen-16 produced in the coolant and the (unshielded) exposure levels 10 m from the turbine inlets.

19. Estimate the tritium inventory in a 1000 MWe PWR reactor after 300 days' operation. Making reasonable assumptions regarding fuel leakages and other leak rates, calculate the effluent water flow needed to hold tritium concentrations below 1% of MPC.

20. A 500 MWe nuclear power plant is located on a river which is utilized for cooling purposes. The average river flow is 1450 cfs;

low river flow 200 cfs. Making reasonable assumptions on coolant flow, estimate the heat load to be dissipated, the temperature rise at the plant outlet and below a "reasonable" mixing zone. If the excess temperature at low flow is to remain below $5°F$, indicate possible alternatives to meet this requirement. (Annual average wind speed 10.6 mph.)

21. A power reactor has been operating at 1200 MWt for six months. Calculate the fission product inventory in the core of iodine-131, ruthenium-106 and cesium-137. In case of a loss of coolant accident, in what form and quantity would you expect these materials to be released from the core?

22. Calculate the amount of krypton and xenon produced in the complete fission of 75 grams of U-235. If this material has been used in a reactor for 1 year before 50% burn-up has occurred, how much makeup air would have to be added to permit free release to the atmosphere of the accumulated active gases? [$(MPC)_{air}$ for Xe-133 and Kr-85 3×10^{-6} μCi/cc for a 168-hr week]

23. In a suspected sample of milk, 80 pCi/l of iodine-131 have been found. Describe a method for extracting this activity and a system for counting it. What minimum size sample would be required for a typical radiation detector efficiency and geometry to give not more than 3% error in the final assay?

24. Evaluate the hazard from potential plutonium release (a) at a nuclear power plant, (b) during spent fuel transportation, (c) at a fuel reprocessing plant, and (d) from a permanent disposal site. Indicate possible pathways of escape and compare inhalation and ingestion doses.

Date Due

O D Olson			
NOV 12 79			
MAR 18 80			
W Nelson Apr 2			
DEC 15 1983			
FEB 08 1984			
NOV 28 1985			